Impacts of Climate Change on Agriculture in the MENA Region

Mohammed El Hadi El Maknouzi
University of Al Dhaid, UAE & Mohammed V University, Rabat, Morocco

Hicham Sadok
https://orcid.org/0000-0003-4402-0954
Mohammed V University, Rabat, Morocco

IGI Global
Scientific Publishing
Publishing Tomorrow's Research Today

Vice President of Editorial Melissa Wagner
Director of Acquisitions Mikaela Felty
Director of Book Development Jocelynn Hessler
Production Manager Mike Brehm
Cover Design Jose Rosado

Published in the United States of America by
IGI Global Scientific Publishing
701 East Chocolate Avenue
Hershey, PA, 17033, USA
Tel: 717-533-8845 | Fax: 717-533-7115
Website: https://www.igi-global.com E-mail: cust@igi-global.com

Library of Congress Cataloging-in-Publication Data

Names: El Maknouzi, Mohammed El Hadi, 1977- editor | Sadok, Hicham, 1976- editor
Title: Impacts of climate change on agriculture in the MENA region / edited by Mohammed El Hadi El Maknouzi, Hicham Sadok.
Description: Hershey, PA : IGI Global Scientific Publishing, 2026. | Includes bibliographical references and index. | Summary: "The book aims to bridge the gap between several disciplines to provide accessible knowledge in addressing a complex issue, namely the vulnerability of agriculture in the MENA region to climate change and adaptation strategies"-- Provided by publisher.
Identifiers: LCCN 2025020998 (print) | LCCN 2025020999 (ebook) | ISBN 9798337333465 hardcover | ISBN 9798337333472 paperback | ISBN 9798337333489 ebook
Subjects: LCSH: Crops and climate--Middle East | Crops and climate--Africa, North
Classification: LCC S600.7.C54 I465 2026 (print) | LCC S600.7.C54 (ebook) | DDC 630.2/5150956--dc23/eng/20250709
LC record available at https://lccn.loc.gov/2025020998
LC ebook record available at https://lccn.loc.gov/2025020999

British Cataloguing in Publication Data
A Cataloguing in Publication record for this book is available from the British Library.

Table of Contents

Detailed Table of Contents

Water scarcity is one of the main concerns troubling nations, as agricultural activity, food security, and sustainable development cannot be conceived without this vital resource. In this context, seawater desalination plants have become one of the most commonly used solutions by national governments and local authorities to provide potable water and water for agricultural irrigation. Through this study, we seek to demonstrate that employing business contracts may represent an optimal means of addressing the impact of climate change on agriculture in the Middle East and North Africa (MENA) region, using a public-private partnership (PPP) contract for a seawater desalination plant as a case study.

Natural resources are essential to the livelihoods of Morocco's poor. Since the 1960s, the average increase in observed temperatures is 0.23°C per decade; nine of the country's 10 hottest years on record have occurred in the past two decades, and precipitation follows a general downward trend. Faced with such observations, this article seeks to determine whether this ongoing climate change has concrete effects on the four main agricultural productions in Morocco. Thus, using the Pettitt test to estimate the standardized precipitation index (SPI) and the evolution of seasonal average temperatures, the study of the Spearman correlation coefficients between these parametric variations of climate change and the yields of four main agricultural productions during the period 2001-2020 did not reveal a significant link, with the exception of cereal yield during the 5 years of severe drought. This resilience can be explained by the intervention of public policies to promote technology, seeds and irrigation to counteract the degradation of climatic factors.

This chapter examines the impacts of climate change on Algerian agriculture. It focused on three main areas: analyzing the impacts on natural resources and production systems; the dynamics of agricultural labour; and assessing agricultural performance and growing food dependency. Main results highlight the degradation of natural resources (water, soil, vegetation), transformations in production systems (cereals, vegetable crops, livestock), and their socioeconomic consequences (rural unemployment, exodus, food dependency). The data and evidence reveal growing vulnerability to climate hazards, exacerbating food security challenges and the need for appropriate agrarian reforms.

This chapter aims to evaluate the effects of climate change on food security in North Africa using an ARDL-PMG model and future climate projections from CMIP6. The results indicate that in the long term, agricultural capital, labor, and climatic conditions (De Martonne index) exert a positive influence on agricultural production. In the short term, the effects vary by country. The integration of Shared Socioeconomic Pathways (SSP) allows for the estimation of future agricultural losses that could reach –4.1% under the SSP5–8.5 scenario, with more pronounced impacts in Morocco and Tunisia. The study highlights the increasing vulnerabilities and emphasizes the need for adapted agricultural and climate policies to ensure regional food security.

This chapter aims to examine the impacts of climate change on agriculture in North Africa using an integrated Ricardian-Production Function approach. Focusing on 5 countries: Morocco, Algeria, Tunisia, Libya, and Egypt, the analysis combines panel data (1980-2020) with climate projections under three Shared Socioeconomic Pathways (SSP1-2.6, SSP2-4.5, and SSP3-7.0), to quantify effects on agricultural value added. The model incorporates key variables such as temperature, precipitation, capital investment, labor, and irrigated area. The main findings reveal that: while optimistic scenarios (SSP1-2.6) show manageable declines of 15-20% in inter-annual rate of change of agricultural added value, pessimistic pathways (SSP3-7.0) precipitate catastrophic losses exceeding 130% in vulnerable countries. Egypt's high adaptive capacity can mitigate up to 90% of potential losses, while Libya and Algeria remain highly vulnerable. Precipitation levels emerge as the most critical factor, although the cumulative effect of rising temperatures could become significant under pessimistic scenarios.

This chapter investigates the impact of climate vulnerability on pre-tax income inequality across MENA countries over 1996–2021. Using random-effects panel models, it examines the direct effect and interactions with agricultural employment and institutional quality. Four econometric models explore these relationships, finding that climate vulnerability is positively associated with inequality, with agricultural dependence and weak institutions intensifying the effect. Food security and ecosystem degradation emerge as key drivers. Graphical analysis shows non-linear country dynamics. Findings emphasize the complex interplay of environmental, economic, and institutional factors, offering insights for targeted policy interventions to mitigate the socio-economic consequences of climate risks.

This chapter explores the transformative potential of agricultural policy synergies between Algeria and Tunisia to address shared climate vulnerabilities, including water scarcity, aquifer depletion, and rising food import dependency. By systematically analyzing five key areas of collaboration—water governance exchange, drought-resistant crop development, cross-border aquifer management, agri-tech knowledge transfer, and climate-smart value chains—the study demonstrates how integrated approaches can enhance resilience, optimize resource use, and reduce economic risks. The chapter argues that bilateral cooperation could unlock ecological, economic, and social dividends, such as slower aquifer depletion, lower cereal import costs, and rural job creation. However, realizing these synergies requires overcoming institutional fragmentation and political inertia. The chapter concludes with a systemic framework for Maghreb agricultural resilience, positioning Algeria and Tunisia as potential leaders in regional climate adaptation.

Drought, salt, high temperatures, pH imbalances in the soil, heavy metal toxicity, and floods are examples of abiotic stresses that pose serious challenges to agricultural sustainability and crop output worldwide. Plant physiological, biochemical, and molecular processes are negatively impacted by these environmental stresses, which are made worse by the rapid rate of climate change. This results in stunted development, decreased photosynthetic efficiency, disturbed nutrient balance, and eventually significant yield losses. The fundamental processes via which abiotic stressors affect plant systems are thoroughly examined in this chapter, with special attention paid to oxidative stress, hormone signaling, osmotic and ionic imbalances, and stress-induced metabolic reprogramming. It also clarifies crop-specific susceptibilities and geographical differences in stress exposure, emphasizing the disproportionate impacts on rainfed agricultural systems and staple food crops in areas like the Middle East, South Asia, and sub-Saharan Africa. The chapter offers an integrative framework of mitigation strategies in addition to evaluating the environmental and economic effects of prolonged abiotic stress, such as soil degradation and biodiversity decline. These consist of microbial inoculants, sophisticated biotechnology treatments, traditional agronomic techniques, and sustainable land management strategies. In the context of creating resilient agroecosystems, the function of precision technology, climate-smart agriculture, and international policy frameworks is also covered. The chapter emphasizes the critical need for flexible, science-based approaches to protect global food security in the face of growing environmental stressors via multidisciplinary insights and case studies tailored to certain regions.

Mian Muhammad Ahmed, College of Life Science and Technology, Tarim University, China
Syeda Maira Hamid, College of Life Science and Technology, Tarim University, China
Zhang Rui, Department of Horticulture and Forestry, Tarim University, China
Pan Zhiyong, College of Horticulture and Forestry Sciences, Huazhong Agricultural University, Wuhan, China
Guo Xingyu, College of Horticulture and Forestry, Tarim University, China
Ye Qian, College of Life Science and Technology, Tarim University, China
Shangqi Yu, College of Life Science and Technology, Tarim University, China
Muhammad Asim, College of Horticulture and Forestry Sciences, Huazhong Agricultural University, Wuhan, China
Xia Qiuhao, College of Horticulture and Forestry, Tarim University, China
Ayaz Ahmad, College of Life Science and Technology, Tarim University, China
Yerzat Yerhazi, College of Horticulture and Forestry, Tarim University, China
Yue Qinyang, College of Horticulture and Forestry, Tarim University, China

Climate change is worsening drought, salinity and extreme heat across MENA, threatening food security. Microbial biotechnology offers a sustainable solution by harnessing beneficial soil microbes like PGPR, AMF and endophytes that enhance crop resilience through nutrient mobilization, stress hormone regulation and antioxidant protection. Field trials in key crops (wheat in Egypt, olives in Morocco, date palms in Saudi Arabia) show 15-28% yield increases, though challenges remain in strain optimization, farmer adoption and policy support. Scaling up requires region-specific microbial consortia, advanced delivery systems like nano-encapsulation, and integration with digital agriculture. By overcoming these barriers, microbial solutions can become vital for climate-smart agriculture in MENA, supporting both food security (SDG 2) and climate action (SDG 13) amid growing environmental stresses.

Citrus farming in the MENA region faces increasing challenges due to climate-induced water scarcity. Ensuring precise and efficient irrigation is essential to sustain production under semi-arid conditions. This chapter presents a remote sensing-based method that combines multispectral (Sentinel-2) and radar (Sentinel-1) imagery to estimate the water requirements (ETc) of citrus trees in southern Morocco over a five-year period. Vegetation indices such as NDVI and radar-based indices like VH-VV were used to compute crop coefficients (Kc), which were then coupled with meteorological data to estimate ETc. Radar imagery, less affected by weather conditions, provided high accuracy (RMSE = 0.050) and continuous monitoring. Spatial and temporal variability in water demand was effectively captured, enabling targeted irrigation strategies. This approach offers a scalable and cost-effective tool to improve water management and resilience in agriculture, contributing to key climate adaptation priorities in the MENA region.

The technoscientific transformation in the agriculture sector has significantly altered the traditional seed system that has profoundly affected the rights of the farmers. The evolution of biotechnology including genetically modified crops, gene editing, hybrid seeds protected under the Intellectual Property regime has reshaped the farmers right to seed sovereignty, sharing, saving and exchanging practices. In the Global South, there exist a wide spectrum of conflict between the technological advancement and the rights of farmers. In light of the above perspective, by tracing the historical shift from community based traditional seed knowledge to corporate based seed and patent technologies, this chapter critically analyzes the innovation

Climate change remains a critical barrier to sustainable development globally, with energy production responsible for approximately 60% of greenhouse gas (GHG) emissions. The United Arab Emirates (UAE), a country with substantial coastal infrastructure in an arid region, is particularly vulnerable to climate-related risks, including rising sea levels and extreme heat events. Despite substantial efforts—including investments in renewable energy, a net-zero carbon target by 2050, and the implementation of comprehensive policies outlined in its Nationally Determined Contribution (NDC)—progress toward Sustainable Development Goal 13 (Climate Action) remains below targeted levels. This study examines the UAE's progress on SDG 13 objectives and employs the Delphi methodology to identify key barriers hindering advancement. These include inadequate public regulatory frameworks, insufficient private sector incentives, and limited climate finance mechanisms. The paper also explores actionable insights by benchmarking international best practices and offering strategic recommendations.

This chapter explores the legal safeguards available to consumers in the UAE's agricultural sector in the face of climate change and increasing reliance on biotechnology. It analyzes the regulatory framework that governs food safety and biosafety, focusing on Federal Law No. (9) of 2020 on Biosafety of Genetically Modified Organisms and its Executive Regulations issued under Cabinet Resolution No. (84) of 2022. These legal instruments are assessed alongside Federal Law No. (10) of 2015 on Food Safety to evaluate how the UAE ensures consumer protection amid environmental pressures and evolving agricultural practices. The chapter highlights the UAE's proactive approach to regulating genetically modified products through labeling, import controls, and environmental risk assessments. By examining the intersection of climate adaptation, technological innovation, and legal regulation,

Preface

The challenge of feeding a growing global population in the face of a rapidly changing climate is one of the most pressing issues of our time. This is especially true for the Middle East and North Africa (MENA) region, an area already defined by its arid and semi-arid conditions, fragile ecosystems, and high dependency on agriculture for both employment and food security. The compounding pressures of climate change—including prolonged droughts, extreme temperatures, desertification, and water scarcity—are not merely environmental issues; they are fundamental threats to economic stability, social equity, and food sovereignty. It is within this critical context that this book, Impacts of Climate Change on Agriculture in the MENA Region, is presented. This collection of chapters represents a concerted effort by leading experts to provide a multifaceted and in-depth examination of the challenges and potential solutions for one of the world's most vulnerable regions.

This book is a comprehensive exploration of the intricate relationship between climate change and agriculture in the MENA region. It is divided into three core thematic sections. The first section establishes the foundational context, examining the biophysical and socioeconomic impacts of climate change on agriculture across the region. Chapters in this section delve into the physiological effects of abiotic stress on crops, use quantitative analysis to model regional agricultural dynamics, and explore the economic and social consequences, including the link between climate change and income inequality.

The second section shifts the focus to technological and institutional responses. It investigates how cutting-edge technologies, such as remote sensing, microbial biotechnology, and digital transformation, can be leveraged to enhance crop resilience and optimize resource management. It also critically analyzes the role of legal frameworks, property rights, and policy synergies in creating a supportive environment for sustainable agricultural development.

The final section is dedicated to forward-looking topics, including specific case studies on marine aquaculture, biodiversity, and the legal implications of new agricultural technologies. It also explores the role of business contracts and legal safeguards in addressing modern challenges like seawater desalination and consumer protection. By weaving together a diverse range of perspectives—from agricultural science and economics to law and policy—this book provides a holistic view of a complex issue, moving beyond a simple description of problems to a robust discussion of innovative and actionable solutions.

The importance of this topic "Impacts of Climate Change on Agriculture in the MENA Region" cannot be overstated, both for the MENA region and for the world at large. The MENA region is a global hotspot for climate vulnerability. Its unique geographic and climatic characteristics, combined with a high reliance on rainfed agriculture and limited freshwater resources, make it particularly susceptible to climate shocks. Rising temperatures, shifting rainfall patterns, and sea-level rise threaten to destabilize food production systems, exacerbating existing social and political tensions. For instance, countries in the Nile Delta, a vital agricultural hub, face a dual threat from saltwater intrusion and reduced water flow, jeopardizing the livelihoods of millions. The impacts, however, are not confined to the region itself; they ripple outward through global food supply chains, impacting international food security and economic stability. By studying the MENA region, we can glean critical insights that are applicable to other arid and semi-arid parts of the world, making this research a vital contribution to the global effort to build climate-resilient food systems. The solutions explored here—from water management technologies to climate-smart policy—can serve as models for global adaptation strategies.

This book is intended for a diverse audience of scholars, practitioners, and policymakers. It will be an invaluable resource for academics and students in fields such as agricultural science, environmental economics, climate studies, and international development. The detailed case studies, quantitative analyses, and multidisciplinary approach will be particularly useful for those conducting research or seeking a deeper understanding of the MENA region's agricultural landscape.

Policymakers and government officials will find the book's policy recommendations and institutional analyses to be highly relevant for formulating effective climate adaptation strategies. The chapters on legal frameworks, policy synergies, and technological adoption provide concrete guidance for creating resilient and sustainable agricultural sectors.

Finally, agricultural professionals, development workers, and private sector stakeholders will benefit from the practical insights into technologies like remote sensing, microbial biotechnology, and marine aquaculture. The book's focus on sustainable practices and legam and economic analysis can inform decision-making on the ground and help bridge the gap between scientific innovation and practical application. The twenty chapters in this book will therefore address:

CHAPTER OVERVIEW

Chapter 1: Utilizing Business Contracts to Address the Impact of Climate Change on Agriculture in the MENA Region: Seawater Desalination Plant Contract as a Mode introduces the book by one of its editors with a practical and legal perspective. This chapter analyzes how business contracts, specifically those for seawater desalination plants, can be structured to address the impacts of climate change. It offers a blueprint for using contractual mechanisms to secure water resources and mitigate risks in a water-scarce region.

Chapter 2: Climate Change and Agricultural Production in Morocco: Assessing Impacts and Resilience offers, by the second editor of this book, a detailed case study of Morocco, a country with a significant agricultural sector highly vulnerable to climate change. This chapter assesses the specific impacts of climate change on Moroccan agricultural production and analyzes the country's existing resilience strategies and adaptation policies, providing a focused, regional perspective on the broader themes of the book.

Chapter 3: The effects of climate change on agriculture in Algeria: Structural constraints and the search for resilience provides a similar deep dive into Algeria, examining the structural constraints, such as land degradation, that limit agricultural resilience. It highlights Algeria's national efforts to combat desertification and its performance review and assessment of implementation systems for climate adaptation.

Chapter 4: Climate change, agriculture, and food security in North Africa a quantitative analysis shifts to a quantitative approach, using models like the de Martonne aridity index and global climate model data (CMIP6) to provide a data-driven analysis of climate change's effects on North African agriculture. This chapter offers a rigorous, statistical perspective on the region's long-term dynamics under different climate scenarios.

Chapter 5: Economic impacts of climate change on agriculture in North African region: A Ricardian Multi-Regional Analysis further expands on the economic implications by applying the Ricardian-production function to a multi-regional panel data set. This chapter analyzes how climate variables, such as precipitation and temperature, influence agricultural income and productivity, providing robust economic evidence for the need for climate-resilient policies.

Chapter 6: Climate Change and Inequality: The Agricultural Channel explores the critical social dimensions of climate change, specifically examining the relationship between climate vulnerability and income inequality. It shows how the agricultural channel can exacerbate existing disparities, particularly for the bottom 50% of the income distribution, and highlights the need for policies that protect vulnerable populations from climate shocks.

Chapter 7: Policy synergies for climate-resilient agriculture: Can Algeria and Tunisia learn from each other? is a comparative policy analysis that examines the climate-resilient agricultural strategies of Algeria and Tunisia. It identifies areas of policy synergy and mutual learning, providing insights into how neighboring countries can cooperate to enhance their collective adaptive capacity.

Chapter 8: Abiotic Stress and Crop Production in a Changing Global Environment sets the stage by providing a foundational overview of the physiological and biochemical mechanisms through which abiotic stressors—such as drought, salinity, and extreme temperatures—impact plant systems. The chapter explores the disproportionate effects on rainfed agricultural systems and staple food crops in regions like South Asia and the African Sahel, offering an integrative framework of mitigation strategies, including microbial inoculants and precision technology.

Chapter 9: Role of Microbial Biotechnology in Mitigating Climate-Induced Agricultural Stress delves into a biological solution to abiotic stress. This chapter examines the potential of microbial inoculants and other biotechnological solutions to enhance crop resilience against drought, salinity, and other environmental stressors, showcasing how biological innovations can contribute to sustainable agriculture.

Chapter 10: Using Multispectral and Radar Imagery to Estimate Citrus Water Requirements in the Context of Climate Change in the MENA Region: Case from Morocco provides a practical application of remote sensing technology. This chapter presents a specific case study from Morocco, detailing how multispectral and radar imagery can be used to accurately estimate water requirements for citrus crops, a critical insight for water-scarce regions.

Chapter 11: Technoscientific Impact of Agriculture on Farmers' Seed Sovereignty focuses on the crucial issue of seed sovereignty. This chapter argues that farmers should be considered "co-creators" of agricultural innovations and have legal ownership of their field data. It advocates for an inclusive policy approach that diversifies intellectual property protections to safeguard farmers' rights.

Chapter 12: Digital transformation, climate change, and agriculture: towards a better sustainability in the MENA region provides a comprehensive look at the role of digital technologies, from precision farming to AI-driven forecasting, in fostering agricultural sustainability. This chapter discusses how digital transformation can bridge knowledge gaps, improve efficiency, and build a more resilient agricultural sector.

Chapter 13: Impact of Biodiversity Loss on Consumer's Right to Food: Sustainable Agriculture as a Solution to Climate Change links biodiversity loss to consumer rights. This chapter explores how the decline in biodiversity affects food security and quality, arguing that sustainable agricultural practices are essential for protecting both ecosystems and the consumer's right to safe and nutritious food. It discusses international frameworks like the Convention on Biological Diversity (CBD) and the Kunming-Montreal Global Biodiversity Framework (GBF).

Chapter 14: Expert's Opinion on the UAE's Progress Toward Sustainable Development Goal 13, Climate Action is a country-specific analysis of the United Arab Emirates. This chapter provides an expert's perspective on the UAE's progress towards achieving SDG 13 (Climate Action), detailing its policies and national determined contributions (NDCs) to reduce emissions and build climate resilience.

Chapter 15: Legal Safeguards for Consumers in the UAE's Agricultural Sector Amid Climate Change delves into the legal frameworks in the UAE that protect consumers from the negative impacts of climate change on the agricultural sector. This chapter examines laws and regulations designed to ensure food safety, quality, and affordability in a changing environment, offering a model for other nations.

Chapter 16: Towards responsible marine aquaculture integrating the blue economy and sustainable development: marine aquaculture introduces marine aquaculture as a promising solution for food production in the MENA region. This chapter concludes the book by addressing an original, complementary, and essential theme in the preservation of life in the face of ongoing climate change. It addresses the principles of a "blue economy" and explores how responsible marine aquaculture can be integrated into sustainable development strategies to provide a reliable food source while minimizing environmental impact.

This book is more than a compilation of research in the three main areas covered by the study of climate change impacts on agriculture in the MENA region, namely: Regional Impacts and Socioeconomic Analysis; Technological and Institutional Solutions; and Law, Policy, and Emerging Solutions. It constitutes an essential resource and a call to action. By synthesizing a wealth of knowledge from various disciplines, it offers a unique and comprehensive understanding of the complex challenges facing the MENA agricultural sector. The chapters collectively demonstrate that building resilience requires a multifaceted approach combining technological innovation, sound legal and institutional frameworks, and a thorough understanding of socioeconomic dynamics.

Our hope is that this book will serve as a foundational text for scholars, a practical guide for policymakers, and a source of inspiration for all those dedicated to building a sustainable and food-secure future for the MENA region and the world. It is a testament to the idea that by working together across disciplines and borders, we can transform vulnerability into resilience and ensure that agriculture can thrive, even in the face of our planet's most profound environmental shifts.

Mohammed El Hadi El Maknouzi

University of Al Dhaid, UAE

Hicham Sadok
Mohammed V University in Rabat, Morocco

Acknowledgment

It is with immense gratitude that we acknowledge the contributions of all the authors who dedicated their time, expertise, and research to the twenty chapters that form the core of this book. Their insightful work and commitment to exploring the complex impacts of climate change on agriculture in the MENA region have made this publication possible.

We would also like to extend our sincere thanks to the University of Al Dhaid in the UAE and Mohammed V University in Rabat for their invaluable support. Their institutional backing and encouragement of academic discourse have been instrumental in the realization of this project.

Introduction

Climate change represents one of the most pressing global challenges of the 21st century, with profound implications for food security, economic stability, and social well-being. While its effects are felt worldwide, the Middle East and North Africa (MENA) region stands at a critical crossroads, facing a perfect storm of climate-related vulnerabilities. Characterized by a high degree of aridity, limited freshwater resources, and significant socioeconomic challenges, the MENA region is disproportionately susceptible to the adverse impacts of a warming planet. This book, a comprehensive compilation of 20 chapters, delves into the intricate relationship between climate change and agriculture in the MENA region, providing a detailed analysis of the threats, a deep exploration of the regional and socioeconomic consequences, and a forward-looking examination of the strategies and policies necessary to build agricultural resilience. By integrating insights from climate science, agronomy, biotechnology, and public policy, this volume aims to serve as a vital resource for researchers, policymakers, and practitioners committed to securing the future of food production in one of the world's most vulnerable areas.

The agricultural sector in the MENA region is inherently susceptible to a variety of abiotic stresses—non-living environmental factors that negatively impact plant growth and productivity. Climate change is acting as a powerful amplifier of these stresses, creating a more hostile environment for crop cultivation. Drought, a perennial challenge in the region, is becoming more frequent, severe, and prolonged. This scarcity of water is a direct threat to rain-fed agriculture, which is a cornerstone of the region's food production. The book's detailed analyses in multiple chapters highlight how reduced precipitation and altered rainfall patterns are diminishing soil moisture and groundwater reserves, forcing farmers to make difficult choices about irrigation, often with limited resources.

In addition to drought, extreme temperatures are rising, leading to heat stress that impairs critical physiological processes in plants. The book explores how high temperatures disrupt photosynthesis, accelerate plant respiration, and cause dehydration, ultimately leading to significant reductions in crop yield. The specific impacts on staple crops, such as wheat, barley, and maize, are examined in detail, revealing how their thermal tolerance thresholds are being breached more frequently, particularly during crucial stages of their growth cycle.

Furthermore, the issue of salinity is explored in depth. As rising temperatures increase evaporation rates, the concentration of salts in both soil and irrigation water intensifies. This phenomenon is exacerbated by unsustainable irrigation practices and the intrusion of saltwater into coastal aquifers. The chapters dedicated to this topic detail the physiological and biochemical mechanisms by which salinity stress inhibits nutrient uptake, causes osmotic stress, and leads to ion toxicity in plants. The combined effect of these abiotic stresses-drought, heat, and salinity-is creating a cumulative burden on the region's agroecosystems, making agricultural production increasingly precarious. The text provides a molecular-level perspective on these challenges, exploring the genetic and biochemical pathways that are compromised, and laying the groundwork for developing more resilient crop varieties.

The consequences of climate-induced agricultural stress extend far beyond crop fields. The chapters in this book provide a granular analysis of the regional and socioeconomic repercussions, painting a picture of a complex and interconnected crisis. The MENA region's diverse geography means that the impacts are not uniform. For instance, the Nile Delta in Egypt faces the twin threats of saltwater intrusion and a reduced flow from the Nile River, while countries in the Maghreb-including Algeria, Morocco, and Tunisia-are grappling with desertification and severe water shortages. The book uses specific case studies to illustrate these varied challenges, providing empirical data on how crop production in these nations is projected to decline under different climate scenarios.

Beyond the immediate agricultural losses, the socioeconomic ripple effects are significant. Decreased crop yields lead to food insecurity, making populations more reliant on food imports, which in turn exposes them to volatile international markets. This dynamic can drive up food prices, disproportionately affecting the most vulnerable populations. The book examines the links between climate change, food price volatility, and social instability, showing how agricultural shocks can become catalysts for broader social and political unrest.

The decline of the agricultural sector also has profound implications for rural livelihoods. Many communities in the MENA region are heavily dependent on farming for their income. The chapters on this topic detail how climate change-induced crop failures and land degradation are leading to poverty, displacement, and rural-to-urban migration. This mass movement of people places an added strain on urban infrastructure and social services, while simultaneously depleting the traditional agricultural knowledge base in rural areas. The book also highlights the long-term ecological consequences, such as the loss of agricultural biodiversity and the degradation of soil health, which create a self-perpetuating cycle of environmental decline and reduced productivity. The detailed analysis in the book's chapters highlights how these interconnected issues demand a holistic and coordinated response that goes beyond simply improving farming techniques.

Recognizing the gravity of these challenges, a significant portion of this book is dedicated to exploring a wide array of strategies and policies designed to enhance agricultural resilience in the MENA region. The solutions proposed range from traditional, time-tested practices to cutting-edge technological innovations. On the policy front, the book advocates for integrated water resource management, which includes promoting efficient irrigation systems like drip and sprinkler irrigation, and investing in rainwater harvesting and desalinization technologies. It also emphasizes the need for comprehensive land-use policies that combat desertification and protect fertile lands from urban sprawl. The book's policy-focused chapters provide a blueprint for institutional reform, advocating for stronger governance, better data collection, and increased investment in agricultural extension services to help farmers adapt to new realities.

In terms of on-the-ground practices, the book examines the importance of climate-smart agriculture. This includes promoting conservation tillage, which helps retain soil moisture; intercropping, which enhances biodiversity and soil health; and the use of drought-tolerant and salt-tolerant crop varieties. The book also delves into the exciting potential of biotechnological innovations. Several chapters are dedicated to the role of genome editing and other genetic engineering techniques in developing new crop varieties that are more resilient to abiotic stresses. The text explores how these technologies can accelerate the development of crops that require less water, can tolerate higher temperatures, and thrive in saline soils, offering a powerful tool in the fight against climate change.

Finally, the book underscores the crucial role of international cooperation and knowledge sharing. It argues that no single country can solve this crisis alone and that collaborative research, technology transfer, and shared best practices are essential for building a resilient regional food system. By weaving together these diverse perspectives-from the molecular biology of plant stress to the geopolitics of food security-this book provides a robust framework for understanding and addressing the complex challenges facing agriculture in the MENA region. It is a call to action, offering not just a diagnosis of the problem but a comprehensive roadmap for a sustainable and food-secure future.

Mohammed El Hadi El Maknouzi

University of Al Dhaid, UAE & Mohammed V University, Rabat, Morocco

Hicham Sadok

Mohammed V University, Rabat, Morocco

Chapter 1

Utilizing Business Contracts to Address the Impact of Climate Change on Agriculture in the MENA Region:

Seawater Desalination Plant Contract as a Mode

Mohammed El Hadi El Maknouzi
https://orcid.org/0000-0003-3396-8985
College of Law, University of Al Dhaid, Sharjah, UAE

ABSTRACT

Water scarcity is one of the main concerns troubling nations, as agricultural activity, food security, and sustainable development cannot be conceived without this vital resource. In this context, seawater desalination plants have become one of the most commonly used solutions by national governments and local authorities to provide potable water and water for agricultural irrigation. Through this study, we seek to demonstrate that employing business contracts may represent an optimal means of addressing the impact of climate change on agriculture in the Middle East and North Africa (MENA) region, using a public-private partnership (PPP) contract for a seawater desalination plant as a case study.

DOI: 10.4018/979-8-3373-3346-5.ch001

INTRODUCTION

There is broad consensus that climate change poses a significant threat to global agri-food systems, with arid and semi-arid regions being the most exposed to such risks. This makes it essential to develop a thorough understanding of the climate-related challenges and the measures needed to address them, in order to ensure the long-term sustainability of agriculture (Dzvene et al., 2025). Undoubtedly, climate change represents one of the major challenges facing agriculture in developing countries, directly affecting the livelihoods of farmers (Naik et al., 2025). This reality, which links the fate of agriculture to climate conditions, has implications for all countries, particularly the MENA region, whose peoples are bound by historical and cultural ties, in addition to facing similar climate-related challenges.

In this context, it is evident that freshwater scarcity in the MENA region gives rise to serious problems, the severity of which is compounded by the effects of climate change (Gabr, 2023). This necessitates a transformation in practices, particularly those used for irrigating agricultural crops, as a necessary step to address the depletion of groundwater resources and to safeguard food security in regions facing the combined pressures of climate change and water scarcity (Darwish et al., 2025).

Most estimates indicate that nearly half of the world's population suffers increasingly from severe water scarcity for at least one month each year, while a quarter of the global population faces extremely high levels of water stress. This situation has made seawater desalination an essential solution for ensuring the sustainability and safety of freshwater, which plays a decisive role in meeting the water needs of residential communities, agriculture, and industry, particularly in arid, dry, and coastal regions (Xu et al., 2025).

Seawater desalination technology has witnessed significant expansion in use, driven by the reduction in the cost of freshwater production. It has emerged as a reliable and efficient solution, thanks to sustained research efforts. Considering the notable advancements in desalination technology, the unit cost of freshwater production has decreased significantly (Ji et al., 2025).

A large part of this development is attributed to advancements in technology and engineering, with the future of desalination expected to incorporate highly permeable and fouling-resistant membranes, along with desalination plants powered by renewable energy sources (Moon et al., 2025). The use of solar-powered evaporation for desalination represents an emerging, cost-effective, and sustainable approach to addressing the global freshwater crisis (Zhou et al., 2025).

At the technical level, reverse osmosis (RO) currently holds the largest share of the global market (Xu et al., 2025) and stands as the most widely used method for seawater desalination. This process requires complex pre-treatment operations, including coagulation, filtration, and chemical cleaning, to reduce membrane fouling (Ji et al., 2025). However, overstating the advantages of seawater desalination through reverse osmosis or other similar technologies should be avoided, as brine discharges represent one of the major challenges due to their potential environmental impact on marine ecosystems (Sola et al., 2025).

All of this confirms that climate-resilient technologies are an effective means of protecting agricultural systems from climate fluctuations (Naik et al., 2025). Indeed, sometimes it is necessary to integrate social data and engineering solutions to develop water infrastructure (Sangsefidi et al., 2025). In this context, the law emerges as a powerful tool to address the impact of climate change on agriculture. A large part of seawater desalination plant projects relies fundamentally on the law, particularly contracts, as the source of parties' obligations.

It should be noted that if the economy is based on exchange, then the realization of such transactions is governed by business contracts. In line with this, contracts appear as agreements between economic actors to regulate the flows between them. Contracts are present at every stage, from the supply of raw materials and the conceptualization of the product to its marketing and eventual acquisition by the final recipient (such as the consumer), not to mention the accompanying operations aimed at securing financing, consulting, and insurance (Testu, 2019).

In practice, business contracts constitute the core framework for the establishment of seawater desalination plants, with a variety of legal provisions and contractual principles—falling within the broad scope of business law—giving effect to these contracts. Attention must also be paid to the fact that the impact of climate change—whether positive or negative—on agriculture is closely tied to the fluctuation of the financial cost of agricultural products and their market prices for consumers.

Against this backdrop, this research seeks to demonstrate that leveraging business contracts provides an optimal means of addressing the impact of climate change on agriculture in the MENA region. The study will use a seawater desalination plant contract as a case study, focusing on the role of project companies in reducing the financial burden borne by states in constructing and operating such facilities. Additionally, the research will discuss how specific contractual provisions contribute to stabilizing the prices of water used for irrigating agricultural products.

SIGNIFICANCE OF THE RESEARCH AND DATA COLLECTION

The significance of analyzing the use of business contracts to address the impact of climate change on agriculture in the MENA region—focusing on a seawater desalination plant contract as a case study—becomes particularly evident when surveying the number of studies published in journals indexed in the Scopus database (See Table 1).

Table 1. Scopus database query.

Conditions	Description
Open access and non-open access documents	**Access**
April 25, 2025	**Query search date**
All years	**Years**
All Subject area	**Subject area**
Article, book chapter, book, conference paper and review	**Document type**
English language only	**Language**
Query nº (1)	
Search within article title, abstract, keywords (The impact of climate change on agriculture)	**Search query**
)20,961) documents found	**Result (Number of documents)**
Query nº (2)	
Search within article title, abstract, keywords (Seawater Desalination Plant Contract)	**Search query**
(51) documents found	**Result (Number of documents)**
Query nº (3)	
Search within article title, abstract, keywords (Seawater Desalination Plant Contract)	**Search query**
(27) documents found	**Result (Number of documents)**

The results of Search Query No. 1 revealed a substantial volume of studies addressing the impact of climate change on agriculture, spanning a wide range of disciplines, particularly in environmental sciences, biology, earth and planetary sciences, engineering, as well as the social sciences. By contrast, Search Query No. 2 identified a total of 51 studies, most of which focused on keywords such as Desalination, Seawater, Reverse Osmosis, Water Supply, Desalination Plant, and Water Filtration.

In this context, the results of Search Query No. 3 confirmed the existence of 27 studies, where the scope was limited to social sciences with a particular emphasis on law as a keyword. Upon careful examination of the titles of these studies, it was observed that most of them address general issues relating to law and contracts, including the compatibility of business contracts with legal systems, particularly in relation to alternative dispute resolution mechanisms, governance, and smart contracts. However, no study was found that explores the use of business contracts specifically as a mechanism for addressing the impact of climate change on agriculture in the MENA region, particularly through an analytical framework centered on seawater desalination plant contracts. This clearly underscores the originality of the present research in terms of its title, scope, anticipated findings, and proposed recommendations.

MATERIALS AND METHODS

To fully address the research problem concerning the utilization of business contracts as a means of mitigating the impact of climate change on agriculture in the MENA region, it is necessary to employ specific tools, foremost among them being practice. It is well established that the complexity of business contracts renders them closely connected to practical implementation. Accordingly, this research will focus primarily on such practices in relation to seawater desalination plant contracts.

Complementing this, an interdisciplinary approach will be adopted, with law and contract to the fore, while remaining open to other relevant fields such as environmental sciences, water, agriculture, climate, and economics, among others. This methodology—which integrates law, contract, and practice while drawing on related disciplines—will make it possible to extract findings and recommendations aimed at identifying the potential contributions of law and contract to addressing the impact of climate change on agriculture.

The Establishment of Seawater Desalination Plants through Project Companies: An Application of Business Contracts to Address the Impact of Climate Change on Agriculture

The establishment of seawater desalination plants through project companies represents one of the applications of business contracts for addressing the impact of climate change on agriculture. However, the adoption of such decisions by a public legal entity—whether the State itself or one of its affiliated bodies—is closely tied to the financial burden that may be incurred. In many cases, this requires the formation of a project company to undertake the construction and operation of a seawater desalination plant within the framework of a PPP contract, which constitutes one of the forms of business contracts in its broader sense. The production of such a facility is to be allocated, wholly or partially, to agricultural irrigation. In this regard, it becomes necessary to examine whether the public legal entity should participate in the share capital of the project company established for the construction and operation of the seawater desalination plant, or whether its role should be limited to exercising oversight over the company's activities and operational management.

The Impact of the Financial Cost of Seawater Desalination Plants on the Decision to Establish a Project Company under a PPP Contract

It is well established that the broad concept of business contracts extends to encompass PPP contracts in their various forms and mechanisms, including project companies established pursuant thereto. In this connection, it is important to emphasize that water scarcity resulting from climate change has negative impacts on agricultural production, inevitably leading to higher production costs. This drives the search for alternatives to rainfall to ensure the continuity of irrigation processes at a reduced—or at least reasonable—financial cost. The reality is that the MENA region, including the Arab world, is one of the world's arid regions, which calls for in-depth knowledge of water management techniques (Jagannathan et al., 2009) as well as mechanisms for reducing the financial cost of producing and supplying water suitable for irrigation.

In a related context, the flexibility of contractual clauses and the adoption of a balanced risk-sharing mechanism between the partners have a positive influence on the private sector party's willingness to engage in PPP projects. It is evident that private sector partners tend to favor public sector entities that demonstrate a high degree of financial stability, with a clear preference for projects situated in more developed areas. Furthermore, governmental financial support serves as an incentive for private sector partners to advance with partnership projects (Kopańska et al., 2024). This is a clear demonstration of the way that financial considerations are one of the decisive indicators for the success of a PPP project.

It must also be noted that the availability of accurate and comprehensive data enables the formulation of an effective strategy for establishing a seawater desalination plant dedicated to agricultural irrigation. For example, as previously clarified, reverse osmosis (RO) technology is considered the most cost-effective due to its energy efficiency. Furthermore, within the framework of developments brought about by artificial intelligence (AI), machine learning enhances the accuracy of estimations significantly by identifying hidden patterns in large datasets.

Here, the main findings indicate that larger desalination plants, specialized technologies—such as thermal desalination—and remote locations contribute to increased costs. Moreover, the choice of procurement model influences financial expenditures, with evidence suggesting that projects financed by the private sector through PPP mechanisms tend to be more cost-efficient than those funded by governments, primarily due to supplier dynamics and competitive advantages.

In addition, with specific reference to the MENA region, geographical analysis reveals a high concentration of seawater desalination plants, accompanied by elevated capital expenditures driven by the intense demand for freshwater (Giwa et al.,

2025). This highlights that improving the financial costs of seawater desalination projects would contribute significantly to addressing the challenges associated with water scarcity.

In analyzing the indicators affecting the financial cost of a seawater desalination plant project, it is necessary to account for the energy component. Fluctuations in energy prices have a direct impact on the project's preparation costs and its operational sustainability (Czipf, 2025), a matter particularly relevant to seawater desalination projects.

In general, and based on the foregoing analysis, the principal indicators that, if neglected, may lead to an increase in the financial cost of constructing and operating seawater desalination plants can be summarized in Table 2.

Table 2. The key indicators to be considered in controlling the financial cost of seawater desalination plants.

Indicator Classification	Impact on Financial Cost
Climatic Indicator	Climate change often results in reduced rainfall levels, leading to water scarcity for irrigation purposes. This scarcity inevitably raises the cost of producing each liter of water, which may increase the final price of agricultural products.
Infrastructure Indicator	The absence or underdevelopment of essential infrastructure in the geographical area designated for the construction of the seawater desalination plant increases financial costs and diminishes its attractiveness to private sector partners.
Data-Related Indicator	The availability of accurate and comprehensive data facilitates the formulation of an effective strategy for developing a seawater desalination plant dedicated to agricultural irrigation. Conversely, incomplete or inaccurate data undermines the prospects of success for the partnership project, consequently increasing its financial burden.
Technical Indicator	Reverse osmosis (RO) technology is widely recognized as the most cost-effective option, due primarily to its superior energy efficiency, which leads to reduced operational costs. By contrast, opting for a technology with higher energy consumption inevitably results in increased operational expenditures.
Energy Indicator	Fluctuations in global energy prices have a direct impact on the financial cost of operating seawater desalination plants.

Table 2 illustrates that the indicators to be adopted in PPP contracts concerning the establishment and operation of seawater desalination plants intended for agricultural irrigation are highly diverse, making their exhaustive enumeration impracticable. These indicators extend to encompass contractual aspects, infrastructure, data accuracy, technology, and energy, among others. Consequently, controlling the aforementioned indicators would contribute to reducing the financial cost associated with the operation of seawater desalination plants, aligning with the negative impact of climate change on agriculture due to the decline in rainfall.

In reality, an excessive focus on the financial aspects of seawater desalination plant projects may lead to the neglect of social considerations and the requirements of sustainable development. The adoption of the Social Return on Investment (SROI) approach assists in identifying projects that yield the greatest environmental benefits for the investment, which enables governments to allocate their funding more effectively to support projects that enhance environmental sustainability. This facilitates the integration of environmental and financial efficiency into a single performance metric (Lee & Jung, 2025).

It is certain that innovation, in its various forms, forms an important means of achieving efficiency; however, it does not represent an inherent feature of partnership projects (Rangel & Galende, 2010). Within this context, it is essential to recognize a fundamental fact: The high financial cost of establishing and operating a seawater desalination plant project—particularly in countries with limited public financial resources—justifies the recourse to PPP contracts.

In effect, the adoption of PPP contracts serves to overcome the complexities associated with conventional public procurement procedures, particularly as such contracts place the operational risk solely upon the private sector party. However, this should not be construed as an attempt to replace traditional public-private cooperation mechanisms, such as public procurement contracts. Rather, the use of PPPs remains contingent upon the specific financial and managerial advantages they provide—a condition that applies to seawater desalination plant contracts.

This demonstrates that the financial dimension is the primary foundation justifying the recourse to PPP contracts. Consequently, such contracts are fundamentally structured as financial arrangements rather than purely legal constructs (Dirnat & Souffir, 2005), which allows for their use as a mechanism to address the impact of climate change on increasing financial costs.

It must also be factored in that the general philosophy underlying PPP contracts rests on two fundamental pillars. The first concerns the existence of a long-term contract concluded between a public legal entity and a private sector partner. The second lies in assigning the private sector partner a comprehensive task—namely, the design, construction, and operation of a public facility, although this may, in some cases, be limited to certain specific stages (Bloch, 2005). However, the success of the overall process depends on the use of various mechanisms, chief among which is the project company.

The involvement of project companies within the framework of PPP contracts constitutes one of the clearest manifestations of the application of private law, as this type of organization often takes the form of a joint-stock company and is therefore subject to the legal provisions governing company contracts. In practice, the process typically begins with the formation of a consortium composed of several enterprises, each specializing in a particular field—such as a construction company, another responsible for repair and maintenance, and a financial institution specialized in project financing. The objective of this consortium is to submit a unified and integrated offer during the legal and regulatory procedures for selecting the contracting party.

It should be noted that the said consortium does not possess any separate legal personality of its own, as it merely comprises an agreement between the group of enterprises brought together to execute the project. In other words, the consortium functions as a contractual instrument, with the agreement establishing it setting forth the rules applicable to its members and the manner in which it operates. The project company does not succeed to the rights and obligations of the selected consortium until after the PPP contract is signed (Bergere et al., 2010). In practical terms, the PPP contract is concluded with the project company (Emery, 2006).

The establishment of a project company is not mandatory, particularly with respect to PPP contracts relating to small-scale projects. As a general rule, the private sector partner is the sole point of contact vis-à-vis the public legal entity concerning all matters related to the performance of the contract (Bergere et al., 2010). In such cases, upon the conclusion of the PPP contract with the project company, the latter will, in turn, enter into several contracts with enterprises responsible for the construction of the facilities and for operation, especially with banks, in order to obtain the necessary financing (Emery, 2006).

Put differently, establishing a project company is not always a mandatory requirement, as the same company that signed the PPP contract with the public legal entity may proceed with its implementation, subsequently concluding contracts with certain suppliers and service providers to ensure the success of the project (See Figure 1).

Figure 1. Scenarios where project companies are mandatory or optional.

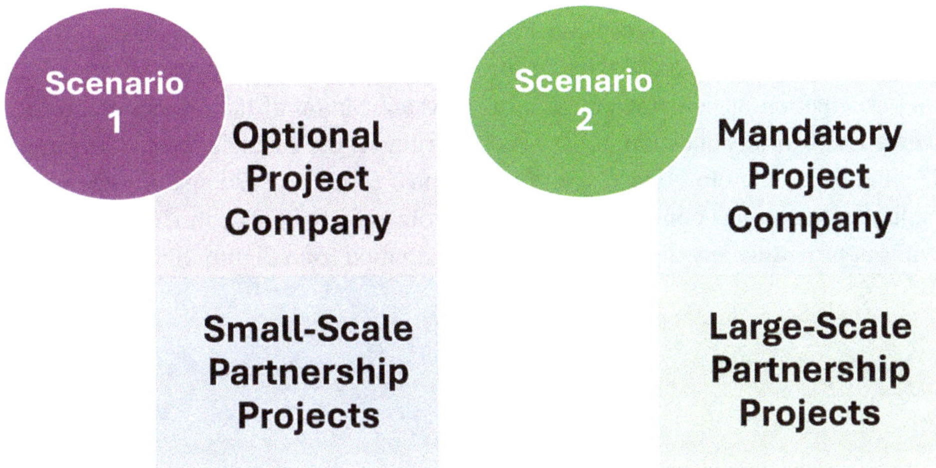

This confirms that the involvement of project companies within the framework of PPP contracts is optional and may not be justified in the case of smaller projects. However, it tends to become necessary—and may even be mandatory—for large-scale projects such as seawater desalination plants, as one of the mechanisms that may be employed to address the impact of climate change on agriculture.

This raises an issue that has a direct bearing on the financial cost of seawater desalination plants carried out under PPP contracts: Is it in the interest of the public legal entity to participate in the share capital of the project company?

EXAMINING THE MERITS OF THE PUBLIC LEGAL ENTITY'S DECISION TO PARTICIPATE IN THE SHARE CAPITAL OF THE PROJECT COMPANY FOR A SEAWATER DESALINATION PLANT

It should first be noted that the private sector partner, under a PPP contract, undertakes responsibility for performing a comprehensive task encompassing the design, construction or repurposing, maintenance, and/or operation of the seawater desalination plant, with an obligation to provide full or partial financing. This should be understood in light of the fact that PPP contracts are not classified as public procurements and are not subject to the legal framework governing such procurements, bearing in mind that in public procurement, it is the public entity that assumes the financial burden, which may include design, construction, maintenance, operation, and other related components (Terneyre, 2003).

Within this context of divergence, a central issue emerges from the moment the PPP contract for the seawater desalination plant is signed—namely, the role that the public legal entity will assume within the share capital structure of the project company. More specifically, what position should the public entity adopt regarding its participation or non-participation in the share capital of the project company? The answer to this question has a direct bearing on the financial cost. The public legal entity is therefore faced with two options: Either to participate in the share capital of the project company or to limit its role to that of the contracting authority without providing any direct or indirect contribution (See Figure 2).

Figure 2. Decisions available to the public legal entity regarding participation in the share capital of the project company.

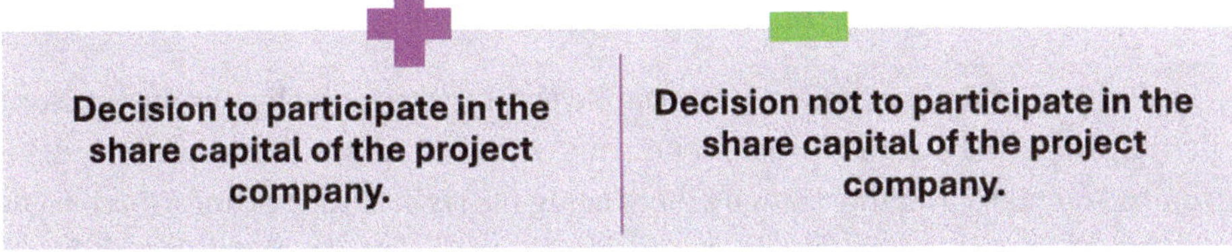

The public legal entity may decide to participate in the share capital of the project company if it deems that such participation serves its interests and contributes to the success of the PPP contract with the private sector partner. However, this approach, which the public legal entity may choose to follow, entails certain advantages and disadvantages, the extent of which varies depending on whether it involves acquiring a participation stake or assuming controlling oversight over the project company (See Figure 3).

Figure 3. Percentages granting either oversight or mere participation in the share capital of the project company.

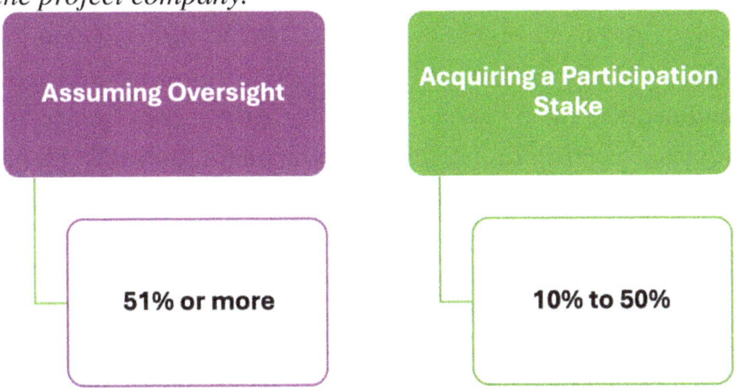

It should be noted that a decision by the public legal entity to acquire a participation stake in the share capital of the project company for a seawater desalination plant—or to extend that participation to acquiring control over its oversight authority—gives rise to a set of effects with both positive and negative dimensions, as follows:

- **Decision by the public legal entity to acquire a participation stake in the project company's share capital**: The participation stake is defined as ownership of a portion of a company's share capital ranging from 10% to 50%. In practice, the concept of participation as defined here reflects the intention to establish influential relations within the company in which shares or stakes have been subscribed or purchased, with the degree of influence varying according to the percentage held.

Regardless of the strategic role the public legal entity may play through owning a participation stake in the project company's share capital, this approach yields several advantages, foremost among them being the right to receive information and a portion of the annual profits. It also facilitates the public legal entity's ability to monitor the activities of the project company, thus enabling participation in decision-making when some practices contradict the terms of the PPP contract or conflict with the objectives underlying the establishment of the seawater desalination plant.

Moreover, owning a significant stake in the project company's share capital allows the public legal entity to make an active contribution to shaping the overall direction of various decisions taken within the internal bodies of the project company (such as general assemblies and the board of directors) in a manner that serves the public interest. This role becomes increasingly important when it concerns the general economic policy of the project company, which affects its financial planning and production quality—factors that have a direct impact on the company's standing, both positive and negative.

In a related context, the public legal entity's ownership of 33% + 1 of the voting rights in a project company structured as a joint-stock company entitles it to block decisions made at an extraordinary general meeting convened to decide on issues such as amendments to the bylaws or capital increases. The concept of a blocking minority is realized when a shareholder owns more than one-third (1/3) of the shares, enabling them to object to decisions passed by a two-thirds (66.66%) majority, regardless of the number of shareholders present or represented.

In certain special cases, some legislations, such as French law, are explicit in authorizing the public legal entity to hold a minority stake in the project company's share capital (Peyrical, 2017). This requires adapting the company's bylaws to comply with public law rules, given that public funds are involved (Sée, 2025).

Similarly, shareholders' agreements (Larrieu, 2007) appear as an effective means to elevate the level of participation held by the public legal entity in the project company to that of exercising supervisory authority within its internal bodies alongside other shareholders. Most of these agreements are concluded among shareholders who form coalitions to create a majority, thus ensuring the stability of shareholding and management, and consequently providing protection against hostile actions that may be taken by the minority (Ripert & Roblot, 2003).

Despite the aforementioned advantages, the public legal entity's ownership of a participation stake in the share capital of the project company is not without significant drawbacks, foremost among which are the financial consequences resulting from the failure of the PPP contract to achieve its intended objectives. Such failure may stem from weak commercial or industrial policies adopted by the project company—particularly in cases where the project company sells each liter of desalinated water at a price equal to or lower than its actual production cost. In such circumstances, the PPP contract may be terminated before its scheduled expiry date, exposing the public legal entity to a share of the financial costs and losses arising from the failure of the seawater desalination plant project.

- **Decision by the public entity to acquire oversight over the project company:** Although legal approaches to the concept of oversight differ (Schmidt, 1998), they converge on the same fundamental idea—that a natural or legal person, whether acting individually or in concert pursuant to an agreement, possesses the authority to manage and direct a company that is otherwise legally independent. Such oversight may take the form of one company exercising authority over another through indirect ownership of shares held by its subsidiary, or through controlling the legal and economic powers arising from share ownership (Laparde, 2007).

One of the most significant consequences of acquiring oversight is the attainment of financial dominance, achieved through the ownership of a majority or substantial portion of shares and voting rights in general assemblies. This, in turn, enables the entity to exercise decisive influence over the company's internal decision-making bodies, including general assemblies and the board of directors (Sortais, 1993).

As a result, and within this general framework, the public legal entity's acquisition of oversight over the project company grants it the authority to appoint members of the company's internal structures, define its financial and commercial policies, and exercise a range of other powers. By exercising such oversight, the public legal entity can direct the company's activities in a manner consistent with the objectives of the public service entrusted to its management and operation. This is one of the key powers enabling the orientation of the seawater desalination project toward addressing the impact of climate change on agriculture.

That said, in operational reality, the exercise of oversight by a public legal entity over the project company suffers from an inherent drawback: Most of the decisions it is expected to make are likely to be predominantly influenced by considerations related to the public service, compelling it to prioritize the public interest over the private interests of the project company. While the decisions in which the public legal entity participates within the project company may indeed produce positive and direct outcomes for the users of the public service under the PPP contract, they may simultaneously undermine the company's performance. This could expose the project company to financial difficulties and, ultimately, lead to the failure of the seawater desalination plant project.

These analyses lead us to a fundamental conclusion: That the exercise of oversight over the project company by a public legal entity would, in most cases, strip the PPP contract of its core essence. Moreover, it runs counter to the general principles underpinning the establishment of such contracts. This scenario could lead to an increase in the financial costs of the seawater desalination project, as a significant portion of the operating expenses would shift to the public entity. As a result, the financial viability of such projects would be undermined, which ultimately weakens efforts to address the impacts of climate change on agriculture in the MENA region. All of this highlights the need for careful consideration of the scenario in which the public legal entity refrains from participating in the share capital of the project company.

In many PPP contracts, the public legal entity initiating the tender procedure often chooses not to participate in the share capital of the project company, limiting its role to overseeing how the company performs the delegated tasks. Before assessing the effectiveness of the public legal entity's decision to refrain from contributing to the project company's share capital, it is important to note that the project company typically enters into multiple contracts—with banking institutions for financing, insurance companies for risk coverage, subcontractors for parts of the project's execution, and technical service providers for equipment supply and operational needs.

Nevertheless, the public legal entity retains the right to require—either through the tender documents or within the framework of the PPP contract with the private sector partner—that it be informed of all agreements the project company intends to conclude with third parties. This condition is intended to allow for prior review and approval before such agreements are signed and become legally effective. It is therefore clear that the public legal entity may exercise oversight over the project company's activities both during the formation of such agreements with suppliers and other partners and throughout their implementation.

This approach is consistent with the underlying rationale for establishing PPP contracts, while also protecting the public legal entity from the financial risks it may incur in the event of project failure, particularly if it holds a stake in the project company's capital. This consideration becomes even more critical given that the larger the production capacity of seawater desalination plants, the higher the associated construction and operational costs.

However, this approach should be subject to certain exceptions: The public legal entity may be permitted to participate in the project company's share capital when the nature of the project warrants such participation or when there is a need for the temporary replacement of a financially distressed shareholder pending the finding of another partner possessing the required financial and technical qualifications. In the same context, the documents constituting the PPP contract, together with the project company's bylaws, should incorporate specific provisions designed to prevent restrictive conditions on the purchase or sale of the project company's shares.

This strategy falls within what may be termed "smart participation," aimed at directing public capital in a manner that ensures the success and sustainability of projects arising from PPP contracts, while addressing the limitations of public financial resources. Such an approach aligns with the requirements of projects designed to mitigate the impacts of climate change on agricultural irrigation, particularly seawater desalination plants, given their complex nature.

It is also important to emphasize, within the context of supporting the ethical dimension of business contracts, that the agreements and charters repeatedly issued by the United Nations—particularly by the Committee on Economic, Social and Cultural Rights (CESCR)—underscore the necessity for individuals to have access to sufficient quantities of water of acceptable quality and at a reasonable cost (Leray, 2023). This principle applies equally to farmers, who have the right to obtain adequate water for irrigating agricultural products, also with acceptable quality and cost, and with sufficient flexibility to allow adaptation to climate change. Special attention must be paid to a fact that has a direct impact on the financial burden borne by farmers: Agricultural irrigation consumes large quantities of water, despite the use of water-saving technologies (Bandarra, 2009).

COMBINING LEGAL PROVISIONS AND CONTRACTUAL CLAUSES TO REGULATE THE PRICING OF SEAWATER DESALINATION PLANT WATER

Once the seawater desalination plant is completed and the project company begins its operation, a key issue arises concerning the prices at which the produced water will be sold to consumers—a concept that extends to include farmers. It is important to note that the regulation of desalinated water prices is governed by legal and regulatory provisions, which are broadly implemented through contractual clauses binding both parties to the PPP contract.

Regulating the Prices of Seawater Desalination Plant Water through Legal and Regulatory Provisions

It is essential at the outset to underscore the mandatory interconnection between the applicable legal and regulatory rules governing water prices, the contractual provisions incorporated into the PPP contract, and the commercial and industrial model adopted for the desalination plant. At this level of analysis, several factors must be considered, as they have a direct impact on the cost of desalinated water before it becomes a commodity within the trade and agricultural consumption cycle.

In reality, improving water-use efficiency contributes to reducing crop losses caused by extreme temperatures (Kürklü et al., 2025). Additionally, the cost of water production may vary depending on the technology used for desalination. For instance, the technical and economic feasibility of membrane distillation differs from that of reverse osmosis (Schwantes et al., 2025).

Within this context, some advocate pricing as an effective tool to encourage water conservation, while others argue that imposing regulations and awareness campaigns may be more effective (Prakash Khedun et al., 2025). Selling desalinated water to farmers at extremely low prices may discourage rational consumption and could even lead to overuse. This underscores the importance of pricing each liter of water in a way that strikes a balance between encouraging agricultural production and promoting daily consumption efficiency, based on pricing regulations that allow for adjustments in response to climate variability.

At the level of decision-making authority, determining water prices remains an open discussion among the relevant entities, whether directly or indirectly involved. In this regard, and for the purpose of cost recovery of water services, Article 9 of Directive 2000/60/EC of the European Parliament and of the Council of 23 October 2000 established a framework for Community action in the field of water policy (European Union. 2000) and clarified the components of calculating the total cost of water. These were identified as follows: financial, environmental, and resource costs, bearing in mind that the latter remains the most controversial in terms of its conceptualization and assessment (Sanabria & Torres, 2020).

All of this confirms that addressing climate change in its relation to agriculture requires the establishment of a structured pricing system for desalinated water. This should be based on an accurate definition of the price components and allow for necessary adjustments to account for the effects of drought and water scarcity on agricultural activities.

In pursuit of this objective, several models have been developed for calculating water pricing which factor in both the human right to water and the imperative of environmental preservation, while also considering certain local costs associated with the regular provision of water services—such as municipal expenses for the construction, operation, maintenance, and management of infrastructure. This approach helps support animal and plant production within the biosphere and regulates issues related to climate change (Banderas Tarabay & González-Villela, 2018).

Furthermore, water prices should reflect the environmental costs resulting from damage to ecosystems and their services, as well as resource costs associated with the loss of social welfare caused by the failure to use water for the most socially beneficial purposes. This necessitates the implementation of environmental performance indicators in agriculture for the recovery of irrigation service costs, to combat pollution, and ultimately reduce environmental impact (Mysiak & Gómez, 2015).

Despite the various constraints affecting water pricing, water market transactions remain viable (Liu & Yan, 2022), turning it into the sort of commercial arena sought by private sector partners for profit-making purposes. This necessitates careful consideration of the fact that establishing a seawater desalination plant within the framework of a PPP contract for agricultural irrigation purposes inevitably requires balancing the imperatives of serving the public interest—safeguarded by the public legal entity—with the private sector partner's inherent pursuit of profit. Added to this is the objective of mitigating the impacts of climate change on agriculture, which calls for pricing mechanisms for the water produced by desalination plants to exhibit a high degree of flexibility and adaptability.

Tracking developments in the water trade market reveals that, over time, there has been a significant increase in the volume of water rights owned by governments, non-governmental organizations, and investors who are not landowners. This shift has made water one of the most valuable assets held by many irrigation users (Seidl et al., 2020). This observation also applies to seawater desalination plants developed under PPP contracts, as such contracts typically span long implementation periods during which the private sector partner retains substantial water rights for the contract's duration. These rights are only transferred to the public legal entity upon the contract's expiration.

In this regard, it should be noted that many countries have long intervened through legislative instruments to encourage private investment in the water sector. France, for instance, permits, under specific legal and regulatory provisions, the management of public services by private sector partners. This approach is similarly adopted in several countries in the MENA region, such as Morocco, where financial resources are sometimes limited.

This option enables the local authorities responsible for potable water services to engage private sector partners to manage public services for periods adequate enough for them to recover their investments, recoup expenditures, and generate profits, subject to necessary periodic administrative reviews. These legal and regulatory provisions ensure that the public authority retains control over the public service while guaranteeing that consumers are fully informed about its characteristics, pricing terms, and the level of service performance provided to the public (Mauvais, 2008).

Accordingly, the pricing of one liter of desalinated water intended for agricultural irrigation should be based on a composite model, provided that one of its components is specifically allocated to covering the cost of climate change impacts on agriculture. In general, the key components of this composite pricing model can be summarized in Table 3.

Table 3. Key components to be considered when determining the price per liter of desalinated water for agricultural irrigation purposes.

Component	Rationale for Impact on Pricing
Climate Change	The geographical location of the desalination plant must be considered, as well as the extent to which it has been exposed to climatic variations over a preceding period (e.g., 10 or 15 years). This will allow for the establishment of regulatory and contractual rules that reflect such variables adequately.
Public Interest	Whether intended for agricultural irrigation or daily drinking consumption, water constitutes a public service provided by the state to individuals. This necessitates incorporating the public interest as a key factor when setting the price per liter.
Private Interest	To incentivize private sector partners to invest in desalination plants, the pricing per liter must enable the generation of net profits and the recovery of investment and operating costs.
Tax Regime	The national tax policy plays a decisive role in the viability of investments in the water sector. It may therefore be necessary to consider exempting desalinated water from taxation or applying reduced rates to achieve a flexible balance between addressing the impact of climate change on agriculture and encouraging profitable private sector investment in areas subject to harsh and changing climatic conditions.
Other Elements	This includes several other factors, such as environmental costs and municipal expenses, among others.

The components outlined in the table above were presented in a deliberate order, underscoring the necessity of prioritizing the impact of climate change when determining the price per liter of desalinated water intended for agricultural irrigation (See Figure 4).

Figure 4. Priority ranking of the key components to be considered when determining the price per liter of desalinated water.

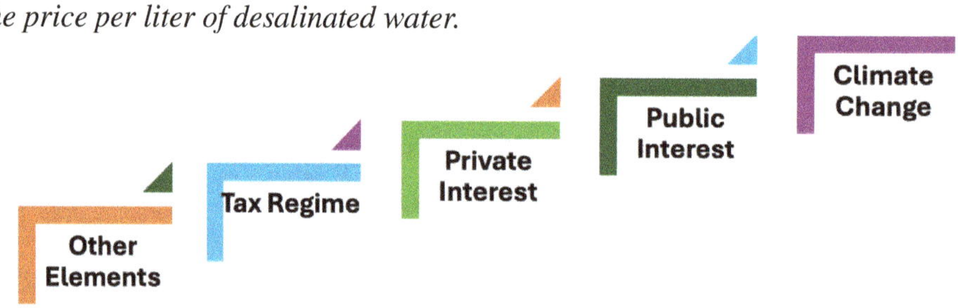

This prioritization is further shaped by additional factors, chief among them being energy considerations. As previously discussed, the national energy policy has a direct influence on the financial cost of operating desalination plants designated for agricultural irrigation. Therefore, a country's progress in the field of renewable energy would facilitate the operation of desalination plants using, for example, solar energy. Such progress would reduce the energy bill directly and, consequently, lower the cost per liter of desalinated water.

All of this confirms that the behavior of the private sector partner and the nature of its decisions regarding investment and trade in water are influenced by pricing (Laurans, 2006) and the profit margins that can be achieved. Equally important is the degree of awareness among farmers regarding the value of investing in technologies that promote water conservation (Zhou et al., 2010), as the adoption of such solutions contributes directly to reducing the overall financial burden associated with using desalinated water for agricultural irrigation.

Regulating the Prices of Seawater Desalination Plant Water through PPP Contract Clauses

As part of a comprehensive approach to regulating the pricing of desalinated water, the clauses of PPP contracts can serve as effective tools for controlling the price of desalinated water. The use of clauses to regulate product pricing is not a novel practice; commercial reality demonstrates that, in many markets, parties resort to contractual clauses to manage price fluctuations, which helps safeguard the contractual balance necessary to ensure the economic viability of the contract.

However, incorporating the element of addressing the impact of climate change on agriculture necessitates drafting a key part of the seawater desalination plant contract for irrigation based on an integrated three-tiered approach (See Figure 5).

Figure 5. Levels of PPP contract clause formulation to address the impact of climate change.

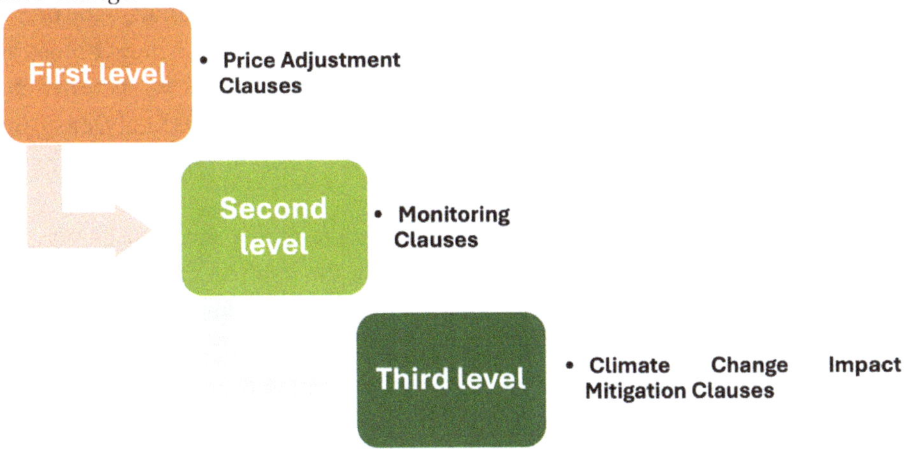

The effectiveness of these three levels of clauses in enabling the PPP contract for a seawater desalination plant designated for irrigation to address the impacts of climate change necessitates the consideration of these factors:

First Level: Price Adjustment Clauses

When unforeseen circumstances beyond the control of the parties occur—such as climate change—the question arises as to whether one party should, in such cases, be required to assist the other in overcoming the resulting difficulty in continuing to fulfill its contractual obligations, or whether the agreement between the parties obliges them to proceed with the contract's execution without modification. In response to this question, some commentators have argued that the anticipation of amending the pricing clauses would undoubtedly provide a safeguard for the parties exposed to the new constraints, particularly given that legal instruments do not address all possible concerns (Fourgoux, 2020).

Therefore, the importance arises of including in a PPP contract a clause regarding the adjustment of the prices of water produced by seawater desalination plants designated for agricultural irrigation. However, such an adjustment must not compromise the balanced economic structure of the contract, which ensures the continued flow of profits for the private sector partner, in addition to the uninterrupted provision of the public service at acceptable prices.

In this context, and in a field closely related to partnership contracts, it is worth noting that the French Public Procurement Code (Code de la commande publique)—considered a historical source for many laws in the MENA region—imposes a specific rule in this regard, which constitutes one of the forms of partnership between the public and private sectors. It requires that, when it becomes apparent during the awarding of the contract to the supplier that the performance of the contract will necessitate the use of significant quantities of materials—particularly raw materials—the prices of which are directly affected by global market fluctuations, the contract must, in such cases, include a clause for adjusting the set price in accordance with the official indices used to determine such prices (Conseil d'Etat, 2009). Accordingly, ensuring the proper implementation of price adjustment clauses requires that they be based on agreed-upon and applicable indices.

Nonetheless, in practice, the application of price adjustment clauses may, in certain cases, encounter difficulties. The refusal of the public legal entity to apply such a clause may give rise to its liability for a contractual breach or unjustified refusal (Linditch, 2024), and this liability may likewise extend to the private sector partner. Consequently, in the event of a dispute concerning the content of the price adjustment clause, jurisdiction shall lie with the judiciary for the purpose of its interpretation (Conseil d'Etat, 2010).

Second Level: Monitoring Clauses on the Performance of the PPP Contract

It is customary for PPP contracts to include monitoring clauses that grant the public legal entity the authority to oversee the performance of the contract. This is achieved by requiring the project company to provide a set of data periodically. Practice demonstrates that these monitoring clauses vary depending on the type of PPP contract and the specific requirements of the project. Their content focuses primarily on auditing, establishing a financial oversight system, technical supervision of the facilities, quality control, and setting performance indicators to ensure proper execution, among other elements.

In this regard, periodic reports on the performance of the contract, as well as the annual report on the price and quality of the public service, are fundamental guarantees for the sustained efficiency of the public service (Mauvais, 2008). It should be noted that in the event the public legal entity holds a share in the capital of the project company, the Companies Law grants it the right to access a range of financial and accounting data.

The submission of periodic data to the public legal entity enables it to make decisions in a timely manner, particularly in the event of climate change that may have adversely or positively affected the seawater desalination plants designated for agricultural irrigation.

Third Level: Climate Change Impact Mitigation Clauses in PPP Contracts

The specific nature of climate change's impact on agriculture necessitates the inclusion of a dedicated clause addressing this type of risk within PPP contracts for seawater desalination plants. One of the key advantages of such a clause is that it allows the execution of the contract to adapt to the adverse effects of climate change on the price of desalinated water. However, the extent of climate change's impact on agriculture must be verified using performance indicators annexed to the PPP contract. This underscores the importance of establishing a measurable climate change impact indicator that specifies the price of one liter of water under normal conditions and in cases where the impact of climate change is confirmed. In this regard, the climate change clause may be formulated as follows:

> *"If, during the execution of the contract, the performance indicators confirm the occurrence of climate change resulting in an increase in the price per liter of water, the agreed prices in this regard shall apply."*

That said, it must be acknowledged that the realization of the climate change risk necessitates supporting farmers by reducing the cost per liter of water produced by seawater desalination plants, which would have an adverse effect on the profit margin achievable by the private sector partner. This situation requires covering such risks through an insurance policy.

In this context, some have emphasized that the ability to forecast water flows during specific periods enhances the feasibility of insuring contracts related to this process (Lu et al., 2017). Insurance contracts based on the historical performance of water flow indicators represent an effective tool for mitigating the risks associated with water shortages. However, uncertainty persists regarding the impact of climate change on water flow behavior and how this may translate into an increase or decrease in risks for insurers and insured parties alike—an area that remains insufficiently explored (Meyer et al., 2017).

In conclusion of this analysis of the three levels of clauses—which, if properly drafted and implemented, would contribute to regulating and adjusting the pricing of seawater desalination plant outputs in response to the impact of climate change on agriculture—a forward-looking approach requires considering the development of an innovative model for managing the water supply services sector: A water supply management contract combined with water rights trading, particularly in the field of agricultural irrigation (Guo, et al., 2019). Additionally, a revenue- and cost-sharing contract could further strengthen the coherence of the project's green supply chain (Wang et al., 2022).

A comprehensive and interdisciplinary approach must also be adopted—integrating technology, policies, and socio-economic factors—to develop resilient agricultural systems (Farah et al., 2025), with all of this reflected in contracts that depart from traditional drafting and implementation models and move toward incorporating the impact of climate change on agriculture and irrigation.

CONCLUSION

This study has demonstrated that employing business contracts represents an effective means of addressing the impact of climate change on agriculture in the MENA region. The challenge of irrigation water scarcity during drought periods can be overcome by leveraging PPP contracts and utilizing project companies that help reduce the financial burden borne by countries in the MENA region to establish and operate such facilities. However, the success of this endeavor rests on the consideration of the following aspects:

- One of the main issues is that the assumption by a public legal entity of oversight authority over the project company, or the acquisition of an ownership stake in its capital, will, in most cases, result in shifting a substantial portion of the financial burden of establishing and operating seawater desalination plants onto the entity itself. Such a situation undermines its capacity to address the impacts of climate change on agriculture in the MENA region. The situation requires the rationalization of financial resources in agricultural areas that are continuously or intermittently affected by climate change, particularly in the form of insufficient or irregular rainfall.

Given these challenges, careful consideration must be given to the possibility that the public legal entity may not contribute capital to the project company to cover additional financial costs related to ensuring the continuity of agricultural activity. However, this approach should be subject to certain exceptions, under which the public legal entity is authorized to contribute capital to the project company in order to address specific circumstances. This flexibility will enable rapid adaptation to the constraints imposed by climate change on agricultural irrigation.

- In parallel, it is necessary to adopt an innovative and multi-element legislative and regulatory model for establishing the price per liter of water produced by seawater desalination plants. This model must incorporate climate change as a measurable and adaptable factor linked directly to pricing. This will provide effective opportunities to adjust the pricing of desalinated water outputs in response to climate change, particularly during drought periods characterized by reduced rainfall, which has an adverse effect on the levels of irrigation dams and groundwater reserves.
- Finally, the drafting of the PPP contract for a seawater desalination plant designated for agricultural irrigation must be structured to accommodate the risk of adverse climate-related impacts on water pricing. This calls for the inclusion of a dedicated clause addressing this issue (the Climate Change Impact on Pricing Clause), provided that its application is tied to a measurable index (the Climate Change Impact Pricing Index).

REFERENCES

Bandarra, N.-J. (2009). Un développement durable pour la Méditerranée. *Revue du Marché commun et de l'Union européenne,* (529).

Banderas Tarabay, A. G., & González-Villela, R. (2018). The environmental added value of water: A proposal. *International Journal of Sustainable Development and Planning, 13*(3), 406–417. DOI: 10.2495/SDP-V13-N3-406-417

Bloch, P. (2005). Présentation synthétique du partenariat Public-Privé. In *Partenariat pubic-privé* (Dossier). *Revue de Droit Bancaire et Financier.*

Czipf, C. (2025). The impact of changing energy prices, interest rates, and investment costs on the net present value and internal rate of return for alternative energy projects. *Discover Sustainability, 6*(1), 161. Advance online publication. DOI: 10.1007/s43621-025-00921-7

Conseil d'État. (2009, 9 décembre). *Motivation de dévolution en marché global et insertion d'une clause de révision de prix.* Recueil Lebon – Recueil des décisions du Conseil d'État, (6).

Conseil d'État. (2010, 9 avril). *Le juge de cassation contrôle l'interprétation d'un cahier des charges.* Recueil Lebon – Recueil des décisions du Conseil d'État, (4).

Darwish, T., Shaban, A., Faour, G., Jomaa, I., Moubarak, P., & Khadra, R. (2025). Sustainable and Traditional Irrigation and Fertigation Practices for Potato and Zucchini in Dry Mediterranean Regions. *Sustainability (Basel), 17*(5), 1860. Advance online publication. DOI: 10.3390/su17051860

Dirnat, C., & Souffir, J. (2005). Approche contractuelle de la dimension financière du contrat de partenariat: un déconcentrant ménage à trois. In *Partenariat pubic-privé* (Dossier). *Revue de Droit Bancaire et Financier.*

Dzvene, A. R., Zhou, L., Slayi, M., & Dirwai, T. L. (2025). A scoping review on challenges and measures for climate change in arid and semi-arid agri-food systems. *Discover Sustainability, 6*(1), 151. Advance online publication. DOI: 10.1007/s43621-025-00945-z

Emery, C. (2006). L'exécution des conventions de délégation de service public (Chap. 3, Folio 3223). In Dalloz (Ed.), *Encyclopédie des collectivités locales.* Dalloz.

European Union. (2000). Directive 2000/60/EC of the European Parliament and of the Council of 23 October 2000 establishing a framework for Community action in the field of water policy. https://data.europa.eu/eli/dir/2000/60/oj

Farah, A. A., Mohamed, M. A., Musse, O. S. H., & Nor, B. A. (2025). The multifaceted impact of climate change on agricultural productivity: A systematic literature review of SCOPUS-indexed studies (2015–2024). *Discover Sustainability*, *6*(1), 397. Advance online publication. DOI: 10.1007/s43621-025-01229-2

Fourgoux, J.-L. (2020). La clause de révision de prix dans les contrats d'affaires. *Actualité Juridique Contrats d'affaires,* (12).

Gabr, M. E. (2023). Impact of climatic changes on future irrigation water requirement in the Middle East and North Africa's region: A case study of upper Egypt. *Applied Water Science*, *13*(7), 158. Advance online publication. DOI: 10.1007/s13201-023-01961-y

Giwa, A., Ademola, H., & Yusuf, A. O. (2025). Machine learning and application for modeling and prediction of desalination cost globally. *Desalination*, *608*, 118829. Advance online publication. DOI: 10.1016/j.desal.2025.118829

Guo, H., Chen, X., Dong, Z., & Zhang, H. (2019). Construction method of water right trading based on water-saving management contract. *Water Resources Protection, 35*(3), 33-38and62. DOI: 10.3880/j.issn.1004-6933.2019.03.005

Jagannathan, N. V., Mohamed, A. S., & Kremer, A. (Eds.). (2009). *Water in the Arab world: Management perspectives and innovations*. World Bank., https://documents.worldbank.org/curated/en/590691468052797602

Ji, H., Lee, H. S., hyun Moon, J., Thu, K., Kim, Y.-D., & Jeon, W.-J. (2025). Experimental and theoretical evaluation of a novel hybrid multi-effect adsorption desalination process. *Separation and Purification Technology*, *361*, 131522. Advance online publication. DOI: 10.1016/j.seppur.2025.131522

Kopańska, A., Osinski, R., & Korbus, B. (2024). Private entities motivations to participate in public-private partnerships. *Socio-Economic Planning Sciences*, *92*, 101841. Advance online publication. DOI: 10.1016/j.seps.2024.101841

Kürklü, A., Pearson, S., & Felek, T. (2025). Climate change impacts on tomato production in high-tech soilless greenhouses in Türkiye. *BMC Plant Biology*, *25*(1), 339. Advance online publication. DOI: 10.1186/s12870-025-06307-1 PMID: 40089667

Larrieu, D. (2007). L'interprétation des pactes extra-statutaires. *Revue des Sociétés,* (4).

Laurans, Y. (2006). Role of economic analysis in the Water Framework Directive. *Houille Blanche*, (4), 75–80. DOI: 10.1051/lhb:200604012

Lee, S. T., & Jung, S. M. (2025). Social return on investment as a tool for environmental impact and strategic assessments: Evidence from South Korea. *Environmental Impact Assessment Review, 112*, 107765. Advance online publication. DOI: 10.1016/j.eiar.2024.107765

Leray, G. (2023). L' ineffectivité du droit à l'eau. *Recueil Dalloz,* (44).

Linditch, F. (2024). Responsabilité contractuelle – Fait générateur In Dalloz (Ed.), *Répertoire de la responsabilité de la puissance publique* (Vol. R), Dalloz: Paris.

Liu, M., & Yan, A. (2022). Feasibility analysis of water resources market transaction of water diversion project based on evolutionary game. *Water Science and Technology: Water Supply, 22*(3), 3061–3071. DOI: 10.2166/ws.2021.423

Lu, M., Lall, U., Robertson, A. W., & Cook, E. (2017). Optimizing multiple reliable forward contracts for reservoir allocation using multitime scale streamflow forecasts. *Water Resources Research, 53*(3), 2035–2050. DOI: 10.1002/2016WR019552

Mauvais, F. (2008). The delegated management: An inventive and efficient way for water and waste water public utilities management. *Revue des Sciences de l'Eau, 21*(2), 143–153. DOI: 10.7202/018463ar

Meyer, E. S., Characklis, G. W., & Brown, C. (2017). Evaluating financial risk management strategies under climate change for hydropower producers on the Great Lakes. *Water Resources Research, 53*(3), 2114–2132. DOI: 10.1002/2016WR019889

Moon, J., Son, S., Kim, J., & Park, K. (2025). Critical challenges in high-salinity seawater reverse osmosis systems: Technical, energy, and environmental reviews. *Desalination, 607*, 118811. Advance online publication. DOI: 10.1016/j.desal.2025.118811

Mysiak, J., & Gómez, C. M. (2015). Water pricing and taxes: An introduction. In Lago, M., Mysiak, J., Gómez, C. M., Delacámara, G., & Maziotis, A. (Eds.), *Use of economic instruments in water policy* (Vol. 14, pp. 15–32). Springer., DOI: 10.1007/978-3-319-18287-2_2

Naik, B. M., Singh, A. K., Venkatesan, P., Maji, S., Sunil, J., & Naik, M. R. (2025). Assessing the contribution of climate resilient agricultural technologies in advancing sustainable development goals in Telangana, India. *Discover Sustainability, 6*(1), 117. Advance online publication. DOI: 10.1007/s43621-025-00870-1

Peyrical, J.-M. (2017). Marché de partenariat: exécution et extinction (Chap. 2, folio n°3230). In *Encyclopédie des collectivités locales* (Vol. M). Paris: Dalloz.

Prakash Khedun, C., Lewis, A. C., & Kaiser, R. A. (2025). Building Resilience into the Public Water Supply System through Targeted Data-Driven Conservation Strategies. *Journal of Irrigation and Drainage Engineering*, *151*(3), 04025011. Advance online publication. DOI: 10.1061/JIDEDH.IRENG-10453

Rangel, T., & Galende, J. (2010). Innovation in public-private partnerships (PPPs): The Spanish case of highway concessions. *Public Money & Management*, *30*(1), 49–54. DOI: 10.1080/09540960903492380

Ripert, G., & Roblot, R. (2003). *Traité de droit commercial (T. 1, V. 2), Les sociétés commerciales.* L.G.D.J.

Sanabria, S., & Torres, J. (2020). Water price: Environment sustainability and resource cost. *Water (Basel)*, *12*(11), 1–18. DOI: 10.3390/w12113176

Sangsefidi, Y., Rios, A., Bagheri, K., Carroll, M. W., & Davani, H. (2025). Integrating social data and engineering solutions for developing resilient water infrastructure against coastal climate change. *Scientific Reports*, *15*(1), 7241. Advance online publication. DOI: 10.1038/s41598-025-91147-0 PMID: 40021917

Schmidt, D. (1998). Les définitions du contrôle d'une sociétés, in, La prise de contrôle d'une société. *Revue de jurisprudence commerciale,* (11).

Schwantes, R., Morales, Y., Pomp, E., Singer, J., Chavan, K., & Saravia, F. (2025). Thermally driven ultrapure water production for water electrolysis – A techno-economic analysis of membrane distillation. *Desalination*, *608*, 118848. Advance online publication. DOI: 10.1016/j.desal.2025.118848

Sée, A. (2025). Les participations minoritaires des personnes publiques au capital des sociétés de production d'énergie renouvelable. *L' Actualité Juridique Droit Administratif (AJDA)*(14).

Seidl, C., Wheeler, S. A., & Zuo, A. (2020). High turbidity: Water valuation and accounting in the Murray-Darling Basin. *Agricultural Water Management*, *230*, 105929. Advance online publication. DOI: 10.1016/j.agwat.2019.105929

Sola, I., Santana-Anticoy, C., Silva-García, R., Pérez-Hernández, G., Pereira-Rojas, J., Blanco-Murillo, F., Díaz, M. J., Sáez, C. A., & Rodríguez-Rojas, F. (2025). Evaluating physico-chemical and biological impacts of brine discharges for a sustainable desalination development on South America's Pacific coast. *Journal of Hazardous Materials*, *489*, 137464. Advance online publication. DOI: 10.1016/j.jhazmat.2025.137464 PMID: 39955987

Sortais, J.-P. (1993). Protection des minoritaires: droit des sociétés. In Dalloz (Ed.), *Répertoire des sociétés* (Vol. P). Dalloz: Paris.

Terneyre, P. (2003). Les nouveaux contrats de partenariat et la typologie des instruments du partenariat public-privé. *Revue de droit Immobilier,* (520).

Testu, F.-X. (2019). Titre 0 - Introduction. In Dalloz (Ed.), *Dalloz référence Contrats d'affaires*. Paris.

Wang, H., Xue, L., Peng, J., Liu, G., & Chen, Z. (2022). Green Operations Management of Interbasin Water Transfer Project: An Extended Framework of Joint Pricing and Inventory Management. *Discrete Dynamics in Nature and Society, 2022*(1), 5927108. Advance online publication. DOI: 10.1155/2022/5927108

Xu, J., Wang, M., Xu, G., Sokolova, E., Vail, T., & Song, D. (2025). A critical review of ISO standards for seawater desalination: Classification, applications, and future development. *Separation and Purification Technology, 367*, 132898. Advance online publication. DOI: 10.1016/j.seppur.2025.132898

Zhou, C. H., He, Y. L., Chen, Z., Fan, Y., Yimingniyazi, A., Ma, P. C., & Kadier, A. (2025). A novel and sustainable invasive plant wood-based solar interfacial evaporator for efficient brackish water desalination. *Desalination, 606*, 118760. Advance online publication. DOI: 10.1016/j.desal.2025.118760

Zhou, H. C., Wu, L., He, B., & Shao, H. S. (2010). Study of agricultural water rights transfer price in Dalian. *Dalian Ligong Daxue Xuebao/Journal of Dalian University of Technology, 50*(2), 251-257.

Chapter 2

Climate Change and Agricultural Production in Morocco:

Assessing Impacts and Resilience

Hicham Sadok
https://orcid.org/0000-0003-4402-0954
Mohammed V University in Rabat, India

ABSTRACT

Natural resources are essential to the livelihoods of Morocco's poor. Since the 1960s, the average increase in observed temperatures is 0.23°C per decade; nine of the country's 10 hottest years on record have occurred in the past two decades, and precipitation follows a general downward trend. Faced with such observations, this article seeks to determine whether this ongoing climate change has concrete effects on the four main agricultural productions in Morocco. Thus, using the Pettitt test to estimate the standardized precipitation index (SPI) and the evolution of seasonal average temperatures, the study of the Spearman correlation coefficients between these parametric variations of climate change and the yields of four main agricultural productions during the period 2001-2020 did not reveal a significant link, with the exception of cereal yield during the 5 years of severe drought. This resilience can be explained by the intervention of public policies to promote technology, seeds and irrigation to counteract the degradation of climatic factors.

DOI: 10.4018/979-8-3373-3346-5.ch002

1- INTRODUCTION

Climate change poses a threat to livelihoods and social well-being and tends to have a detrimental effect on the economic development of African countries. The World Bank estimates that 100 million people worldwide, mainly in South Asia and Africa, are at risk of falling back into poverty due to the effects of climate change (Hallegatte et al., 2016; Baarsch et al., 2020). African countries are most widely affected by this phenomenon due to the weakness of their production systems based essentially on subsistence agriculture. This is dependent on climatic vagaries, low yields and the absence of adequate adaptation strategies. Dwindling harvests therfore expose rural households to food insecurity, poverty and widening inequalities.

The risks associated with climate change can therefore lead to increased poverty and place poor households in a kind of poverty trap that can annihilate development efforts. In Africa, variations in rainfall and temperature induced by climate change are weakening food systems and exposing an average of 30 million people to food insecurity (Epule et al., 2017; Verpoorten et al., 2013; Sissoko et al., 2011). The absence of an insurance mechanism to cover against risks also limits the ability of households to adapt to climate shocks.

Indeed, it is this conception of climate change, measured by variations in precipitation and temperature, which we will retain in this article to estimate the impact of the latter on the evolution of rural poverty in Morocco through the channel of agricultural production. Certainly, the economic growth recorded over the last two decades has certainly led to a drop in the poverty rate, but according to the latest report on poverty from the High Commission for Planning (HCP, 2022) nearly 3.5 million people are still below one of the selected poverty thresholds, i.e., 9% of the total population (2.3 million people suffer from multidimensional poverty and 1.2 million from monetary poverty). However, an increasing trend in poverty in Morocco and several developing countries is recently noticeable. It would be mainly attributable, temporarily, to the negative effects of the health crisis and inflation, and more structurally to the effects of climate change on agriculture (Angelsen and Dokken, 2018; Muller et al., 2011).

Faced with this observation, the objective of this article is to characterize, firstly, the parameters of this climate change through precipitation and the increase in temperature, to evaluate, secondly, whether this ongoing climate change has concrete effects on rural poverty through production of the four main agricultural crops in Morocco, namely Cereals, Fruits, Vegetables, Market Gardening and Sugar (MGSC), to which the majority of rural people are devoted.

From a methodological point of view, the article uses the Pettitt test to characterize the parameters of ongoing climate change, then the Spearman correlation coefficients to assess the nature of the relationship between climate change and rural poverty through the channel transmission which is the dominant agricultural production in this country.

Thus, the rest of the article is organized as follows: section 2 presents a review of the literature on the impact of climate change on poverty through its effects on agriculture; section 3 describes the analysis method and the stylized facts of the variables studied; section 4 presents the impact of climate change on agriculture and rural poverty; and section 5 is devoted to the results and analysis before concluding

2- THE IMPACT OF CLIMATE CHANGE ON POVERTY THROUGH ITS EFFECTS ON AGRICULTURE: A REVIEW

The first contributions on the links between climate and poverty date back to the 1990s. Thus, Udry (1995) measures the potential impact of climate variability on poverty. Paxson (1992) found that rainfall shocks have serious consequences for the welfare of farming households, especially when they do not have savings as a means of mitigating the shocks. However, the absence of hedging against risk can lead to poverty mainly among the poorest farmers due to credit or liquidity constraints (Dercon and Christiaensen, 2011; Dercon and Krishnan, 2005;Fafchamps et al., 1998).

In more recent work, several studies have focused on the impacts of climate shocks on poverty. Because of their location, poor countries are the most exposed to various effects of climate change, whether water stress, the intensity of droughts or heat waves, loss of agricultural yields or the degradation of natural habitats. It is thus estimated, using indicators taking into account these effects of climate change, that exposure to climate risks is around 90% in Africa and Southeast Asia (Byers et al., 2018), and it is the poorest people within these regions who are most at risk (Abidoye and Odusola, 2015).

The most important transmission channels of climate variability on vulnerability to poverty are agriculture (Auffhammer and Schlenker, 2014), health (Deschenes, 2014) or conflicts (Burke et al., 2015).In the case of agriculture, several studies have highlighted the relationship between crop yields and climate variability (Rosenzweig et al., 2014), hypothesizing that extreme temperatures and low levels of precipitation can reduce yields, increase the risk of food insecurity and increasing levels of poverty. A significant proportion of the population of developing countries depends directly on activities likely to be affected by climate change, particularly the most modest. The survival of the latter depends more on natural capital than on human capital (Dasgupta et al., 2010).), due to the fact that their economic activities depend exclusively on the benefits provided by nature (Noack et al., 2015), or even threatened their existence: for the period 2007-2016, the mortality rate due to natural disasters is approximately 4 times higher in poor countries (Formetta and Feyen, 2019).

The most modest also have to deal with indirect impacts such as the increase in the price of agricultural commodities resulting from lower yields or extreme weather events (Hallegatte and Rozenberg, 2017). They are particularly sensitive to variations in these prices since they spend a large part of their income on food. Rising prices could threaten food security in certain regions, notably in Africa or South Asia (Hertel, 2015). Another mechanism involving indirect impacts can affect income when labor productivity declines due to high temperatures (Deryugina and Hsiang, 2014; Heal and Park, 2016), especially for outdoor work.

For these different types of impacts, the adaptation capacity of the most modest is less and climate change exacerbates pre-existing difficulties. Most of the time, they do not benefit from insurance mechanisms or access to basic health services that could mitigate price or income shocks. These living conditions forces them, in the event of damage caused by a natural disaster such as a storm or flood, to draw on their own assets. With fewer assets, it is more difficult to cope with risk. Their assets are also less diversified: for poor urban households, housing constitutes the bulk of their assets (Moser, 2007), while for low-income rural households, herds represent the bulk of their capital, likely to be lost in the event of drought (Nkedianye et al., 2011). In the event of climatic hazards, the poorest are also the most affected by diseases such as malaria or water-borne diseases (Hallegatte, 2015). Thus, a severe environmental shock results in long-term effects for the poorest, thus increasing their risks of falling into the poverty trap (Barret et al., 2007). It is in these types of circumstances that climate change acts as an amplifier of risks for the poorest.

In a prospective study on the relationship between the poverty and climate change, Hallegatte and Rozenberg (2017) use a global database to model this impact by 2030 through five channels: agricultural productivity, food prices, natural disasters, labor productivity and health. The study finds that the impact of climate change on poverty depends, on the one hand, on the fraction of the population working in agriculture, its productivity and the diversification of income within households, and on the other hand, the fraction of the income of each household budget devoted to food consumption.

The study by Hallegatte and Rozenberg (2017) also highlights that the number of people living in extreme poverty in 2030 would be estimated between 35 million and 122 million, depending on the climate impact scenario as stipulated in figure below.

Figure 1. Scenarios of additional people in extreme poverty in 2030 due to climate change

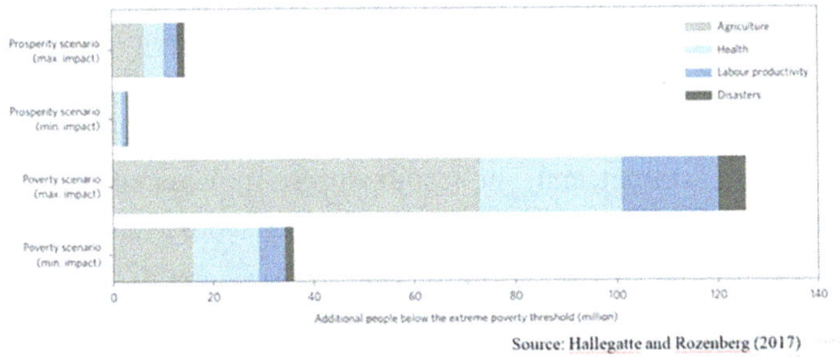

Source: Hallegatte and Rozenberg (2017)

In another study involving a panel of households from 21 regions in eight African countries, Dzanku et al. (2015) find that poverty increases with yield gaps due to climate variability and production conditions. Azzarri and Signorelli (2020) use cross-sectional data from 24 African countries, linked to geospatial information on agro-climatic and market access conditions, to measure the impact of climate variability and extreme weather shocks on poverty. They find that climate shocks are associated with a 35% decrease in total per capita consumption and a 17% increase in extreme poverty.

Thus, most of the studies dealing with the relationship between climate change and rural poverty converge on the need to make agriculture in poor countries more resilient to the harmful effects of climate change (Devarajan et al., 2015). In the absence of implementing decarbonization policies on a global scale, support through local public policies is likely to mitigate the degree of deterioration in the living conditions of the most vulnerable households (Hallegatte, 2016).

3- METHOD AND MATERIALS

The correlation between precipitation and agricultural production has been the subject of several studies. Rain is certainly an essential determinant of agricultural production but it is not exclusive (Sambou et al., 2021). To study the relationship between these two phenomena, it is interesting to first analyze the effects of climate change in Morocco through precipitation and temperature over recent decades. To do this, we will study the time series of the precipitation index to determine the water supply in Morocco, and the evolution of the annual and seasonal average temperature. Thus, using the Pettitt test (1979), we will calculate the annual change in precipitation, the deviation from the mean of the series and finally calculate the Standardized Precipitation Index (SPI).

The choice of this statistical test, called nonparametric, is adapted to the problem addressed because it makes no assumptions about the underlying distribution of the data. Especially,it will apply to assumed data not having a Gaussian distribution. The Pettitt test is thus adapted to the hydrometeorological data for which the distributions are often asymmetrical.The test examines the existence of a break at an unknown time (t) in the series using a formula derived from Mann-Whitney (Servat et al., 1998). The absence of a break in the series (xi) of size N constitutes the null hypothesis. The Pettitt test (1979) defines the variable Ut,N:

$$U_{t,N} = \sum_{i=1}^{t} \sum_{j=t+1}^{N} D_{ij} \quad \text{où } D_{ij} = \text{sgn}(X_i - X_j) \text{ tel que } \begin{cases} \text{sgn}(X) = 1 & \\ \text{sgn}(X) = 0 & \text{pour} \\ \text{sgn}(X) = -1 & \end{cases} \begin{cases} X > 0 \\ X = 0 \\ X < 0 \end{cases}$$

Wher eDij = sgn (xi xj) with sgn (Z) = 1 if Z > 0, 0 if Z = 0 and -1 if Z < 0.

Pettitt test proposes testing the null hypothesis using the Knstatistic defined by the maximum absolute value of Ut, N for (t) varying from 1 to N -1. If the null hypothesis of homogeneity of the series is rejected, the test offers an estimate of the break date.

SPI (Standardized Precipitation Index) (McKee et al., 1993) is calculated as follows: IPS = (Pi Pm)/S with:

Pi = cumulative rainfall for a year i

Pm = annual average over the series considered

S = standard deviation of annual totals over the same period.

With regard to agricultural productivity, we will analyze the areas sown for each crop as well as the yields obtained.

3.1 The stylized facts of climate change

In the study of rainfall anomalies in Morocco over a century, and from the mapping of the three homogeneous rainfall regions of Morocco developed by Knippertz et al. (2003), it appears from the time series of the precipitation index calculated for each hydrological year that the interannual and decadal variability is considerable. The figure below reveals that in the Mediterranean region (région méditerranéenne MED), precipitation is below the value that have prevailed since the late 1970s, while in the Atlantic region (région de l'Atlantique, ATL), precipitation was low from the 1970s through the early 1990s, but with some wet years during the late 1990s; precipitation in the southern Atlas region (Region du Sud de l'Atlas, SOA) was above average in most years after the mid-1980s, especially during the early this millennium before sinking during the last decade.

Figure 2. Rainfall anomalies in the regions of Morocco

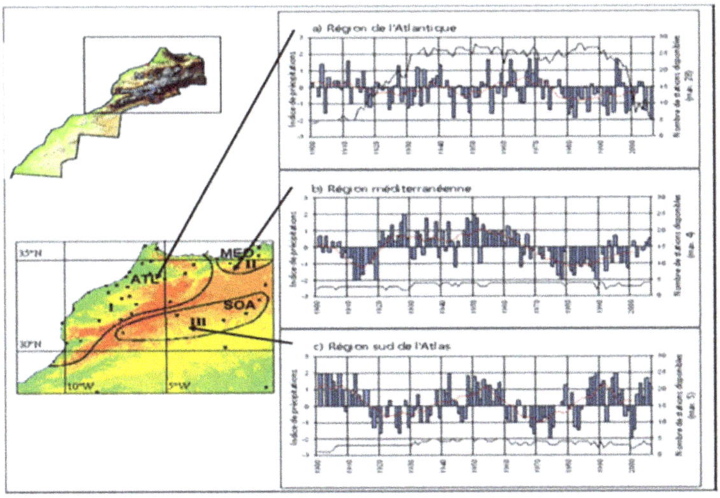

Source: UNDP Climate Change Country Profiles

Source: UNDP Climate Change Country Profiles

Generally speaking, and beyond the variability depending on the region, the average annual rainfall in Morocco over a century has evolved with a constant and marked negative trend since 1960. The overall data presented below in the form of an anomaly by compared to the norm (regional average) varies with an increase reaching a maximum of 12%, and a maximum decrease of 49%. The median overall change is -9 to -31%. Changes in the proportion of total precipitation falling during severe events tend to decrease. The rate varies between -21% and +11%; and variations in rainfall maxima over 1 and 10 days tend to decrease: changes in projections range from -11 mm to +17 mm of annual maximum for 1 day, and from -27 mm to +26 mm of annual maximum for 10 days.

These data therefore help to illustrate why Morocco is considered one of the world countries most affected by water stress: since the end of the 1970s, water reserves in terms of precipitation have decreased, from annual average of 22 billion m^3 between 1945 and 1978 (yellow line in the figure below) to an annual average of 15 billion m^3 between 1979 and 2020 (dark blue line). The decline in inputs mentioned above, combined with the increase in demand due to demographic growth and economic development, has put Morocco in a situation of water stress: between 1960 and 2020, the per capita availability of water resources renewable water increased from 2,560 m^3 to around 620 m^3 per person per year, placing Morocco in a situation of structural water stress (less than 1,000 m^3), and quickly approaching the absolute shortage threshold of 500 m^3 per person and per year.

Figure 3. Morocco's water inflows (in billion m³/year)

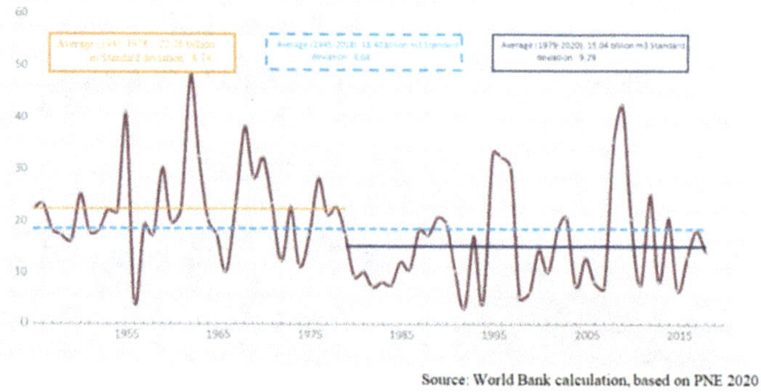

Source: World Bank calculation, based on PNE 2020

Another decisive criterion for estimating the degree of climate change and its consequences on the most vulnerable populations consists in noting the tendencies of the average annual and seasonal temperature of the recent past and those projected for the future.

Based on a panel of 126 low- and middle-income countries over 1960-2017, the study conducted by the Banque de France reveals that positive and lasting temperature differences from their historical norm have a non-linear negative effect on economic and per capita growth. In Morocco, a sustained increase in temperature of 1°C reduces the annual growth of real GDP per capita by 0.74% to 1.52%. In all the countries studied, the results suggest that global warming constitutes a poverty trap, complicating adaptation to climate change in developing countries, particularly in countries with the lowest income levels due to their low resilience and their high socio-economic vulnerability.

Thus, the figure below summarizes the average annual and seasonal temperature of Morocco, as well as the trends projected in the decades to come according to 3 scenarios (A2: a temperature increase of 1.5°C; A1B: a temperature increase of 0.7° C and B1: an increase in temperature according to the trend observed since 1960).

Figure 4. Annual and seasonal average temperature trends

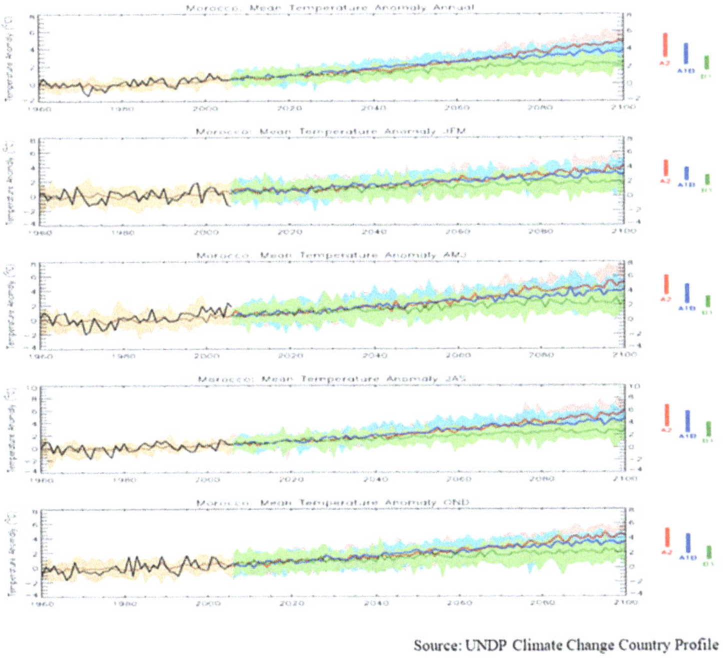

Source: UNDP Climate Change Country Profile

Analysis of the graph reveals that the average annual temperature in Morocco has increased by 1.2°C since 1960, an average rate of 0.23°C per decade. This upward trend varies by season and is statistically significant only in the season (April-May-June, AMJ) and (September-October-November, SON). The fastest rate of temperature increase is observed during the AMJ season with an average rate of temperature increase of 0.37°C per decade.

The frequency of days qualified as hot, defined as the temperature exceeding by 10% the average temperature of the climate of the region during the season considered, has increased significantly since 1960 in all seasons except during the SON season. Thus, the average number of "hot" days per year increased by 21.5 days (5.9%), especially during the JJA season, when this number increased by 3.8 days per month (i.e., 12.4% of additional days during the JJA season).

The average number of "hot" nights per year increased by 40 (i.e., 10.8% more nights). The most strongly observed rate of increase is during the SON season, where the average number of warm nights increased by 6.3 (i.e., 2.2% more nights during this season).

The frequency of cold days and nights has decreased considerably in all seasons: they have decreased by 21 (5.8% of days) during the period studied. This rate of decline is most rapid in summer (JJA), where the average number of cold summer days decreased by 2.2 days per month (i.e., 7.2% of summer days). Regarding the average number of "cold" nights, the latter decreased by 29 (i.e., 7.9% of summer nights), and the fastest decrease was observed in summer (JJA) where the average number of cold nights decreased by 2.9 nights per month (i.e., 9.4% of nights in the JJA season).

Climate change in Morocco, captured from rainfall and temperature anomalies recorded by season, as well as the trend of the latter developed under 3 scenarios (Figures 2 and 3), indicates that Morocco experienced in 2020 compared to 1960:

- An increase in the average annual temperature between +0.5°C and +1.2°C depending on the season and the region;
- An average reduction in the annual volume of precipitation of around -6%;
- An intensification and increase in the frequency of extreme events (drought, heavy rainfall) and a disruption of seasonal rainfall.

3.2 Agricultural production and poverty

Given its structural dependence on weather conditions, agriculture would be the sector most directly vulnerable to climate change. The sector is expected to experience strong variations in the volume and quality of its production, including fishing and livestock activities. The quantification of the impacts is nevertheless particularly uncertain due to the heterogeneity of the effects according to the different agricultural activities, and the many existing threshold effects on yields.

Climate change (increase in temperature and change in precipitation) would decrease agricultural yields in most regions. Losses on the hottest days would explain most of this decrease (Carleton & Hsiang, 2016). This effect would be exacerbated by the new proliferation of insects or bacteria attacking crops (Deutsch, et al., 2018). Thus, under a business-as-usual scenario, approximately 90% of the rural population would be exposed to simultaneous losses of production in agriculture by 2100 according to the study by Hallegatte and Rozenberg (2017). The sector should see great diversity in yield developments.

While economies can adapt in the medium to long term by changing crop types, this development takes time and, in the short term, average yield losses can be massive. The cost of climate change on the agricultural sector is therefore largely linked to the costs of the gradual adaptation of the sector (via the need to make substitutions between crops).

Based on household survey data from 89 developing countries, the World Bank defines extreme and moderate poverty as people living on less than \$1.90 and between \$1.90 and \$3.10 a day. Based on this methodology, the High Commission for Planning (HCP), the main institution for the production of official statistics in Morocco, measures and maps poverty according to several thresholds[1]. We have selected the threshold that seem the most significant in relation to the issue raised in this article, namely absolute poverty calculated as the food poverty line plus the cost of non-food items enabling households to meet their food needs.Taking these criteria into consideration, the latest HCP report (2022) on poverty and vulnerability in Morocco notes that the absolute poverty line, after having experienced a decrease between 2001 and 2010, increased from 1.7% in 2011 to 3% in 2021 at the national level, from 3.9% to 6.8% in rural areas and from 0.5% to 1% in urban areas. By aggregating this percentage, there are therefore 3.2 million additional people who have been engulfed in poverty.This increase in poverty in Morocco, whose economy has suffered enormous losses over the last decade, is mainly due to a series of shocks whose effects are cumulative, and which led to a sharp slowdown in economic activity over the last 10 years, as shown in the figure below. This decline is not fading; deflated GDP growth fell in the first quarter of 2023 by 0.3% compared to 2019, and the reasons for this slowdown, apart from the Covid crisis and inflation, are mainly due to the consequences of the drought which is hitting the country severely, leading to a sharp contraction in agricultural GDP compared to 2019, and which represents 13% of overall GDP.

Figure 5. Economic growth in Morocco slowing over past 20 years

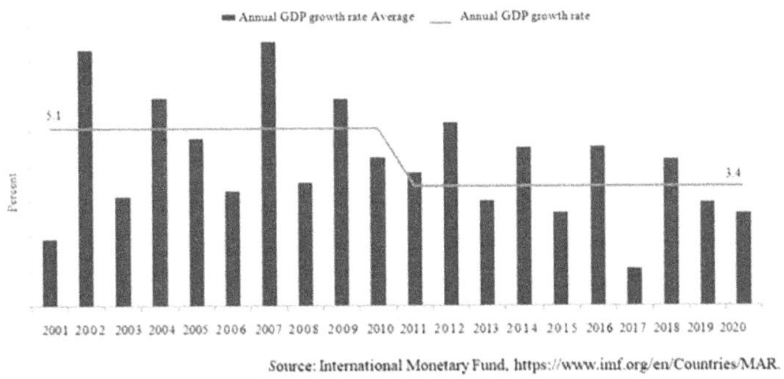

Source: International Monetary Fund, https://www.imf.org/en/Countries/MAR.

The slowdown in growth seems to occur mainly in agriculture (primary sector). The deceleration in GDP growth before and after 2010 was particularly striking in the primary sector, as shown in the figure below. This is largely due to the slowdown in the agricultural sector, which represents around 95% of primary sector production, and which became more volatile between 2001 and 2020. Under these conditions, it is estimated that almost ten years of progress in reducing poverty and vulnerability have been lost (HCP report, 2022) due, perhaps, to climate change, a relationship we seek to address in this work through agricultural production.

Figure 6. Big slowdown in agricultural growth

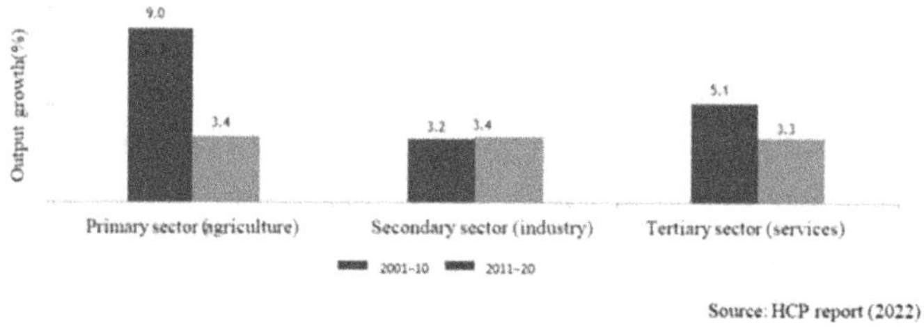

Source: HCP report (2022)

But if climate change observed over the past two decades poses a threat to livelihoods and social well-being and tends to have a detrimental effect on poverty and economic development, then what are the salient signs of this relationship between climate change and agricultural production thus contributing to worsening poverty in Morocco?

4- IMPACT OF CLIMATE CHANGE ON AGRICULTURE AND RURAL POVERTY

Climate change would not just have an impact on the quantity produced but would also change the quality of the products. For a warming of 2.5°C by 2050, the zinc, iron and protein content of many food crops should experience a decrease of between 3 and 17% (Myers-Smith et al., 2019).In addition, climate change is expected to reduce the quality of livestock products due to the proliferation of diseases and increased heat stress to which animals would be subjected. This would be characterized by a decrease in protein content and an increase in fat content (Gregory, 2010).

Agricultural losses would be much greater in the hottest regions of Morocco, which today represent nearly 51% of national agricultural production. These are regions where economies much more dependent on agricultural activities are concentrated (e.g, the agricultural sector represents 22% of GDP and 60% of export revenues in Haouz region) and, therefore, more sensitive to rising temperatures and lack of rainfall.

Global warming would have an impact on labor productivity via its effects on the health of farmers (due to heat stress, humidity levels) and on the labor market (loss of jobs and cumulative agrarian knowledge, know-how and adaptations to the new type of agriculture).The combination of climate change, including increased temperatures and humidity levels, would induce labor productivity losses. According to a 2019 report by the International Labor Organization (ILO), high temperatures create health risks that will translate into lost productivity (ILO, 2019).Labor productivity could fall by up to 1.7% for each degree of warming above 1.5°C mean daily temperature (Deryugina & Hsiang, 2014) and a drop in worker productivity of 50% from 33-34°C. According to the IPCC, by 2100, climate change could reduce labor productivity by up to 60% during the hottest months compared to the rest of the year.

Since the 2010s, public policies have focused on improving water efficiency in the agricultural sector, which is by far the largest consumer of water in Morocco. These policies focused on the modernization of the irrigation system and subsidies for investments in equipment on farms. Thus, between 2008 and 2018, the area irrigated by drip irrigation was multiplied by 3.5, leading to a 92% increase in the real agricultural value added. And although they represent only 20% of agricultural land, irrigated areas contribute more than half of the total agricultural value added. This transformation has contributed to reducing the volatility of agricultural sector growth (orange lines in the Figure below).However, as the yields of rainfed crops (which cover 80% of agricultural land) are strongly correlated with rainfall, the agricultural sector thus remains a major source of macroeconomic volatility, and rainfall shocks explain nearly 37% of the variance of GDP in the medium term, even if agriculture represents only 13% of GDP (World Bank, 2022). Droughts can also have repercussions on the national budget and the balance of payments due to the emergency support measures provided to farmers during exceptionally dry years and the need to import larger volumes of cereals when the national production fall. These impacts are amplified when droughts coincide with international food price shocks, as seen with the war in Ukraine and its impact on the global grain market.

Figure 7. Precipitation vs Agricultural Value Added

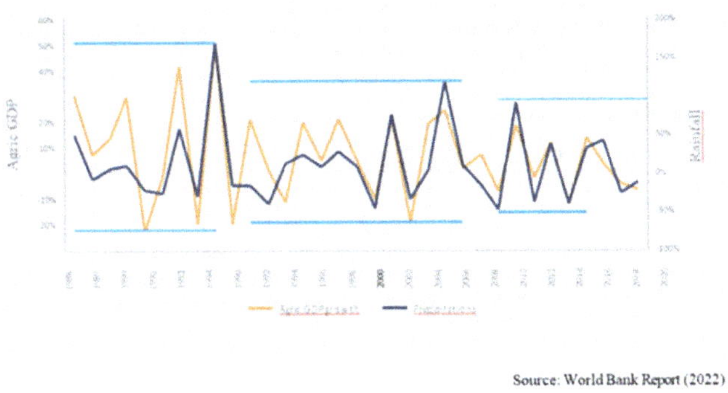

Source: World Bank Report (2022)

Droughts and water scarcity disproportionately affect poor farmers. In total, 79% of Morocco's poor live in rural areas, and most of them depend on rain-fed agriculture as a source of food and income. In this context, the increase in the incidence, severity and duration of droughts has become a key vulnerability factor for the rural population. The penetration of agricultural insurance schemes remains very limited for most small producers[2]. It must be said that, despite several decades of a massive and selective "dam policy", irrigated areas cover barely 17% of the country's useful agricultural land, so that more than 80% of cultivated land remains subject to climatic hazards. Thus, when the vegetative cycle of plants is favoured by appropriate rainfall, the cereal harvest[3] (nearly two thirds of cultivated areas) is good, bringing in its wake agricultural GDP, which in turn stimulates other agricultural sectors and thus reduces rural poverty. Unfortunately, this dynamic also works in the other direction: a rainfall deficit, or even a drought, leads to a poor cereal season and therefore a drop in agricultural GDP accompanied by an increase in rural poverty. The transformation of the Moroccan agricultural sector brought about by the Green Morocco Plan strategy (Plan Maroc Vert, PMV) has created few jobs and has not provided sufficient opportunities for poor rural populations[4]. As a result, migration remains one of the few adaptation measures to climate shocks available to many vulnerable smallholder farmers, a trend that could accelerate in the coming decades.

5- RESULTS AND ANALYSIS

Analysis of the figure below shows that there is a large dispersion of points for the returns of Cereals, Fruits and Vegetables, Legumes, Market gardening and sugar crops (MGSC). The correlation coefficients of the main agricultural productions of Morocco are everywhere close to the zero value. This explains that the relationships between total annual precipitation and crop yields are weak. In other words, annual rainfall does not alone determine the low or high level of agricultural yields. Other factors, such as technology, seeds or the spatio-temporal distribution of precipitation, influence the evolution of yields.The first result to highlight during this analysis is, for better agricultural production, the regularity of rains is much more important than the quantity of falling rain. A good spatio-temporal distribution of precipitation is an essential element for the growth and productivity of cultivated plants.

Figure 8. Rainfall and crop yield ratio

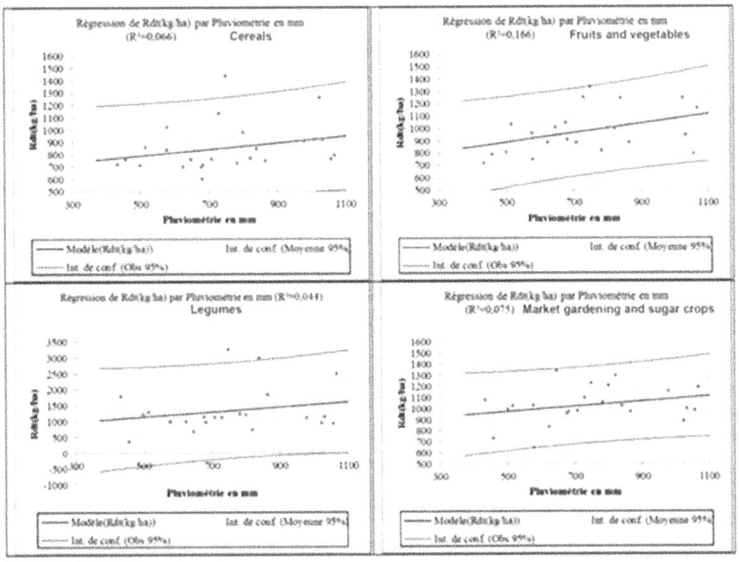

Table 1 below shows that, over a 20-year period (2001-2020), a good yield for one crop does not necessarily mean a good yield for all crops. However, if we exclude from the Spearman correlation calculation the 5 largest differences in rank with the two variables (the yields of the 2 crops, in this case cereals and legumes, which are the most exposed to lack of rain during drier years), the correlations are in many cases significant. This is explained by the fact that in exceptional years in terms of rainfall, irregularity affects crops differently, even if certain non-climatic factors may intervene (technology, seeds, irrigation, etc.). This observation underlines a second result consisting of encouraging farmers to modernize their processes and especially to practice polyculture so that a poor harvest of one crop is partially compensated by a better harvest of another crop.

Table 1. Spearman correlation coefficients between the yields of 4 crops for the period 2001-2020

	Fruits and Vegetables	Cereals	Legumes	Market gardening and Sugar crops
Fruits and Vegetables	-	0,43 (0,03)	0,56 (0,01)	0,83 (0,02)
Cereals		-	0,84 (0,06)	0,76 (0,05)
Legumes			-	0,69 (0,04)
Market Gardening and sugar crops				-

Figure 8 above shows that there is, at first glance, a lack of correlation between annual rainfall and yields of different crops. For fruits and vegetables, there is a definite correlation with all other crops, but it is weak. This is also visible in Table 2A below for market gardening and sugar crops. This observation (line A of Table 2) goes against the grain of the elements noted during the literature review. The basic hypothesis states that seasonal climatic irregularities, by definition different from one year to the next due to ongoing climate change, could be responsible for a correlation between production volatility and annual precipitation. However, when we exclude the five years where the difference in rainfall is the greatest, we obtain significant correlations (line B of table 2). But when we take into account the 20 years of the study, the correlation turns out to be very weak, or even non-existent for certain crops such as fruits and vegetables, market gardening and sugar crops.

Thus, this result highlights a third observation which demonstrates that, despite ongoing climate change, agricultural production of the main crops demonstrates certain resilience. Public policies to promote irrigation, appropriate seeds and technology in general may be factors explaining this result, even if this policy raises other issues related to the sustainability of this agricultural model.

Table 2. Spearman correlation between yields of 4 crops and annual precipitation per season

	Fruits and vegetables				Cereals				Legumes				MGSC			
Yield and precip	**Rain JFM**	**Rain AMJ**	**Rain JAS**	**Rain OND**	**Rain JFM**	**Rain AMJ**	**Rain JAS**	**Rain OND**	**Rain JFM**	**Rain AMJ**	**Rain JAS**	**Rain OND**	**Rain JFM**	**Rain AMJ**	**Rain JAS**	**Rain OND**
A.spearm-an. Corr N=20	0,17 0,08	0,15 0,05	0,16 0,03	0.07 0.02	0,34 0,15	0,39 0,11	non sig	0,28 0,13	0,47 0,19	0,56 0,05	0,45 0,05	0, 400, 10	0,06 0,08	0,05 0,03	non sign	0,09 0,01
B.spearm-an.Corr N=15	0,35 0,05	0,41 0,04	0,63 0,03	0,54 0,04	0,72 0,06	0,68 0,04	0,12 0,01	0,53 0,03	0,68 0,04	0,81 0,06	0,21 0,02	0,57 0,03	0,49 0,04	0,59 0,05	0,48 0,04	0,67 0,09

The application of the Pettitt test, and the analysis of deviations from the mean of the series and the SPI highlight a rainfall disturbance in Morocco during this millennium. Recorded volumes have fluctuated considerably over the past two decades. Even though we noted interruptions in rainfall due to drought on five occasions during the period studied 2001-2020, the yield of agricultural production was not generally deficit, with the exception of the yield of cereals during years of drought.

However, the drop in precipitation and increase in temperatures observed during the period studied disrupted the activities of the most vulnerable populations and the crops least served by irrigation and agricultural investments: drought increases soil vulnerability and yield of non-irrigated crops; the intra- and inter-seasonal variability of precipitation, the occurrence of extreme phenomena (droughts and heatwaves) and dry sequences limit the yields of these crops and push the population who occupy these activities towards immigration.

This analysis made it possible to draw two major results:

1- The volatility of agricultural yields does not create more rural poverty as stipulated in the literature review, which confirms the data from the poverty census carried out by the HCP (2022). Rural poverty is decreasing as those most vulnerable to climate change immigrate to cities, and those who have the means to adapt, or who are eligible for help and support provided by public policies to invest in technology, seeds and irrigation, manage to perpetuate their activities and improve the yields of their productions.

There is therefore a sort of selection and adaptation which takes place gradually in rural areas and which does not allow the poverty rate to worsen. This is explained by the fact that, when they are not constrained by immigration, farmers equipped with more modern means tend to intensify their production by turning to crops with higher added value but on the other hand more intensive in the use of water and soil. This phenomenon is already known in the economic literature by the "Jevons paradox"[5], which, unfortunately, is confirmed once again in this study;

2- The amount of rain that falls and warming temperatures do not alone determine agricultural production. This explains why some years rainfall is low while production is not. Overall, there is even a lack of correlation between yield and average annual rainfall, with the exception of cereal production, where it is demonstrated that the correlations during the 5 years of drought during the period studied are significant.

Technical progress and supporting public policies free the most vulnerable farmers from environmental constraints. The effects of this policy lead them to intensify the production of crops, considered more profitable, and which are becoming increasingly insufficient with ongoing climate change: this is the indirect rebound effect which has been taking place for two decades in the agricultural policy in Morocco; The efficiency of agricultural production is not enough to preserve the environment and circumvent its effects, it is also the behaviors that must be modified. Sobriety is far from being encouraged, or even absent, in the agricultural development model in Morocco; quite the contrary.

6-CONCLUSION

Climate change is certainly a global phenomenon caused essentially by the savagery of capital, but its harmful effects are felt more harshly by poor countries. They pose a serious risk to poverty reduction by threatening to undo decades of development efforts, which depend primarily on agriculture and natural resources, and which are likely to decline due to the limited capacity of the countries concerned to cope with climate variability and extreme weather events in the long term (Muller et al., 2011; Angelsen et al. Dokken, 2018).

The study of the impacts of climate variability on agricultural production is still relevant in Morocco. Regarding the evolution of temperature and the Standardized Rainfall Index (SPI), these parameters reveal a fluctuation of these variables in the time series. We can thus distinguish an average increase in temperatures of 0.23°C per decade since 1960, and a downward trend in precipitation which varies depending on the region.

However, the study of Spearman's correlation coefficients between these parametric variations of climate change and the yields of four main agricultural productions during the period 2001-2020 did not reveal a significant link, with the exception of cereal yield during the 5 years of severe drought.

Furthermore, we can conclude at the end of this work that yields are linked on the one hand to the variability and the uneven spatio-temporal distribution of rainfall which has a downward trend, and on the other hand to structural factors linked to public policy such as encouraging the use of the most suitable techniques, fertilizers and seeds.But it seems that this last aspect manages to compensate for the first to make agricultural yield higher despite the ongoing climate deterioration, with the exception of cereals during the 5 years of severe drought noted during the period studied.But it seems that this last aspect manages to compensate for the first to make agricultural yield higher despite the ongoing climate deterioration, with the exception of cereals during the 5 years of severe drought noted during the period studied. This observation explains the decrease in rural poverty in Morocco noted by the latest HCP report (2022). If this reduction is a fact achieved to the detriment of ongoing climate change, it does not take into consideration the effect of exclusion pushing the rural population towards immigration, and especially the sustainability of public agricultural policies to cope with the poverty and climate change.

REFERENCES

Abidoye, B. O., & Odusola, A. F. (2015). Climate change and economic growth in Africa: An econometric analysis. *Journal of African Economies*, *24*(2), 277–301. DOI: 10.1093/jae/eju033

Angelsen, A., & Dokken, T. (2018). Climate exposure, vulnerability and environmental reliance: A cross-section analysis of structural and stochastic poverty. *Environment and Development Economics*, *23*(3), 257–278. DOI: 10.1017/S1355770X18000013

Auffhammer, M., & Schlenker, W. (2014). Empirical studies on agricultural impacts and adaptation. *Energy Economics*, *46*, 555–561. DOI: 10.1016/j.eneco.2014.09.010

Azzarri, C., & Signorelli, S. (2020). Climate and poverty in Africa South of the Sahara. *World Development*, *125*, 104691. DOI: 10.1016/j.worlddev.2019.104691 PMID: 31902973

Baarsch, F., Granadillos, J. R., Hare, W., Knaus, M., Krapp, M., Schaeffer, M., & Lotze-Campen, H. (2020). The impact of climate change on incomes and convergence in Africa. *World Development*, *126*, 104699. DOI: 10.1016/j.worlddev.2019.104699

Burke, M., Hsiang, S. M., & Miguel, E. (2015). Global non-linear effect of temperature on economic production. *Nature*, *527*(7577), 235–239. DOI: 10.1038/nature15725 PMID: 26503051

Byers, E., Gidden, M., Leclère, D., Balkovic, J., Burek, P., Ebi, K., Greve, P., Grey, D., Havlik, P., Hillers, A., Johnson, N., Kahil, T., Krey, V., Langan, S., Nakicenovic, N., Novak, R., Obersteiner, M., Pachauri, S., Palazzo, A., & Riahi, K. (2018). Global exposure and vulnerability to multi-sector development and climate change hotspots. *Environmental Research Letters*, *13*(5), 055012. DOI: 10.1088/1748-9326/aabf45

Carleton, T. A., & Hsiang, S. M. (2016). Social and economic impacts of climate. *Science*, *353*(6304), aad9837. DOI: 10.1126/science.aad9837 PMID: 27609899

Dasgupta, S., Huq, M., Khan, Z. H., Ahmed, M. M. Z., Mukherjee, N., Khan, M., & Pandey, K. D. (2010). Vulnerability of Bangladesh to cyclones in a changing climate: Potential damages and adaptation cost. *World Bank Policy Research Working Paper*, (5280).

Dercon, S., & Christiaensen, L. (2011). Consumption risk, technology adoption and poverty traps: Evidence from Ethiopia. *Journal of Development Economics*, *96*(2), 159–173. DOI: 10.1016/j.jdeveco.2010.08.003

Deryugina, T., & Hsiang, S. M. (2014). Does the environment still matter? *Daily temperature and income in the United States* (No. w20750). National Bureau of Economic Research.

Deschenes, O. (2014). Temperature, human health, and adaptation: A review of the empirical literature. *Energy Economics*, *46*, 606–619. DOI: 10.1016/j.eneco.2013.10.013

Deutsch, C. A., Tewksbury, J. J., Tigchelaar, M., Battisti, D. S., Merrill, S. C., Huey, R. B., & Naylor, R. L. (2018). Increase in crop losses to insect pests in a warming climate. *Science*, *361*(6405), 916–919. DOI: 10.1126/science.aat3466 PMID: 30166490

Devarajan, S., Go, D. S., Maliszewska, M., Osorio-Rodarte, I., & Timmer, H. (2015). Stress-testing Africa's recent growth and poverty performance. *Journal of Policy Modeling*, *37*(4), 521–547. DOI: 10.1016/j.jpolmod.2015.04.006

Dzanku, F. M., Jirström, M., & Marstorp, H. (2015). Yield gap-based poverty gaps in rural Sub-Saharan Africa. *World Development*, *67*, 336–362. DOI: 10.1016/j.worlddev.2014.10.030

Epule, T. E., Ford, J. D., Lwasa, S., & Lepage, L. (2017). Climate change adaptation in the Sahel. *Environmental Science & Policy*, *75*, 121–137. DOI: 10.1016/j.envsci.2017.05.018

Fafchamps, M., Udry, C., & Czukas, K. (1998). Drought and saving in West Africa: Are livestock a buffer stock? *Journal of Development Economics*, *55*(2), 273–305. DOI: 10.1016/S0304-3878(98)00037-6

Formetta, G., & Feyen, L. (2019). Empirical evidence of declining global vulnerability to climate-related hazards. *Global Environmental Change*, *57*, 101920. DOI: 10.1016/j.gloenvcha.2019.05.004 PMID: 31417231

Gregory, N. G. (2010). How climatic changes could affect meat quality. *Food Research International*, *43*(7), 1866–1873. DOI: 10.1016/j.foodres.2009.05.018

Hallegatte, S. (2015). The indirect cost of natural disasters and an economic definition of macroeconomic resilience. *World Bank Policy Research Working Paper*, •••, 7357. DOI: 10.1596/1813-9450-7357

Hallegatte, S., Rogelj, J., Allen, M., Clarke, L., Edenhofer, O., Field, C. B., Friedlingstein, P., van Kesteren, L., Knutti, R., Mach, K. J., Mastrandrea, M., Michel, A., Minx, J., Oppenheimer, M., Plattner, G.-K., Riahi, K., Schaeffer, M., Stocker, T. F., & Van Vuuren, D. P. (2016). Mapping the climate change challenge. *Nature Climate Change*, *6*(7), 663–668. DOI: 10.1038/nclimate3057

Hallegatte, S., & Rozenberg, J. (2017). Climate change through a poverty lens. *Nature Climate Change*, *7*(4), 250–256. DOI: 10.1038/nclimate3253

Hallegatte, S., & Rozenberg, J. (2017). Climate change through a poverty lens. *Nature Climate Change*, *7*(4), 250–256. DOI: 10.1038/nclimate3253

HCP (Haut Commissariat au Plan). (2022)available at:https://www.hcp.ma/Evolution-des-inegalites-sociales-dans-un-contexte-marque-par-les-effets-de-la-COVID-19-et-de-la-hausse-des-prix_a3588.html

Heal, G., & Park, J. (2016). Reflections—temperature stress and the direct impact of climate change: A review of an emerging literature. *Review of Environmental Economics and Policy*, *10*(2), 347–362. DOI: 10.1093/reep/rew007

Hertel, T. W. (2015). The challenges of sustainably feeding a growing planet. *Food Security*, *7*(2), 185–198. DOI: 10.1007/s12571-015-0440-2

ILO (International Report Organization). (2019) "Annual Report 2029" https://www.ilo.org/wcmsp5/groups/public/---ed_emp/documents/publication/wcms_739991.pdf (Accessed30 October 2023)

Knippertz, P., Christoph, M., & Speth, P. (2003). Long-term precipitation variability in Morocco and the link to the large-scale circulation in recent and future climates. *Meteorology and Atmospheric Physics*, *83*(1), 67–88. DOI: 10.1007/s00703-002-0561-y

McKee, T. B., Doesken, N. J., & Kleist, J. (1993, January). The relationship of drought frequency and duration to time scales. In *Proceedings of the 8th Conference on Applied Climatology* (Vol. 17, No. 22, pp. 179-183).

. Moser, S. C. (2007). More bad news: the risk of neglecting emotional responses to climate change information.

Müller, C., Cramer, W., Hare, W. L., & Lotze-Campen, H. (2011). Climate change risks for African agriculture. *Proceedings of the National Academy of Sciences of the United States of America*, *108*(11), 4313–4315. DOI: 10.1073/pnas.1015078108 PMID: 21368199

Myers-Smith, I. H., Grabowski, M. M., Thomas, H. J., Angers-Blondin, S., Daskalova, G. N., Bjorkman, A. D., Cunliffe, A. M., Assmann, J. J., Boyle, J. S., McLeod, E., McLeod, S., Joe, R., Lennie, P., Arey, D., Gordon, R. R., & Eckert, C. D. (2019). Eighteen years of ecological monitoring reveals multiple lines of evidence for tundra vegetation change. *Ecological Monographs*, *89*(2), e01351. DOI: 10.1002/ecm.1351

Nkedianye, D., de Leeuw, J., Ogutu, J. O., Said, M. Y., Saidimu, T. L., Kifugo, S. C., Kaelo, D. S., & Reid, R. S. (2011). Mobility and livestock mortality in communally used pastoral areas: the impact of the 2005-2006 drought on livestock mortality in Maasailand. *Pastoralism: Research. Pastoralism, 1*(1), 1–17. DOI: 10.1186/2041-7136-1-17

Noack, F., Wunder, S., Angelsen, A., & Börner, J. (2015). Responses to weather and climate: a cross-section analysis of rural incomes. *World Bank Policy Research Working Paper*, (7478).

Paxson, C. H. (1992). Using weather variability to estimate the response of savings to transitory income in Thailand. *The American Economic Review*, •••, 15–33.

Pettitt, A. N. (1979). A non-parametric approach to the change-point problem. *Journal of the Royal Statistical Society. Series C, Applied Statistics, 28*(2), 126–135.

Rosenzweig, C., Elliott, J., Deryng, D., Ruane, A. C., Müller, C., Arneth, A., Boote, K. J., Folberth, C., Glotter, M., Khabarov, N., Neumann, K., Piontek, F., Pugh, T. A. M., Schmid, E., Stehfest, E., Yang, H., & Jones, J. W. (2014). Assessing agricultural risks of climate change in the 21st century in a global gridded crop model inter-comparison. *Proceedings of the National Academy of Sciences of the United States of America, 111*(9), 3268–3273. DOI: 10.1073/pnas.1222463110 PMID: 24344314

Sambou, M. J. G., Pohl, B., Janicot, S., Landry Famien, A. M., Roucou, P., Badiane, D., & Gaye, A. T. (2021). Heat waves in spring from Senegal to Sahel: Evolution under climate change. *International Journal of Climatology, 41*(14), 6238–6253. DOI: 10.1002/joc.7176

Servat, D., Perrier, E., Treuil, J. P., & Drogoul, A. (1998). When agents emerge from agents: Introducing multi-scale viewpoints in multi-agent simulations. In *Multi-Agent Systems and Agent-Based Simulation: First International Workshop, MABS'98, Paris, France, July 4-6, 1998.* [Springer Berlin Heidelberg.]. *Proceedings, 1*, 183–198.

Sissoko, K., van Keulen, H., Verhagen, J., Tekken, V., & Battaglini, A. (2011). Agriculture, livelihoods and climate change in the West African Sahel. *Regional Environmental Change, 11*(S1), 119–125. DOI: 10.1007/s10113-010-0164-y

Udry, C. (1995). Risk and saving in Northern Nigeria. *The American Economic Review, 85*(5), 1287–1300.

Verpoorten, M., Arora, A., Stoop, N., & Swinnen, J. (2013). Self-reported food insecurity in Africa during the food price crisis. *Food Policy, 39*, 51–63. DOI: 10.1016/j.foodpol.2012.12.006

World Bank. (2022) “Maroc: Rapport de suivi de la situation économique (Hiver 2022-23): Le Maroc face aux chocs d'offre” a. https://documents1.worldbank.org/curated/en/099332002132325417/pdf/IDU0e3742c55010a3044730a80a0ad6062acd9db.pdf (Accessed 25 October 2023)

ENDNOTES

1 On this subject, see the link to the “poverty & vulnerability” Methodology on the institutional site of the High Commission for Planning in Morocco.https://www.hcp.ma/methodologie-pauvrete-vulnerabilite_r497.html

2 The penetration rate of agricultural insurance in Morocco is estimated at 17%, although it is highly variable: 40% for large farms (> 50 ha), 24% for medium-sized farms (between 10 and 50 ha) and less than 3% for small farms (< 10 ha).

3 The cereal harvest is so important that it constitutes one of the main hypotheses on the basis of which the architecture of any finance bill is constructed, often taking as reference an annual average of cereal production of around 75 million quintals.

4 Indeed, the jobs diagnosis recently carried out by the World Bank in collaboration with the HCP shows that the strong growth in agricultural value added recorded over the past two decades has had a negative multiplier effect on jobs.

5 In his essay published in 1865 entitled “The Coal Question” the British economist William Stanley Jevons emphasizes that the idea that more economical use of a resource would be equivalent to less consumption is a total confusion. It's the opposite that is true.

Chapter 3

The Effects of Climate Change on Agriculture in Algeria:

Structural Constraints and the Search for Resilience

Amine M. Benmehaia
https://orcid.org/0000-0001-8110-1601
University of Biskra, Algeria

Slimane Bédrani
ENSA Algiers, Algeria

ABSTRACT

This chapter examines the impacts of climate change on Algerian agriculture. It focused on three main areas: analyzing the impacts on natural resources and production systems; the dynamics of agricultural labour; and assessing agricultural performance and growing food dependency. Main results highlight the degradation of natural resources (water, soil, vegetation), transformations in production systems (cereals, vegetable crops, livestock), and their socioeconomic consequences (rural unemployment, exodus, food dependency). The data and evidence reveal growing vulnerability to climate hazards, exacerbating food security challenges and the need for appropriate agrarian reforms.

DOI: 10.4018/979-8-3373-3346-5.ch003

INTRODUCTION

Algeria today faces an unprecedented agricultural challenge, intersecting simultaneously food security imperatives, growing water constraints, and the impacts of climate change. This chapter analyzes the complex dynamics linking water, labor, and agricultural performances in a profoundly changing environmental context.

Since its independence, Algeria has undergone several land reforms aimed at modernizing its agricultural sector. Yet, the country remains structurally dependent on imports to cover its basic food needs, with a bill that now exceeds $10 billion annually.[1] This paradoxical situation—significant agricultural potential but insufficient production—has worsened under the combined effects of demographic pressure and climate change.

The manifestations of climate change in Algeria are now undeniable. Data from the National Meteorological Office, World Bank and IPCC reports, show an average temperature increase of 1.5°C since 1960, accompanied by a 20 to 30% reduction in precipitation in inland regions. These climatic disturbances have led to a gradual northward shift of isohyets, reducing arable land without irrigation and increasing pressure on already limited water resources. At the same time, the domestic agricultural sector is facing profound transformations in its workforce dynamics (Chaami et al., 2025). The massive rural exodus, particularly marked among young people, has caused an aging of the active agricultural population. Several studies (such as Bouchakour & Saad, 2020; Benmehaia & Benmehaia, 2018) reveal that the average age of farmers exceeds 55, raising the crucial question of generational continuity.

This chapter will focus on three main areas: analyzing the impacts of climate change on natural resources and production systems; studying transformations in the rural world and the dynamics of agricultural labor; and assessing agricultural performance and growing food dependency. Through this analysis, we will attempt to answer a central question: how is climate change reconfiguring the relationships between water, labor, and agricultural production in Algeria, and what adaptations does this imply for current agricultural policies?

This chapter adopts an interdisciplinary analytical approach that draws from the fields of political ecology, agroecological resilience, and climate vulnerability frameworks. The relationships between climate change, agricultural production, and rural labor are examined through the lens of socio-ecological systems, wherein environmental stressors and institutional responses are co-determinants of outcomes (Folke et al., 2016; Scoones, 1998). Rather than treating climate impacts as exogenous shocks, we consider them as interacting with structural vulnerabilities—notably land degradation, water stress, and labor informality—that are rooted in historical policy choices and institutional lock-ins (North, 1990; Pierson, 2000). This aligns with recent scholarship in environmental political economy, which emphasizes how path-dependent development trajectories shape the adaptive capacity of agrarian systems in the Global South (Clapp, 2021; Ribot, 2014).

The data used come primarily from national (National Office of Statistics, Ministry of Agriculture, National Agency for Hydraulic Resources) and international (FAO, World Bank, IPCC) institutions, supplemented by recent studies. This approach will shed light on the complex interactions between natural, economic, and social factors shaping the future of Algerian agriculture.

1. CLIMATE CHANGE: SIGNS AND AMPLIFICATION OF RISKS IN ALGERIA

Algeria is experiencing the full impact of climate change, with an increase in extreme events. This section analyzes the evolution of droughts, floods, and other climatic hazards that are gradually reshaping the country's natural and agricultural balance.

1.1. Increased droughts and isohyet shifts

As in all North African countries, there are numerous signs of climate change in Algeria: rising temperatures, increasing extreme events (droughts, floods, sandstorms, and forest fires), and a profound shift in its rainfall patterns. The National Climate Change Agency and the Renewable Energy Development Center, among others, report that the frequency of extreme weather events has doubled since the early 2000s. Heat waves have become more frequent and intense, with temperatures regularly exceeding 50°C in Saharan regions. At the same time, torrential rain events are increasing, particularly in coastal areas and highlands (Sahabi-Abed et al., 2023; Fellag et al., 2021; Hamitouche et al., 2024; Benmehaia et al., 2020; Benmehaia, 2023).

According to the National Meteorological Office, from 1991 to 2024, only these seven weather stations in the entire country reported continuous temperature readings.[2] From these weather reports, the ONM established a long-term trend representing monthly average temperatures. The hottest month over this entire period was July 2023 with 33.1°C. January 2005 was the coldest month with an average of 8.9°C. The annual average temperature was around 19.8°C for the years after 1991 and around 21.4°C for the last years before 2024. In addition, temperature variability (inter-annual difference) shows a clear upward trend since the 1990s (Figure 1), according to FAO (2025) data.

Figure 1. Change in annual temperatures 1961-2024 in Algeria (Source: Based on FAO data: fao.org/faostat/en/#data)

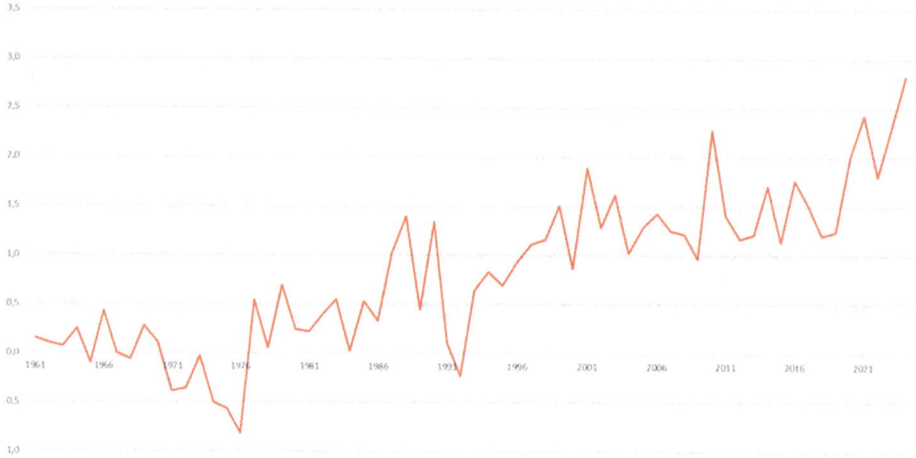

Fires in Algeria have intensified in recent years, exacerbated by heatwaves and prolonged droughts linked to climate change. According to the European Forest Fire Information System (EFFIS, 2023), Algeria lost more than 100 000 hectares of vegetation cover during the 2022 summer season alone, double the ten-year average. The northern regions (Tizi Ouzou, Béjaïa, Jijel) are the most affected, where Mediterranean forests (cork oak, cedar) are burning at an unprecedented rate. The Algerian General Directorate of Forests (DGF, 2023) emphasizes that a large proportion of fires are of anthropogenic origin (negligence or criminal acts), but their spread is favored by extreme weather conditions.

Furthermore, data from the World Meteorological Organization (WMO, 2023) confirm a trend toward accelerated aridification since the 1990s, with an average decrease in precipitation of 15 to 30% in inland regions. This decline, coupled with a temperature increase of 1.2°C since the pre-industrial era (IPCC, 2022), has caused a significant shift in isohyets, permanently altering the geography of arable land. Nedjraoui & Bédrani (2008) indicate that at the steppe station of Mécheria (Southern Oran), the decrease in precipitation is around 18 to 27% and the dry season has increased by 2 months during the last century.

The retreat of the 400 mm/year isohyet towards the north is particularly revealing. According to data from the Ministry of Water Resources (2022), this rainfall line, which traditionally delimited the rainfed crop zone, has migrated nearly 80 km in four decades, reducing the areas suitable for unirrigated cereals. The Tiaret region, once considered a cereal granary, illustrates this decline: its average rainfall has fallen from 380 mm/year in the 1980s to less than 280 mm/year over the last decade

(Haouari, 2024; WMO, 2023). This progressive aridification is accompanied by an intensification of acute drought episodes. The period 1990-2023 saw five years of drought deficit, with the last 4 years (from 2020-2023) are the hottest ever observed (Haouari, 2024).

Soils, subject to increased wind erosion in the absence of vegetation cover, are losing up to 20 tons of fertile soil per hectare per year in the most affected areas (UN-CCD, 2020, 2023). This phenomenon is exacerbated by the gradual disappearance of steppe rangelands, where natural vegetation is receding in the face of the encroaching desert. Satellite images from the Copernicus program show a reduction of nearly 25% in pastoral areas in steppe regions such as Naâma and El Bayadh since 2000.

This transformation of the water regime poses major challenges for agricultural policies. Irrigated areas, concentrated on only 8% of the useful agricultural area, are becoming production sanctuaries, but their sustainability is threatened by the overexploitation of groundwater. The Mio-Plio-Quaternary aquifer in the northern Sahara has seen a drawdown of 32 m in 43 years in the northern Oued Rhir, and nearly 5 m in 30 years in Ouargla (OSS, 2003). However, alternative solutions, such as the mobilization of unconventional water, are struggling to compensate for these losses: the desalination plants planned by 2030 (like the 25 already installed) will only cover 60% of agricultural needs (APS, 2023a). The shift in isohyets in Algeria is not simply a matter of climate fluctuation. This is a territorial restructuring that requires an overhaul of production models, calling into question the viability of traditional rain-fed agriculture. Ongoing reforms, such as the revision of the National Water Plan, must integrate this new geo-climatic situation to avoid a collapse of agro-pastoral systems in the coming decades.

These observations are consistent with broader scientific evidence documenting the advance of aridification across the Mediterranean and North African regions. Recent work by Oubadi & Faci (2025) confirms a statistically significant increase in the frequency and duration of heatwaves across Algeria's climate zones from 1979 to 2023, with notable acceleration after the year 2000. Similarly, Sahabi-Abed et al. (2023), using CMIP6 climate models, project a continued northward migration of aridity lines and a sharp contraction of humid and semi-humid agroecological zones by mid-century, particularly under high-emission scenarios. This trajectory threatens to permanently reconfigure Algeria's agricultural geography, echoing the concerns raised by Nedjraoui & Bédrani (2008) more than a decade ago.

Recent studies also point to the compounding effects of temperature anomalies and precipitation shifts, which together reduce effective growing days and disrupt planting calendars (Benmehaia, 2023; Fellag et al., 2021). From a policy perspective, these transformations underscore the urgency of transitioning from static agricultural zoning to dynamic agroecological planning, as advocated in resilience-based agricultural frameworks (Altieri & Nicholls, 2025). Without such anticipatory adaptation strategies, Algeria risks locking in production systems that are increasingly mismatched to the evolving bioclimatic realities.

1.2. Increased flooding and inundation

Algeria presents a marked climate paradox, where prolonged droughts and violent floods coexist with increasing frequency. According to official data, the number of major flood events increased significantly between the 1990s and 2000s and the 2010s and 2020s. This trend is consistent with IPCC (2021) projections for the Mediterranean region, which predict an intensification of extreme rainfall events despite an overall decrease in annual precipitation.

The accelerated urbanization of flood-prone areas is significantly exacerbating the impact of these phenomena. Several areas identified as flood-prone in the 1980s have been urbanized, often illegally. This observation is particularly striking in the Oued El Harrach valley, where soil permeability decreased considerably between 2000 and 2022. During the torrential rains of October 2022, this same Oued recorded a record flow rate of 1 200 m^3/second, compared to 300 m^3/second for a comparable episode in 1966. As for the consequences for peri-urban agriculture, the Directorate of Agricultural Services of Algiers (MADR, 2022) reports that more than half of the farms in the green belt suffered damage during the last three major floods. Vegetable crops, despite being vital to the capital's water supply, is paying a heavy price for these repeated disasters.

Faced with this growing threat, protective infrastructure appears insufficient. The Ministry of Water Resources acknowledges that only 12% of at-risk waterways have containment systems that meet international standards. The situation is particularly worrying for dams: 43% have not benefited from technical draining since their commissioning, according to the National Agency for Dams and Transfers (2022), reducing their capacity to regulate floods.

The intensification of flood risks in Algeria mirrors broader patterns observed across the Mediterranean Basin, where hydrological extremes are becoming more frequent and erratic. Recent climatological research confirms a paradoxical trend of decreasing mean annual rainfall alongside increasing intensity of precipitation events, which overwhelms local hydrological systems (Hamitouche et al., 2024; IPCC, 2022). This "precipitation intensification effect" is particularly acute in urbanized watersheds, where impermeable surfaces amplify runoff and flash flooding risks (Alfieri et al., 2015). Studies focusing on Algeria's flood-prone zones, such as Merniz et al. (2019), highlight the increasing variability of rainfall across seasons and the inadequacy of existing drainage infrastructure to cope with peak flows. Moreover, Algeria's exposure is magnified by institutional and spatial planning failures, including informal urban expansion in flood basins, insufficient dam desiltation, and underinvestment in early warning systems (Mozas & Ghosn, 2013). From a risk governance perspective, this reflects what scholars refer to as compound vulnerability, where climate hazards interact with systemic governance gaps (Ribot, 2014). Anticipatory urban planning, ecological flood management, and the retrofitting of hydraulic infrastructure are critical if Algeria is to shift from reactive disaster response to proactive resilience building.

2. EFFECTS OF CLIMATE CHANGE ON NATURAL RESOURCES

Accelerated soil degradation, water scarcity, and vegetation depletion are all signs of these changes. Here, we will examine how climate change is compromising the very foundations of agricultural production in Algeria.

2.1. Soil degradation: Water and eolian erosion

Soil degradation in Algeria is reaching high levels, with both water and eolian erosion processes accelerating under the combined effects of climate change and inappropriate agricultural practices. MADR (2022) reports estimate that nearly 45% of the country's usable agricultural land shows advanced signs of degradation, with soil losses reaching up to 30 tons per hectare per year in the most affected areas.

Water erosion primarily affects the northern regions of the country. A report by the World Bank (2023) reveals that in certain watersheds, soils lose an average of 20 mm of thickness per year due to torrential rains. These phenomena are exacerbated by deforestation, with forest cover falling from 12% in the 1960s to less than 7% today, according to MADR (2022) data, as also confirmed by several studies and official reports. Traditional agricultural terraces, once effective at retaining soil, are often abandoned or poorly maintained, reducing their protective capacity.

The progressive desertification of the steppe, exacerbated by overgrazing and illegal land clearing, creates favorable conditions for the formation of increasingly frequent sandstorms. The national environmental monitoring network recorded a 60% increase in days with sandstorms between 2000 and 2020. The consequences of this dual erosion are particularly worrying. Studies, such as those by Morsli et al. (2013) and Benderradji et al. (2006), have shown that soils in affected areas significantly lose organic matter over twenty years, thus reducing their fertility. This degradation is irreversible on a human scale, as highlighted in a recent study by Merabtene et al. (2021) on dryland soils. According to the FAO (2011) report, cereal lands lose an average of 10 to 15 tons of fertile soil per hectare each year, reducing their water retention capacity and organic fertility. Unsuitable agricultural practices, such as monoculture and intensive tillage, exacerbate this degradation.

Faced with this situation, erosion control measures remain insufficient. The National Soil Conservation Program, launched in 2018, has only covered 15% of the identified priority areas, according to an evaluation report by the General Directorate of Forests (DGF, 2024). Modern conservation techniques, such as grass strips or contour farming, are struggling to become widespread due to the fragmentation of farms and a lack of farmer training.

The severity of Algeria's soil erosion crisis is confirmed by a growing body of scientific literature, which highlights the accelerating pace of both hydric and eolian degradation across arid and semi-arid landscapes. Recent studies using remote sensing and field-based erosion models have quantified annual topsoil losses exceeding 20 to 30 tons/ha/year in the Tell and High Plateaux regions, particularly on sloped terrains lacking vegetation cover (Morsli et al., 2013; Benderradji et al., 2006). A 2021 study by Merabtene *et al.* on Tessala soils shows sharp declines in organic carbon stocks and microbial activity, particularly in fallow or over-cultivated parcels. These findings are not isolated: comparative research across the Maghreb indicates that Algeria exhibits some of the highest land degradation rates in North Africa, largely due to land-use mismanagement and delayed policy interventions (Benabdelkader et al., 2021; UNEP, 2023).

Erosion dynamics are further amplified by feedback loops between land degradation and climate change: loss of vegetation increases albedo and evapotranspiration, which in turn accelerates desertification processes (UNCCD, 2023). The expansion of sandstorms and wind erosion fronts in Algeria's steppe margins, as documented by Chenchouni & Neffar (2022), illustrates this spiraling degradation. Eolian processes are no longer confined to traditionally vulnerable zones like El Bayadh or Naâma; they are now encroaching into agricultural frontiers in southern Sétif, Djelfa, and M'sila. The depletion of vegetative barriers—particularly alfa grass and natural shrublands—further exposes bare soils to multidirectional wind flows and sand encroachment, as confirmed by satellite analysis from the Copernicus program.

Moreover, these forms of degradation are not only ecological but also deeply institutional. The limited diffusion of conservation techniques—such as contour farming, cover cropping, and agroforestry—is often tied to fragmented land tenure, under-resourced extension services, and insufficient farmer training. As emphasized by Altieri & Nicholls (2021), addressing erosion in dryland contexts requires more than technical fixes; it demands a shift toward agroecological design principles, which treat soil as a living system embedded in a wider socio-ecological context. The erosion crisis in Algeria is thus emblematic of a broader structural failure: a disconnect between centralized soil policy and localized land-use realities. Without a paradigm shift toward integrated landscape management, Algeria's soil capital—already fragile—risks permanent depletion, threatening food security and rural livelihoods in the decades to come.

2.2. Depletion of vegetation

Algeria, in some regions, is experiencing a sharp decline in its vegetation cover, a phenomenon that is part of a dynamic of accelerated degradation of natural ecosystems. According to MADR (2022) data, the national forest area has declined from 4.1 million hectares in 1962 to approximately 2.5 million hectares today, a reduction of nearly 40%. This trend is confirmed by satellite images from the Copernicus program, which reveal an average loss of 25 000 hectares of vegetation cover per year since 2010.

Steppe formations, which historically covered nearly 20 million hectares, are undergoing particularly severe degradation. The study by Macheroum & Chenchouni (2022) in the semi-arid Algerian regions shows that the Alfa plant density of the steppe has declined significantly over the last three decades. Overgrazing, combined with recurring droughts, has transformed vast areas into near-barren lands, and with their capacity for natural regeneration severely compromised.

The Mediterranean forests in the north of the country are not spared. The World Bank (2023b) report highlights that 45% of forest areas show signs of advanced degradation. Endemic species such as the cork oak and the Atlas cedar are seeing their ranges drastically reduced. Recurring fires—more than 30 000 hectares burned each year on average, according to World Bank (2023b) data—accelerate this process of disappearance. Several studies have shown that the reduction in vegetation cover leads to a significant decline in biodiversity, with the disappearance of numerous animal and plant species. Furthermore, the carbon storage capacity of Algerian ecosystems has declined by 35% since 1990, according to estimates from the United Nations Environment Programme (UNEP, 2023) and the Ministry of the Environment and Renewable Energy (MEER, 2023).

The socioeconomic impacts are equally worrying. The General Directorate of Forests (DGF, 2023) estimates that nearly 500 000 people directly dependent on forest resources are seeing their livelihoods threatened. Traditional activities such as beekeeping, the gathering of aromatic and medicinal plants, and extensive livestock farming are particularly affected by this ongoing degradation.

Given this situation, reforestation efforts remain insufficient. The National Reforestation Program, which aims to plant 1.3 million hectares by 2030, has only met 15% of its mid-term objectives according to the MADR (2022) reports. Plant survival rates, often below 30% due to harsh climatic conditions and lack of maintenance, significantly limit the effectiveness of these initiatives.

The degradation of Algeria's vegetation cover is not only a visible symptom of climate change but also a key driver of broader ecological collapse. Recent satellite-based studies confirm a marked decline in Normalized Difference Vegetation Index (NDVI) values across the Tellian Atlas, High Plateaux, and northern Sahara over the past two decades, with average losses exceeding 15–25% in high-stress zones (Thi et al., 2023; Zeng et al., 2024). This downward trend is most severe in areas of overlapping pressures—such as drought frequency, overgrazing, fire recurrence, and agricultural expansion—illustrating what UNEP (2023) defines as "synergistic degradation". In the steppe regions, for instance, Chenchouni & Nefar (2022) demonstrate that alfa grass (*Stipa tenacissima*) has not only declined in density but also in biomass productivity and root integrity, impairing its regenerative capacity. The collapse of these perennial species destabilizes the soil-vegetation-climate equilibrium and accelerates desertification feedback loops.

In forest ecosystems, tree mortality and forest fragmentation are rising. Mediterranean species such as cork oak (*Quercus suber*) and Aleppo pine (*Pinus halepensis*) exhibit increased canopy dieback and reduced recruitment rates due to hotter, drier summers (Vennetier, 2020; World Bank, 2023b). These shifts are further intensified by the recurrence of wildfires, whose frequency and magnitude have sharply increased under extreme weather patterns (EFFIS, 2023). In some cases, post-fire regeneration has failed entirely due to seedbed desiccation and invasive species proliferation, particularly in degraded peri-urban forests. Meanwhile, reforestation efforts remain fragmented and largely ineffective. As Bouzid et al. (2022) note, many public plantation projects suffer from low survival rates, poor species selection, and a lack of community involvement, resulting in what scholars refer to as "institutional reforestation without ecological restoration" (Reynolds et al., 2007).

The implications extend beyond ecology. Vegetation loss has direct socioeconomic consequences for rural livelihoods, especially those dependent on non-timber forest products (NTFPs), beekeeping, and pastoral mobility. According to Slimani & Aidoud (2004), the decline in wild forage availability has forced transhumant herders to reduce herd mobility and adopt more sedentary, input-intensive systems. This not only undermines traditional knowledge systems but also increases dependence on imported fodder and concentrates. Moreover, vegetation loss weakens the carbon sink capacity of Algerian ecosystems. Estimates by Bastin et al. (2017) suggest that the Maghreb region has already lost significant part of its pre-industrial vegetative carbon storage potential. Without urgent intervention, Algeria risks entering a "vegetation tipping point" where ecosystem services such as rainfall regulation, erosion control, and biodiversity support collapse irreversibly.

In response, many scholars now advocate for a transition from narrow afforestation schemes to landscape-scale ecological restoration, integrating water harvesting, native species replanting, and community-led governance structures (Aronson et al., 2020; Mansourian et al., 2017). Such approaches would require rethinking land-use rights, aligning environmental goals with local livelihoods, and moving beyond symbolic plantation targets toward resilience-oriented planning.

2.3. Water stress: Between structural shortage and persistent waste

The water crisis in Algeria reveals a striking paradox: while resources are drastically diminishing, their management remains marked by systemic inefficiencies. Official figures from the Ministry of Water Resources (2017) are telling: with less than 300 m^3 of freshwater available per capita per year, the country is well below the shortage threshold defined by the United Nations. This critical situation is not inevitable, however, but the result of questionable political and technical choices.

Groundwater, the country's main reserves, are being exploited well beyond their replenishment capacity. The recent study by Kherbache & Molle (2021) demonstrates that the northern aquifers are experiencing worrying rates of overexploitation. The Mitidja aquifer, the main water resource for the entire central region, has seen its level drop by an average of 3 meters over the past decade. In Saharan regions, uncontrolled water withdrawals threaten fossil fuel reserves that are thousands of years old.

The dilapidated and poorly maintained distribution network generates significant losses. Data from MRE and studies such as Mozas & Ghosn (2013) indicate that 40% of the drinking water produced never reaches users, evaporating through leaks and illegal connections. Algiers, the capital, holds the national record with loss rates approaching 50%. This waste stands in stark contrast to the difficulties millions of citizens face in accessing water, particularly in rural and peri-urban areas.

Agriculture, the largest consumer with 70% of available resources, primarily uses gravity-fed irrigation systems,[3] which exhibit low water yields (see Table 1 below). However, there has been a significant increase in water-efficient irrigation systems. However, this remains insufficient compared to what should be achieved. The water issue is the thorny issue in the expansion of vegetable crops. Vegetable crops consumes 40% of the country's irrigation water, with varying efficiencies (Oulmane et al., 2020). While large public schemes (such as those in the Lower Soummam Plain) achieve acceptable water yields (75%), private farms often experience significant waste. A study by Belaidi et al. (2022) in the Mitidja region revealed that 60% of farms continue to use gravity-fed irrigation methods, with losses of up to 50%. Faced with this situation, technical solutions exist but are struggling to gain traction. Seawater desalination, presented as a panacea, only covers 18% of drinking water needs. Projects for the reuse of treated wastewater remain almost at the experimental stage, with only 5% of their potential being exploited. As for water pricing, it in no way reflects its true value or scarcity, thus perpetuating a culture of waste.

Table 1. Changes in irrigated areas (2000-2010)

Years	Irrigated area	Irrigation system (ha)		
	Total (ha)	Gravity	Sprinkler	Drip
2000	350 000	275 000	70 000	5 000
2001	617 427	458 421	102 978	56 028
2002	644 978	433 531	127 570	83 877
2003	722 320	485 019	138 301	99 000
2004	793 334	516 108	159 739	117 487
2005	825 206	524 503	153 006	147 697
2006	835 590	481 046	175 056	179 488
2007	907 293	557 327	183 182	166 784
2008	906 174	566 971	183 584	155 619
2009	936 862	573 667	211 594	196 475
2010	981 736	573 667	211 594	196 475

Source: MADR (2022)

This multifaceted crisis calls for a complete overhaul of water governance. As the World Bank (2023) report highlights, Algeria urgently needs to reconcile its food security imperatives with the preservation of its water resources. The time for half-measures and cosmetic solutions is over: an entire management model must be rethought before the shortage turns into a humanitarian catastrophe.

Algeria's water crisis is symptomatic of broader challenges faced by arid and semi-arid countries undergoing rapid demographic and agricultural transitions under climate stress. Peer-reviewed studies confirm the severity of aquifer depletion across major production zones, particularly in the Mitidja plain and Saharan regions, where water tables are dropping by several meters per decade (Kherbache & Molle, 2021). According to Abouelmagd & Ahmed (2024), groundwater overdraft in the northern Sahara has reached unsustainable levels, risking long-term salinization and collapse of fossil reserves. Algeria's per capita water availability has now fallen below 300 m^3/year—a threshold categorized as "absolute scarcity" by the FAO (FAO AQUASTAT, 2022)—with projections indicating further declines due to population growth and reduced precipitation under RCP4.5 and RCP8.5 scenarios (Mokhtari *et al.*, 2025). Comparative analyses show that Algeria lags behind Morocco and Tunisia in terms of treated wastewater reuse and water governance reforms, despite having higher hydroclimatic stress (Qadir et al., 2010). The continued dominance of gravity-fed irrigation (in over 60% of irrigated farms) remains a critical inefficiency, exacerbating evaporative losses and placing pressure on already overdrawn aquifers.

Beyond technical limitations, the persistence of Algeria's water crisis is largely institutional. As noted by Ouessar et al. (2021), governance fragmentation, poor pricing incentives, and weak enforcement mechanisms contribute to systemic inefficiencies in allocation, conservation, and infrastructure maintenance. Water pricing remains heavily subsidized and disconnected from actual scarcity, leading to a "tragedy of the commons" dynamic where users lack incentives to adopt conservation practices.

3. IMPACTS ON THE MAIN AGRICULTURAL PRODUCTIONS

Faced with new climatic constraints, crop and livestock systems are undergoing radical transformations. This section assesses the evolution of cereal yields, the expansion of irrigated market gardening, and the difficult adaptation of pastoral activities.

3.1. Cereals: Decreasing yields in dry areas

Grain yields in Algeria have been highly uneven, particularly in semi-arid and arid regions. The Figure 2 below illustrates the stagnation of wheat yields (in tons per hectare) since 2010. While in the 1970s and 1980s, yields occasionally exceeded 2 tons per hectare, recent decades have shown increased volatility, with yields often below 1.5 tons per hectare, and marked collapses during years of severe drought, such as 2021.

Figure 2. Changes in wheat yields (in quintals/hectare)

This phenomenon is mainly explained by the deterioration of agro-climatic conditions. Several studies confirm that traditional cereal production areas, such as the High Plateaux, are experiencing accelerated aridification. In the regions of Sétif and Batna, water deficit significantly reduced durum wheat yields between 2000 and 2020. In the High Plateaux, precipitation has decreased by 20 to 30% since 1990 (Benmehaia et al., 2020; Benmehaia, 2023), while average temperatures have increased by nearly 1.5°C, increasing evapotranspiration and water stress in crops.

Public policies are struggling to halt this decline. Despite subsidies for inputs (fertilizers, seeds), productivity is stagnating. The National Agricultural Development Plan (PNDA) has certainly enabled increased mechanization, but without significant yield improvements in rainfed agricultural areas. According to the MADR (2022), less than 15% of cereal-growing areas benefit from supplemental irrigation, while climate models predict worsening droughts.

The decline in cereal yields in drylands is not only a climate issue, but also the result of unsuitable production systems and fragile resource management. Without a transition to resilient practices (agroecology, diversification, efficient irrigation), dependence on wheat imports – already at 70% – will only increase.

However, yield stagnation cannot be attributed to climate alone. Structural inefficiencies in Algeria's cereal support schemes have created what some analysts call a "low productivity trap." Despite input subsidies (for seeds and fertilizers) and state-guaranteed purchase prices, most farmers remain locked into low-input, low-output production systems (Bekkis & Benmehaia, 2025). Fragmented landholdings, poor access to certified seeds, and limited extension services constrain the adoption of improved practices, especially among smallholders. Compared with neighboring Morocco and Tunisia, Algeria has made slower progress in expanding climate-smart cereal practices, such as conservation agriculture, early-maturing drought-resistant varieties, and integrated soil fertility management (FAO, 2019; Karrou et al., 2011). These gaps are increasingly significant in the context of volatile international markets, where over 70% of Algeria's wheat needs are met through imports. Without a structural shift toward diversified and climate-adapted cereal systems, Algeria's strategic food autonomy remains at risk, and future climate shocks could severely destabilize local grain markets and farmer livelihoods.

3.2. Expansion of vegetables crops under irrigation

Algerian agriculture has undergone a major transformation over the past two decades, with the spectacular expansion of irrigated market gardening. This phenomenon, driven by proactive agricultural policies and growing urban demand, is radically reshaping the national agricultural landscape. Statistics from the Ministry of Agriculture (MADR, 2022) reveal that the area devoted to irrigated market gardening has more than tripled since 2000, now reaching nearly 450,000 hectares, or 12% of the total usable agricultural area.

This expansion can be explained by several cyclical factors. First, urban demand for fresh vegetables has exploded with changing consumption patterns. The FAO statistics (based on the Faostat 2025 data on production and trade) notes about 60% increase in per capita consumption of vegetables since 2010. Subsequently, public policies, particularly through the National Agricultural Development Plan (PNDA), have massively subsidized irrigation equipment (drip, sprinkler) and tunnel greenhouses. The National Agricultural Development Fund (FNDA) thus financed more than 80 000 localized irrigation projects between 2010 and 2020.

Geographically, this expansion follows contrasting patterns. In the coastal plains (Mitidja, Bas Chélif), there is an intensification of existing systems, with increasing specialization in high-value crops such as sheltered tomatoes and peppers. Studies by Belaidi et al. (2022) and Belaidi & Benmehaia (2024) shows that 75% of vegetable crops in the region have adopted drip irrigation. At the same time, peripheral areas (High Plateaus, South) are experiencing an expansion of irrigated areas, often to the detriment of steppe rangelands or marginal cereal lands.

On a technical level, this shift is accompanied by a partial modernization of production systems. Benabdelkader et al. (2021) notes that 45% of vegetable crops now use selected seeds and 30% use integrated pest management. However, these technical advances mask practices that are often unsustainable. According to the FAO (2005), mineral fertilizer applications have quadrupled in market gardening areas since 2000, leading to some soil salinization. Economically, this sector generates mixed results. On the one hand, it now represents 28% of the value of national agricultural production (MADR, 2022) and employs more than 1.2 million permanent and seasonal workers. On the other, it creates a growing dependence on imported inputs (seeds, pesticides, equipment), which weighs on the trade balance. Finally, this expansion poses major environmental challenges. The proliferation of drilling (according to reports from the ABHS, the Sahara Watershed Agency) threatens groundwater, particularly in Saharan areas where resources are fossil fuels. Furthermore, the expansion of plastic greenhouses generates problems of visual pollution and agricultural waste management. Faced with these challenges, the public authorities are attempting to rectify the situation. The new 2024-2030 Agricultural Plan provides for a vast program of conversion to agroecological systems, with increased subsidies for water conservation and biological control. However, the effectiveness of these measures will depend on a profound change in practices among producers, as well as on reinforced control of water resources.

The expansion of irrigated vegetable crops in Algeria is not unique; it reflects a broader trend observed across semi-arid and arid regions in North Africa, where farmers have shifted toward high-value horticulture in response to climate stress, land fragmentation, and rising urban demand. Studies in comparable agroecological zones confirm that this transformation is largely driven by profitability differentials, government subsidies for irrigation infrastructure, and changes in food consumption patterns (Schreinemachers et al., 2018; Ringler et al., 2022; Hanjra & Williams, 2020). In the Algerian context, however, this shift has been particularly rapid: between 2000 and 2020, vegetable production increased by over 150%, while cereals declined or stagnated (FAO, 2025). Belaidi et al. (2022) demonstrated that access to drip irrigation systems significantly boosts yield and water productivity for key crops such as tomatoes and peppers in the Mitidja plain. Yet this technical transition has occurred unevenly across regions: while the Mitidja and Lower Chélif have benefitted from large-scale support programs, peripheral areas such as the High Plateaux and semi-arid interior are still reliant on subsurface aquifers and informal boreholes, often without regulatory oversight (Benmehaia & Bekkis, 2024).

These developments raise critical questions about sustainability, equity, and long-term viability. Despite the widespread adoption of irrigation, recent studies show that only 32–45% of water-use efficiency potential is realized across Algeria's horticultural farms, due to inappropriate irrigation scheduling, poor maintenance, and limited technical training (Oulmane et al., 2020; Belaidi et al., 2022). Moreover, the expansion of market gardening has been accompanied by intensified agrochemical use, including synthetic fertilizers and pesticides, with negative implications for soil salinization, aquifer contamination, and biodiversity loss.

3.3. Transformations of livestock farming

The livestock sector in Algeria is going through a period of profound change, marked by technical, spatial, and economic restructuring. FAO (2025) data indicate that the national sheep and goats population has experienced growth of 5.3% per year on average since 1980 (from 16 millions in 1980 to 36 millions in 2023), whereas cattle heads by 3.2% (as it was 1.3 million in 1980, and became 1.8 in 2023). These are masking contrasting trends depending on the species. Sheep, which represent nearly 60% of the total livestock population, are seeing their numbers stagnate in traditional livestock farming areas such as the High Plateaus, while they are increasing in peri-urban regions as a result of intensification. However, the production levels show trends more explicitly. The Figure 3 below represents chicken, sheep and cattle meats production levels based on FAO statistics.

Figure 3. Evolution of beef, sheep, and poultry production in Algeria (in 1000 tons of production)

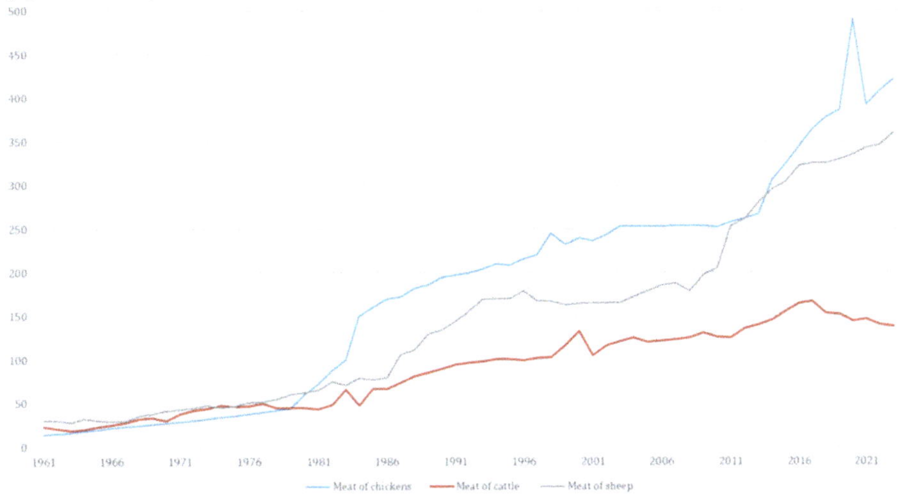

Traditional pastoral systems, once dominant, are undergoing a structural crisis linked to the scarcity of natural rangelands. The study by Boussaada et al. (2022) reveals that the area of steppe rangelands has decreased significantly since 1990 due to the combined effects of desertification and cultivation. This situation is forcing many livestock farmers to change their practices, with increased reliance on stall feeding and imported feed supplements. Production costs have thus increased by more than 35% in ten years, according to estimates from the National Chamber of Agriculture (MADR, 2022).

Cattle farming is undergoing a more radical transformation, with the emergence of semi-industrial units around major urban centers. The MADR (2022) data counted 1200 modern dairy farms, producing nearly 40% of the milk collected. This partial modernization, however, hides disparities: 80% of cattle farms remain family-run, with fewer than five cows. The average milk productivity of high-dairy cattle does not exceed 2 500 liters per cow per year, far below European yields.

Poultry (and rabbit) farming have experienced the most spectacular growth, driven by sharply rising domestic demand. Poultry production has doubled since 2010, reaching 421 000 tons in 2023, enabling the country to achieve relative self-sufficiency. This performance, however, masks an almost total dependence on imported livestock feed, which represents 70% of production costs.

Spatial transformations in the sector are equally striking. There has been a gradual migration of livestock farming activities to peri-urban areas, facilitated by improved road infrastructure and proximity to consumer markets. This geographic concentration, however, creates growing environmental problems, particularly in terms of livestock manure management.

Supporting public policies are struggling to keep pace with these changes. Livestock subsidies, which represented 25% of the agricultural budget in 2022, remain focused on the purchase of imported feed rather than on genetic improvement or training. The 2020-2025 National Livestock Development Program has only achieved 30% of its mid-term objectives. Faced with these challenges, some livestock farmers are innovating by developing mixed systems combining fodder crops and livestock production. Several studies have documented several successful experiments in using agricultural by-products for animal feed, thereby reducing dependence on imports. These local initiatives, while still marginal, could inspire a broader transition towards more sustainable and climate-resilient livestock systems.

Algeria's livestock sector illustrates a dual transformation: the decline of extensive pastoral systems and the parallel rise of semi-intensive and peri-urban livestock production. This bifurcation mirrors broader trends in North Africa, where climate change, rangeland degradation, and shifting consumption patterns have pushed traditional systems toward a crisis of viability (Boussaada et al., 2022; Moussaoui et al., 2021). In steppe regions, satellite-based assessments and scenarios show a significant decline in palatable biomass availability over the last two decades, largely driven by the collapse of herbaceous cover and increased rainfall variability (Mokhtari *et al.*, 2025; Martínez-Valderrama et al., 2018). The result has been a marked reduction in herd mobility, a growing dependence on commercial feed, and a rise in herd sedentarization—all of which increase costs and reduce resilience.

4. EFFECTS ON AGRICULTURAL EMPLOYMENT AND RURAL DYNAMICS

Climate change is reshuffling the social organization of rural Algeria. Here, we will analyze its consequences for employment, internal migration, and the emergence of new forms of agricultural labor.

4.1. Unemployment and rural exodus

The agricultural employment crisis in Algeria is worsening due to the combined effects of climate change and structural changes in the sector. The World Bank also noted that the youth unemployment rate is projected to be 29.3% in 2024. This situation particularly affects rural youth, among whom unemployment exceeds approximately 35%, widening a generational gap in rural Algeria.

The rural exodus is accelerating in response to this growing poverty. FAO data shows that nearly 1.3 million people have left rural areas over the past twenty years. The regions of the High Plateaus and the South are the most affected by this phenomenon, with depopulation rates reaching 2.5% per year in some municipalities. This migration fuels the expansion of urban peripheries, where infrastructure struggles to keep up. The underlying causes of this relative human depopulation are multiple. Agricultural land degradation, which affects 45% of arable land according to the National Observatory of Agricultural Land (MADR, 2022), reduces local employment opportunities. Smallholder farmers, faced with declining yields and erratic rainfall, are gradually abandoning their land. MADR (2022) reports highlight that 40% of farmers with less than 5 hectares have ceased farming since 2015.

Public policies are struggling to stem this trend. Rural employment support programs, such as the ANSEJ agricultural program, have created only 50,000 jobs since their launch in 2010, according to MADR (2022) reports. This exodus has socioeconomic consequences. Algerian cities are struggling to absorb this migratory flow, with an already high urban unemployment rate (12.25% in 2023 according to World Bank data). In rural areas, the aging of the agricultural workforce—the average age of farmers now exceeds 55—threatens the sustainability of traditional production systems. A few local initiatives are attempting to reverse this trend. In regions of Tlemcen and Sétif, agricultural cooperatives have successfully retained young people through projects promoting local products. However, these initiatives remain isolated and would require stronger institutional support to have a national impact.

4.2. New forms of agricultural labor

Algerian agriculture is undergoing a profound transformation in its employment and work organization, marked by the emergence of hybrid and precarious forms of activity. World Bank data reveals that nearly 28.5% of agricultural workers lack a stable employment status, oscillating between seasonal work, micro-entrepreneurship, and multiple job opportunities. This widespread precariousness is paradoxically accompanied by the emergence of new professions linked to the partial modernization of the sector.

At the same time, the development of the agri-food industry has given rise to specialized profiles such as processing technology or quality control technicians. The Ministry of Vocational Training currently offers 21 specializations in these new sectors. Furthermore, digital technology is tentatively beginning to penetrate the agricultural sector, creating new intermediaries. As mobile phones become available to the most isolated rural households, platforms such as *Fellah-Tech* and *Agri-Emploi* have been connecting thousands of casual workers with farmers seeking temporary labor since 2021. While these innovations facilitate the matching of supply and demand, they also accentuate the fragmentation of labor.

Agricultural cooperatives, encouraged by the public authorities, are attempting to organize this new workforce. According to the APS (2023), there are 204 active cooperatives, including 162 newly created ones. However, these structures often struggle to reconcile profitability and social protection, with many functioning as mere purchasing groups rather than true cooperatives—in the legal sense of the term (Brabez & Bédrani, 2015).

In peri-urban areas, a complex informal economy has developed around short supply chains. Thousands of micro-farmers sell their produce directly via social media or pop-up outlets, escaping any regulatory framework. It is estimated that a significant portion of fresh produce transactions now fall outside formal channels. This restructuring of agricultural labor poses major challenges in terms of social protection. The new status of self-employed agricultural workers, created in 2022, attempts to remedy this situation. Faced with these changes, agricultural unions are struggling to adapt their actions. The majority continue to focus on owner-operators, neglecting the new profiles of precarious workers. Yet, it is precisely in the articulation between these different forms of work that the future of the Algerian agricultural sector lies.

5. WORSENING FOOD DEPENDENCE

It should be noted that the population's food situation has improved significantly since the country's independence. Indeed, the average food intake increased from 1 577 calories during the period 1963-1967 to 3 219 kilocalories for the period 2009-2013 (Bouzid et al., 2022). Despite agricultural development efforts, however, Algeria is seeing its dependence on food imports increase. This final section explores the mechanisms and implications of this growing vulnerability.

5.1. Increasing imports

Algeria's dependence on food imports is reaching high levels, despite official rhetoric about food sovereignty. Figures from Algerian Customs are unequivocal: the agri-food import bill exceeded $8 billion in 2022, representing nearly 23.5%

of the country's total import value (USDA, 2022). This trend has intensified over the past decade, with an average annual growth of 8% in external food purchases.

Cereals constitute the largest component of this dependence (Bekkis & Benmehaia, 2025). According to data from the Ministry of Commerce (2023), Algeria imported 7.2 million tons of wheat in 2022, or 75% of its needs. The country thus ranks among the world's top five wheat importers, with a cereal bill exceeding $4 billion annually. This situation is all the more worrying given that local production is stagnating at around 3 million tons, barely covering 30% of growing national demand.

Dairy products and vegetable oils are exacerbating food dependency. Domestic milk production (3.6 billion liters) in 2019 only covered 60% of consumption (around 6 billion liters) (Abdelli et al., 2021). As for edible oils (mainly soybean oil), foreign purchases represent 46.3% of national production, according to FAO (2025) statistics, while the Minister of Trade and Export Promotion claims that Algeria now covers all of its needs through its own production (Grar, 2023), without emphasizing that all refined oil is produced from imported raw materials.

This increased dependency can be explained by several structural factors. First, the relative failure of agricultural policies to significantly boost local production. The National Agricultural Development Plan (PNDA) has failed to reverse this trend, as MADR (2022) figures displayed. Subsequently, population growth and changing consumption patterns have increased domestic demand faster than local supply.

The economic consequences of this dependence are significant. The World Bank (2024) highlights that food imports now account for 15% of the country's trade deficit. This situation makes the national economy vulnerable to fluctuations in international markets, as demonstrated by the Ukrainian crisis with the surge in wheat prices in 2022. Politically, this food dependence creates recurring tensions. Public subsidies for basic products (flour, milk, oil) represent a growing burden on public finances, estimated at $3 billion annually (World Bank, 2023). At the same time, periodic shortages of certain food products cause social discontent, as during the milk and oil crises of recent years. Faced with this situation, the authorities are trying to develop local alternatives. The dairy production support program has enabled a 20% increase in national harvests since 2020. Similarly, the ambitious olive oil plan aims to reduce dependence on imported oils. However, these efforts remain insufficient given the scale of the needs, and Algeria seems doomed, for a long time to come, to massively import a large portion of its basic food.

5.2. Growing share of imports in the food ration

Algeria's food security increasingly relies on imported products, with foreign foodstuffs increasingly making up the Algerian household's food basket. FAO (2025) data reveals that imported products now represent a significant portion of the daily calories consumed by Algerians. This nutritional dependence has increased for basic products. Imported wheat largely dominates the daily diet. According to Bekkis et al. (2022, 2023), 85% of the bread consumed in Algeria is made from imported soft wheat, mainly from France and Ukraine. Semolina, the basis of couscous, contains 60% foreign durum wheat. This situation places Algeria in a position of extreme vulnerability to fluctuations in international markets, as demonstrated by the 2022 Ukrainian crisis. Dairy products also illustrate this growing dependence. Various MADR (2022) reports from different agencies estimate that 70% of dairy products consumed in Algeria are made from imported milk powder, mainly from Europe and New Zealand. Even traditional yogurt, a staple of the local diet, is often made from imported powdered milk. Attempts to revive local production are struggling to reverse this trend, with only 30% of collected milk meeting industrial standards (Bekkis & Benmehaia, 2025).

Cooking oils constitute a third worrying source of dependency. Despite the country's significant olive oil potential, vegetable oil imports have increased by an average of 12% per year since 2015. This invasion of imported products is gradually changing traditional eating habits. It should be noted that the share of processed products (biscuits, ready meals, etc.) in the diet of urban households has increased from 15% to 35% in ten years. These products, mostly imported or manufactured locally from foreign raw materials, contribute to the emergence of new public health problems such as obesity and diabetes.

The consequences of this food dependence are multiple. Economically, it places a heavy burden on the trade balance, with a food bill representing 25% of total imports. It makes the country vulnerable to supply crises, such as during the milk and oil shortages of 2018 and 2022.

Faced with this situation, public policies oscillate between liberalization and protectionism. The new 2024-2030 Agricultural Plan provides for significant investments in strategic sectors (cereals, milk, oilseeds), but sector experts doubt their effectiveness without a profound reform of the agricultural production system. As noted in a World Bank (2023) report, the main challenge facing the Algerian economy lies in its heavy dependence on hydrocarbon revenues and public spending.

Indeed, Algeria's food dependence has deepened over the past two decades, despite successive agricultural development plans. The country currently imports over 70% of its cereal needs—mainly wheat—and large portions of its dairy and oil products (USDA, 2022; Bekkis & Benmehaia, 2025). According to FAO (2025), imported items now account for a significant share of daily caloric intake: over 85% of bread is made from imported soft wheat, and most dairy products are produced from foreign milk powder. These dependencies expose Algeria to price volatility and geopolitical shocks, as evidenced during the COVID-19 crisis and the 2022 Ukraine conflict, which caused wheat prices to spike and supply contracts to falter.

Despite recent government rhetoric around "food sovereignty," the agro-industrial system remains heavily reliant on imported inputs, including seeds, fertilizers, feed, and machinery—creating a structural vulnerability that monetary reserves and hydrocarbon revenues can only temporarily offset.

This growing dependency is not just economic; it reflects a deeper disconnect between national consumption patterns and local production systems. The National Agricultural Development Plan (PNDA), while expanding irrigated horticulture, has failed to meaningfully increase strategic food crop output (cereals, pulses, oilseeds). Public support remains skewed toward large-scale input subsidies and state food procurement, often bypassing smallholder production and agroecological adaptation (Benmehaia & Bekkis, 2024). Comparative experiences from Morocco and Egypt show that targeted investments in domestic wheat, legumes, and traditional grains—paired with local seed systems and consumer awareness—can mitigate dependency over time. In Algeria, transitioning toward resilient and diversified food systems will require shifting focus from short-term import substitution to long-term structural reforms, including crop diversification, agro-processing development, and stronger rural–urban market linkages. Without such transformation, the country's food security will remain at the mercy of external markets, undermining its broader economic and political resilience.

6. CONCLUSIONS

Climate change is reconfiguring the material foundations of Algerian agriculture. It is no longer a distant or abstract concern, but a systemic force disrupting the biophysical and institutional equilibria that underpin food production, rural employment, and land use. This chapter has examined how prolonged droughts, isohyetal shifts, increased frequency of extreme events, and rising evapotranspiration are intersecting with internal structural weaknesses—most notably the degradation of soil and water resources, the fragmentation of labor regimes, and the failure to stabilize domestic food production. These climate-induced transformations are exposing the fragility of Algeria's agricultural model, which remains anchored in input-driven expansionism, policy centralization, and a dual structure separating irrigated capital-intensive zones from extensive, low-productivity hinterlands.

The persistence of environmental degradation—desertification, groundwater overdraft, and vegetation collapse—is not simply the product of ecological stress. It reflects the inertia of institutional lock-in, where policy instruments such as subsidies, price guarantees, and public procurement have failed to evolve in response to ecological thresholds or demographic shifts. Similarly, the structural exodus of rural youth is not merely economic; it indicates a failure to generate viable agrarian futures, as land, capital, and technical support remain inaccessible to new entrants. The current policy architecture rewards scale and access to state networks rather than resilience, sustainability, or inclusion.

In this context, the concept of socio-ecological resilience (Folke et al., 2016) provides a useful framework for rethinking Algeria's agricultural future. Resilience is not only the capacity to absorb shocks but also the ability to reorganize systems in ways that reduce vulnerability, enhance adaptive capacity, and sustain ecosystem services. Algeria's current trajectory reveals high exposure and sensitivity to climate stress, but limited institutional and technical adaptive capacity—particularly in dryland cereal zones, steppe livestock systems, and degraded watersheds.

What is needed, therefore, is not incremental reform but a strategic shift toward adaptive transformation. We outline four priority axes for a revised agrarian strategy grounded in both theoretical insight and empirical urgency: *Territorialized agroecological planning*: Move beyond national production targets to regionally tailored adaptation plans. This includes investing in agroecological infrastructure (e.g., soil conservation, small-scale rainwater harvesting), restoring vegetative cover using native species, and supporting diversified cropping systems that stabilize yields under variable climates. *Decentralized water governance and pricing reform*: Transition from infrastructure-heavy, top-down water management to Integrated Water Resource Management (IWRM) frameworks. Introduce differentiated water pricing based on scarcity and efficiency, strengthen basin-level institutions, and regulate informal groundwater extraction through community-based monitoring and sanctions. *Reconfiguring support systems for smallholders and rural youth*: Current public spending disproportionately supports large commercial farms. A redirection is needed toward land access reform, rural vocational training, and inclusive innovation platforms. Policies should target marginalized zones where labor reserves exist but productivity gains are constrained by institutional exclusion. *Building food system sovereignty through input autonomy and market relinking*: Reduce reliance on imported seeds, fertilizers, and feed by supporting local input production chains. Link small and medium farms to urban markets through public procurement, consumer cooperatives, and decentralized agro-processing, thereby creating circular value chains that stabilize both production and demand.

Without these structural corrections, Algeria's rural and food systems risk entering a state of climate-driven hysteresis, where damage becomes irreversible, and adaptation windows close. A coherent agrarian transition must treat climate adaptation not as a technical fix but as an opportunity to realign agricultural systems with ecological realities, social equity, and long-term sovereignty.

REFERENCES

Abdelli, R., Yasmine, S., Kaouche, S., & Benhacine, R. (2021). Etat des lieux de la filière laitière en Algérie et perspectives de développement. *Journal Algérien des Milieux Arides, 11*(1), 4–14.

Abouelmagd, A., & Ahmed, M. (2024). Groundwater in North Africa: Effects of climatic and anthropogenic pressures on groundwater availability. In *The Geology of North Africa* (pp. 305–321). Springer International Publishing. DOI: 10.1007/978-3-031-48299-1_11

Alfieri, L., Burek, P., Feyen, L., & Forzieri, G. (2015). Global warming increases the frequency of river floods in Europe. *Hydrology and Earth System Sciences, 19*(5), 2247–2260. DOI: 10.5194/hess-19-2247-2015

Altieri, M. A., Nicholls, C. I., Montalba, R., Vieli, L., & Vazquez, L. L. (2025). Agroecology and the limits to resilience: Extending the adaptation capacity of agroecosystems to drought. *Frontiers in Agronomy, 7*, 1534370. DOI: 10.3389/fagro.2025.1534370

APS. (2023a). Les stations de dessalement contribueront à hauteur de 60% à l'alimentation en eau potable d'ici 2030. Publié le 27 février. https://www.aps.dz/economie/152381

APS. (2023b). Agriculture: création de plus de 160 nouvelles coopératives depuis 2020. Publié le 2 juillet. https://www.aps.dz/economie/142312

Aronson, J., Goodwin, N., Orlando, L., Eisenberg, C., & Cross, A. T. (2020). A world of possibilities: Six restoration strategies to support the United Nation's Decade on Ecosystem Restoration. *Restoration Ecology, 28*(4), 730–736. DOI: 10.1111/rec.13170

Bastin, J. F., Berrahmouni, N., Grainger, A., Maniatis, D., Mollicone, D., Moore, R., & Castro, R. (2017). The extent of forest in dryland biomes. *Science, 356*(6338), 635–638. DOI: 10.1126/science.aam6527

Bekkis, S., Benmehaia, M. A., & Kaci, A. (2022). Les enjeux de la dépendance de la filière de blé en Algérie: Analyse par asymétries de réponses de l'offre dans la chaîne de valeur. *New Medit, 21*(1), 133–147. DOI: 10.30682/nm2201h

Bekkis, S., Benmehaia, M. A., & Kaci, A. (2023). Price transmission in the wheat market in Algeria: Threshold cointegration approach. *International Journal of Food and Agricultural Economics, 11*(1), 17–32.

Belaidi, S., & Benmehaia, A. M. (2024). The selection of modern irrigation technologies within a rural community in Algeria: Stated preferences approach. *Journal of Rural and Community Development, 19*(2).

Belaidi, S., Chehat, F., & Benmehaia, M. A. (2022). The adoption of water-saving irrigation technologies in the Mitidja plain, Algeria: An econometric analysis. *New Medit, 21*(1), 53–73. DOI: 10.30682/nm2201d

Benabdelkader, M., Saifi, R., & Saifi, H. (2021). Sustainable Agriculture in Some Arab Maghreb Countries (Morocco, Algeria, Tunisia). In *Agro-environmental sustainability in MENA regions* (pp. 233–261). Springer International Publishing. DOI: 10.1007/978-3-030-78574-1_10

Benderradji, M. E. H., Alatou, D., Arfa, A. M. T., & Benachour, K. (2006). Problèmes de dégradation de l'environnement par la désertification et la déforestation. Impact du phénomène en Algérie. *New Medit.* N 4.

Benmehaia, A. M., & Bekkis, S. (2024). Agri-food policy trends in Algeria: Selected explorations. *New Medit, 23*(3), 49–62. DOI: 10.30682/nm2403d

Benmehaia, A. M., Merniz, N., & Oulmane, A. (2020). Spatiotemporal analysis of rainfed cereal yields across the eastern high plateaus of Algeria: An exploratory investigation of the effects of weather factors. *Euro-Mediterranean Journal for Environmental Integration, 5*(3), 54. DOI: 10.1007/s41207-020-00191-x

Benmehaia, M. A. (2023). Assessing asymmetrical effects of climate change on cereal yields in Algeria: The NARDL-AEC approach. *Environment, Development and Sustainability, 27*(2), 1–22. DOI: 10.1007/s10668-023-04079-y

Benmehaia, M. A., & Benmehaia, R. (2018, March). Socioeconomic analysis of date palm sector: The case of Biskra region (Algeria). *Sixth International Date Palm Conference.* Abu Dhabi, United Arab Emirates (pp. 19-21).

Bouchakour, R., & Saad, M. (2020). Farm and farmer characteristics and off-farm work: Evidence from Algeria. *The Australian Journal of Agricultural and Resource Economics, 64*(2), 455–476. DOI: 10.1111/1467-8489.12349

Boussaada, D., Yerou, H., Benabdelli, K., & Djelailia, S. (2022). Evaluation des potentialités pastorales des parcours steppiques algériennes: Cas de M'sila (Algérie). *Livestock Research for Rural Development, 34*(4).

Bouzid, A., Lazereg, M., Bédrani, S., Behnassi, M., & Baig, M. B. (2022). Facteurs naturels et réglementaires sous-jacents à la dépendance alimentaire en Algérie. In Behnassi, M., Baig, M. B., Sraïri, M. T., Alsheikh, A. A., & Abu Risheh, A. W. A. (Eds.), *Sécurité alimentaire et systèmes alimentaires climato-intelligents*. Springer., DOI: 10.1007/978-3-030-92738-7_17

Brabez, F., & Bédrani, S. (2015). Les coopératives agricoles de services: Étude de cas. *Les cahiers du CREAD*, Volume 31. *Numéro*, *113*, 193–217.

Chaami, M., Bédrani, S., Bessaad, A., & Lambarraa-Lehnhardt, F. (2025). Why is there a shortage of agricultural labor in Algeria? Evidence from M'zirâa, southern Algeria's leading vegetable market. *New Medit*, *25*(1). Advance online publication. DOI: 10.30682/nm2501h

Chenchouni, H., & Neffar, S. (2022). Soil organic carbon stock in arid and semi-arid steppe rangelands of North Africa. *Catena*, *211*, 106004. DOI: 10.1016/j.catena.2021.106004

Clapp, J. (2021). Food: Governance challenges for a hot and hungry planet. In: *Global Governance Futures* (pp. 207-221). Routledge.

DGF. (2024). Conservation des eaux et des sols et lutte contre la désertification. http://www.dgf.org.dz/fr/content/conservation-des-eaux-et-des-sols-et-lutte-contre-la-d%C3%A9sertification

EFFIS. (2023). Forest fires in Europe, Middle East and North Africa 2023. https://forest-fire.emergency.copernicus.eu

FAO. (2005). Utilisation des engrais par culture en Algérie. Service de la gestion des terres et de la nutrition des plantes. Division de la mise en valeur des terres et des eaux. Organisation des nations unies pour l'alimentation et l'agriculture, Rome.

FAO. (2011). *La pratique de la gestion durable des terres. Directives et bonnes pratiques en Afrique subsaharienne. TerrAfrica, Panorama mondial des approches et technologies de conservation (WOCAT) et Organisation des Nations Unies pour l'alimentation et l'agriculture*. FAO.

FAO. (2019). *Proceedings from the Regional Workshop on Climate-Smart Agriculture in the Near East and North Africa*. FAO, Cairo.

FAO. (2025). Statistics on productions and trade. FAO: https://fao.org/faostat/en/#data

Fellag, M., Achite, M., & Walega, A. (2021). Spatial-temporal characterization of meteorological drought using the Standardized precipitation index. Case study in Algeria. *Acta Scientiarum Polonorum. Formatio Circumiectus*, *20*(1), 19–31. DOI: 10.15576/ASP.FC/2021.20.1.19

Folke, C., Biggs, R., Norström, A. V., Reyers, B., & Rockström, J. (2016). Social-ecological resilience and biosphere-based sustainability science. *Ecology and Society*, *21*(3), art41. DOI: 10.5751/ES-08748-210341

Grar K, 2023. L'Algérie atteint son autosuffisance en huile de table. *L'économiste maghrébin*, du 23-3-2023.

Hamitouche, Y., Zeroual, A., Meddi, M., Assani, A. A., & Alkama, R. (2024). Changes in extreme precipitation indices across Algeria climate zones. *International Journal of Climatology*, *44*(8), 2537–2560. DOI: 10.1002/joc.8467

Hanjra, M. A., & Williams, T. O. (2020). Global change and investments in smallholder irrigation for food and nutrition security in Sub-Saharan Africa. In *The role of smallholder farms in food and nutrition security* (pp. 99–131). Springer International Publishing. DOI: 10.1007/978-3-030-42148-9_6

Haouari, M. (2024). Évolution récente des températures moyennes en Algérie. [On onm-blog.meteo.dz.]. *Journal of the American Medical Association*, *8*, 1–09.

IPCC. (2022). AR6 Synthesis Report. *Climatic Change*, *2023*, •••. https://www.ipcc.ch/assessment-report/ar6/

Karrou, M., Oweis, T., & Bahri, A. (2011). *Improving water and land productivity in rainfed systems*. ICARDA Policy Briefs.

Kherbache, N., & Molle, F. (2021). *La surexploitation des ressources en eau: le cas du bassin de la Macta en Algérie*. IRD.

Macheroum, A., & Chenchouni, H. (2022). Short-term land degradation driven by livestock grazing does not affect soil properties in semiarid steppe rangelands. *Frontiers in Environmental Science*, *10*, 846045. DOI: 10.3389/fenvs.2022.846045

MADR. (2022). *Statistiques agricoles de l'Algérie: Série B*. Ministère de l'Agriculture et du Développement Rural.

Mansourian, S., Dudley, N., & Vallauri, D. (2017). Forest landscape restoration: Progress in the last decade and remaining challenges. *Ecological Restoration*, *35*(4), 281–288. DOI: 10.3368/er.35.4.281

Martínez-Valderrama, J., Ibáñez, J., Del Barrio, G., Alcalá, F. J., Sanjuán, M. E., Ruiz, A., & Puigdefábregas, J. (2018). Doomed to collapse: Why Algerian steppe rangelands are overgrazed and some lessons to help land-use transitions. *The Science of the Total Environment, 613*, 1489–1497. DOI: 10.1016/j.scitotenv.2017.07.058

MEER. (2023). *Premier rapport biennal actualise de l'Algérie: à la convention cadre des nations unies sur le changement climatique*. Ministère de l'Environnement et des Énergies Renouvelables.

Merabtene, M. D., Faraoun, F., Mlih, R., Djellouli, R., Latreche, A., & Bol, R. (2021). Forest soil organic carbon stocks of Tessala Mount in North-West Algeria-Preliminary estimates. *Frontiers in Environmental Science, 8*, 520284. DOI: 10.3389/fenvs.2020.520284

Merniz, N., Tahar, A., & Benmehaia, A. M. (2019). Statistical assessment of rainfall variability and trends in northeastern Algeria. *Journal of Water and Land Development, 40*(1), 87–96. DOI: 10.2478/jwld-2019-0009

Ministère des Ressources en Eau. (2017). *Plan National de l'Eau: PNE*. MRE.

Morsli, B., Habi, M., & Meddi, M. (2013). Dynamique de l'érosion en zone méditerranéenne algérienne: Facteurs explicatifs de variation du ruissellement et de l'érosion sous différentes occupations du sol. *Revue des Sciences de l'Eau, 26*(2), 89–105. DOI: 10.7202/1016061ar

Moussaoui, A., Abbas, K., Bottalico, F., & Ottomano, G. (2021). Pastoralism in Algeria: transformation, challenges and prospects. *Pastoralism and Sustainable Development*, 189.

Mozas, M., & Ghosn, A. (2013). *État des lieux du secteur de l'eau en Algérie*. Études & Analyses, Ipemed.

Nedjraoui, D., & Bédrani, S. (2008, April). La désertification dans les steppes algériennes: Causes, impacts et actions de lutte. *Vertigo, 8*(1), •••. http://vertigo.revues.org/index448.html

North, D. C. (1990). *Institutions, Institutional Change and Economic Performance*. Cambridge University Press. DOI: 10.1017/CBO9780511808678

OSS. (2003). *Système aquifère du Sahara septentrional: Une conscience de bassin. 2e édition. Hydrogéologie* (Vol. II). Observatoire du Sahara et du Sahel.

Oubadi, M., & Faci, M. (2025). Analysis of the evolution of heatwaves recordings in Algeria during the period 1979–2023. *Theoretical and Applied Climatology, 156*(4), 1–9. DOI: 10.1007/s00704-025-05456-5

Ouessar, M., Sghaier, A., Frija, A., Sghaier, M., & Baig, M. B. (2021). Impacts of climate change on agriculture and food security in Tunisia: challenges, existing policies, and way forward. In *Emerging Challenges to Food Production and Security in Asia, Middle East, and Africa: Climate Risks and Resource Scarcity* (pp. 65–99). Springer International Publishing. DOI: 10.1007/978-3-030-72987-5_3

Oulmane, A., Chebil, A., Frija, A., & Benmehaia, M. A. (2020). Water-saving technologies and total factor productivity growth in small horticultural farms in Algeria. *Agricultural Research*, *9*(4), 585–591. DOI: 10.1007/s40003-019-00446-2

Pierson, P. (2000). Increasing returns, path dependence, and the study of politics. *The American Political Science Review*, *94*(2), 251–267. DOI: 10.2307/2586011

Qadir, M., Bahri, A., Sato, T., & Al-Karadsheh, E. (2010). Wastewater production, treatment, and irrigation in Middle East and North Africa. *Irrigation and Drainage Systems*, *24*(1), 37–51. DOI: 10.1007/s10795-009-9081-y

Reynolds, J. F., Smith, D. M. S., Lambin, E. F., Turner, B. L., Mortimore, M., Batterbury, S. P., & Walker, B. (2007). Global desertification: Building a science for dryland development. *Science*, *316*(5826), 847–851. DOI: 10.1126/science.1131634

Ribot, J. (2014). Cause and response: Vulnerability and climate in the Anthropocene. *The Journal of Peasant Studies*, *41*(5), 667–705. DOI: 10.1080/03066150.2014.894911

Ringler, C., Agbonlahor, M., Barron, J., Baye, K., Meenakshi, J. V., Mekonnen, D. K., & Uhlenbrook, S. (2022). The role of water in transforming food systems. *Global Food Security*, *33*, 100639. DOI: 10.1016/j.gfs.2022.100639

Sahabi-Abed, S., Ayugi, B. O., & Selmane, A. N. E. I. (2023). Spatiotemporal projections of extreme precipitation over Algeria based on CMIP6 global climate models. *Modeling Earth Systems and Environment*, *9*(3), 3011–3028. DOI: 10.1007/s40808-023-01716-3

Schreinemachers, P., Simmons, E. B., & Wopereis, M. C. (2018). Tapping the economic and nutritional power of vegetables. *Global Food Security*, *16*, 36–45. DOI: 10.1016/j.gfs.2017.09.005

Scoones, I. (1998). Sustainable rural livelihoods: A framework for analysis. IDS Working Paper 72.

Slimani, H., & Aidoud, A. (2004). Desertification in the Maghreb: A case study of an Algerian high-plain steppe. In: Environmental Challenges in the Mediterranean 2000–2050: Proceedings of the NATO Advanced Research Workshop on Environmental Challenges in the Mediterranean 2000–2050 Madrid, Spain 2–5 October 2002 (pp. 93-108). Dordrecht: Springer Netherlands. DOI: 10.1007/978-94-007-0973-7_6

Thi, N. Q., Govind, A., Le, M. H., Linh, N. T., Anh, T. T. M., Hai, N. K., & Ha, T. V. (2023). Spatiotemporal characterization of droughts and vegetation response in Northwest Africa from 1981 to 2020. *The Egyptian Journal of Remote Sensing and Space Sciences*, *26*(3), 393–401. DOI: 10.1016/j.ejrs.2023.05.006

UNCCD. (2020). Neutralité en matière de dégradation des terres: Mettre le concept en pratique. https://www.unccd.int/sites/default/files/ldn_targets/algeria-ldn-country-report-final-draft.pdf

UNCCD (2023). Report from Algeria. Performance review and assessment of implementation system. Seventh reporting process.

USDA (2022). Food and agricultural import regulations and standards country report: Algeria. FAIRS Annual Country Report. AG2022-0013. USDA.

Vennetier, M. (2020). Forêts et changement climatique. Le constat en région méditerranéenne. *Sciences Eaux & Territoires*, *33*(3), 18–25. DOI: 10.3917/set.033.0018

WMO. (2023). *State of the Climate in Africa 2024. WMO-No. 1370.* World Meteorological Organization.

World Bank (2023a). Diagnostic sur la gestion des risques climatiques et de catastrophe en Algérie. Banque internationale pour la reconstruction et le développement, Groupe de la Banque mondiale.

World Bank (2023b). Note sur les forêts algériennes: Gestion durable des forêts pour lutter contre les feux de forêts. Banque internationale pour la reconstruction et le développement.

World Bank. (2024). Rapport de suivi de la situation économique en Algérie: Investir dans les données pour une croissance diversifie. Banque Internationale pour la Reconstruction et le Développement. Banque Mondiale. Bekkis S. & Benmehaia M.A (2025). Céréales et lait en Algérie: Sujétions de service public et résultats économiques. *Maghreb, Machrek*, •••, 261.

Zeng, H., Wu, B., Elnashar, A., & Fu, Z. (2024). Dryland dynamics in the Mediterranean region. In *Dryland Social-Ecological Systems in Changing Environments* (pp. 243–271). Springer Nature Singapore. DOI: 10.1007/978-981-99-9375-8_8

1 See Benmehaia & Bekkis, 2024 for more detailed explorations on food security aspects in Algeria.

2 See Merniz et al. (2019) and Oubadi & Faci (2025) for more details on the precipitation trends and patterns in Algeria

[3] See Belaidi et al. (2022) for detailed analysis of irrigation systems adoption in Algeria.

Chapter 4

Climate Change, Agriculture, and Food Security in North Africa:
A Quantitative Analysis

Abdallahi H'moide
https://orcid.org/0009-0000-8266-1715
Mohammed V University in Rabat, Morocco

Raouf Radouane
Mohammed V University in Rabat, Morocco

ABSTRACT

This chapter aims to evaluate the effects of climate change on food security in North Africa using an ARDL-PMG model and future climate projections from CMIP6. The results indicate that in the long term, agricultural capital, labor, and climatic conditions (De Martonne index) exert a positive influence on agricultural production. In the short term, the effects vary by country. The integration of Shared Socioeconomic Pathways (SSP) allows for the estimation of future agricultural losses that could reach –4.1% under the SSP5–8.5 scenario, with more pronounced impacts in Morocco and Tunisia. The study highlights the increasing vulnerabilities and emphasizes the need for adapted agricultural and climate policies to ensure regional food security.

DOI: 10.4018/979-8-3373-3346-5.ch004

INTRODUCTION

Climate change is one of the major challenges facing humanity in the 21st century. Its devastating consequences on economic activities, natural resources and ecosystems are prompting political decision-makers to implement numerous programs and strategies. The impacts of climate change manifest themselves in extreme weather events, such as prolonged droughts, water shortages and floods, compromising global food security and threatening the stability of agricultural systems. This climate change, combined with pollution from greenhouse gas emissions, is mainly the result of human activities. Indeed, CO2 concentrations in the atmosphere have risen from 280 ppm in 1750 to over 413 ppm in 2018 (Nordhaus, 2019).

Developing countries, including those in North Africa, although low contributors to global greenhouse gas emissions, are the hardest hit by the impacts of climate change. These countries suffer from increased problems of access to water, food security and loss of biodiversity (United Nations Development Programme [UNDP], 2023). Rising temperatures and increased intensity of extreme weather events are likely to reduce agricultural yields, particularly in vulnerable regions (Intergovernmental Panel on Climate Change [IPCC], 2022; UNDP, 2023). Moreover, the impact on agriculture is particularly severe, affecting yields and increasing food prices, exacerbating the economic situation in this region (Stern, 2007).

Climate impacts on food security take several forms. On the one hand, they directly affect agricultural production through changes in growing conditions, infestation and disease, and disruptions to the seasonal cycles on which harvests depend (Ngure et al., 2021). On the other hand, climate change exerts a strong influence on human health, agricultural livelihoods, food distribution channels, household purchasing power and market flows (Berhanu & Wolde, 2019). Similarly, the economic implications are considerable. Climate change, by threatening agricultural production, leads to higher food prices. In this sense, economic losses are unevenly distributed, with consumers in low-latitude regions bearing the brunt of the impacts. If international trade becomes more restricted, these global losses could increase considerably (Stevanović et al., 2016).

This real phenomenon represents a growing threat to Africa, a continent which, despite being responsible for a small share of global greenhouse gas emissions, is nonetheless experiencing devastating impacts on food security, access to water and natural resources. According to the World Meteorological Organization WMO (2023), the African continent has warmed faster than the rest of the world over the past 60 years. In 2023, one of its three hottest years, Africa experienced extreme weather events such as deadly heat waves, floods, heavy rains and prolonged droughts. The average temperature was 0.61°C above the 1991-2020 average and 1.23°C above the 1961-1990 reference period. This represents a warming of around +0.3°C per decade between 1991 and 2023, compared with the global average.

In light of these findings, the central question addressed in this chapter is how climate change affects food security, defined here as agricultural production, in North Africa over a long period of time. While most existing studies either rely on econometric analyses that do not explicitly incorporate future climate scenarios or focus on climate simulations that overlook the economic and food security dimensions, this research adopts an integrated framework that bridges these two approaches. In particular, it integrates an advanced econometric technique, notably the ARDL-PMG model, with data from future climate projection models (CMIP6) and socio-economic scenarios (SSPs)[1]. This approach establishes a direct link between projected climate change and economic and food dynamics, but also takes into account the complex interactions between biophysical and socio-economic variables. To our knowledge, the North African context—which is marked by a strong reliance on agriculture and a notable susceptibility to climatic fluctuations—has not yet seen the use of such an integrated technique. In order to improve food security and climate resilience in the North African agricultural sector, this study intends to not only offer a novel methodological contribution but also, based on its findings, suggest adaptation measures.

The rest of the chapter is structured as follows: the next section outlines the regional and national context, highlighting the climatic, agricultural, and socio-economic specificities of the countries studied. This is followed by a literature review that summarizes previous work on the relationship between climate change, agriculture, and food security from a perspective similar to that of our study. This section also highlights the research gap that our study aims to fill. The following section describes the data used and the approach employed, before presenting and analyzing the empirical results. Finally, the chapter concludes with a presentation of the main policy implications and a discussion of the study's limitations.

REGIONAL CONTEXT

Of all the African sub-regions, North Africa has experienced the most rapid global warming, with an average rise of around 0.4°C per decade between 1991 and 2023, compared with just 0.2°C per decade between 1961 and 1990. This region is one of the most sensitive to climate change due to its geographical location and dry environment (Salami and Babatunde, 2024; K. Waha et al. 2017).

The North African region is also facing declining rainfall and increasing water stress, which is directly affecting agricultural production and, consequently, food security. Indeed, rainfall anomalies were slightly above the 1991-2020 average in northeast Africa, and the western part of North Africa is experiencing a marked rainfall deficit. A large part of North Africa and North-West Africa recorded a below-normal annual rainfall deficit, particularly in Morocco, Algeria, Tunisia and western Libya.

As a result, climate change is primarily responsible for the region's environmental problems, which threaten the region's long-term stability, particularly in terms of employment, poverty, inequality and underdevelopment. Food insecurity affects 7.8% or 20.7 million of the region's population (Food and Agriculture Organization of the United Nations [FAO], 2024). This food insecurity is mainly caused by climate change. This is because it affects the main dimensions of food security, i.e. availability, stability, utilization and access (A. Feher et al., 2022; C. Hofisi et al., 2013).

According to the IPCC climate zones, the North African region is characterized by a dry tropical and dry warm temperate climate. Although several vital sectors in the North African region are also vulnerable to the impacts of climate change, the sector most affected remains agriculture. The water, energy, infrastructure and tourism sectors are also under increasing pressure. However, agriculture, which employs a large proportion of the population - Mauritania (33%), Morocco (31%) and Algeria (10%)- and is an essential source of income and food security, is particularly exposed (International Labour Organization [ILO], 2023). This sector, which is therefore an economic and social pillar in North Africa, finds itself at the heart of the challenges posed by climate change. Because of its dependence on climatic conditions and water resources, it bears the brunt of rising temperatures, increased drought and soil degradation (Deeb et al., 2024).

Variations in rainfall and temperature, extreme heat waves and drought pose a direct threat to agricultural production and hence to food security. Studies show that with global warming at the 3°C, 2°C and even 1.5°C threshold set by the Paris agreement, average annual temperatures in the region will exceed the 1994-2005 average. The same applies to mean annual precipitation, which has already fallen over much of the region (IPCC, 2022).

AGRICULTURAL PATTERNS AND CLIMATE VARIABILITY

Cartographic analysis, in Figures 1 and 2, of the spatial and temporal distribution of agricultural production and the climatic variables, temperature and precipitation, reveals contracted results. These results are presented for the six countries of the North African region - Morocco, Mauritania, Algeria, Tunisia, Libya and Egypt - and for four distinct periods over the last thirty years: 1995-2001, 2002-2008, 2009-2015 and 2016-2023. In terms of agricultural production, Morocco and Egypt have seen continuous growth in output, particularly over the last two periods. This evolution can be explained by the implementation of certain targeted agricultural policies and the expansion of irrigation systems, such as the Green Morocco Plan for Morocco (Ouraich and Tyner, 2014). However, Libya and Tunisia are characterized by relatively constant or slightly declining production levels. This highlights the structural and climatic constraints that hamper their ability to adapt and intensify their agricultural activities.

Figure 1. Contrasting trends in agricultural production and precipitation in North Africa over the period 1995-2023

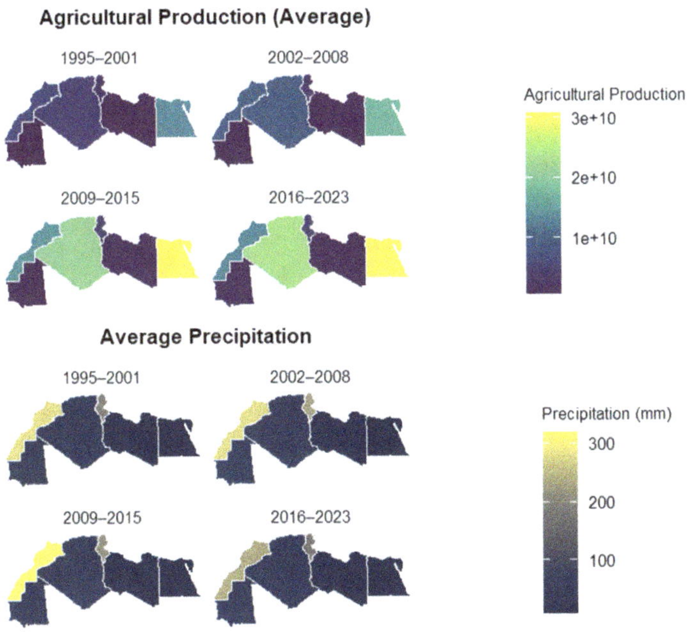

Source: authors' calculations

Figure 1 shows the evolution of agricultural production in parallel with that of rainfall. The results show that, at the start of the period, Morocco, Tunisia and Mauritania benefited to some extent from moderately higher levels of precipitation (>300 mm annual average) than the other countries, although trends over the last three decades show no significant increase. On the other hand, countries on the eastern seaboard - notably Libya, Egypt and central Algeria - suffer from climatic conditions with a tendency to aridity (<100 mm/year), confirming the findings of some studies that these countries are among the areas most vulnerable to increasing aridity (Taabni and Jihad, 2012 ; IPCC, 2022; Waha et al., 2017; World Bank, 2018). Indeed, the decline recorded in the region from the year 2002 indicates a climate change that compromises the stability of the water supply needed for agriculture. However, this decline in rainfall levels, combined with an increase in production in certain countries, notably Algeria and Egypt, over the two periods 2009-2015 and 2016-2023, raises fears of overexploitation of water resources, likely to lead to environmental degradation and increased food insecurity in the long term. Moreover, these observations underline the importance of implementing agricultural and climate policies and access adaptation strategies based on sustainable water management, the development of drought-resistant crops or the improvement of agricultural techniques (El-Malky M. et al., 2016; Wahid et al., 2022).

Generally speaking, the absence of a favorable trend in rainfall, combined with the phenomenon of global warming, is accentuating the vulnerability of agricultural systems, particularly in semi-arid regions.

Figure 2. Contrasting trends in agricultural production and temperature in North Africa over the period 1995-2023

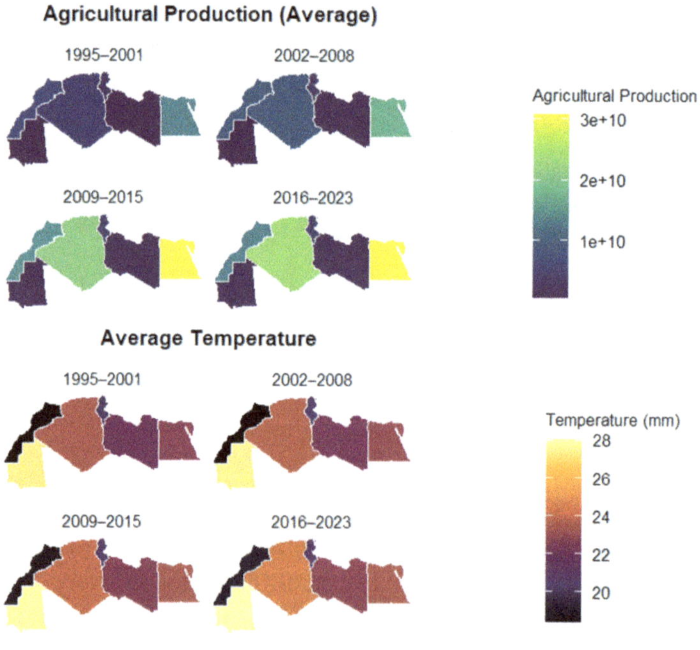

Source: authors' calculations.

Figure 2 shows a global trend towards rising temperatures. All countries are recording a significant rise in mean annual temperatures over the period 1995-2023, which is in line with climate projections (IPCC, 2022; Almazroui et al., 2020). In turn, rising temperatures will exert increasing pressure on water resources, aggravating the risk of water stress, particularly in Morocco, Algeria and Tunisia (Schilling et al., 2012).

With a population of 213 million, these dynamics highlight the importance of the challenges posed by climate change for food security in North Africa. In this context, the IPCC (2022) and the FAO (2024) advocate an in-depth overhaul of agricultural systems to make them more resilient to climate change. This includes promoting climate-adapted agricultural practices, improving water resource management, as well as strengthening regional cooperation for a better response to the interconnected challenges of sustainability and food security (FAO, 2019).

This heightened vulnerability of agriculture to climate variability or weather conditions has cascading repercussions on the economy as a whole, exacerbating poverty and inequality. Autonomous and/or planned adaptation measures and mitigation strategies will therefore be crucial to mitigate these impacts and support populations, particularly rural populations, who depend on agriculture. This issue demands particular attention, both for the sustainability of existing agricultural systems and for the food security of populations. The challenge is all the greater in a region characterized by chronic water scarcity and growing demand for natural resources.

LITERATURE REVIEW

Studies on the relationship between climate change, agriculture, and food security have grown significantly in recent decades, both globally and regionally. The studies by Rosenzweig and Parry (1994) and Nelson et al. (2010) have shown that rising temperatures and changing precipitation patterns pose a direct threat to agricultural productivity and the ability to feed a rapidly growing global population, particularly in low-income countries. According to these authors' simulations, adaptation measures taken by farmers could mitigate some impacts, but would not substantially reduce the disparity in vulnerability between developed and developing countries. These observations highlight the scale of the challenge posed by climate change, particularly for vulnerable regions, and call for an analysis of regional impact dynamics (Waha et al. 2017).

In the African context, several studies have assessed the impact of climate change on various aspects of food security. Affoh et al. (2022) studied the effect of climate variability and change on the three dimensions of food security—availability, accessibility, and utilization—using a panel of 25 countries in sub-Saharan Africa over the period 1985–2018. Using an ARDL-PMG approach, they demonstrate that rainfall has a significant positive impact on these three aspects in the long term. On the other hand, rising temperatures penalize availability and accessibility, without affecting utilization. These results confirm the vulnerability of food systems in Africa to climate variability and the importance of distinct adaptation measures based on the dimensions of food security. Fusco (2022) examines the influence of climate change on food security in North and East Africa for the period 2000-2012. The results indicate that fluctuations in temperature and precipitation have an adverse impact on food security in these areas. This reaffirms the structural vulnerability of their food systems to climate variability. Similarly, Pickson et al. (2024) analyze the effects of climate variables on cereal farming in Africa using data covering the period 1970Q1 to 2017Q4. The results reveal that rising temperatures have a negative effect on cereal crops, while increased precipitation and harvested area have a positive effect.

However, despite the significant contribution of these studies on the relationship between climate change and food security, they suffer from certain methodological constraints. On the one hand, they do not include climate projections and future scenarios, which limits their ability to predict the future impacts of climate change. On the other hand, the use of large panels tends to minimize national particularities and mask regional heterogeneities (Affoh et al., 2023; Fusco, 2022). Although these approaches are useful for identifying global trends, they struggle to reflect country-specific situations, especially in contexts where agricultural and food systems are strongly determined by local factors.

From a country-level perspective, Attiaoui and Boufateh (2019) study the impact of climate change on cereal production in Tunisia for the period 1975-2014, using an ARDL panel approach. The results show that rainfall deficits have a persistent negative impact on yields. Similarly, Jatuporn and Takeuchi (2023), in assessing the impact of climate change on economic growth in the agricultural sector and its variability in Thailand, indicate that extreme weather events are detrimental to the agricultural economy, while an increase in total rainfall is positively related to it.

Food availability is directly threatened by climate change, which reduces the supply capacity of natural ecosystems. This leads to a decline in food production due to rising temperatures and changing rainfall patterns (Xie et al., 2021).

In a study conducted in India, researchers examined the influence of climate change on the main crop, based on data spanning from 1970 to 2020. The analyses show that, in the long term, high temperatures have an adverse impact on agriculture (Choudhary and Gupta, 2024).

Other authors, meanwhile, have examined food security, agricultural productivity, and CO_2 emissions together. They conclude that although agriculture and food availability are crucial for economic growth and the resilience of agricultural systems, the impact of CO_2 emissions remains ambivalent—sometimes beneficial in the short term due to the carbon fertilization effect, but harmful in the long term due to its cumulative effects on production (Frimpong et al., 2024; Ozdemir, 2022).

While previous studies have used econometric methods to assess the effect of climate change on agriculture, other studies have distinguished themselves by combining climate models, agricultural models, and economic models in order to directly link climate projections to impacts on food security. With this in mind, Khan et al. (2020), Vatankhah et al. (2019) and Ouraich and Tyner (2014) examine the economic consequences of agricultural productivity losses due to climate change by 2050, integrating climate, agronomic and economic models. Their results indicate a significant estimated loss in GDP, a notable increase in agricultural prices, and a significant decrease in consumption, thereby causing multiplier effects on the industrial and commercial sectors.

However, most current research, whether conducted at the regional or national level, focuses either on econometric analyses that do not explicitly integrate future climate scenarios, or on climate simulations that do not directly take into account food security dimensions. Our study addresses this gap by adapting an integrated framework to the North African situation, characterized by high dependence on agriculture and significant vulnerability to climate fluctuations.

METHODOLOGY

Over the past two decades, the literature on the economics of climate change has undergone significant development in terms of modeling approaches (Castle and Hendry, 2020). Regional climate projections point to declining rainfall, rising temperatures and continued loss of arable land, exposing the region to systemic agricultural and food shocks. Faced with these challenges, there is a need for robust economic evaluation tools to simulate the effects of climate on food security, and possibly inform adaptation decisions. The ultimate goal is to contribute to a better understanding of climate issues in a particularly exposed and strategic region.

Data

Econometric Variables and Data Sources

The study is based on panel data covering six North African countries: Algeria, Egypt, Libya, Mauritania, Morocco and Tunisia, for the period 1995 to 2023. The ultimate aim is to quantify the impacts of climate change on agricultural production, using a variety of economic and climatic factors. The variables taken into account include agricultural production as the variable of interest, agricultural labor, agricultural capital (or gross fixed capital formation), climatic conditions, temperature and precipitation, and a synthetic aridity index (De Martonne index). The temporal dimension was determined by the homogeneous and reliable availability of statistical series for all the variables used. Table 1 lists the variables, their units of measurement, abbreviations and data sources.

Agricultural economic data are taken from the FAOSTAT database. Real agricultural production, measured here in constant 2015 USD, is the foundation of food security, notably through its principal role in the food availability dimension (Ardakani et al., 2020; Jiang & Chen, 2020). This choice makes it possible to directly capture changes in the performance of the agricultural sector, while neutralizing the effects of inflation. The agricultural labor force, i.e. the number of people employed in the agricultural sector, and agricultural gross fixed capital formation (in constant 2015 USD) also come from the same source, respectively through the data series on the agricultural labor force and agricultural investment. The inclusion of these variables in the model makes it possible to control for the effect of economic inputs in the agricultural production fabric. The annual climate variables, namely mean annual temperature and mean annual precipitation, are taken from the Copernicus Climate Change Service and the Our World in Data database. These sources are widely recognized in the literature for their reliability.

Finally, the missing data encountered at variable level were treated by a rigorous method of imputation by the intra-country mean, taking into account, of course, the temporal structure and the low rate of missing data. This imputation approach is widely used in panel studies, particularly in the presence of continuous data with low correlation to random absences (Missing Completely At Random [MCAR]). Nevertheless, the robustness of this approach was verified by additional sensitivity analyses, (i) comparing descriptive statistics before and after imputation, (ii) re-estimating the models using other imputation techniques (e.g., median substitution), and (iii) checking that the signs, magnitudes, and significance of the estimated coefficients remained unchanged. The analyses revealed negligible variation between techniques, confirming that the imputation process had no significant impact on the econometric results. As a result, the data set was structured according to a balanced panel structure, with 6 cross-sectional units (countries) and 29 annual periods, ensuring the statistical consistency of the econometric models subsequently implemented.

Climatic Index and Projection Scenarios

To channel the combined effects of heat and humidity, we have also introduced a unitless De Martonne index, based on the classic formula combining precipitation and temperature. This index, frequently used in the literature, characterizes the degree of annual climatic aridity for each country (Andrade et al., 2021; Aieb et al., 2020; Pellicone et al., 2019).

It is calculated according to the following formula:

$$IM = \frac{P}{T + 10}$$

Where *P* represents the average annual precipitation (in mm) and *T* the average annual temperature (in °C).

A high value is indicative of more humid climatic conditions favorable to agriculture, while a low value is indicative of more arid conditions.

For climate projection data, we are using the results of phase 6 of the Coupled Model Intercomparison Project (CMIP6) for the North African region. Recent IPCC scenarios (2022), i.e. Shared Socioeconomic Pathways (SSP), will be used in this sense, to discover possible long-term dynamics according to different emission and development contexts.

The Coupled Model Intercomparison Project Phase 6 (CMIP6) is the latest generation of global climate models. It improves the accuracy of climate projections by offering higher geographical resolution and improved dynamic processes compared to previous phases (Arfasa et al., 2023). The integration of CMIP5 Representative Concentration Pathways (RCPs) with Shared Socioeconomic Pathways (SSPs) is a key innovation of CMIP6, enabling a more comprehensive assessment of future climate outcomes by simultaneously considering socioeconomic and biophysical factors. The SSPs cover factors such as population, GDP, urbanization, technological progress, energy demand, and land use changes[2]. They also describe five different narrative trajectories of global socioeconomic development, ranging from sustainable development (SSP1) to fossil fuel-intensive growth (SSP5) (Supharatid & Nafung, 2021; Arfasa et al., 2023).

Table 1. List of variables and data source

Variable name	Unit of measure	Abbreviation	Data sources
Agricultural production	Constant USD (2015)	Yagri	FAOSTAT, Gross Production Value (current thousand US$)
Agricultural labor	Number of people	Lagri	FAOSTAT (Agricultural labor force)
Gross agricultural fixed capital formation	Constant USD (2015)	Kagri	FAOSTAT (Agricultural Investment)
Average annual temperature	Degrees Celsius (°C)	Temp	Copernicus Climate Change Service, Our World in Data,
Average annual precipitation	Millimeters (mm)	Precip	Copernicus Climate Change Service, Our World in Data
De Martonne climatic index	Index without unit	IM	Calculated from climatic data (Temp, Precip)

The Econometric Approach

Model Specification

To explore the existing relationship between agricultural production and other endogenous variables, notably climate, we will deploy an advanced econometric approach that is quite widespread in the literature for studying the impacts of climate change. Thus, the variable of interest, agricultural production, will be explained by the endogenous variables: agricultural capital, agricultural labor and the De Martonne index. The equation is written as follows:

Agricultural production
=f(Agricultural capital, Agricultural labor, De Martonne index)

$$Y_{AGRit} = f\left(K^{\beta_1}_{AGRit}, L^{\beta_2}_{AGRit}, IM^{\beta_3}_{AGRit}\right)$$

So the logarithmic form of our equation can be presented as follows:

$$Ln\ Y_{AGRi,t} = \beta_0 + \beta_1 Ln K_{AGRi,t} + \beta_2 Ln L_{AGRi,t} + \beta_3 Ln IM_{AGRi,t} + \mu$$

$Y_{agri,it}$ is the country's agricultural value added i for the period. $tK_{agri,it}$ is the agricultural capital stock (gross fixed agricultural capital formation). $L_{agri,it}$ is the agricultural workforce. And IM_{it} is De Martonne's climatic index.

We will deploy autoregressive staggered lag models (ARDL) in panel (ARDL-Panel). These models extend the ARDL framework from time series to panel data, making it possible to simultaneously study cross-sectional attributes between individuals while capturing interactions between time series (Gaies et al., 2021). The potential of these models lies in their ability to analyze variables with different orders of integration, since unlike conventional cointegration approaches these models can efficiently estimate short- and long-term relationships between variables (Pesaran et al., 2001; Htike et al., 2021). This flexibility makes these models particularly suitable for studying the economic impacts of climate change, where economic and climate variables do not generally share the same level of integration (Jebli et al., 2022). It should be noted that ARDL-Panel also adapts to panel data structures where the sample size of two dimensions, temporal and individual, is relatively small. Thus it provides rigorous estimates even with limited data availability (Htike et al., 2021; Haug, 2002; Yu et al., 2022).

Introduced by Pesaran, Shin and Smith (1999), the Pooled Mean Group (PMG) approach will be deployed to carry out our ARDL-Panel modeling. Several authors have deployed this approach to address issues related to the impacts of climate change on agriculture and food security (Attiaoui and Boufateh, 2019; Yaqoob et al., 2023; Zhang et al, 2024). The PMG-ARDL approach makes it possible to simultaneously estimate short-term relationships (which may vary from one country to another) and a long-term relationship that is identical for all the countries studied.

Thus our model is mathematically formulated as follows:

$$\Delta \ln Y_{agr\,i,it} = \phi_i\big[\ln Y_{agr\,i,i,t-1} - \beta_1 \ln K_{agr\,i,i,t-1} - \beta_2 \ln L_{agr\,i,i,t-1}$$

$$-\beta_3 \ln IM_{i,t-1} - \beta_4 trend_i - \alpha_i\big] + \sum_{j=1}^{p-1} \gamma_{ij} \Delta ln\, Y_{agr\,i,i,t-j} + \sum_{j=0}^{q-1} \delta_{1ij} \Delta ln\, K_{agr\,i,i,t-j}$$

$$+ \sum_{j=0}^{r-1} \delta_{2ij} \Delta ln\, L_{agr\,i,i,t-j} + \sum_{j=0}^{s-1} \delta_{3ij} \Delta lnIM_{i,t-j} + \mu_i + \varepsilon_{it}$$

Where:

Δ is the first difference operator.

$trend_i$ is a time trend specific to each country.

ϕ_i represents the speed of adjustment towards long-term equilibrium (error correction term, COINTEQ).

$\beta_1, \beta_2, \beta_3$ are the long-term coefficients common to the entire panel.

$\gamma_{ij}, \delta_{1ij}, \delta_{2ij}, \delta_{3ij}$ represent short-term effects specific to each country.

μ_i is the country-specific fixed effect.

ε_{it} represents the random error term.

Statistical Tests

A set of statistical tests must precede ARDL-PMG modeling to ensure the validity and robustness of the results. Firstly, we will ensure stationarity by deploying the panel unit root test, namely CADF (Cross-sectionally Augmented Dickey-Fuller) and CIPS (Cross-sectionally Augmented Im-Pesaran-Shin), but also the CSD (cross-sectional dependence test) in order to verify the independence of the panel units and/or detect the potential presence of unobserved common shocks i.e. omitted variables such as global climate shocks, by way of example (Pesaran, 2014, 2006, 2007). The equations for these three are presented respectively as follows:

- CADF (Cross-sectionally Augmented Dickey-Fuller)

For each i, the test is written:

$$\Delta y_{it} = a_i + b_i y_{i,t-1} + c_i \bar{y}_{t-1} + \sum_{p=1}^{P} d_{ip} \Delta y_{i,t-p} + \sum_{p=1}^{P} e_{ip} \Delta \bar{y}_{t-p} + \varepsilon_{it}$$

Where $\bar{y}_t = \frac{1}{N}\sum_{i=1}^{N} y_{it}$ is the cross-sectional mean. This test is estimated for each country *i*.

- CIPS (Cross-sectionally Augmented IPS test)

This is the average of CADF statistics:

$$CIPS = \frac{1}{N}\sum_{i=1}^{N} t_i^{CADF}$$

with t_i^{CADF} the t-statistic associated with the b_i coefficient in the CADF test

- CSD (cross-sectional dependence test)

$$CD = \sqrt{\frac{2}{N(N-1)}}\sum_{i=1}^{N-1}\sum_{j=i+1}^{N}\hat{\rho}_{ij}$$

where $\hat{\rho}_{ij}$ represents the correlation of residuals between units *i* and *j*. Under the null hypothesis of no cross-sectional dependence, $CD \sim N(0,1)$ when $T \to \infty$ and $N \to \infty$.

Secondly, other complementary tests before and after estimation will also be used to guarantee the validity of the model and ensure that basic econometric assumptions are not violated. These tests are Hsiao's (1986) structural homogeneity test to determine the appropriate structure for our model, Kao's co-integration test (based on Engle-Granger) to verify the existence of a co-integration relationship between variables, and Dumitrescu and Hurlin's (2012) panel causality test to study unidirectional or bidirectional relationships between variables. These conditions must be taken into account to ensure the reliability and robustness of the estimates of the Pooled Mean Group (PMG) or Mean Group (MG) model.

Climate Change Impact Assessment

The results of long-term estimates from the ARDL-PMG model will then be used to quantitatively assess the future impact of climate change on agricultural production in North African countries, and hence on regional food security. For this purpose, we integrate climate projections of temperature and precipitation from the CMIP6 model calculated by Almazroui et al. (2020). In their study, the authors analyzed 27 global climate models from the sixth phase of the Coupled Model Intercomparison Project (CMIP6) under three future SSP scenarios, varying from low-forcing SSP1-2.6, medium-forcing SSP2-4.5 and high-forcing SSP5-8.5 (O'Neill et al.,

2017). They examined projected changes in temperature and precipitation over the African continent and its eight sub-regions by 2100.

In our analysis, we specifically exploit the projections for the North Africa region (NAF as abbreviated in the study), on the three distinct SSP scenarios: SSP1-2.6 (low emissions), SSP2-4.5 (intermediate emissions), and SSP5-8.5 (high emissions). Table 2 shows temperature and precipitation projections for the North African region over the period 2030-2099.

Table 2. Projected trends in mean temperature (in °C) and mean precipitation (in %) by decade for the period 2030-2099.

Scenario	NAF temperature (°C)	Precipitation NAF (%)
SSP1-2.6	+0.04	-1.00
SSP2-4.5	+0.26	-1.82
SSP5-8.5	+0.69	-4.11

These climate projections will be incorporated into the calculation of the future value of the De Martonne climate index, which jointly captures the influence of temperature and precipitation. Consequently, predicted fluctuations in this index will be used to estimate the impacts of climate variability on agricultural production by directly applying the elasticities obtained from the ARDL model. As a result, the future variation in agricultural production can be presented as follows:

$$\Delta Y_{AGRcc} = \beta_{IMLT} \times \Delta IM$$

Where ΔY_{AGRcc} is the variation in agricultural production due to climate change, β_{IMLT} is the long-term coefficient of the logarithm of the climate index and ΔIM is the variation of this index between its initial and future values. This approach will make it possible to rigorously quantify the potential impacts of climate change on the agricultural sector, highlighting future food security issues in the North African region (Emadodin et al., 2021; Tayebeh et al., 2019).

RESULTS AND DISCUSSION

Descriptive Analysis and Preliminary Tests

Descriptive Analysis

Table 3 presents the descriptive analysis of all the logarithmic variables in our panel, based on a sample of 174. The position measures suggest a strong central consistency of the variables, notably with almost equal means and medians. In terms of dispersion measures, the standard deviation shows low values, indicating that the observations are relatively clustered around their means and therefore less dispersed, testifying to a certain stability both between countries and over time. The skewness coefficients show that the distributions are mainly symmetrical, although there are slight asymmetries in some variables, notably LNL_AGRI and LN_TEMP. Kurtosis, which measures the flatness or sharpness of the distribution compared with a normal distribution, reveals distributions that are relatively flat compared with a normal distribution, with values generally below 3, indicating a flatter distribution with lighter tails than normal. To refute or affirm normality, the results of the Jarque-Bera (JB) test reveal that, for most variables, the data distribution is asymmetric and differs significantly from a classic bell-shaped normal distribution, as shown by the low probability values. However, only the LN_TEMP variable has a probability above the 5% significance level, indicating that the distribution of this variable is normal. Figure 3 illustrates the evolution of the variables in logarithm for the different countries in the panel for the period from 1995 to 2023.

Table 3. Descriptive statistics

Statistics	LNY_AGRI	LNL_AGRI	LNK_AGRI	LN_ TEMP	LN_ PRECIP	LNIM
Average	22.23	13.86	20.21	3.12	4.09	0.61
Median	22.28	13.75	20.43	3.13	4.19	0.63
Maximum	24.42	16.50	21.97	3.36	6.20	2.87
Minimum	18.03	12	17.80	2.90	1.80	-1.80
Standard deviation	1.36	1.39	1.24	0.13	1.19	1.23
Skewness	-0.41	0.32	-0.45	0.07	-0.20	-0.08
Kurtosis	2.49	1.82	1.90	2.30	1.88	1.86
Jarque-Bera	6.84	12.90	14.76	3.74	10.29	9.60
Probability	0.0327	0.0016	0.0006	0.1542	0.0058	0.0082

Figure 3. Variable trends over the 1995-2023 period (source: authors' calculations)

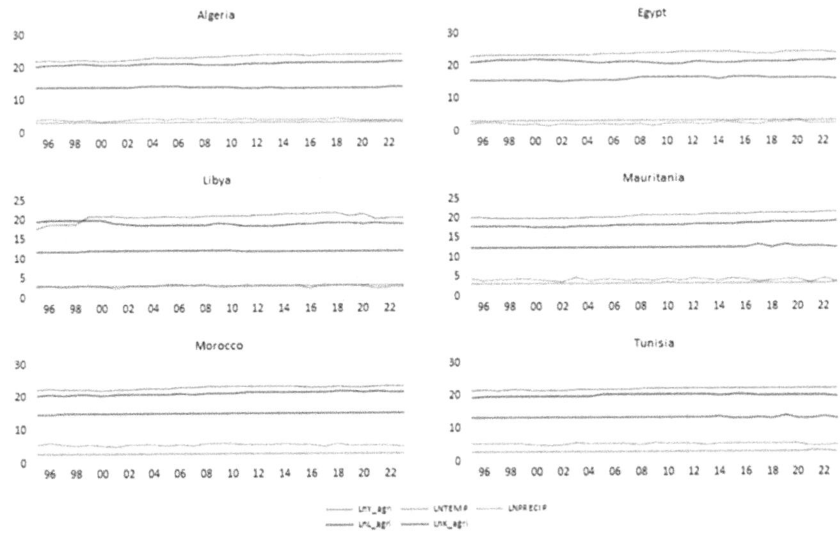

Preliminary Tests

Table 4 shows the results of the cross-sectional dependence test (CSD). This test, including the Breusch-Pagan LM, Pesaran scaled LM, Bias-corrected scaled LM and Pesaran CD statistics, shows that all the LNY_AGRI, LNK_AGRI, LNL_AGRI and LNIM variables exhibit statistically significant cross-dependence. Agricultural production (LNY_AGRI) shows a very high Pesaran CD statistic, confirming a strong interdependence of agricultural shocks between countries. Similarly, the other economic and climatic variables reveal high levels of statistically significant cross-sectional dependence. The LNIM variable, which captures the combined effect of climate change on soil moisture, also shows significant dependence, underlining the commonality of regional climate shocks in North Africa. This is consistent with the geographical proximity, agroclimatic similarities, and regional climate trends often shared across the region.

These results of the CSD test explicitly highlight that the six North African countries share, to some extent, substantial common shocks, most likely linked to climatic, economic and regional agricultural policy conditions. It is essential to explicitly incorporate this dependence into the econometric methodology to prevent estimation bias and ensure robust conclusions.

Table 4. Cross-sectional dependence (CSD) test results

Variable	Breusch-Pagan LM	Pesaran scaled LM	Bias-corrected scaled LM	Pesaran CD
LNIM	60.46 ***	8.30 ***	8.19 ***	5.49 ***
LNK_AGRI	134.97 ***	21.90 ***	21.80 ***	1.97 **
LNY_AGRI	274.49 ***	47.38 ***	47.27 ***	16.13 ***
LNL_AGRI	63.92 ***	8.93 ***	8.82 ***	1.19 ***

***** describes the 5% and 10% significance levels respectively.

Series stationarity is a fundamental assumption in the estimation of Pesaran (2007) ARDL-PMG models. The CIPS test is used to assess the presence of unit roots, taking into account the interconnections between the countries in the panel. This test is of particular importance in our setting, marked by significant regional interdependence (validated by Pesaran's CD test) (See table 5). The variable LNY_AGRI (agricultural production) shows a CIPS value of -3.98, statistically significant with a specification including a constant and a trend. This result implies that the series is stationary in level, i.e. it is integrated of order zero: I(0). In addition, agricultural capital has a statistically significant CIPS value of -3.43, which also leads to rejection of the unit root null hypothesis at the 1% significance level, confirming stationarity at level, i.e. I(0). Martonne's climatic index also indicates stability at the stationary level. Examination of the CADF results by country nevertheless reveals some heterogeneity in the stationarity of individual variables, with series that are clearly stationary in some countries and non-stationary in others (see Table 6). This combination of zero-order and first-order integrated variables, with no second-order integrated variables, confirms that the essential methodological preconditions for robust estimation of an error-correction ARDL panel model, as advocated by Pesaran et al. (1999), are met in this empirical analysis.

Table 5. CIPS and CADF panel unit root test results

Variable	CIPS t-stat	Order of integration	CADF by country
LNL_AGRI	-2.45*	I(1)	ALG: 1.10, EGY: -2.20, LIB: 0.76, MAU: -5.53*, MOR: -2.66, TUN: -6.18*
LNY_AGRI	-3.98*	I(0)	ALG: *-3.61*, EGY: *-3.65*, LIB: -3.38, MAU: -3.20, MOR: -2.65, TUN: -7.39*
LNK_AGRI	-3.43*	I(0)	ALG: -3.08, EGY: -2.02, LIB: -5.80*, MAU: -3.25, MOR: -4.55, TUN: -1.90
LNIM	-3.68*	I(0)	ALG: 0.59, EGY: -0.98, LIB: -6.06*, MAU: -8.06*, MOR: -5.30*, TUN: -2.28

,,*** describes 1%, 5% and 10% significance levels respectively. ALG = Algeria, EGY = Egypt, LIB = Libya, MAU = Mauritania, MOR = Morocco, TUN = Tunisia

In addition, it is essential to test the structural homogeneity of the coefficients in order to highlight the diversity of short-term dynamics between countries, which fully justifies the adoption of the PMG approach. To this end, the structural homogeneity test of the parameters (slopes and constants) proposed by Hsiao (1986) is used (see Table 6). The test results explicitly reject the hypothesis of total homogeneity (identical slopes and constants). This rejection suggests that economic and climatic dynamics vary considerably from country to country, calling into question the assumption of a uniform relationship for the entire sample. Furthermore, testing of the partial homogeneity hypothesis (identical constants but different slopes) also results in a categorical rejection. The same applies to H3, where the slopes are identical but the constants different. These results show that the coefficients linked to the explanatory variables (slopes) vary significantly from one country to another, highlighting significant structural heterogeneity. These results underline the importance of applying an econometric methodology capable of taking this heterogeneity into account. Consequently, this justifies the use of the ARDL panel model with Mean Group (MG) or Pooled Mean Group (PMG) specifications, which enable country-specific dynamics to be effectively incorporated into a common framework (Pesaran, 1999, 2007).

Table 6. Structural homogeneity test by Hsiao (1986)

Hypothesis tested	**Specification**	**F-statistics**	**P-value**
H_1: Total homogeneity vs. total heterogeneity	All slopes and constants are identical	15.312	< 0.0001
H_2: Partial homogeneity vs. total heterogeneity	Identical constants, but different slopes	15.349	< 0.0001
H_3: Total homogeneity vs. partial homogeneity	Identical slopes, but different constants	6.596	0.0000127

Table 7 shows the results of Kao and Pedroni's test for the existence of a long-term relationship (co-integration), based on the Engle-Granger methodology adapted to panel data. The null hypothesis of no co-integration was rejected, indicating that the variables studied share a statistically significant long-term relationship. These results empirically justify the application of an error correction model (ECM) in the ARDL framework. In addition, the robust variance estimation (HAC method with Bartlett kernel) confirms the stability and robustness of the results in the face of potential heteroscedasticity and autocorrelation. These results reinforce the empirical validity of the ARDL panel framework used, confirming that the selected economic and climate variables are linked by stable long-term dynamics. This methodological validation is essential before interpreting the long-term coefficients of the estimated model (Gengenbach et al., 2006).

Table 7. Results of the Kao co-integration test (based on Engle-Granger)

Statistics	Value	P-value
ADF (t-stat)	-1.827766	0.0338
Residual variance	0.054153	-
HAC variance (Bartlett)	0.054136	-

As a final step in our pursuit of statistical diagnostics, we will check the causal relationship between the variables using the Dumitrescu and Hurlin (2012) test. The results, as illustrated in Table 8, reveal a complex interaction between agricultural and climatic variables. At first glance, a significant bidirectional relationship is observed between agricultural gross fixed capital formation (LNK_AGRI) and agricultural value added (LNY_AGRI). This indicates a mutual feedback between investment in agricultural capital and agricultural production, underlining the importance of simultaneously strengthening these two dynamics to boost production. Secondly, a unidirectional causality from agricultural labor (LNL_AGRI) to production (LNY_AGRI) is also detected, showing that the evolution of employment in the sector directly influences agricultural performance. The relationships between climatic and agricultural variables are equally significant. Temperature influences agricultural production, highlighting the vulnerability of agricultural performance to climate change. There is also a unidirectional relationship between agricultural production and precipitation, as well as between temperature and precipitation. These results can be interpreted with caution, as dynamic correlations linked to the spatio-temporal variability of the region rather than a direct influence. On the other hand, bidirectional causality is observed between temperature (LNTEMP) and labor (LNL_AGRI), as well as between temperature and agricultural capital (LNK_AGRI. These results reflect the mutual influence between climatic conditions and the structural dynamics of the agricultural sector, which may reflect adjustments in terms of investment and labor in response to climatic variations.

Table 8. Summary of Dumitrescu and Hurlin's test for causality between variables.

Relationship tested	p-value	Type of causality
LNK_AGRI LNY_AGRI	0.0055	Bidirectional
LNY_AGRI LNK_AGRI	0.0007	Bidirectional
LNL_AGRI LNY_AGRI	0.0243	Unidirectional
LNTEMP LNL_AGRI	0.0069	Bidirectional
LNL_AGRI LNTEMP	0.0224	Bidirectional
LNTEMP LNK_AGRI	0.0207	Bidirectional
LNK_AGRI LNTEMP	0.0022	Bidirectional
LNY_AGRI LNTEMP	0.00004	Unidirectional
LNY_AGRI LNPRECIP	0.0220	Unidirectional
LNTEMP LNPRECIP	0.0325	Unidirectional

DISCUSSION

After these statistical diagnostics, which we considered important for the success of our modeling, we present and discuss in this subsection the short- and long-term results of the ARDL-PMG model, chosen from several models estimated on the basis of Akaike's information criteria (AIC). The results of the short-term estimation of the ARDL panel model (PMG), as presented in Table 9, show a statistically significant error correction coefficient (-0.302). This coefficient indicates the speed of adjustment towards long-term equilibrium, i.e. that around 30% of the imbalance between actual agricultural production and its long-term equilibrium value is corrected each year. This result strongly confirms the existence of a mechanism for returning to long-term equilibrium. As for the short-term explanatory variables, most of them show insignificant coefficients. This indicates that the immediate annual variations in agricultural capital, the climatic index and agricultural labour have no statistically significant impact on agricultural growth in the very short term. However, the first lag of farm labor (D(LNL_AGRI(-1))) displays a negative effect with marginal significance. This indicates that a fall in agricultural employment tends to produce, with a certain time lag, a negative adjustment in agricultural output in the short term. This could reflect the effects of structural adjustments or productivity constraints when agricultural employment changes abruptly. These results show that the short-term agricultural situation in North Africa is complex and slow to change. Adjustments are made gradually towards long-term stability, rather than reacting rapidly to changes in production conditions or climate.

Table 9. Short-term results of the ARDL-PMG model

Variable	Coefficient	Std. Error	t-Statistic
COINTEQ	-0.301757 *	0.098599	-3.060462
D(LNY_AGRI(-1))	0.138671	0.152508	0.909274
D(LNK_AGRI)	0.060669	0.081351	0.745768
D(LNK_AGRI(-1))	-0.135709	0.133649	-1.015417
D(LNL_AGRI)	-0.068287	0.252955	-0.269957
D(LNL_AGRI(-1))	-0.713076 ***	0.415581	-1.715854
D(LNIM)	0.048647	0.080618	0.603427
D(LNIM(-1))	0.044491	0.073384	0.606278
C (constant)	0.562100 *	0.140477	4.001367

,,*** describes the 1%, 5% and 10% significance levels, respectively.

So as not to limit ourselves solely to aggregate short-term results. Table 10 presents the short-term results of the ARDL-PMG model for each of the six panel countries. The results show large differences between countries for the error correction coefficient (COINTEQ). Egypt, Libya, Morocco and Tunisia show statistically significant coefficients (-0.346, -0.224, -0.658 and -0.479), indicating rapid adjustment to equilibrium after economic or climatic shocks. However, Morocco represents the fastest adjustment, despite the fact that its food security faces various climatic challenges (Akesbi, 2024). By contrast, Algeria and Mauritania have very low coefficients, -0.049 and -0.055 respectively. This shows that there is little or no movement towards long-term equilibrium, which could be due to structural problems or inefficiencies in their agricultural economies. The climate index shows statistical significance only for Egypt and Libya. The results for the other variables reveal heterogeneous effects. The effects of agricultural capital and agricultural labor differ both between countries and between the lags considered in the model. Overall, agricultural labor exerts a negative effect on agricultural production in all countries, which corroborates with the overall results of the short-term model for the same variable (see table 9). In contrast, agricultural capital generally has a positive effect in all countries. This indicates the need to step up investment in the agricultural sector in these countries. This can be seen as an essential lever for developing effective adaptation strategies in the face of climate change, and for guaranteeing the long-term food security of North African countries.

Table 10. Short-term results of the ARDL-PMG model by country

Variable	Algeria	Egypt	Libya	Mauritania	Morocco	Tunisia
COINTEQ	-0.0491	-0.3455 *	-0.2235 *	-0.0546	-0.6584 *	-0.4793 *
D(LNY_AGRI(-1))	-0.1850	0.3257	-0.1838	0.3280	0.7104 *	-0.1632
D(LNK_AGRI)	0.1730	-0.1484	-0.1619	0.1549	0.34408 *	0.0023
D(LNK_AGRI(-1))	-0.3218 ***	-0.1193	0.3144	0.1935	-0.4855 *	-0.3955 *
D(LNL_AGRI)	-0.0662	-0.1109	1.0772	-0.1958	-0.7932	-0.3207***
D(LNL_AGRI(-1))	-0.2370	-0.1421	-1.0411	-0.0910	-2.6553*	-0.1114
D(LNIM)	0.2082 ***	0.0173	0.3407	-0.0087	-0.2167 *	-0.0485
D(LNIM(-1))	0.1428	0.0330	0.3248	-0.0263	-0.21373*	0.0062
C (constant)	0.2092***	0.8240 *	0.8124 *	0.1551	0.4205 *	0.9502 *

,,*** describes the 1%, 5% and 10% significance levels, respectively.

Table 11 summarizes the long-term results of the ARDL- PMG model. They explicitly confirm the existence of a statistically significant long-term relationship between agricultural production and the explanatory variables included in our analysis. The De Martonne index (LNIM), which takes into account both precipitation and temperature, has a positive and significant effect on agricultural production. The coefficient value of the index (0.5) confirms that more favorable climatic conditions, marked by increased humidity and moderate temperatures, favor agricultural production. This dimension is of crucial importance in a context of pronounced regional climate change, calling for adaptation strategies specifically focused on reducing climatic constraints (Pellicone et al., 2019). Similarly, Agricultural capital with a positive and significant coefficient of 0.338 suggests that investment in agricultural fixed capital will have a positive impact on long-term growth in agricultural output, i.e. a 0.33% increase in agricultural capital will cause an average 1% increase in agricultural output. More pronounced is the significantly greater impact of agricultural labor (LNL_AGRI), characterized by a coefficient of 0.912. In other words, a 0.91% increase in farm labor will increase agricultural output by 1%. This underlines the major importance of human capital in the agricultural sector in North Africa, highlighting that the availability and efficient use of agricultural labor in a context of climate change adaptation are key elements in increasing regional agricultural production (Huang et al., 2023; Kurukulasuriya et al., 2006).

These long-term results provide a robust empirical basis for developing policy recommendations aimed at sustainably enhancing food security in North Africa. This could be achieved through specific investments in agricultural capital, the development of agricultural human resources, and the implementation of adequate climate adaptation measures.

Table 11. Long-term results of the ARDL-PMG model

Variable	Coefficient	Std. Error	t-Statistic
LNK_AGRI	0.337986 *	0.145871	2.317021
LNL_AGRI	0.911735 *	0.206874	4.407197
LNIM	0.490858 *	0.124190	3.952471
@TREND	0.015383 *	0.005645	2.725286

,,*** describes the 1%, 5% and 10% significance levels, respectively.

As mentioned in the previous section, the coefficient value of the climatic index will be used as the elasticity of crop variations based on temperature and precipitation projections from Almazroui et al. (2020). In line with the study's approach, we use the SSP scenarios presented in Table 2. After integrating these projections and calculating the variations in agricultural production. Table 12 shows the results of this variation. The climate projections applied to North Africa indicate a general decline in agricultural production by 2100, with average annual losses intensifying according to the emission scenarios.

Table 12. Projected change in agricultural production under different SSP climate scenarios (in %)

	Algeria	Egypt	Libya	Mauritania	Morocco	Tunisia
ΔY_{AGRcc} **SSP 1-2.6**	-0,8	-0,7	-0,8	-0,7	-0,9	-0,8
ΔY_{AGRcc} **SSP 2-4.5**	-1,7	-1,6	-1,7	-1,5	-1,8	-1,7
ΔY_{AGRcc} **SSP 5-8.5**	-3,8	-3,7	-3,9	-3,7	-4,1	-4,0

In Algeria, projections show that agricultural production will decline with climate change. In the SSP1-2.6 scenario, the decline is fairly slight, estimated at -0.8%. This shows that there is a certain capacity to adapt to current climate impacts. On the other hand, with SSP2-4.5, the loss increases almost twofold (-1.7%), showing a growing fragility in the face of more severe water and heat problems. The SSP5-8.5 scenario predicts a drop of 3.8%, which would lead to a major reduction in production over the long term. These results show that Algeria needs to improve its adaptation policies, especially in terms of water management, crop diversification and, to some extent, increased productivity in irrigated areas (Bouznit and Aïssaoui, 2023).

Egypt's trajectory is relatively similar to that of Algeria, although slightly less marked under the SSP1-2.6 scenario (-0.7%). This resistance can be explained by the importance of irrigation in Egyptian agriculture. However, from scenario SSP2-4.5 onwards, the decline is -1.6%, showing greater sensitivity to very high temperatures and to water availability, particularly from the Nile. In the SSP5-8.5

scenario, the loss is -3.7%, which threatens food security, especially in a country that relies heavily on imports. It is therefore becoming urgent to implement adaptation strategies, particularly technological ones such as efficient irrigation, greenhouses and resistant plants, to maintain production in the face of climate change (Ramadan, 2024; Abdel-Maksoud, 2018; Roushdi, 2024).

Libya is very fragile, with losses estimated at -0.8%, -1.7% and -3.9% according to three different scenarios. This country, which has harsh climatic conditions and relies heavily on fossil water for irrigation, could lose much of its ability to produce food if major investments are not made to adapt (Abunnour et al., 2016). The country also faces extreme climatic events. These figures show that it is important to consider agriculture in Libya in terms of water security, climate resilience and long-term strategic planning (Zaki et al., 2022).

Mauritania shows the smallest losses in each situation (-0.7%, -1.5%, -3.7%), which may give the impression that it is less exposed. In fact, this is mainly because agriculture is still of little importance to the country's economy, which is mainly concentrated in the Senegal River valley. However, a drop of almost 4% in a country where food insecurity already affects a large proportion of the population would represent a considerable challenge. Improving irrigation and implementing adaptation measures is therefore essential.

Morocco loses the most in all scenarios, peaking at -4.1% under SSP5-8.5. This vulnerability is due to the agricultural sector's high dependence on rainfall, especially for cereals and food crops. Even under a sustainable scenario, the decline is still significant (-0.9%). Aware of these impacts, the country has put in place strategies to adapt, such as the Green Morocco Plan and the Generation Green Strategy, but these efforts will need to be intensified to limit future losses (Ouraich and Tyner, 2014; Benabdelkader et al., 2021). Agriculture in Morocco requires structural transformations and adaptation of production systems to enhanced arid conditions.

Tunisia's trajectory is very similar to Morocco's, with losses estimated at -0.8%, -1.7% and -4.0% depending on the scenario. This situation highlights the country's high exposure to climatic risks, compounded by overexploitation of water resources and low efficiency of traditional irrigation. The Central and Southern regions, already vulnerable, could experience alarming yield declines. In this context, Tunisia needs to accelerate the integration of adaptation policies into its agricultural planning, combining technical innovation, water governance and targeted support for the most exposed farmers (Khalifa, 2025).

Finally, at the regional level, the results indicate that agricultural production in North Africa is expected to decline in the future due to climate change. The projected declines average –0.75% in a sustainable development scenario (SSP1–2.6), reach –1.65% in an intermediate scenario (SSP2–4.5), and peak at –3.86% in a climate inaction scenario marked by substantial dependence on fossil fuels (SSP5–8.5).

The data indicates that the region's agricultural systems are inherently vulnerable to escalating climatic stressors, necessitating the implementation of robust, diverse, and sustainable adaptation strategies to safeguard food security.

CONCLUSION AND POLICY IMPLICATIONS

Throughout this chapter, we have focused on the impacts of climate change on agriculture and food security in the North African region. Using a panel ARDL model with a Pooled Mean Group (PMG) specification, we were able to establish clear and robust long-term relationships between agricultural production and its main economic and climatic determinants. Our results show that agricultural fixed capital, agricultural labor and climatic conditions, as captured by the Martonne index, all exert a significant positive long-term influence on regional agricultural production. This confirms what previous studies have found about the links between climate, agricultural resources and food security.

In the short term, the analysis shows major differences between countries. Morocco and Tunisia adapt quickly to change, thanks in part to their agricultural reforms. By contrast, Algeria and Mauritania show slower dynamics, indicating economic or structural rigidities to be overcome to improve their resilience to climate shocks. These dynamics were supported by tests that verified cointegration, stationarity (CADF, CIPS) and dependence between the different parties, proving that the ARDL method is appropriate for a regional group with structural links and national characteristics. On the other hand, Dumitrescu-Hurlin tests show significant links between economic and climatic variables. For example, in some countries, temperature affects agricultural production, just as agricultural capital influences labor. These links show that there are interactions between human and climatic factors in the agricultural system in North Africa. In addition, Hsiao's (1986) structural homogeneity test indicated heterogeneity in the agricultural and economic dynamics of countries. This structural heterogeneity fully justifies the use of ARDL-PMG, effectively capturing these country specificities while maintaining a consistent methodological framework for the panel as a whole.

The use of Almazroui et al.'s (2020) CMIP6 climate scenarios, such as SSP1-2.6, SSP2-4.5 and SSP5-8.5, helps link economic results to future projections. The application of these projections in the De Martonne climate index and their combination with the econometric results of the ARDL model provide robust and valuable estimates of the magnitude of future climate change impacts on agricultural production and, by extension, on food security in the North African region.

Long-term simulations based on projections by Almazroui et al. (2020) show an average decline in agricultural production. This drop depends on the level of emissions: it varies from -0.7% to -0.9% with SSP1-2.6, and can reach -3.7% to -4.1% with SSP5-8.5. Morocco and Tunisia are hard hit by climate change, with annual losses of -4.1% and -4.0%. This shows the urgent need to improve their adaptation policies. Egypt and Mauritania have smaller, but still significant losses (-3.7%). This shows that they are less directly affected, but still very vulnerable. Algeria and Libya lost between -3.8% and -3.9%, due to intense heat and lack of water. These projected losses call into question the region's food security, particularly with limited natural resources, a growing population and increased dependence on food imports.

In conclusion, this study offers important contributions to the existing literature on climate change, agriculture and food security in North Africa by addressing three important points: (i) relevant and robust econometric modeling, (ii) country disaggregation highlighting national contrasts, and (iii) integration of forward-looking climate scenarios. The results argue in favor of differentiated agricultural and climate policies, focusing on structural adaptation, modernization of agricultural practices, water governance, and the integration of climate science into food system planning. Faced with the predicted worsening of climatic conditions, the transformation of North African agricultural systems is no longer an option, but a strategic necessity.

Given the scale of climate change's impact on food security, particularly in terms of availability, it is essential to radically reevaluate agricultural strategies in the region. It is imperative that the most vulnerable countries begin the transition to more resilient production systems without delay.

Based on the findings of this study and the lessons learned from the literature review, various adaptation measures could be implemented to mitigate the vulnerability of food systems:

- To reduce dependence on a limited set of crops that are particularly vulnerable to climate variability, crop diversification is essential (Fekad et al., 2020).
- Small-scale irrigation systems should be developed to maximize water use efficiency and reduce dependence on rainfall patterns.
- Sustainable soil and water management methods are being implemented to maintain their quality and availability in the long term (Akinyi et al., 2022).
- Mobilization of specific funds for adaptation—whether through independent or programmed measures—with a particular focus on the regions and sectors most vulnerable to the impacts of climate change.
- Sustainable soil and water management methods are being implemented to maintain their quality and availability in the long term (Akinyi et al., 2022).
- Mobilization of specific funds for adaptation—whether through independent or programmed measures—with a particular focus on regions and sectors most vulnerable to the impacts of climate change.

• The implementation of innovative agricultural practices with the aim of increasing productivity and mitigating the impact of climate change, such as climate-smart agriculture (Chevallier, 2023; Fusco et al., 2020).

LIMITATIONS

Like all studies, this study has its own limitations. First, climate variability was measured using the De Martonne aridity index, which, although widely used in the literature, remains a relatively simple indicator that does not take into account key climate factors such as evapotranspiration or seasonal rainfall distribution. Second, the analysis focused exclusively on a single dimension of food security, namely availability, represented here by total agricultural production. This approach does not take into account other equally important dimensions, such as access, utilization, and stability, which could provide a better understanding of the impact of climate change on food security in the region. Third, the study did not include certain potentially relevant explanatory variables, such as CO_2 emissions or other indicators of environmental stress, which could influence agricultural productivity. Fourth, the analysis was conducted at an aggregate level for the entire agricultural sector, without distinguishing between specific crops. Given the heterogeneity of crop responses to climate shocks, future research could examine the impacts of climate change on key staple or strategic crops, taking into account each country's comparative advantages in production. Taking these limitations into account in future work would allow for a more nuanced and multidimensional assessment of the impacts of climate change on food security in North Africa.

REFERENCES

Abunnour, M. A., Hashim, N. M., & Bin Jaafar, M. (2016). Agricultural water demand, water quality and crop suitability in Souk-Alkhamis Al-Khums, Libya. *IOP Conference Series. Earth and Environmental Science*, *37*, 012045. Advance online publication. DOI: 10.1088/1755-1315/37/1/012045

Affoh, R., Zheng, H., Dangui, K., & Dissani, B. M. (2022). The Impact of Climate Variability and Change on Food Security in Sub-Saharan Africa: Perspective from Panel Data Analysis. *Sustainability (Basel)*, *14*(2), 759. DOI: 10.3390/su14020759

Aieb, A., Lefsih, K., Scarpa, M., Bonaccorso, B., Cicero, N., Mimeche, O., & Madani, K. (2020). Statistical modeling of monthly rainfall variability in Soummam watershed of Algeria, between 1967 and 2018. *Natural Resource Modeling*, *33*(4), e12288. Advance online publication. DOI: 10.1111/nrm.12288

Akinyi, D. P., Ng'ang'a, S. K., Ngigi, M., Mathenge, M., & Girvetz, E. (2022). Cost-benefit analysis of prioritized climate-smart agricultural practices among smallholder farmers: Evidence from selected value chains across sub-Saharan Africa. *Heliyon*, *8*(4), e09228. DOI: 10.1016/j.heliyon.2022.e09228

Almazroui, M., Saeed, F., Saeed, S., Islam, M. N., Ismail, M., Klutse, N. B., & Siddiqui, M. H. (2020). Projected Change in Temperature and Precipitation Over Africa from CMIP6. *Earth Systems and Environment*, *4*(3), 455–475. DOI: 10.1007/s41748-020-00161-x

Andrade, C., Contente, J., & Santos, J. A. (2021). Climate change projections of aridity conditions in the Iberian peninsula. *Water (Basel)*, *13*(15), 2035. DOI: 10.3390/w13152035

Ardakani, Z., Bartolini, F., & Brunori, G. (2020). New evaluation of small farms: Implication for an analysis of food security. *Agriculture*, *10*(3), 74. DOI: 10.3390/agriculture10030074

Arfasa, G. F., Sekyere, E. O., & Doke, D. A. (2023). Temperature and precipitation trend analysis using the CMIP6 model in the Upper East region of Ghana. *All Earth*, *36*(1), 1–14. DOI: 10.1080/27669645.2023.2290966

Attiaoui, I., & Boufateh, T. (2019). Impacts of climate change on cereal farming in Tunisia: A panel ARDL-PMG approach. *Environmental Science and Pollution Research International*, *26*(13), 13334–13345. DOI: 10.1007/s11356-019-04867-y

Benabdelkader, M., Saifi, R., & Saifi, H. (2021). Sustainable agriculture in some Arab Maghreb countries (Morocco, Algeria, Tunisia). In *Springer water* (pp. 233–261). https://doi.org/DOI: 10.1007/978-3-030-78574-1_10

Berhanu, M., & Wolde, A. O. (2019). Review on Climate Change Impacts and its Adaptation strategies on Food Security in Sub-Saharan Africa. *Agricultural Social Economic Journal, 19*(3), 145–154. DOI: 10.21776/ub.agrise.2019.019.3.3

Bouznit, M., & Aïssaoui, R. (2023). The impacts of climate change factors and innovative capabilities on food production in Algeria: Evidence from ARDL model. *Environment, Development and Sustainability, 26*(9), 23889–23908. DOI: 10.1007/s10668-023-03627-w

Castle, J. L., & Hendry, D. F. (2020). Climate Econometrics: An Overview. *Foundations and Trends® in Econometrics, 10*(3–4), 145–322. https://doi.org/DOI: 10.1561/0800000037

Chevallier, R. (2023). Strengthening Africa's climate-smart agriculture and food systems through enhanced policy coherence and coordinated action. *The South African Journal of International Affairs, 30*(4), 595–618. DOI: 10.1080/10220461.2024.2318712

Choudhary, T. F., & Gupta, M. (2024). Impact of Climate Change on Agriculture: Evidence from Major Crop Production in India. The Indian Economic Journal, 72(3), 442-459. (Original work published 2024)DOI: 10.1177/00194662241230655

Data and statistics. (2023). International Labour Organization. https://www.ilo.org/data-and-statistics

Deeb, M., Smagin, A. V., Pauleit, S., Fouché-Grobla, O., Podwojewski, P., & Groffman, P. M. (2024). The urgency of building soils for Middle Eastern and North African countries: Economic, environmental, and health solutions. *The Science of the Total Environment, 917*, 170529. DOI: 10.1016/j.scitotenv.2024.170529

Dumitrescu, E.-I., & Hurlin, C. (2012). Testing for Granger Non-Causality in Heterogeneous Panels. *Economic Modelling, 29*(4), 1450–1460. DOI: 10.1016/j.econmod.2012.02.014

El-Malky, M. G., Abdelwahab, M. M., Bryant, C., & Tantawi, S. (2016). VULNERABILITY AND ADAPTATION ASSESSMENT OF AGRICULTURE TO CLIMATE CHANGE IN EGYPT USING MATHEMATICAL MODELS. *Journal of Environmental Sciences (China), 35*(1), 1–19. DOI: 10.21608/jes.2016.27758

Emadodin, I., Corral, D. E. F., Reinsch, T., Kluß, C., & Taube, F. (2021). Climate Change Effects on Temperate Grassland and Its Implication for Forage Production: A Case Study from Northern Germany. *Agriculture*, *11*(3), 232. DOI: 10.3390/agriculture11030232

FAO. (2019). Proceedings from the Regional Workshop on Climate-Smart Agriculture in the Near East and North Africa. Cairo. 68 pp. Licence: CC BY-NC-SA 3.0 IGO. https://openknowledge.fao.org/items/a7cad295-ed2d-47b7-8c27-52450f4a7ac0

FAO, IFAD, WHO, WFP and UNICEF. (2024). The State of Food Security and Nutrition in the World 2024. Financing to eliminate hunger, food insecurity and all forms of malnutrition. Rome. https://doi.org/DOI: 10.4060/cd1254fr

Feher, A., Raicov, M., Brad, I., Toader, C. S., Sirbulescu, C. E., & Ben Saleh, R. (2022). Food security in a changing climate world. *Review on Agriculture and Rural Development*, *11*(1-2), 32–38. DOI: 10.14232/rard.2022.1-2.32-38

Fekad, M., & Bekalu, Y. (2020). Climate change impacts on household food security and its adaptation options in rural Ethiopia: A systematic review. [IJRD]. *EPRA International Journal of Research & Development*, *5*, 2455–7838. DOI: 10.36713/epra4142

Frimpong, T. D., Atchadé, M. N., & Tona Landu, T. (2024). Assessing the impact of CO2 emissions, food security and agriculture expansion on economic growth: A panel ARDL analysis. *Discover Sustainability*, *5*(1), 424. DOI: 10.1007/s43621-024-00630-7

Fusco, G. (2022). Climate Change and Food Security in the Northern and Eastern African Regions: A Panel Data Analysis. *Sustainability (Basel)*, *14*(19), 12664. DOI: 10.3390/su141912664

Fusco, G., Melgiovanni, M., Porrini, D., & Ricciardo, T. M. (2020). How to Improve the Diffusion of Climate-Smart Agriculture: What the Literature Tells us. *Sustainability (Basel)*, *12*(12), 5168. DOI: 10.3390/su12125168

Gaies, B., Nakhli, M. S., & Sahut, J. (2021). Is financial globalization polluting? – Symmetric and asymmetric effects of external liabilities on CO2 emissions in the MENA region. *Research Square (Research Square)*. https://doi.org/DOI: 10.21203/rs.3.rs-467034/v1

Gengenbach, C., Palm, F. C., & Urbain, J. (2006). Cointegration Testing in Panels with Common Factors*. *Oxford Bulletin of Economics and Statistics*, *68*(s1), 683–719. DOI: 10.1111/j.1468-0084.2006.00452.x

Hofisi, C., Chigavazira, B., Mago, S., & Hofisi, M. (2013). "Climate Finance Issues": Implications for climate change adaptation for food security in Southern Africa. *Mediterranean Journal of Social Sciences*. Advance online publication. DOI: 10.5901/mjss.2013.v4n6p47

Huang, W., Liu, Q., & Abu Hatab, A. (2023). Is the technical efficiency green? The environmental efficiency of agricultural production in the MENA region. *Journal of Environmental Management, 327*, 116820. DOI: 10.1016/j.jenvman.2022.116820

IPCC. 2022: *Climate Change 2022: Impacts, Adaptation, and Vulnerability.* Contribution of Working Group II to the Sixth Assessment Report of the Intergovernmental Panel on Climate Change [H.-O. Pörtner, D.C. Roberts, M. Tignor, E.S. Poloczanska, K. Mintenbeck, A. Alegría, M. Craig, S. Langsdorf, S. Löschke, V. Möller, A. Okem, B. Rama (eds.)]. Cambridge University Press. Cambridge University Press, Cambridge, UK and New York, NY, USA, 3056 pp., DOI: 10.1017/9781009325844

Jatuporn, C., & Takeuchi, K. (2023). Assessing the impact of climate change on the agricultural economy in Thailand: An empirical study using panel data analysis. *Environmental Science and Pollution Research International, 30*(3), 8123–8132. DOI: 10.1007/s11356-022-22743-0

Jiang, X., & Chen, Y. (2020). The potential of absorbing foreign agricultural investment to improve food security in developing countries. *Sustainability (Basel), 12*(6), 2481. DOI: 10.3390/su12062481

Khalifa, J. (2025). The Impacts of Climate Change, Agricultural Productivity, and Food Security on Economic Growth in Tunisia: Evidence from an Econometrics Analysis. *Research on World Agricultural Economy*, 577–599. DOI: 10.36956/rwae.v6i1.1457

Khan, M. A., Tahir, A., Khurshid, N., Husnain, M. I., Ahmed, M., & Boughanmi, H. (2020). Economic Effects of Climate Change-Induced Loss of Agricultural Production by 2050: A Case Study of Pakistan. *Sustainability (Basel), 12*(3), 1216. DOI: 10.3390/su12031216

Kurukulasuriya, P., Mendelsohn, R., Hassan, R., Benhin, J., Deressa, T., Diop, M., Eid, H. M., Fosu, K. Y., Gbetibouo, G., Jain, S., Mahamadou, A., Mano, R., Kabubo-Mariara, J., El-Marsafawy, S., Molua, E., Ouda, S., Ouedraogo, M., Séne, I., Maddison, D., & Dinar, A. (2006). Will African agriculture survive climate change? *The World Bank Economic Review, 20*(3), 367–388. DOI: 10.1093/wber/lhl004

Nelson, G. C., Rosegrant, M. W., Palazzo, A., Gray, I., Ingersoll, C., Robertson, R. D., Tokgoz, S., Zhu, T., Sulser, T. B., Ringler, C., Msangi, S., & You, L. (2010). Food Security, Farming, and Climate Change to 2050: Scenarios, Results, Policy Options. Washington, DC: International Food Policy Research Institute (IFPRI). https://doi.org/10.2499/9780896291867

Ngure, M. W., Wandiga, S. O., Olago, D. O., & Oriaso, S. O. (2021). Climate change stressors affecting household food security among Kimandi-Wanyaga smallholder farmers in Murang'a County, Kenya. *Open Agriculture*, *6*(1), 587–608. DOI: 10.1515/opag-2021-0042

Nordhaus, W. (2019). Climate Change: The Ultimate Challenge for Economics. *The American Economic Review*, *109*(6), 1991–2014. DOI: 10.1257/aer.109.6.1991

O'Neill, B. C., Kriegler, E., Ebi, K. L., Kemp-Benedict, E., Riahi, K., Rothman, D. S., van Ruijven, B. J., van Vuuren, D. P., Birkmann, J., Kok, K., Levy, M., & Solecki, W. (2017). The roads ahead: Narratives for shared socioeconomic pathways describing world futures in the 21st century. *Global Environmental Change*, *42*, 169–180. DOI: 10.1016/j.gloenvcha.2015.01.004

Ouraich, I., & Tyner, W. E. (2014) Climate Change Impacts on Moroccan Agriculture and the Whole Economy: An Analysis of the Impacts of the Plan Maroc Vert in Morocco. WIDER Working Paper 2014/083. Helsinki: UNU-WIDER. https://doi.org/DOI: 10.35188/UNU-WIDER/2014/804-9

Ozdemir, D. (2022). The impact of climate change on agricultural productivity in Asian countries: A heterogeneous panel data approach. *Environmental Science and Pollution Research International*, *29*(6), 8205–8217. DOI: 10.1007/s11356-021-16291-2

Rosenzweig, C., & Parry, M. L. (1994). Potential impact of climate change on world food supply. Nature, 367(6459), 133–138. https://doi.org/10.1038/367133a0

Pellicone, G., Caloiero, T., & Guagliardi, I. (2019). The De Martonne aridity index in Calabria (Southern Italy). *Journal of Maps*, *15*(2), 788–796. DOI: 10.1080/17445647.2019.1673840

Pesaran, M. H. (2006). Estimation and Inference in Large Heterogeneous Panels with a Multifactor Error Structure. *Econometrica*, *74*(4), 967–1012. DOI: 10.1111/j.1468-0262.2006.00692.x

Pesaran, M. H. (2007). A simple panel unit root test in the presence of cross-section dependence. *Journal of Applied Econometrics*, *22*(2), 265–312. DOI: 10.1002/jae.951

Pesaran, M. H. (2014). Testing Weak Cross-Sectional Dependence in Large Panels. *Econometric Reviews*, *34*(6-10), 1089–1117. DOI: 10.1080/07474938.2014.956623

Pesaran, M. H., Shin, Y., & Smith, R. P. (1999). Pooled Mean Group Estimation of Dynamic Heterogeneous Panels. *Journal of the American Statistical Association*, *94*(446), 621–634. DOI: 10.1080/01621459.1999.10474156

Pickson, R. B., Boateng, E., Gui, P., & Chen, A. (2024). The impacts of climatic conditions on cereal production: Implications for food security in Africa. *Environment, Development and Sustainability*, *26*(7), 18333–18360. DOI: 10.1007/s10668-023-03391-x

Salami, A. A., & Babatunde, O. R. (2024). Environmental Challenges, The Impacts of Climate Change in North Africa Region: A review. *Earth and Environmental Sciences Library*, 281–294. https://doi.org/DOI: 10.1007/978-3-031-58315-5_14

Schilling, J., Freier, K. P., Hertig, E., & Scheffran, J. (2012). Climate change, vulnerability and adaptation in North Africa with focus on Morocco. *Agriculture, Ecosystems & Environment*, *156*, 12–26. DOI: 10.1016/j.agee.2012.04.021

State of the Climate in Africa 2023. (2024, November 26). World Meteorological Organization. https://wmo.int/publication-series/state-of-climate-africa-2023

Stern, N. (2007). *The economics of climate change*. https://doi.org/DOI: 10.1017/CBO9780511817434

Stevanović, M., Popp, A., Lotze-Campen, H., Dietrich, J. P., Müller, C., Bonsch, M., Schmitz, C., Bodirsky, B. L., Humpenöder, F., & Weindl, I. (2016). The impact of high-end climate change on agricultural welfare. *Science Advances*, *2*(8), e1501452. Advance online publication. DOI: 10.1126/sciadv.1501452

Supharatid, S., & Nafung, J. (2021, November 1). Projected drought conditions by CMIP6 multimodel ensemble over Southeast Asia. *Journal of Water and Climate Change*, *12*(7), 3330–3354. DOI: 10.2166/wcc.2021.308

Taabni, M., & Jihad, M. E. (2012). Eau et changement climatique au Maghreb : Quelles stratégies d'adaptation? *Les Cahiers d'Outre-Mer*, *65*(260), 493–518. DOI: 10.4000/com.6718

The Economics of Climate Change: The Stern Review - Grantham Research Institute on climate change and the environment. (2024, August 19). Grantham Research Institute on Climate Change and the Environment. https://www.lse.ac.uk/granthaminstitute/publication/the-economics-of-climate-change-the-stern-review/

The World Bank. (2018). Beyond scarcity: water security in the Middle East and North Africa. (2018). In *Washington, DC: World Bank eBooks*. https://doi.org/DOI: 10.1596/978-1-4648-1144-9

United Nations Development Programme (UNDP). (2023). Annual Report 2022. New York: UNDP. https://annualreport.undp.org/2022/assets/Annual-Report-2022-FR.pdf

Vatankhah, T., Moosavi, S. N., & Tabatabaei, S. M. (2019). The economic impacts of climate change on agriculture in Iran: A CGE model analysis. *Energy Sources. Part A, Recovery, Utilization, and Environmental Effects*, *42*(16), 1935–1949. DOI: 10.1080/15567036.2019.1604903

Waha, K., Krummenauer, L., Adams, S., Aich, V., Baarsch, F., Coumou, D., Fader, M., Hoff, H., Jobbins, G., Marcus, R., Mengel, M., Otto, I. M., Perrette, M., Rocha, M., Robinson, A., & Schleussner, C.-F. (2017). Climate change impacts in the Middle East and Northern Africa (MENA) region and their implications for vulnerable population groups. *Regional Environmental Change*, *17*(6), 1623–1638. DOI: 10.1007/s10113-017-1144-2

Wahid, N., Bouazzama, B., Yousfi, H. E., & Zascerinska, J. (2022). A HISTORICAL OVERVIEW OF EVOLUTION OF THE IRRIGATED AGRICULTURE SECTOR IN MOROCCO. *Journal of Regional Economic and Social Development*, *14*, 109–118. DOI: 10.17770/jresd2022vol14.6896

Xie, H., Wen, Y., Choi, Y., & Zhang, X. (2021). Global Trends on Food Security Research: A Bibliometric Analysis. *Land (Basel)*, *10*(2), 119. DOI: 10.3390/land10020119

Yaqoob, N., Ali, S. A., Kannaiah, D., Khan, N., Shabbir, M. S., Bilal, K., & Tabash, M. I. (2023). The effects of Agriculture Productivity, Land Intensification, on Sustainable Economic Growth: A panel analysis from Bangladesh, India, and Pakistan Economies. *Environmental Science and Pollution Research International*, *30*(55), 116440–116448. DOI: 10.1007/s11356-021-18471-6

Zaki, N. A., Kløve, B., & Haghighi, A. T. (2022). Expanding the irrigated areas in the MENA and Central Asia: Challenges or opportunities? *Water (Basel)*, *14*(16), 2560. DOI: 10.3390/w14162560

Zhang, Y., Ahmad, M. I., Shen, Q., Boota, M. W., Zhao, C., & Ma, H. (2025). Unveiling the environmental impact of corn production in China: Evidence from panel ARDL approach. *Frontiers in Sustainable Food Systems*, *8*, 1492262. Advance online publication. DOI: 10.3389/fsufs.2024.1492262

KEY TERMS AND DEFINITIONS

Climate vulnerability: is the degree to which a system is likely to be hurt by the bad effects of climate change, which includes changes and peaks.

The ARDL-PMG Model: is an econometric tool used to look at both long-term and short-term changes in panel data models with different short-run factors for each country.

Food Security: A condition in which all individuals have consistent access to food that is safe, nutritious, and sufficient to meet their dietary needs.

De Martonne Index: A climatic metric that is used to assess aridity and is derived from temperature and precipitation data.

Socioeconomic Pathways (SSPs): Scenarios that delineate various global trajectories that influence greenhouse gas emissions and global climate outcomes.

Agricultural Capital: The tangible and monetary resources that are utilized in agricultural production, including livestock, infrastructure, and machinery.

Adaptation Strategies: Strategies that are implemented to enhance the resilience of agricultural systems and mitigate the negative consequences of climate change.

ENDNOTES

1 See Climatic Index and Projection Scenarios, subsection 2, in the "Data" section

2 Refer to the next section (Climate Change Impact Assessment) for details on the different SSP narrative trajectories.

Chapter 5
Economic Impacts of Climate Change on Agriculture in the North African Region:
A Ricardian Multi-Regional Analysis

Amine M. Benmehaia
https://orcid.org/0000-0001-8110-1601
University of Biskra, Algeria

Ali Chebil
University of Carthage, Tunisia

Amine Oulmane
https://orcid.org/0000-0001-6145-2965
CREAD, Algiers, Algeria

ABSTRACT

This chapter aims to examine the impacts of climate change on agriculture in North Africa using an integrated Ricardian-Production Function approach. Focusing on 5 countries: Morocco, Algeria, Tunisia, Libya, and Egypt, the analysis combines panel data (1980-2020) with climate projections under three Shared Socioeconomic Pathways (SSP1-2.6, SSP2-4.5, and SSP3-7.0), to quantify effects on agricultural value added. The model incorporates key variables such as temperature, precipitation, capital investment, labor, and irrigated area. The main findings reveal that: while optimistic scenarios (SSP1-2.6) show manageable declines of 15-20%

DOI: 10.4018/979-8-3373-3346-5.ch005

in inter-annual rate of change of agricultural added value, pessimistic pathways (SSP3-7.0) precipitate catastrophic losses exceeding 130% in vulnerable countries. Egypt's high adaptive capacity can mitigate up to 90% of potential losses, while Libya and Algeria remain highly vulnerable. Precipitation levels emerge as the most critical factor, although the cumulative effect of rising temperatures could become significant under pessimistic scenarios.

1. INTRODUCTION

Climate change represents one of the most significant challenges to sustainable development in the 21st century, with profound implications for agricultural systems worldwide (Ghadge *et al.*, 2020). The agricultural sector, being inherently dependent on climatic conditions, faces unprecedented risks from rising temperatures, shifting precipitation patterns, and increasing frequency of extreme weather events (Gomez-Zavaglia *et al.*, 2020). These challenges are particularly acute in the North African region, which stands out as one of the most climate-vulnerable regions globally, with projections indicating that it will experience temperature increases exceeding the global average and significant reductions in precipitation (IPCC, 2022). Indeed, climate models consistently predict that rising temperatures and declining rainfall will reduce crop yields, diminish water availability, and exacerbate land degradation, thereby threatening food production systems and rural economies (Sowers *et al.*, 2011). The region's heavy reliance on agriculture for livelihoods, employment, and food security -approximately 14-30% of employment and 10-15% of GDP in North African countries (World Bank, 2021)- makes it especially susceptible to these climatic shifts. This highlights the importance of protecting the sector to ensure economic and social stability.

The economic assessment of climate change impacts on agriculture has been a focal point of environmental and resource economics, with the Ricardian approach emerging as a prominent methodological framework. Developed by Mendelsohn *et al.* (1994), the Ricardian model estimates the economic value of climate impacts by examining how variations in climatic conditions affect agricultural land values or revenues. Unlike traditional models that focus on crop-specific yield responses, the Ricardian approach captures farmers' adaptive behaviors by analyzing revealed preferences in land markets, thereby providing an ample assessment of climate impacts (Mendelsohn & Dinar, 2009). This method has been widely applied in diverse contexts, including Sub-Saharan Africa (Kurukulasuriya *et al.*, 2006; Kabubo-Mariara & Karanja, 2007; Serdeczny *et al.*, 2017; Kurukulasuriya & Mendelsohn, 2007), South Asia (Sanghi & Mendelsohn, 2008; Abidoye *et al.*, 2017; Hossain *et al.*, 2019), and Latin America (Seo & Mendelsohn, 2008). However, its application in North Africa remains limited, with few studies addressing the region's unique agro-climatic and socio-economic conditions (Eid *et al.*, 2007; Meyghani *et al.*, 2021; Alboghdady & Hendawy, 2016; Migliore *et al.*, 2019). Existing research on climate change and agriculture in North Africa has primarily focused on biophysical impacts, such as crop yield projections (Schilling *et al.*, 2012), or qualitative assessments of vulnerability (Verner, 2012), others addressed the economic valuation of these impacts and the role of adaptation strategies (Frija et al., 2021).

There is a growing body of research on the impacts of and adaptation to climate change. However, these impacts vary greatly depending on the local context. To better understand these disparities, this study compares the effects of climate change across different regions in North Africa, highlighting how geographic and socioeconomic differences shape vulnerabilities and adaptation needs. To do so, we employ an integrated Ricardian-Production Function approach in order to quantify the economic impacts of climate change on agriculture across North Africa. The analysis extends the traditional Ricardian model by incorporating production function theory, which explicitly accounts for the role of inputs such as capital, labor, and irrigation in moderating climate effects. Furthermore, a composite index that measures national adaptive capacities, drawing on indicators such as governance quality, irrigation infrastructure, and technological innovation was introduced to capture the disparities in terms of adaptive potential.

This hybrid approach enables a more nuanced understanding of how climatic factors interact with agricultural inputs to influence productivity, thereby offering insights into potential adaptation pathways. The study utilizes a panel dataset spanning five North African countries from 1980 to 2022, combining climate variables (temperature and precipitation) with economic data on agricultural value added, capital formation, labor, and irrigated area. By leveraging this panel dataset, the analysis captures both the direct effects of climate change and the adaptive capacity of agricultural systems. Furthermore, the study projects future impacts under three Shared Socioeconomic Pathways (SSPs)—optimistic (SSP1-2.6), central (SSP2-4.5), and pessimistic (SSP3-7.0)—to assess how different warming trajectories may affect agricultural productivity by 2050. These projections are complemented by scenario analyses that evaluate the potential of adaptation measures, such as expanded irrigation and improved institutional capacity, to mitigate climate damages.

The implications of this research extend beyond academia, offering actionable recommendations for stakeholders at local, national, and regional levels. By quantifying the economic costs of climate inaction, the study underscores the urgency of investing in water-efficient technologies, institutional reforms, and cross-border collaboration. The chapter also contributes to the broader literature on climate-economy interactions by refining the Ricardian model to better capture regional specificities, such as the role of irrigation in mitigating heat stress and the diminishing returns of capital investment under extreme climate conditions. Ultimately, this chapter aims to bridge the gap between scientific knowledge and policy implementation, providing an evidence-based roadmap for safeguarding North Africa's agricultural sector in an increasingly uncertain climate future.

This chapter is organized as follows: Section 2 is a regional overview of the climate change and agriculture in North Africa. Section 3 presents the research methodology, detailing the integrated Ricardian-Production Function model, data sources, and empirical specifications. Section 4 analyzes the results, including descriptive statistics, regression estimates, scenario-based projections of climate impacts under varying adaptation strategies, and policy implications, emphasizing region-specific recommendations for enhancing agricultural resilience. Finally, Section 5 concludes by synthesizing key findings, addressing limitations, and proposing avenues for future research.

2. CLIMATE CHANGE AND AGRICULTURE IN NORTH AFRICA: A REGIONAL OVERVIEW

North Africa's agricultural sector stands at the frontline of climate change impacts, with its predominantly arid and semi-arid ecosystems facing unprecedented environmental stress. The region, comprising Morocco, Algeria, Tunisia, Libya, and Egypt, has experienced significant climatic shifts since the mid-20th century, marked by declining precipitation, rising temperatures, and increased frequency of extreme weather events (Schilling *et al.*, 2012; Waha et al., 2017; Sowers *et al.*, 2011; Salami & Babatunde, 2024). These changes have profound implications for food security, water resources, and rural livelihoods across the region.

Since the 1970s, North Africa has undergone rapid demographic and economic transformations that have intensified pressure on its fragile agro-ecological systems. Population growth, urbanization, and agricultural expansion have strained already limited water resources, while climate change has exacerbated these pressures. Historical climate data indicate a steady decline in annual rainfall, particularly in the Maghreb, where precipitation has become more erratic, disrupting traditional farming cycles (Hamed *et al.*, 2024; Pickson & Boateng, 2022; World Bank, 2021). Temperature increases have also been significant, with the Intergovernmental Panel on Climate Change (IPCC) projecting a rise of 2–4°C by 2100 for the region, further accelerating evaporation rates and soil moisture loss (IPCC, 2022). The following Figure 1 demonstrate the upward trend (starting from 1991) in temperatures changes in the five North African countries (1961-2024) according to FAO statistics.

Figure 1. Temperature changes (°C) in the five North African countries (1961-2024) (Source: Based on data from FAO statistics, 2025).

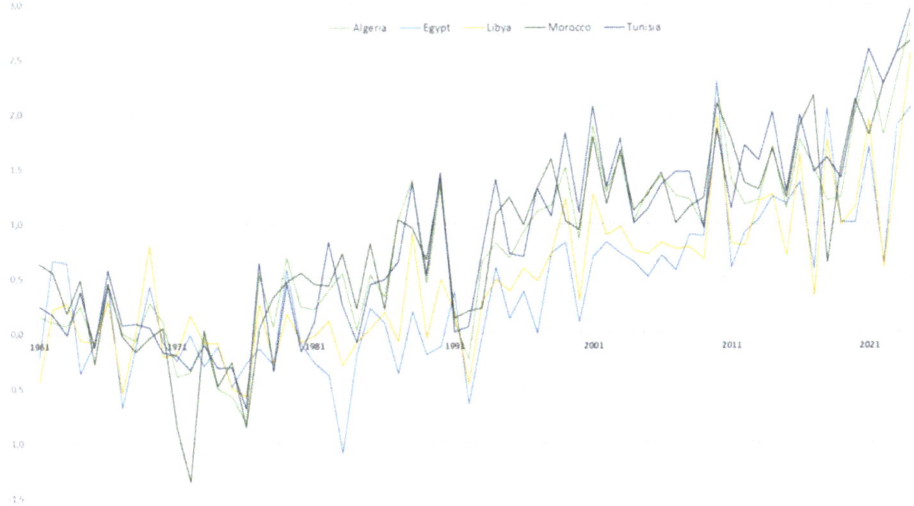

Water scarcity is the most pressing challenge. Renewable water resources per capita have plummeted from approximately 1 200 m³/year in the 1960s to just 548 m³/year by 2010—well below the threshold of 1 000 m³/year deemed necessary for sustainable development by the World Health Organization (Falkenmark et al., 1989). Algeria, for instance, faces a critical situation, with projections suggesting its water availability could drop to 282 m³ per capita without substantial intervention (Oxford Analytica, 2024). Egypt, heavily dependent on the Nile, is also under strain due to upstream dam developments and rising demand from its rapidly growing population, which is expected to double by 2075 (FAO, 2022).

Agriculture remains a cornerstone of North Africa's economy, employing a significant portion of the workforce and contributing to food security. However, the sector is highly sensitive to climatic fluctuations. Rain-fed farming, which dominates in Morocco, Tunisia, and parts of Algeria, is particularly vulnerable to prolonged droughts and irregular rainfall. Studies indicate that cereal yields could have significant decline in Algeria and Tunisia by 2050 under current warming trends (Le Mouël *et al.*, 2023; Benmehaia & Bekkis, 2024). Similarly, olive production—a key export for Tunisia and Morocco—faces significant yield reductions due to heat stress and water shortages (Uzundumlu & Ateş, 2025; Fernández-Lobato *et al.*, 2022; World Bank, 2021). Irrigation, though critical for stabilizing production, is increasingly unsustainable. Groundwater depletion is rampant, especially in Egypt's Nile Delta and Morocco's Souss-Massa region, where over-extraction has led to saltwater intrusion and declining aquifer levels (FAO, 2022). Libya's agricultural collapse amid ongoing conflict further illustrates how political instability exacerbates climate vulnerabilities, with farmers abandoning lands due to water shortages and damaged infrastructure (ICRC, 2022).

North African governments have adopted varying approaches to climate adaptation, reflecting divergent economic and political contexts. Morocco has emerged as a regional leader, implementing the Green Morocco Plan in 2008 to promote climate-smart agriculture and renewable energy, with ambitions to derive 52% of its electricity from renewables by 2030 (CAT, 2023). Tunisia, meanwhile, has integrated climate resilience into its national development strategy, emphasizing drought-resistant crops and water-efficient irrigation (Verner et al., 2018).

However, structural barriers persist. Algeria's irrigation potential remains underutilized, with only 0.62 million hectares irrigated despite a theoretical capacity of 2.2 million hectares—a stark contrast to Morocco's 1.98 million hectares (FAO, 2022). Egypt's focus on large-scale, capital-intensive agricultural projects has drawn criticism for neglecting smallholder farmers, who are disproportionately affected by climate shocks (Perez *et al.*, 2021; Gabr, 2023).

2.1. Morocco: A regional leader facing persistent challenges

Morocco's agricultural sector has undergone significant transformations since the 1970s, shaped by both climatic shifts and evolving policy approaches. In the post-independence era, the government prioritized cereal self-sufficiency through state-led irrigation projects and input subsidies, particularly in fertile regions like the Gharb and Haouz plains. However, these efforts were hampered by volatile rainfall and inefficient water use, leading to recurring droughts in the 1980s and 1990s (World Bank, 2021). The structural adjustment programs of the 1990s further disrupted the sector by reducing subsidies, exposing smallholder farmers to market instability (Akesbi, 2012).

By the 2000s, Morocco recognized the need for a more resilient agricultural model. The launch of the Plan Maroc Vert (Green Morocco Plan, PMV) in 2008 marked a turning point, aiming to modernize agriculture while addressing climate vulnerabilities. The PMV's dual-pillar approach combined large-scale agribusiness development with support for small-scale farmers, emphasizing drip irrigation, drought-resistant crops, and value-chain integration (CAT, 2023). Morocco's climate is increasingly characterized by rising temperatures and declining precipitation, particularly in the south and interior regions (Filahi *et al.*, 2016 ; Lachgar *et al.*, 2022). The 2015–2016 drought—the worst in 30 years—reduced cereal yields by 70%, slashing GDP growth by 1.5% and underscoring the sector's exposure to climatic shocks (World Bank, 2021).

Water scarcity is acute: renewable resources have dwindled to 500 m^3 per capita annually, with groundwater overexploitation causing aquifer depletion in critical areas like the Souss-Massa basin (Attar *et al.*, 2024 ; ABHSM, 2021). The Al Wahda and Bin El Ouidane dams, key to irrigation, often operate below capacity due to recurrent droughts (OECD, 2020). Despite progress, disparities persist. Smallholder farmers—constituting 80% of the sector—often lack access to finance and technology, limiting their adaptive capacity (FAO, 2022). Groundwater depletion continues unabated in regions like the Saïss Plain, where illegal wells proliferate (Closas & Villholth, 2016). Climate models predict yield declines of 10–30% for staples like wheat by 2050, threatening rural livelihoods (IPCC, 2022).

2.2. Algeria: The permanent struggling with water scarcity and policy gaps

Algeria's agricultural sector has faced systemic challenges since the 1970s, shaped by post-independence land reforms, erratic climate conditions, and fluctuating economic policies. During the 1970s–1980s, the government prioritized food self-sufficiency through state-run farms and cereal subsidies, particularly for durum wheat—a staple crop. However, these efforts were undermined by recurrent droughts, inefficient irrigation, and bureaucratic mismanagement (Bessaoud, 2016; Benmehaia & Bekkis, 2024). The 1986 oil price collapse and subsequent Structural Adjustment Programs (SAPs) in the 1990s further weakened the sector, as subsidies for fertilizers and equipment were slashed, leaving small farmers vulnerable (Oulmane *et al.*, 2022; World Bank, 2021).

Algeria's climate is among the most arid in North Africa, with 80% of its territory classified as desert or semi-desert. Since the 1970s, rainfall has decreased by 20–30% in key agricultural regions like the High Plains, while temperatures have risen by 1.8°C (Mokhtari *et al.*, 2025). The 2018–2021 drought—the worst in 50 years—devastated cereal production, forcing Algeria to import over 70% of its wheat needs (Benmehaia & Bekkis, 2024; FAO, 2022).

Water scarcity is critical: renewable water resources stand at about 11 billion m^3/year, but only 3.8 billion m^3 are mobilizable due to infrastructure limitations (FAO, 2022). Per capita water availability has plummeted to 180 m^3/year—far below the 500 m^3 threshold for absolute scarcity (FAO, 2022). Overexploitation of groundwater has caused aquifer levels to drop significantly in many regions (Kherbache & Molle, 2022; Oulmane *et al.*, 2022).

The government's response has been slow and fragmented. Algeria ratified the UNFCCC and submitted National Communications on climate change, but implementation of adaptation measures has lagged. Subsidies for agricultural inputs, once a pillar of state support, were scaled back under IMF structural adjustment programs in the 1990s, leaving farmers exposed to market fluctuations. State-led irrigation projects have struggled with inefficiency (Oulmane et al., 2022), while groundwater depletion worsens in key farming regions like the High Plains. Unlike Morocco, Algeria has not pursued large-scale renewable energy integration in agriculture, missing opportunities to reduce reliance on fossil-fuel-powered water pumps.

Political and economic instability further complicates climate adaptation. Hydrocarbon revenues, which dominate Algeria's economy, have historically overshadowed agricultural development. Rural areas suffer from underinvestment, pushing small farmers toward informal water pumping or abandonment of land. The 2021 forest fires, exacerbated by drought, highlighted the country's vulnerability to extreme climate events. Without significant reforms in water governance and agricultural modernization, Algeria's ability to adapt to worsening climatic conditions remains uncertain.

2.3. Tunisia: A struggle against aridification

Tunisia's agricultural sector sits at the crossroads of climate vulnerability and economic necessity. The country's Mediterranean climate, historically favorable for diversified farming, has been undergoing rapid aridification since the late 20th century. Between 1970 and 2023, average temperatures increased by 1.4°C while annual rainfall decreased by 15-20% (World Bank, 2024b), particularly in critical cereal-growing regions of the central and southern interior (Attiaoui & Boufateh, 2019; Ouessar *et al.*, 2021). This climatic shift has compressed Tunisia's already narrow window for rain-fed agriculture, forcing farmers to confront unprecedented challenges in water management and crop viability.

The country's water crisis presents the most immediate threat to agricultural sustainability. Tunisia's renewable water resources have dwindled to 400 m^3 per capita annually, well below the 500 m^3 threshold for absolute water scarcity (FAO, 2022). Traditional water harvesting systems like the Jessour and Tabias that sustained agriculture for centuries are proving inadequate against prolonged droughts. The Siliana and Kairouan governorates, once breadbaskets of cereal production, have seen groundwater tables drop significantly due to over-extraction for irrigation (Bachta & Ben Nasr, 2020; Gannouni *et al.*, 2023). This unsustainable water use has led to increasing soil salinity, particularly in coastal areas where seawater intrusion affects 40% of irrigated lands (FAO, 2022).

Olive oil production, Tunisia's agricultural crown jewel accounting for 20% of total agricultural exports, faces existential threats from climate change (Zaied & Zouabi, 2016). Research indicates potential yield reductions of 32% by 2050 under current warming trends. The 2022 drought caused olive production to plummet by 60%, dealing a severe blow to rural economies where over 1 million families depend on olive cultivation (FAO, 2022). Cereal crops, particularly durum wheat, show even greater vulnerability with projected declines by mid-century (Gannouni *et al.*, 2023; World Bank, 2021). These impacts are disproportionately affecting smallholder farmers who lack access to irrigation and climate adaptation technologies.

Tunisia's policy response has evolved through successive national strategies. The 2012 Climate Change Adaptation Strategy marked the first comprehensive attempt to address agricultural vulnerabilities, followed by the 2015 Nationally Determined Contribution (NDC) (Shaheen, 2025) that prioritized climate-smart agriculture. The 2021 updated NDC introduced more ambitious targets, including the rehabilitation of 200,000 hectares of degraded agricultural land and the adoption of drought-resistant crop varieties. However, implementation has been hampered by bureaucratic inefficiencies and limited funding. The Agricultural Investment Promotion Agency (APIA) reports that only 35% of planned adaptation measures have been fully executed as of 2023 (APIA, 2023).

Regional disparities in climate vulnerability have become increasingly pronounced. While the northern coastal belt benefits from relatively stable rainfall, central and southern Tunisia face accelerating desertification. The Tataouine and Kebili regions have lost 18% of arable land since 2000 due to combined effects of drought and soil degradation (FAO, 2022). This geographical imbalance is exacerbating rural-urban migration, with agricultural labor force participation dropping from 30% in 1990 to 15% in 2024 (World Bank, 2024b). The social consequences are particularly acute in marginalized interior regions where agriculture remains the primary livelihood.

Tunisia's experience highlights the complex interplay between climate change, water governance, and agricultural sustainability. While the country has demonstrated awareness of climate risks through policy frameworks, the translation of these commitments into on-ground adaptation remains inconsistent. The growing frequency of climate shocks - from the unprecedented 2015 drought to the 2021 heatwaves that reduced fruit yields by 30% (World Bank, 2024b) – continues to test the resilience of Tunisia's agricultural systems. Without significant improvements in water use efficiency and targeted support for small-scale farmers, the sector faces mounting pressures that could undermine both food security and rural stability.

2.4. Libya: A collapsing sector amid conflict and aridification

Libya's agricultural sector, already operating at the margins of viability in one of the world's most arid countries, has been pushed to the brink by the combined pressures of climate change and prolonged conflict. The country's arable land, constituting less than 1% of its total territory, has faced progressive degradation since the 1970s due to declining rainfall and unsustainable irrigation practices. Annual precipitation, historically concentrated in the northern coastal Jefara plain and Jabal al-Akhdar highlands, has decreased by 20-30% since 1980, with drought frequency doubling in recent decades (World Bank, 2024a). This climatic shift has devastated rain-fed agriculture, which previously accounted for 65% of Libya's cereal production (FAO, 2022).

The country's water crisis epitomizes its agricultural challenges. Libya's renewable water resources stand at a critically low 126 m^3 per capita annually, forcing near-total reliance on fossil groundwater from the Nubian Sandstone Aquifer System (AFDB, 2025). The Great Man-Made River project, initiated in the 1980s to transport fossil water from southern aquifers to coastal farms, has suffered severe damage during the 2011 revolution and subsequent conflicts, with pipeline ruptures reducing significantly water delivery (AFDB, 2025). Farmers in traditional agricultural zones like Awiniya now face impossible choices, as destroyed infrastructure and soaring fuel prices make water pumping prohibitively expensive. Olive oil production, a cultural and economic mainstay, has seen prices triple since 2020 due to drought-induced yield collapses (Razek & Othman, 2023).

Conflict has multiplied climate vulnerabilities. The 2011 uprising and ensuing instability disrupted agricultural input supply chains, with the number of functional tractors and harvesters falling by 70% between 2010-2020. The abandonment of farms has become widespread, particularly in the western mountain regions where terraced agriculture dating to Roman times is being lost to erosion. The International Committee of the Red Cross documents cases of farmers returning after years of displacement to find orchards dead and topsoil blown away (ICRC, 2022). This breakdown has forced Libya to import 75% of its food needs, creating dangerous dependencies on global markets (World Bank, 2024a). Coastal agriculture faces distinct climate threats. Sea level rise of 3.2 mm annually is causing saltwater intrusion in the vital Jefara plain, where large parts of irrigated lands show elevated salinity levels (Atkinson, 2023; Zurqani, 2021). Paradoxically, when rare heavy rains occur, they trigger destructive floods due to degraded watershed management - the 2019 Al-Bayda floods destroyed 14 000 hectares of crops (FAO, 2022). These compounding stressors have collapsed agricultural employment from 24% of the workforce in 2000 to just 8% in 2023.

Libya's institutional collapse has left climate adaptation virtually nonexistent. The country ranks among the world's least prepared nations for climate hazards, with no functioning national agricultural policy since 2014 (IFAD, 2022). Localized efforts by international organizations, such as the ICRC's support for soil analysis training in Misrata, provide minimal relief against systemic collapse (ICRC, 2022). The agricultural research system, once robust with 22 experimental stations, now operates at 15% capacity. This institutional vacuum leaves Libya uniquely vulnerable as climate projections indicate temperature increases of 3-4°C by 2100 - among the highest in the Mediterranean region (IPCC, 2022). The case of Libya demonstrates how conflict amplifies climate impacts to catastrophic effect. Where neighboring countries have implemented adaptation strategies, Libya's agricultural sector exists in a state of managed decline, with traditional knowledge systems eroding faster than new solutions can emerge. The 2023 olive harvest, yielding just 30% of pre-2011 levels, stands as a stark indicator of this downward trajectory. Without political stability and massive investment in water infrastructure, Libya's agricultural future appears increasingly untenable under current climate trends.

2.5. Egypt: A looming crisis in the Nile Basin

Egypt's agricultural system, historically dependent on the Nile's predictable rhythms, faces unprecedented strain from climate change and competing water demands. The country's hot desert climate, with average annual rainfall below 20mm, makes it one of the world's most water-scarce nations, relying on the Nile for 97% of its freshwater needs (FAO, 2023). Since the 1970s, Egypt's population has nearly tripled while its per capita water share has plummeted from 1 200 m^3 to 582 m^3 annually - well below the water poverty line. This demographic pressure coincides with climate shifts, including temperature rises of 1.8°C since 1960 and more frequent heatwaves that reduce crop yields while increasing evapotranspiration (IPCC, 2022).

The Nile Delta, Egypt's agricultural heartland producing 60% of the country's food, is under multiple climate threats. Sea level rise of 4.6 mm/year is causing saltwater intrusion that has already rendered 15% of Delta soils unsuitable for traditional crops (FAO, 2022). Combined with the reduced sediment flow from upstream dams (including Ethiopia's GERD), this has accelerated Delta subsidence, with projections suggesting 30% of this vital region could be lost by 2050 (FAO, 2022). Staple crops like wheat, which occupies 1.43 million hectares, face yield reductions of 15-20% under current warming trends (FAO, 2022). Rice cultivation, a key export, has been restricted in northern Delta governorates due to water scarcity, shrinking production by 35% since 2018.

Groundwater depletion compounds surface water shortages. The Western Desert's fossil aquifers, tapped for reclamation projects like Toshka, are being depleted at rates 300% faster than recharge (El-Agha *et al.*, 2024). In Fayoum, over-pumping has caused water tables to drop 2 meters annually, forcing farmers to abandon centuries-old irrigation practices (FAO, 2022). The government's response has prioritized large-scale infrastructure, including 125 desalination plants and wastewater reuse projects, but these cover less than 5% of agricultural needs (Elsaie *et al.*, 2023).

Policy approaches reflect these tensions. The 2014 Constitution's Article 44 enshrined the right to Nile water, while the 2030 Sustainable Development Strategy emphasizes vertical farming and crop pattern modifications. However, implementation remains inconsistent, with smallholder farmers (constituting more than 80% of agricultural holdings) lacking access to climate adaptation technologies (Sattar *et al.*, 2024). The 2021 National Climate Change Strategy identified agriculture as highly vulnerable, but allocated only 12% of adaptation funding to the sector (FAO, 2022). Structural inequalities exacerbate climate vulnerabilities. Historic land concentration - where 5% of farmers control 47% of arable land - means adaptation resources are unevenly distributed. The 1992 tenancy law reforms displaced 1 million small farmers, pushing them onto marginal lands more exposed to climate risks. This legacy continues to shape adaptation disparities, with corporate agribusinesses investing in climate-smart technologies while smallholders resort to unsustainable groundwater use.

Egypt's agricultural future hinges on addressing these intersecting challenges. The country has demonstrated technical capacity through projects like the 1.5 million feddan reclamation initiative and salt-tolerant crop research at the Agricultural Genetic Engineering Research Institute (ICARDA, 2018). However, without equitable water governance and climate-informed land use planning, even these measures may prove inadequate against the accelerating impacts of climate change on the Nile Basin ecosystem.

3. RESEARCH METHODOLOGY

3.1. Ricardian analysis: The theoretical framework

The Ricardian approach offers a powerful economic framework for evaluating how climatic conditions shape agricultural productivity by examining their influence on land values and farm revenues. Originally developed by Mendelsohn *et al.* (1994), this method revolutionized climate impact studies by moving beyond crop-specific assessments to capture farmers' adaptive behaviors through revealed preferences in land markets. In this study, we enrich this framework by integrating it with production function theory, in order to create a more detailed analytical tool that addresses the region's challenges.

The theoretical foundation rests on two complementary economic principles. First, the Ricardian principle posits that farmland values reflect the discounted stream of future profits, which are inherently climate-dependent. Second, neoclassical production theory asserts that agricultural output depends on inputs like capital and labor interacting with environmental conditions. By synthesizing these perspectives, our integrated model accounts for both climate effects and input productivity simultaneously.

The approach builds on several key assumptions supported by empirical evidence: Farmers optimize land use based on prevailing climatic and economic conditions (Mendelsohn & Dinar 2009); Input adjustments (e.g., irrigation expansion) partially mitigate climate stresses (Dinar *et al.*, 1998); and Market prices reflect the marginal productivity of land under different climate regimes. However, the model can predict mainly three following key hypotheses in our context: H_1: Temperature increases exhibit nonlinear effects, with accelerating yield losses beyond optimal thresholds, H_2: Irrigation significantly moderates heat stress impacts, H_3: Capital investments show diminishing returns under extreme climate conditions

The integrated Ricardian-Production Function takes the form

$$y_{it} = \alpha + \beta_i \sum \mathbb{C}_{it} + \gamma_i \sum \mathbb{X}_{it} + \theta_i \sum \mathbb{I}_{it} + \epsilon_{it}$$

where y_{it} represents the agricultural productivity (or a proxy for it), $\mathbb{C}_{it}$ for climate variables, $\mathbb{X}_{it}$ for inputs variables, $\mathbb{I}_{it}$ for interactions terms, and ϵ_{it} is an i.d.d. error term. The model uniquely combines the strengths of two established economic approaches: The *Ricardian component* (climate variables) captures how long-term climatic conditions affect land productivity through farmer adaptation, and the *production function component* (input variables) accounts for how conventional factors of production mediate these climate impacts.

3.2. Data structure and sources

The study employs a yearly country-level panel data for five North African countries, namely: Morocco, Algeria, Tunisia, Libya, and Egypt. The time frame is spanning from 1980 to2022. The dataset combines economic variables (value added of agriculture, gross fixed capital formation, farm labor force, area equipped for irrigation), and two main climate variables (temperature and precipitation), as aggregated annual mean. Table 1 display the main variables and their respective sources, units, and respective symbols.

Table 1. Used variables and sources

Variable label	*Sources*	*Unit*	*Symbol*
Value Added of Agriculture	FAO (2024)	Billion US$, 2015 prices	*y*
Gross Fixed Capital Formation	FAO (2024)	US$, 2015 prices	*C*
Farm Labor Force	AOAD (2024)	Million persons	*L*
Area equipped for irrigation	FAO (2024)	Million Ha	*I*
Temperature	WorldClim (2024)	°C	*T*
Precipitation	WorldClim (2024)	mm	*P*

3.3. Model specification

This study employs an integrated Ricardian-Production Function approach to estimate the economic impacts of climate change on agriculture in North Africa. The Ricardian model assesses how variations in climate (temperature and precipitation) affect farmland productivity, proxied by value added of agriculture. The production function component explicitly links agricultural output to key inputs (capital, labor, and irrigation), capturing both direct climate effects and adaptive capacity. The baseline Ricardian model is specified as

$$y_{it} = \alpha + \beta_1 T_{it} + \beta_2 T_{it}^2 + \beta_3 P_{it} + \beta_4 P_{it}^2 + \Gamma X_{it} + \delta_i + \epsilon_{it}$$

where y_{it} represents the value added of agriculture in country i and year t, T_{it} and P_{it} are the mean temperature, and total precipitation respectively, quadratic terms (T_{it}^2, P_{it}^2) capture nonlinear thresholds, $\boldsymbol{X}_{it}$ represents a vector of production inputs, namely the capital (C_{it}) proxied by the gross fixed capital formation, farm labor force (L_{it}), and the area equipped for irrigation (I_{it}). The term δ_i reflects the country fixed effects (controls for time-invariant heterogeneity, mainly soil quality), and ϵ_{it} is the i.d.d. error term. To evaluate how irrigation mitigates climate risks, we include an interaction term:

$$\beta_5 (T_{it} \times I_{it})$$

as a negative β_5 would suggest that irrigation reduces heat-related losses. The study uses many variants of panel regression estimations. For model validation, multitude of fitness and performances will be assessed, along with two specification tests, namely: Hausman test, multicollinearity, along with a Pooled OLS estimation as it will serve as a baseline but likely biased due to unobserved heterogeneity.

The analysis employs five specifications namely: the Pooled OLS as a naive baseline, the Fixed Effects (FE) to control for unobserved country heterogeneity, the Random Effects (RE) for efficiency under orthogonal individual effects, the Panel-Corrected Standard Errors (PCSE) to address cross-sectional dependence, and the Fixed Effects with Robust Standard Errors (FER) as the preferred specification that simultaneously accounts for country fixed effects, heteroskedasticity, and autocorrelation.

3.4. Scenario simulation procedure

This study employs a structural scenario simulation approach to project climate change impacts on agricultural value added across North African countries. The simulation framework combines econometric estimates from the FER panel regression estimations with climate projection data to generate forward-looking scenarios under different warming pathways.

The simulation processes three core components through sequential computational steps: climate scenario inputs, production function calibration, and adaptation module. For *climate scenario inputs*, we have:

– *Temperature projections*: Derived from CMIP6 ensemble means for SSP1-2.6 (optimistic), SSP2-4.5 (central), and SSP3-7.0 (pessimistic) scenarios (Table 3), linearly interpolated to annual country-level changes from 2020-2050.
– *Precipitation multipliers*: Applied as percentage changes to historical baselines (1990-2020) from WorldClim v2.1, with region-specific scaling factors (optimistic: -3%; central: -7%; pessimistic: -12%) (Table 3).

For the *production function calibration*, the simulation implements the validated Fixed Effects with Robust Standard Errors (FER) specification, where variables are transformed per their empirical treatment. Finally, as an *adaptation module*, the country-specific impacts are adjusted by the formula:

$$y_{t\_adapt} = y_{t\_baseline} \times [1 - (\varphi_i \times I_i)]$$

where φ_i represents the assigned adaptation capacity, and I_i is the irrigation modifier $(1/(1+\ln(I_i + 1)))$ to reflect diminishing returns.

The adaptation capacity scores assigned to each country (ranging from 0 to 1) reflect a composite assessment of institutional, infrastructural, and technological factors that influence their ability to mitigate climate impacts on agriculture. These weights were derived from a structured evaluation of three key dimensions: (1) institutional governance, (2) irrigation infrastructure coverage, and (3) agricultural innovation capacity. Table 2 provides the scoring system in weighting adaptation capacity scores for each country.

Table 2. Scoring assumptions of adaptation capacity

Country	*Score*	*Key rationale and determinants*
Morocco	0.7	Large-scale irrigation projects (Green Morocco Plan), uneven rural adoption, low R&D spending
Algeria	0.5	Limited irrigation (<10% of arable land), slower policy implementation, lower innovation investment
Tunisia	0.8	Advanced water conservation policies (e.g., wastewater reuse), moderate irrigation (14% of arable land), active research system
Libya	0.4	Political instability, degraded infrastructure, minimal adaptive governance
Egypt	0.9	High irrigation coverage (100% Nile-dependent), strong policy frameworks for water management, significant investments

The study employs a scenario-based approach to capture the range of potential climate impacts on North African agriculture by 2050. The three scenarios (Table 3)—Optimistic (SSP1-2.6), Central (SSP2-4.5), and Pessimistic (SSP3-7.0)—are constructed using a combination of global climate model (GCM) projections, regional downscaling, and empirical adjustments to align with historical agricultural responses.

Table 3. Scenario definitions for temperature and precipitation

Scenario	*Global Warming Trajectory*	*Precipitation Change*
Optimistic (SSP1-2.6)	+1.0°C (range: 0.7-1.3°C)	−3% (range: −1% to −5%)
Central (SSP2-4.5)	+1.5°C (range: 1.2-1.8°C)	−7% (range: −5% to −10%)
Pessimistic (SSP3-7.0)	+2.5°C (range: 2.0-3.0°C)	−12% (range: −9% to −15%)

The simulations of "without adaptation" strategy is performed through the baseline climate impacts are computed using the retained model coefficients (precipitation and temperature effects) but exclude the adaptation capacity adjustment, i.e. a pure climate effects. Whereas, "with adaptation" strategy adjusts impacts using country-specific capacity scores and irrigation efficacy, i.e. scaling damages by institutional and technical capacity.

The overall computational workflow is as follows:

1. Baseline calibration: Initializes 2020 values.
2. Monte Carlo Propagation: Runs 10,000 iterations per scenario, sampling from: Coefficient distributions (FER variance-covariance matrix), and climate projection uncertainty ($\pm 1\sigma$ of CMIP6 spreads).
3. Aggregation: Computes weighted regional averages using countries' 2015-2020 y shares: Egypt (30%), Morocco (25%), Algeria (20%), Tunisia (15%), and Libya (10%).
4. Validation Checks: The procedure was verified through:
    - Backcasting tests (1990-2020) showing <5% MAE versus observed y changes
    - Sensitivity analyses confirming robustness to alternative climate downscaling methods, irrigation efficacy parameters, and adaptation capacity thresholds

This implementation procedure provides an empirically grounded yet computationally tractable framework for assessing climate risks to North African agriculture, to the extent that it balances theoretical rigor with policy-relevant outputs.

4. RESULTS AND DISCUSSION

4.1. Preliminary analysis of variables and patterns

The data (Table 4) reveals Egypt as the productivity leader (max $y = 42.42$) with intensive irrigation (max $I = 3.93$) and large labor inputs (max $L = 9.26$), while Libya lags significantly (min $y = 0.97$) due to minimal irrigated area (0.23) and labor force (0.04). Algeria shows the highest capital investment (569.52), though with less efficient productivity conversion than Egypt. Climate variables indicate that Morocco records the highest annual rainfall (max $P = 556$mm) and Egypt's aridity (min $P = 10.88$mm), with temperatures ranging from Morocco's moderate 17.04°C to Egypt's subtropical 24.56°C. The wide standard deviations (y: ±9.47; P: ±128.50) underscore the region's sharp disparities, necessitating country-specific analysis in the Ricardian model to capture distinct climate adaptation pathways.

Table 4. Aggregated agricultural and climate variables across North Africa from 1980-2020

Variables	*Mean*	*S.D.*	*Min (Country)*	*Max (Country)*
VAA (y)	9,06	9,47	0,97 (Libya)	42,42 (Egypt)
Capital (C)	142,11	120,69	2,87 (Tunisia)	569,52 (Algeria)
Irrigated area (I)	1,24	1,10	0,23 (Libya)	3,93 (Egypt)
Labor (L)	2,46	2,24	0,04 (Libya)	9,26 (Egypt)
Temperature (T)	21,50	2,07	17,04 (Morocco)	24,56 (Egypt)
Precipitation (P)	144,36	128,50	10,88 (Egypt)	556,00 (Morocco)

Table 5 reveals distinct agricultural and climatic profiles across North African countries. Findings highlight Egypt's dominance in agricultural productivity ($y = 23.97 \pm 9.21$) supported by intensive irrigation ($I = 3.21 \pm 0.50$) and labor inputs ($L = 5.92 \pm 1.23$). Algeria follows with substantial capital investment ($C = 253.50 \pm 140.49$) but achieves only moderate productivity ($y = 9.83 \pm 5.83$), reflecting potential inefficiencies. Morocco's relatively high precipitation ($P = 307.37 \pm 77.56$) contrasts with Egypt's aridity ($P = 20.53 \pm 6.02$), underscoring divergent climate vulnerabilities. Libya's minimal agricultural activity ($y = 1.43 \pm 0.34$) aligns with its limited labor ($L = 0.17 \pm 0.12$) and irrigation ($I = 0.40 \pm 0.06$), while Tunisia occupies an intermediate position with modest but stable outputs ($y = 2.86 \pm 0.96$). Temperature ranges show Algeria as the warmest ($T = 23.47 \pm 0.36$) and Morocco the coolest ($T = 18.06 \pm 0.46$), with Egypt exhibiting notable variability ($T = 22.91$

± 0.61). These patterns emphasize the need for a detailed investigation on country-specific climate adaptation strategies in Ricardian modeling.

Table 5. Disaggregated agricultural and climate variables across North Africa 1980-2020

Variables	*Morocco*	*Algeria*	*Tunisia*	*Libya*	*Egypt*
y	7.20 (2.48)	9.83 (5.83)	2.86 (0.96)	1.43 (0.34)	23.97 (9.21)
C	106.29 (74.32)	253.50 (140.49)	48.76 (23.73)	99.61 (27.03)	202.39 (138.52)
I	1.44 (0.23)	0.75 (0.39)	0.39 (0.07)	0.40 (0.06)	3.21 (0.50)
L	3.61 (0.71)	1.92 (1.05)	0.70 (0.16)	0.17 (0.12)	5.92 (1.23)
T	18.06 (0.46)	23.47 (0.36)	20.44 (0.49)	22.65 (0.44)	22.91 (0.61)
P	307.37 (77.56)	83.88 (13.28)	270.76 (59.55)	39.27 (6.48)	20.53 (6.02)

In order to examine the stationarity of used variables, the Levin-Lin-Chu (LLC) test is used to determine whether panels (time series) contain unit roots, which implies non-stationarity check. Table 6 below displaying the results of the LLC unit-root tests for used variables.

Table 6. The results of the LLC unit-root tests for used variables

	Adjusted t Statistic	*p-value*	*Interpretation*
Variables in levels			
y	7.1751	1.0000	Fails to reject H_0 (strong evidence of non-stationarity).
C	1.2413	0.8928	Fails to reject H_0 (non-stationary).
I	-0.4663	0.3205	Fails to reject H_0 (non-stationary).
L	-1.4338	0.0758	Fails to reject H_0 at 5% (possible weak stationarity).
T	-2.7132	0.0033	Rejects H_0 (strong evidence of stationarity).
P	-4.3873	0.0000	Rejects H_0 (strong evidence of stationarity).
T×I	-0.70	0.2420	Fails to reject H_0 (non-stationary).
Variables in first difference			
y	-8.1887	0.0000	Rejects H_0 (strong evidence of stationarity).
C	-7.6065	0.0000	Rejects H_0 (strong evidence of stationarity).
I	-4.7279	0.0000	Rejects H_0 (strong evidence of stationarity).
L	-16.1676	0.0000	Rejects H_0 (strong evidence of stationarity).
Variables in logarithm			
T×I	-4.466	0.0000	Rejects H_0 (strong evidence of stationarity).

The LLC unit root tests reveal that y, C, I, and L exhibit non-stationarity in their levels (p-values > 0.05), meaning they contain unit roots and require first-differencing to avoid spurious regression results, while T and P are stationary (p-values < 0.05) and can be used in their raw form (in levels). After first-differencing, y, C, I, and L all show strong evidence of stationarity (p-values $\approx$ 0.000), confirming that the transformation successfully removed unit roots and making them suitable for econometric analysis. This distinction is crucial—non-stationary variables must be differenced before modeling, whereas stationary variables can be used directly without transformation. However, the interaction term $T \times I$, which appears non-stationary failing to reject the stationarity null hypothesis at $p = 0.24$, a logarithmic transformation is performed (due to the high variability). It became also stationary with strong evidence (p-values $\approx$ 0.000). Panel unit root tests were considered but deemed unnecessary for first-differenced production variables and interaction term, while climate variables are already stationary.

4.2. Analysis of regression estimations results

Table 7 presents panel regression estimates of the Ricardian-production function using panel data across five North African countries. The progression from POLS to FER demonstrates how model selection critically influences coefficient estimates, with the Hausman test ($\chi^2 = 8.11, p = 0.087$) favoring FE over RE, while robust standard errors in FER provide the most reliable inference given observed heteroskedasticity (White's $p \approx 0$) and autocorrelation (Wooldridge $p = 0.043$). This multi-model approach ensures robustness checks against specification biases while highlighting the dominance of country-specific effects ($\rho = 0.615$) in shaping agricultural productivity outcomes.

It should be noted that the underlying empirical specification requires distinct interpretation consideration for each variable type: First-differenced terms represent marginal changes influencing subsequent changes in outcome. Variables in levels capture how absolute quantities affect changes. Logarithmically transformed variable require elasticity interpretations, where percentage variations relate to unit changes in the dependent variable.

In terms of production inputs, it seems that capital investment (C) shows a positive coefficient across specifications, though its statistical significance varies considerably. The FE model yields a coefficient of 0.0086 (with no statistical significance, $p = 0.365$), which becomes marginally significant (with statistical significance, $p = 0.042$) when robust standard errors are applied in the FER specification. This pattern indicates that while capital likely contributes to productivity, its effect is modest and sensitive to model specification. The 95% confidence interval [0.0005, 0.0167] in FER suggests a potentially meaningful economic impact despite statistical ambiguity.

Table 7. Panel regression estimations of Ricardian production function for North African countries

Variables	POLS		FE		RE		PSCE		FER	
	Coef (t)	p	Coef (t)	p	Coef (z)	p	Coef (z)	p	Coef (t)	p
const.	-11.15 (-1.35)		-9.61 (-0.71)		-11.15 (-1.35)		-11.15 (-1.09)		-9.61 (-0.76)	
C	0.0048 (0.49)		0.0086 (0.82)		0.0048 (0.49)		0.0048 (0.63)		0.0086 (2.94)	***
I	-0.2385 (-0.17)		-0.1702 (-0.12)		-0.2385 (-0.17)		-0.2385 (-0.25)		-0.1702 (-0.16)	
L	-0.0507 (-0.59)		-0.0505 (-0.59)		-0.0507 (-0.59)		-0.0507 (-1.15)		-0.0505 (-0.58)	
T	0.7540 (0.95)		0.4422 (0.35)		0.7540 (0.95)		0.7540 (0.80)		0.4422 (0.44)	
P	0.0049 (2.38)	**	0.0199 (3.42)	***	0.0049 (2.38)	**	0.0049 (2.26)	**	0.0199 (1.94)	*
T^2	-0.0136 (-0.71)		-0.0037 (-0.13)		-0.0136 (-0.71)		-0.0136 (-0.63)		-0.0037 (-0.19)	
P^2	-0.0007 (-1.80)	*	-0.0002 (-3.30)	***	-0.0007 (-1.80)	*	-0.0007 (-1.23)		-0.0002 (-2.25)	**
$T \times I$	0.3803 (4.19)	***	0.0886 (0.37)		0.3803 (4.19)	***	0.3803 (5.81)	***	0.0886 (0.30)	
σ_u			0.983		0				0.983	
σ_e			0.777		0.777				0.777	
rho			0.615		0				0.615	
LL	−230		−226				−226		−226	
Adj. R^2	0.134		0.204				0.169		0.204	
R^2 within			0.092		0.055		0.092		0.092	
R^2 between			0.169		0.981				0.169	
R^2 overall	0.169		0.002		0.169				0.002	
Wald $\chi^2(5)$	6658	***	5801	***	38.88	***	154	***	5801	***
F(p)	4.859	***	2.39	**			3.70			

Notes: For asterisks, ***, **, and * denote statistical significance at the 1%, 5%, and 10% levels, respectively. The t-ratios (and z-scores) are reported in parentheses below each coefficient. The Hausman test $\chi^2 = 8.11$ (p = 0.087). All models are estimated using balanced panel data (N = 5, T = 41).

The results for irrigated land area (*I*) present a striking contrast between models. Pooled OLS and RE show positive coefficients (1.176 with no statistical significance, $p > 0.4$), while the FE and FER specifications yield negative but insignificant estimates (-0.088 with no statistical significance, $p > 0.9$). This complete reversal suggests that simpler models may be capturing spurious correlations, with the true relationship likely absorbed by country-specific fixed effects. However, labor input (*L*) maintain a consistently negative coefficient across all models (around -0.05), though never statistically significant ($p > 0.5$). This persistent pattern warrants investigation into potential measurement error or omitted variable bias in the labor-productivity relationship. A plausible explanation lies in the partial substitutability between labor and capital. In some farms, particularly those with higher levels of capital investment, modern equipment and technologies reduce the reliance on human labor.

On the other hand, climate variables exhibit model-dependent significance. Precipitation (*P*) shows strong positive effects in FE (0.019, with high statistical significance) that weaken to marginal significance in FER (with $p = 0.08$), while its quadratic term becomes marginally significant (-0.00003, with lower statistical significance, $p = 0.072$) in FER, suggesting possible diminishing returns. Temperature variable (*T*) remain insignificant across all specifications ($p > 0.3$), which indicates that the effects may be captured by time-invariant country characteristics.

The interaction term $T \times I$ shows striking differences across specifications. The POLS, RE, and PCSE models yield large, statistically significant coefficients (about 0.38 with $p < 0.01$), suggesting strong synergistic effects between temperature and irrigation. However, the FE and FER models—which control for unobserved country heterogeneity—show near-zero, insignificant coefficients (with 0.088, $p > 0.7$). This divergence reveals that POLS, RE, and PCSE estimations likely suffer from omitted variable bias, conflating true interaction effects with unobserved country-specific. However, FE absorption explains most of the apparent interaction, implying the adaptation benefit of irrigation is largely time-invariant.

The variance decomposition reveals that 61% of total variance (rho = 0.615) stems from country-specific effects in the FE model, strongly justifying its use over RE specifications. This substantial share of unobserved heterogeneity confirms that regional characteristics – such as persistent soil quality differences, institutional frameworks, or agricultural traditions – significantly influence agricultural outcomes. The FER model's notably narrower confidence intervals, particularly for capital (0.008 ± 0.01 in FE vs. 0.009 ± 0.002 in FER), demonstrate its conservative adjustment for both heteroskedasticity (White's test with $p \approx 0$) and autocorrelation (Wooldridge's test with $p = 0.043$). The PCSE model maintains significance for precipitation variables ($p = 0.024$) by accounting for cross-sectional dependence (Pesaran CD test with $p = 0.004$), but this comes at the cost of ignoring crucial country-specific effects. This trade-off is evident in the precipitation coefficient's drop from 0.019 (FE) to 0.001 (PCSE), suggesting nearly half of the apparent climate effect actually reflects unobserved regional characteristics.

Three key findings emerge from comparing specifications: The Hausman test ($\chi^2 = 8.11, p = 0.087$) decisively rejects RE, confirming FER as the only consistent estimator; POLS severely overestimates irrigated land's impact (-0.238 vs. -0.170 in FE) due to omitted variable bias; and climate variable significance in PCSE ($p < 0.01$) disappears in FER ($p > 0.1$), revealing their sensitivity to model choice.

For scenario simulation, the FER specification offers the most credible estimates despite two constraints: the limited country sample ($N = 5$) reduces statistical power, as seen in capital's marginal significance (coefficient = 0.0086 with $p = 0.042$), and robust error correction yields wider confidence intervals (e.g., capital's CI [0.0005, 0.0167]). Nevertheless, these conservative estimates are methodologically superior to the potentially biased results from POLS or RE specifications, particularly for the temperature-irrigation interaction term (FER: 0.0886 with $p = 0.76$ vs. POLS: 0.3803 with $p < 0.01$). Three FER findings are critical for scenario design: (1) precipitation's robust nonlinear effects (quadratic term = -0.00002 with $p = 0.022$), (2) capital's consistently positive coefficient (0.0086) despite its borderline significance, and (3) temperature variables' statistical insignificance (linear term $p = 0.66$; quadratic $p = 0.85$), suggesting precipitation dominates climate risks in this region. These results justify using FER as the benchmark while acknowledging that its conservative nature reflects the empirical challenges of small-N panel analysis.

These findings highlight the fact that cross-country heterogeneity dominates the climate-agriculture relationship, model choice dramatically affects both coefficient magnitude and significance, and FER's robustness adjustments provide more credible estimates for simulation. The analysis conclusively demonstrates that only FER adequately addresses the triad of econometric challenges: unobserved heterogeneity, error dependence, and cross-sectional correlation. As its results appear more precise, they also avoid the severe bias of alternative specifications, making FER essential for meaningful scenario analysis in North Africa's agricultural sector.

4.3. Scenario projections analysis

The country-level projections reveal stark differences in climate vulnerability and adaptive capacity across North Africa (Table 8). It should be noted that the negative value represents the cumulative percentage decline in y relative to the baseline year (2020) by 2050. In technical terms, if τ is the percentage result projection, than, it should be interpreted as $y_{2025} = y_{2020} e^{\tau/100}$. Accordingly, three key patterns emerge: First, baseline climate impacts vary substantially by country, with Egypt facing the most severe losses (-148.5% without adaptation in the pessimistic scenario) due to its heavy dependence on Nile River irrigation systems. The logarithmic temperature-irrigation interaction term in the model exacerbates damages in highly irrigated systems under extreme warming. In contrast, Libya shows relatively lower, similarly serious, impacts (-120.6%) due to limited existing irrigation infrastructure reducing its exposure to water-related climate shocks.

Second, adaptation effectiveness strongly correlates with assumed capacity parameters, but with diminishing returns. Egypt's high adaptation capacity (0.9) reduces impacts by approximately 90% across all scenarios (e.g., -148.5% to -14.8% in pessimistic case), following the theoretical specification where adapted impact is the base impact × (1 - capacity). Meanwhile, Algeria and Libya, with lower capacities (0.5 and 0.4 respectively), show proportionally smaller mitigations (50-60% of base impacts remain after adaptation).

Table 8. Scenarios results on country-specific climate impact projections (2050)

Country	*Adaptation capacity*	*Impact without adaptation (%)*			*Impact with adaptation (%)*		
		Optimistic	*Central*	*Pessimistic*	*Optimistic*	*Central*	*Pessimistic*
Morocco	0.7	-54.2	-81.7	-137.1	-16.3	-24.5	-41.1
Algeria	0.5	-52.3	-78.8	-132.2	-26.1	-39.4	-66.1
Tunisia	0.8	-48.2	-72.7	-122.0	-9.7	-14.5	-24.4
Libya	0.4	-47.6	-71.8	-120.6	-28.6	-43.1	-72.4
Egypt	0.9	-58.8	-88.5	-148.5	-5.9	-8.8	-14.8

Third, the damage progression across scenarios reveals nonlinear acceleration, particularly for countries with significant irrigation infrastructure. The pessimistic scenario generates impacts about 2.5 times greater than the optimistic scenario, exceeding simple linear scaling due to: (1) quadratic precipitation terms amplifying water stress, and (2) the temperature-irrigation interaction becoming increasingly dominant at higher warming levels. This is especially evident in Morocco's results, where damages grow from -54.2% (optimistic) to -137.1% (pessimistic).

In the aggregated level, the results in Table 9 (and Figure 2) display a summary of the scenarios with and without adaptation in 2050. The regional projections reveal a stable proportional relationship between adaptation and no-adaptation impacts across all climate scenarios, with adaptation consistently mitigating 28-30% of potential losses. This proportional reduction holds despite the growth in absolute damages from optimistic (-53.2% without adaptation) to pessimistic (-134.4%) scenarios, indicating the model's adaptation mechanism operates as a constant multiplier rather than exhibiting scenario-dependent effectiveness. The near-uniform mitigation ratio (adaptation/no-adaptation ≈ 0.28) arises mechanically from the weighted combination of country-specific adaptation capacities (ranging from 0.4 in Libya to 0.9 in Egypt) with their respective agricultural productivity shares, though this masks substantial sub-regional heterogeneity in adaptive performance.

Table 9. Summary of the scenarios with and without adaptation in 2050

Scenarios	*With adaptation*	*Without adaptation*
Optimistic	-15.0	-53.2
Central	-22.6	-80.1
Pessimistic	-37.9	-134.4

The damage progression displays marked convexity, where pessimistic scenario impacts exceed linear expectations relative to central scenarios. The 68% escalation from central (-80.1%) to pessimistic (-134.4%) no-adaptation impacts compared to the 51% increase from optimistic (-53.2%) to central scenarios reflects the compounding influence of quadratic precipitation terms and logarithmic irrigation-temperature interactions in the underlying FER specification. This convexity manifests through two channels: the explicit quadratic precipitation term (-0.00002) accelerates water stress under extreme drying, while the implicit curvature in the irrigation interaction term (0.0886×ln(I + 1)) amplifies temperature effects in high-irrigation systems under severe warming.

Figure 2. Plot of the projected scenarios of the impact of climate change on agricultural value added in North African countries (2021-2050)

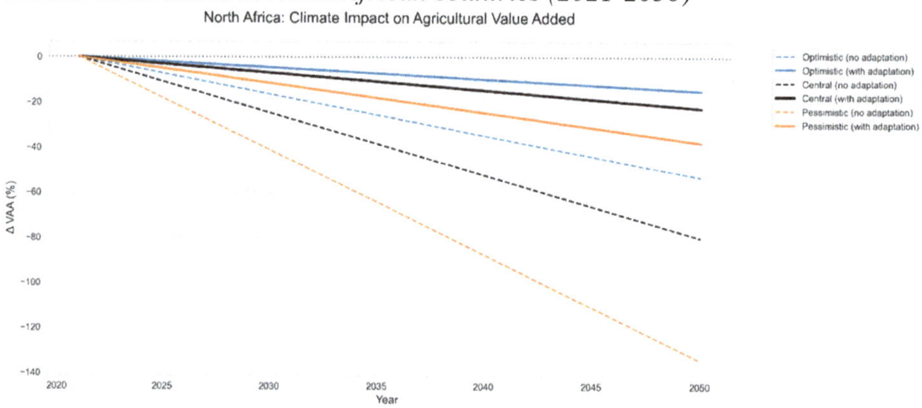

The regional averages conceal critical disparities in national vulnerability profiles. While aggregate results suggest moderate mean adaptation efficacy (28-30% mitigation), the underlying country-level data reveals a bimodal distribution: high-capacity Egypt achieves 90% impact reduction (from -148.5% to -14.8% in pessimistic scenario) through its combination of superior institutional capacity (0.9) and irrigation infrastructure, whereas low-capacity Libya's adaptation only offsets 40% of impacts (from -120.6% to -72.4%) due to constrained institutional resources

(0.4) and limited irrigation coverage. This divergence confirms that regional means systematically underrepresent the exposure of low-capacity states while overstating the adaptability of the region as a whole. The convex damage curve further exacerbates these disparities under extreme scenarios, as seen in Libya's disproportionate deterioration from -47.6% (optimistic) to -120.6% (pessimistic) versus Egypt's -58.8% to -148.5% progression.

The stability of aggregate adaptation efficacy stems from offsetting compositional effects: Egypt's dominant agricultural weight (30% regional share) pulls the mean toward its high-efficiency outcomes, while the logarithmic form of irrigation interactions ensures diminishing marginal returns to adaptive capacity. This creates the apparent paradox where regional aggregates show stable proportional mitigation even as absolute cross-country differences widen under more severe scenarios. The results are fundamentally shaped by the FER model's treatment of climate variables in levels while production variables are differenced - a specification choice that makes impacts sensitive to baseline climate conditions but relatively insensitive to inter-annual variability.

4.4. Policy implications

The empirical findings reveal critical policy challenges for North African agriculture under climate change. Three key areas demand urgent attention. In terms of water-centric adaptation strategies, it seems that the strong precipitation sensitivity (coefficient of 0.019, with high statistical significance in FER) coupled with diminishing returns (quadratic term with -0.00002) necessitates prioritizing water efficiency investments. Accordingly, policies should scale drip irrigation and water harvesting in rainfed systems (particularly Morocco and Tunisia), reform groundwater allocation to prevent aquifer depletion, and develop early warning systems for drought extremes.

For targeted capital investments, while capital shows positive marginal returns (0.0086, with high statistical significance), its modest effect size and sensitivity to specification suggest a focused spending on climate-smart technologies (e.g., drought-resistant seeds) rather than broad mechanization, promoting credit programs tailored to smallholders' adaptive capacity gaps, and establishment of monitoring systems to prevent maladaptation in water-intensive crops. Moreover, in terms of institutional capacity building, the 61.5% variance explained by country fixed effects ($\rho = 0.615$) highlights that Libya and Algeria's low adaptation scores (0.4-0.5) require institutional strengthening, Egypt's irrigation advantage (mitigating 90% of impacts) offers regional learning potential, and promoting cross-border knowledge transfer programs could reduce disparities.

However, there will be many implementation challenges. The convex damage curve (pessimistic impacts 2.5 × optimistic) creates nonlinear policy demands: less than 1.5°C warming means that incremental adaptation suffices, whereas the 2°C warming scenario will requires systemic transformation (crop switching, water rationing). Additionally, the pronounced sensitivity of agricultural value added to precipitation changes, coupled with diminishing marginal returns to increased rainfall, highlighted water scarcity as the central challenge for regional food security. This finding suggests that policies focusing solely on expanding irrigation may prove inadequate without parallel investments in water-use efficiency. The region would benefit most from integrated approaches that combine modern irrigation technologies with institutional reforms in water governance, particularly given the nonlinear acceleration of damages under severe climate scenarios. The convex damage curve reveals that conventional adaptation measures may maintain effectiveness under moderate warming but could become insufficient beyond 2°C, necessitating proactive planning for systemic agricultural transformation. It should be noted that data limitations (N = 5 countries), and model specifications caution against overgeneralization, but the consistent FER results provide a relatively robust basis for climate-resilient agricultural planning.

The analysis reveals critical disparities in adaptive capacity across countries that demand differentiated policy responses. Egypt's demonstrated ability to mitigate nearly 90% of climate impacts through its advanced irrigation systems and institutional capacity offers important lessons for regional cooperation. However, the same results expose alarming vulnerabilities in low-capacity countries like Libya, where adaptation mechanisms only offset 40% of potential losses. This divergence calls for targeted international support to strengthen institutional frameworks and technology transfer programs, particularly for nations with limited adaptive infrastructure. The predominance of country-specific effects in the variance decomposition suggests that successful policies must account for local agro-ecological conditions and governance structures rather than relying on standardized regional approaches.

The complete insignificance of temperature variables in all specifications, contrasted with the strong precipitation effects, suggests that water management rather than heat tolerance should dominate research and development priorities. However, rising temperatures should not be overlooked, as climate change amplifies water scarcity not only through declining precipitation but also by increasing evapotranspiration and the overall demand for water.

The paradoxical results for irrigation effects – showing apparent benefits in naive specifications but null effects in robust models – carry important policy lessons. They caution against overreliance on irrigation expansion as a standalone adaptation strategy, instead emphasizing the need to combine physical infrastructure with institutional and managerial innovations. The results imply that Egypt's irrigation success stems from its comprehensive water governance system rather than irrigation coverage alone, suggesting that other countries must develop complementary policies around water pricing, maintenance regimes, and user participation to achieve similar results.

The consistent superiority of FER in model performance comparisons underscores the importance of context-specific policy design. The large share of variance explained by country fixed effects confirms that effective interventions must be grounded in local realities rather than imported templates. This argues for decentralized policymaking frameworks that empower subnational actors while maintaining strategic coordination at the national level. The findings particularly highlight the need for improved agricultural data systems to better capture localized climate impacts and adaptation potentials, as the current analysis was constrained by limited country coverage.

Ultimately, the results paint a picture of North African agriculture at a critical juncture. While the region possesses some capacity to adapt to moderate climate changes, particularly through water management innovations, the accelerating damages under severe warming scenarios reveal fundamental limits to conventional adaptation strategies. This suggests the need for parallel investments in agricultural diversification, food system resilience, and social safety nets to complement production-focused adaptation measures. The analysis provides empirical support for policies that recognize both the urgent need for climate action and the profound institutional and geographical constraints that shape adaptation potentials across the region.

5. CONCLUSION

This chapter embarked on an ambitious endeavor to rigorously assess the multifaceted economic impacts of climate change on agricultural across five North African countries, while simultaneously evaluating the potential efficacy of various adaptation strategies to mitigate these impacts. Our investigation was motivated by the urgent need to understand how one of the world's most climate-vulnerable regions can sustain its agricultural productivity in the face of escalating environmental challenges. The research adopted an innovative integrated Ricardian-Production Function framework that synergistically combined the strengths of traditional Ricardian analysis with production function theory, creating a robust analytical tool specifically tailored to the unique conditions of North African agriculture. We used a four-decade panel dataset (1980–2022), integrating climate, agricultural, and economic variables. We further enriched the analysis with three scenario projections to 2050 and introduced innovative country-specific adaptation capacity indices capturing governance, technology, and infrastructure.

The findings of this extensive analysis revealed several crucial insights with profound implications for both academic understanding and policy formulation. First, we identified striking regional disparities in climate vulnerability and adaptive capacity across North Africa. Egypt demonstrated remarkable resilience, with its sophisticated Nile-based irrigation systems and robust policy frameworks enabling mitigation of up to 90% of potential climate-induced agricultural losses even under the most pessimistic warming scenarios. This resilience stands in stark contrast to the situation in countries like Libya and Algeria, where political instability, underdeveloped agricultural infrastructure, and weaker governance systems resulted in significantly higher vulnerability to climate impacts. Our results particularly highlighted how these disparities are likely to intensify under more severe climate scenarios, potentially exacerbating existing regional inequalities in food security and economic development.

A particularly significant finding was the dominant role of precipitation variability compared to temperature changes in determining agricultural outcomes across the region. While much of the global climate-agriculture literature focuses on temperature effects, our analysis demonstrated that water availability remains the paramount climatic factor for North African agriculture. This finding has crucial implications for adaptation prioritization, suggesting that water management strategies should form the cornerstone of climate resilience efforts in the region. The study also revealed important nonlinearities in climate impacts, with damage curves becoming progressively steeper under more severe warming scenarios, particularly in countries with limited adaptive resources.

Our analysis of adaptation strategies yielded several nuanced insights. While irrigation expansion showed potential as a climate adaptation tool, we found evidence of diminishing returns, particularly under extreme warming scenarios. This suggests that conventional approaches to irrigation development may need to be rethought and complemented with more efficient water-use technologies. Similarly, while capital investment generally showed positive effects on agricultural production, its effectiveness was significantly moderated by climatic conditions, indicating that investment strategies must be carefully tailored to anticipated future climate regimes.

The policy implications emerging from this research are both far-reaching and urgent. For high-capacity countries like Egypt and Tunisia, our findings suggest that maintaining and enhancing existing institutional frameworks while investing in next-generation water-efficient technologies should be immediate priorities. The success of initiatives like Morocco's Green Plan and Tunisia's water conservation policies demonstrates the potential of proactive, well-designed adaptation strategies. For lower-capacity countries such as Libya and Algeria, the situation demands more fundamental interventions, including conflict resolution, governance strengthening, and substantial international support for infrastructure development. At the regional level, our analysis strongly advocates for enhanced cooperation mechanisms, particularly in the areas of knowledge transfer, technology sharing, and transboundary water management.

Several important limitations of the current study must be acknowledged, each pointing to valuable directions for future research. The relatively small country sample (N = 5) necessarily limits the statistical power of our analysis and calls for caution in generalizing findings. Our reliance on national-level aggregate data, while necessary for regional comparison, inevitably masks important subnational variations in climate exposure and adaptive capacity. The adaptation capacity indices, though carefully constructed, represent static assessments that may not fully capture dynamic socio-political changes. Perhaps most significantly, our focus on production-based adaptation strategies may overlook important non-production dimensions of climate resilience, such as livelihood diversification or market-based risk management tools.

Future research should address these limitations through several promising avenues. First, expanding the analytical framework to incorporate micro-level data from farm households would provide crucial insights into local-scale adaptive behaviors and constraints. Second, developing dynamic modeling approaches that can simulate the evolution of institutional capacities over time would significantly enhance the policy relevance of findings. Third, investigating non-production-based adaptation strategies, including crop diversification, agricultural insurance schemes, and value chain development, could reveal important complementary pathways to resilience. Additionally, incorporating micro-behavioral dimensions of climate vulnerability and accounting for informal labor dynamics would help ensure that adaptation policies are equitable and efficient.

REFERENCES

ABHSM. (2021). *Water stress in Morocco's basins*. Agence du Bassin Hydraulique du Souss-Massa.

Abidoye, B. O., Kurukulasuriya, P., Reed, B., & Mendelsohn, R. (2017). Structural Ricardian analysis of south-east Asian agriculture. *Climate Change Economics (Singapore)*, *8*(03), 1740005. DOI: 10.1142/S201000781740005X

AFDB. (2025). *Defining a new approach to water management in Libya: Issues & options*. AFDB.

Akesbi, N. (2012). Une nouvelle stratégie pour l'agriculture marocaine: Plan Maroc Vert. *New Medit*, *11*(2), 12.

Alboghdady, M., & El-Hendawy, S. E. (2016). Economic impacts of climate change and variability on agricultural production in the Middle East and North Africa region. *International Journal of Climate Change Strategies and Management*, *8*(3), 463–472. DOI: 10.1108/IJCCSM-07-2015-0100

AOAD (2024). Agricultural Statistics Yearbook. Arab Organization for Agricultural Development. Arab League.

APIA. (2023). *Rapports statistiques*. APIA, Ministère de l'Agriculture.

Atkinson, K. (2023). Libya's Soil Resources and Their Potential for Sustained Agricultural Production. In *The Economic Development of Libya* (pp. 134–156). Routledge. DOI: 10.4324/9781003420552-8

Attar, O., Brouziyne, Y., Bouchaou, L., El Bilali, A., Ait Brahim, Y., & Chehbouni, A. (2024). Understanding the trade-offs between climate change-induced aridity and agricultural water demand in the Souss basin, Morocco. *Frontiers in Water*, *6*, 1270078. DOI: 10.3389/frwa.2024.1270078

Attiaoui, I., & Boufateh, T. (2019). Impacts of climate change on cereal farming in Tunisia: A panel ARDL–PMG approach. *Environmental Science and Pollution Research International*, *26*(13), 13334–13345. DOI: 10.1007/s11356-019-04867-y

Bachta, M. S., & Ben Nasr, J. (2020). *Water policy in Tunisia. Water policies in MENA countries*, 161-184.

Benmehaia, A. M., & Bekkis, S. (2024). Agri-food policy trends in Algeria: Selected explorations. *New Medit*, *23*(3), 49–62. DOI: 10.30682/nm2403d

CAT. (2023). Policies & action: Morocco. Policy overview. Climate Action Tracker. https://climateactiontracker.org/countries/morocco/policies-action/

Closas, A., & Villholth, K.G. (2016). Aquifer contracts: a means to solving groundwater over-exploitatin in Morocco?. Internatinal Water Management Instiute (IWMI). Groundwater Solutins Initītie for Policy and Practie – GRIPP. Case Profie Series 01).

Eid, H. M., El-Marsafawy, S. M., & Ouda, S. A. (2007). Assessing the economic impacts of climate change on agriculture in Egypt: a Ricardian approach. World Bank Policy Research Working Paper, (4293).

El-Agha, D. E., Molle, F., Metwally, M. I., Emara, S. R., Shalby, A., Armanuos, A. M., & Gado, T. A. (2024). Toward sustainable management of groundwater in the deserts of Egypt. *Hydrogeology Journal*, *32*(3), 663–678. DOI: 10.1007/s10040-023-02738-y

Elsaie, Y., Ismail, S., Soussa, H., Gado, M., & Balah, A. (2023). Water desalination in Egypt; literature review and assessment. *Ain Shams Engineering Journal*, *14*(7), 101998. DOI: 10.1016/j.asej.2022.101998

Falkenmark, M., Lundqvist, J., & Widstrand, C. (1989). Macro-scale water scarcity requires micro-scale approaches: Aspects of vulnerability in semi-arid development. [). Oxford, UK: Blackwell Publishing Ltd.]. *Natural Resources Forum*, *13*(4), 258–267. DOI: 10.1111/j.1477-8947.1989.tb00348.x

FAO. (2022). *Water efficiency, productivity and sustainability in the NENA regions*. FAO-UN.

FAO. (2024). Statistics on the agricultural sector. FAO-UN. https://fao.org/faostat/en/#data

Fernández-Lobato, L., López-Sánchez, Y., Baccar, R., Fendri, M., & Vera, D. (2022). Life cycle assessment of the most representative virgin olive oil production systems in Tunisia. *Sustainable Production and Consumption*, *32*, 908–923. DOI: 10.1016/j.spc.2022.06.002

Filahi, S., Tanarhte, M., Mouhir, L., El Morhit, M., & Tramblay, Y. (2016). Trends in indices of daily temperature and precipitations extremes in Morocco. *Theoretical and Applied Climatology*, *124*(3-4), 959–972. DOI: 10.1007/s00704-015-1472-4

Frija, A., Oulmane, A., Chebil, A., & Makhlouf, M. (2021). Socio-Economic implications and potential structural adaptations of the Tunisian agricultural sector to climate change. *Agronomy (Basel)*, *11*(11), 2112. DOI: 10.3390/agronomy11112112

Gabr, M. E. (2023). Impact of climatic changes on future irrigation water requirement in the Middle East and North Africa's region: A case study of upper Egypt. *Applied Water Science, 13*(7), 158. DOI: 10.1007/s13201-023-01961-y

Gannouni, S., Riahi, R., Kefi, M., Fehri, R. M., & Rebai, N. (2023). Identification and monitoring of groundwater potential zones using geospatial techniques (Siliana, Tunisia). *Euro-Mediterranean Journal for Environmental Integration, 8*(4), 1015–1034. DOI: 10.1007/s41207-023-00395-x

Ghadge, A., Wurtmann, H., & Seuring, S. (2020). Managing climate change risks in global supply chains: A review and research agenda. *International Journal of Production Research, 58*(1), 44–64. DOI: 10.1080/00207543.2019.1629670

Gomez-Zavaglia, A., Mejuto, J. C., & Simal-Gandara, J. (2020). Mitigation of emerging implications of climate change on food production systems. *Food Research International, 134*, 109256. DOI: 10.1016/j.foodres.2020.109256

Hamed, M. M., Al-Hasani, A. A. J., Nashwan, M. S., Sa'adi, Z., & Shahid, S. (2024). Assessing the growing threat of heat stress in the North Africa and Arabian Peninsula region connected to climate change. *Journal of Cleaner Production, 447*, 141639. DOI: 10.1016/j.jclepro.2024.141639

Hossain, M. S., Arshad, M., Qian, L., Zhao, M., Mehmood, Y., & Kächele, H. (2019). Economic impact of climate change on crop farming in Bangladesh: An application of Ricardian method. *Ecological Economics, 164*, 106354. DOI: 10.1016/j.ecolecon.2019.106354

ICARDA. (2018). *Contributing to Egypt's 1.5 Million Feddan project.* ICARDA Communication Team.

ICRC. (2022). Libya: Conflict weakens the abilities of farmers to mitigate climate risks. International Committee of the Red Cross.

IFAD (2022). The State of Libya: Country strategy note 2023 – 2024, Main report and appendices. No. 5683-LY.

IPCC. (2022). *Climate change 2022: Impacts, adaptation, and vulnerability. Contribution of Working Group II to the Sixth Assessment Report of the Intergovernmental Panel on Climate Change.* Cambridge University Press.

Kabubo-Mariara, J., & Karanja, F. K. (2007). The economic impact of climate change on Kenyan crop agriculture: A Ricardian approach. *Global and Planetary Change, 57*(3-4), 319–330. DOI: 10.1016/j.gloplacha.2007.01.002

Kherbache, N., & Molle, F. (2022). Causes and consequences of the Macta basin closure, Algeria. *International Journal of Water Resources Development, 0*, 1–22.

Kurukulasuriya, P., Mendelsohn, R., Hassan, R., Benhin, J., Deressa, T., Diop, M., Eid, H. M., Fosu, K. Y., Gbetibouo, G., Jain, S., Mahamadou, A., El-Marsafawy, S., Ouda, S., Ouedraogo, M., Sène, I., Maddison, D., Seo, S. N., & Dinar, A. (2006). Will African agriculture survive climate change? *The World Bank Economic Review, 20*(3), 367–388. DOI: 10.1093/wber/lhl004

Kurukulasuriya, P., & Mendelsohn, R. O. (2007). A Ricardian analysis of the impact of climate change on African cropland. *World Bank Policy Research Working Paper,* •••, 4305. DOI: 10.1596/1813-9450-4305

Lachgar, R., Badri, W., & Chlaida, M. (2022). Assessment of future changes in downscaled temperature and precipitation over the Casablanca-Settat region (Morocco). *Modeling Earth Systems and Environment, 8*(2), 2123–2133. DOI: 10.1007/s40808-021-01213-5

Le Mouël, C., Forslund, A., Marty, P., Manceron, S., Marajo-Petitzon, E., Caillaud, M. A., Dumas, P., & Schmitt, B. (2023). Can the Middle East-North Africa region mitigate the rise of its food import dependency under climate change? *Regional Environmental Change, 23*(2), 52. DOI: 10.1007/s10113-023-02045-y

Mendelsohn, R., & Dinar, A. (2009). *Climate change and agriculture: An economic analysis of global impacts, adaptation, and distributional effects.* Edward Elgar Publishing. DOI: 10.4337/9781849802239

Mendelsohn, R., Nordhaus, W. D., & Shaw, D. (1994). The impact of global warming on agriculture: A Ricardian analysis. *The American Economic Review, 84*(4), 753–771. https://www.jstor.org/stable/2118029

Meyghani, S., Khodaparast Mashadi, M., & Salehnia, N. (2021). The Impacts of Climate Change on Value-Added Agriculture in the MENA Region. *Journal of Economics and Regional Development, 27*(20), 129–158.

Migliore, G., Zinnanti, C., Schimmenti, E., Borsellino, V., Schifani, G., Di Franco, C. P., & Asciuto, A. (2019). A Ricardian analysis of the impact of climate change on permanent crops in a Mediterranean region. *New Medit, 18*(1), 41–52. DOI: 10.30682/nm1901d

Mokhtari, S., Benmehaia, A. M., Nourani, A., & Hadjeb, A. (2025). Future water requirements of the wheat crop in Northern Algeria under climate change scenarios. *Regional Environmental Change, 25*(2), 1–14. DOI: 10.1007/s10113-025-02378-w

OECD (2020). Climate Impacts on African Hydropower. IEA report.

Ouessar, M., Sghaier, A., Frija, A., Sghaier, M., & Baig, M. B. (2021). Impacts of climate change on agriculture and food security in Tunisia: challenges, existing policies, and way forward. *Emerging Challenges to Food Production and Security in Asia, Middle East, and Africa: Climate Risks and Resource Scarcity*, 65-99.

Oulmane, A., Kechar, A., Benmihoub, A., & Amine Benmehaia, M. (2022). Assessment of surface water management institutions: A case of public irrigation schemes in northern Algeria. *Water Policy*, *24*(2), 229–241. DOI: 10.2166/wp.2022.025

Oxford Analytica (2024). Algeria's water challenges will grow. Emerald Expert Briefings, (oxan-db).

Perez, N. D., Kassim, Y., Ringler, C., Thomas, T. S., ElDidi, H., & Breisinger, C. (2021). *Climate-resilience policies and investments for Egypt's agriculture sector: Sustaining productivity and food security*. Intl Food Policy Res Inst.

Pickson, R. B., & Boateng, E. (2022). Climate change: A friend or foe to food security in Africa? *Environment, Development and Sustainability*, *24*(3), 1–26. DOI: 10.1007/s10668-021-01621-8

Razek, A., & Othman, S. (2023). Some Economic Aspects of the Olive Crop in Libya. *Alexandria Science Exchange Journal*, *44*(4), 1109–1122. DOI: 10.21608/asejaiqjsae.2023.326999

Salami, A. A., & Babatunde, O. R. (2024). Environmental challenges, the impacts of climate change in North Africa region: A review. *Natural resources deterioration in MENA region: land degradation, soil erosion, and desertification*, 281-294.

Sanghi, A., & Mendelsohn, R. (2008). The impacts of global warming on farmers in Brazil and India. *Global Environmental Change*, *18*(4), 655–665. DOI: 10.1016/j.gloenvcha.2008.06.008

Sattar, A., Brown, C., Rounsevell, M., & Alexander, P. (2024). Typology analysis of Egyptian agricultural households reveals increasing income diversification and abandonment of agricultural activities. *Agricultural Systems*, *218*, 104000. DOI: 10.1016/j.agsy.2024.104000

Schilling, J., Freier, K. P., Hertig, E., & Scheffran, J. (2012). Climate change, vulnerability, and adaptation in North Africa with focus on Morocco. *Agriculture, Ecosystems & Environment*, *156*, 12–26. DOI: 10.1016/j.agee.2012.04.021

Seo, S. N., & Mendelsohn, R. (2008). Measuring impacts and adaptations to climate change: A structural Ricardian model of African livestock management. *Agricultural Economics*, *38*(2), 151–165. DOI: 10.1111/j.1574-0862.2007.00289.x

Serdeczny, O., Adams, S., Baarsch, F., Coumou, D., Robinson, A., Hare, W., & Reinhardt, J. (2017). Climate change impacts in Sub-Saharan Africa: From physical changes to their social repercussions. *Regional Environmental Change*, *17*(6), 1585–1600. DOI: 10.1007/s10113-015-0910-2

Shaheen, H. M. (2025). The Role of the Paris Agreement in Addressing Gender-Based Violence through Nationally Determined Contributions (NDC s): A Comparative Study of Egypt and Tunisia. *Studi Magrebini*, *22*(2), 201–217. DOI: 10.1163/2590034x-20240110

Sowers, J., Vengosh, A., & Weinthal, E. (2011). Climate change, water resources, and the politics of adaptation in the Middle East and North Africa. *Climatic Change*, *104*(3–4), 599–627. DOI: 10.1007/s10584-010-9835-4

Uzundumlu, A. S., & Ateş, T. (2025). Olive Production Forecasts in the Leader Countries for 2023–2027. *Applied Fruit Science*, *67*(2), 56. DOI: 10.1007/s10341-025-01268-z

Verner, D. (Ed.). (2012). *Adaptation to a changing climate in the Arab countries: A case for adaptation governance and leadership in building climate resilience.* World Bank., DOI: 10.1596/978-0-8213-9459-5

Verner, D., Treguer, D., Redwood, J., Christensen, J., McDonnell, R., Elbert, C., & Konishi, Y. (2018). *Climate variability, drought, and drought management in Tunisia's agricultural sector. World Bank.* World Bank.

Waha, K., Krummenauer, L., Adams, S., Aich, V., Baarsch, F., Coumou, D., & Schleussner, C. F. (2017). Climate change impacts in the Middle East and Northern Africa (MENA) region and their implications for vulnerable population groups. *Regional Environmental Change*, *17*(6), 1623–1638. DOI: 10.1007/s10113-017-1144-2

World Bank. (2021). *Climate-smart agriculture in the Middle East and North Africa: Opportunities and constraints*. World Bank Group. https://openknowledge.worldbank.org/handle/10986/35296

World Bank (2024a). Climatology: Libya. Climate Change Knowledge Portal (CCKP). The World Bank Group.

World Bank. (2024b). *Trends and Significant Change against Natural Variability in Tunisia. Climate Change Knowledge Portal (CCKP).* The World Bank Group.

WorldClim. (2024). Records on the bioclimatic variables: Historical weather data. *WorldClim*, *2*, 1. worldclim.org/data/monthlywth.html

Zaied, Y. B., & Zouabi, O. (2016). Impacts of climate change on Tunisian olive oil output. *Climatic Change*, *139*(3-4), 535–549. DOI: 10.1007/s10584-016-1801-3

Zurqani, H. A. (Ed.). (2021). *The soils of Libya*. Springer. DOI: 10.1007/978-3-030-66368-1

Chapter 6
Climate Change and Inequality:
The Agricultural Channel

Salwa Ech-charfi
https://orcid.org/0009-0000-0710-6228
Mohammed V University, Morocco

Idriss El Abbassi
Mohammed V University, Morocco

ABSTRACT

This chapter investigates the impact of climate vulnerability on pre-tax income inequality across MENA countries over 1996–2021. Using random-effects panel models, it examines the direct effect and interactions with agricultural employment and institutional quality. Four econometric models explore these relationships, finding that climate vulnerability is positively associated with inequality, with agricultural dependence and weak institutions intensifying the effect. Food security and ecosystem degradation emerge as key drivers. Graphical analysis shows non-linear country dynamics. Findings emphasize the complex interplay of environmental, economic, and institutional factors, offering insights for targeted policy interventions to mitigate the socio-economic consequences of climate risks.

1. INTRODUCTION

Climate change represents one of the defining challenges of the 21st century, with far-reaching implications that extend beyond environmental degradation to profound social and economic inequalities (IPCC, 2021; IPCC, 2022; IPCC, 2023).

DOI: 10.4018/979-8-3373-3346-5.ch006

While a growing body of research has explored the macroeconomic consequences of climate change and disparities between countries, considerably less attention has been devoted to the mechanisms through which climate vulnerability influences income inequality within countries, especially in the Global South. This analytical gap is particularly salient in the Middle East and North Africa (MENA) region, a zone characterized by acute ecological fragility, deep-rooted socio-economic divides, and strong dependence on climate-sensitive sectors such as agriculture (World Bank, 2018; World Bank, 2020; OECD/FAO, 2018).

The MENA region is among the most vulnerable worldwide to climate change due to its arid to semi-arid climate, low adaptive capacity, and persistent socio-economic challenges (IEMed, 2021; Chancel, Bothe, & Voituriez, 2023). Rising temperatures, the increasing frequency of extreme weather events, and widespread water scarcity exacerbate existing environmental and social stresses, threatening food security, economic development, and social cohesion. Moreover, structural pressures such as high population growth, urbanization, and political instability intensify these risks and limit the scope for effective adaptation (IEMed, 2021). Climate change also acts as a "threat multiplier" in the region, deepening territorial disparities, worsening environmental degradation, and placing additional strain on already fragile economies and institutions (United Nations Economic and Social Commission for Western Asia, United Nations Population Fund Arab States Regional Office, & Save the Children, 2022).

Notably, the Arab region is the only world region where poverty rates have been rising in recent years, contributing to widening inequality (United Nations Economic and Social Commission for Western Asia [ESCWA], 2024). The richest 10 percent of the population in the Arab world hold, on average, 44 times as much wealth per person as the poorest 40 percent. Furthermore, wealth inequality has sharply increased since 2020: in that year, the wealthiest 1 percent held 43 percent of all wealth, rising to 44 percent by 2022. These trends have been particularly acute in low- and middle-income countries across the region. At the same time, climate-associated risks have become the fastest-growing threat to development in the Arab region, compounding structural inequalities and increasing exposure to shocks (United Nations Economic and Social Commission for Western Asia [ESCWA], 2024). To add to that, the MENA region has the highest per capita emissions in the world when considering both domestic and imported emissions, with the top 10% of emitters are responsible for over 60% of total emissions and the bottom 50% for less than 10% (Chancel, Bothe, & Voituriez, 2023).

Agriculture in the MENA region faces significant adaptation challenges due to these climatic pressures, compounded by limited financial resources, water scarcity, and inefficient farming practices (del Pozo et al., 2019). Smallholder and subsistence farmers are particularly vulnerable, as they often lack access to modern technologies, irrigation infrastructure, and institutional support systems necessary for climate resilience. These factors increase the risk of reduced agricultural productivity, income loss, and food insecurity, which in turn exacerbate socio-economic inequalities. Rural communities in fragile areas, especially in water-scarce, conflict-affected, or poorly connected regions, are increasingly exposed to the cumulative effects of climate and economic stress (United Nations Economic and Social Commission for Western Asia, United Nations Population Fund Arab States Regional Office, & Save the Children, 2022).

This chapter therefore seeks to address the underexplored relationship between climate vulnerability and within-country income inequality, with a particular focus on agriculture as a key transmission channel. Climate-induced agricultural disruptions serve as multipliers of socio-economic inequality by impacting rural and low-income populations, often asset-poor, informally employed, and insufficiently covered by social safety nets. Declines in agricultural productivity heighten economic precarity in these communities, perpetuating cycles of vulnerability and marginalization.

To add to that, agriculture remains vital to the MENA region's economy. According to World Bank data, in 2022 the average share of agricultural employment across 18 MENA countries was 9.61 percent, with significant variation, ranging from 30.75 percent in Morocco to just 0.9 percent in Bahrain. Particularly in arid and semi-arid zones, agriculture is highly vulnerable to climate hazards such as extreme heat, irregular precipitation, and chronic water scarcity, risks projected to intensify under current climate trajectories (IPCC, 2022). These vulnerabilities threaten not only food security but also the livelihoods of millions who depend on agriculture.

This chapter builds on the conceptual framework proposed by Islam and Winkel (2017), which argues that climate change tends to amplify pre-existing inequalities by disproportionately affecting vulnerable groups. Islam and Winkel (2017) develop a comprehensive analytical framework highlighting three primary channels through which inequality exacerbates climate vulnerability. First, the exposure channel explains that disadvantaged groups are disproportionately situated in high-risk areas, such as flood-prone slums or drought-affected rural zones, due to limited access to safer environments. This spatial inequality intensifies their exposure to climate hazards (Islam & Winkel, 2017, pp. 12–14). Second, the susceptibility channel demonstrates that even when exposed equally, poorer populations suffer more severe damage owing to substandard housing, fragile livelihoods, and inadequate health infrastructure (Islam & Winkel, 2017, pp. 15–17). For example, in Nepal and MENA countries, the poor reported twice the income losses from climate-related events as the rich (Gentle, P., & Maraseni, T. N., 2012; Bougnoux et al., 2014).

Third, the coping and recovery channel underscores the limited capacity of marginalized populations to withstand and rebound from climate shocks due to scarce savings, restricted insurance access, and insufficient public support (Islam

& Winkel, 2017, pp. 17–21). These channels operate through both economic and political mechanisms, whereby inequality not only shapes vulnerability but is also reinforced by climate-related losses. In this feedback loop, wealth disparities increase exposure, while elite influence may delay inclusive responses to climate risks.

Using empirical data spanning 1996 to 2021, this study fills a critical void in the literature by quantifying the link between climate vulnerability and income inequality within MENA countries. To our knowledge, this is the first empirical study to focus on this relationship in the MENA region, whereas prior research has largely examined related themes such as migration (Bougnoux et al., 2014), food security (Elkhalfi, O. et al, 2025) etc. or the impact of the agricultural sector, specially the productivity of the sector (see Drine, I., 2011 and Alboghdady, M. & El-Hendawy, S. E., 2016) even though they did not look at inequality, they use data up to 2007 and 2009. All this with what was discussed earlier, make this study an addition to be further complemented with future research in the subject.

The analysis examines how this relationship is mediated by the structure of agricultural employment and the strength of institutional governance (United Nations Economic and Social Commission for Western Asia [ESCWA], 2024; World Bank, 2020). By doing so, the chapter aims to inform more inclusive and climate-resilient policy interventions targeting both environmental and social vulnerabilities.

The remainder of the chapter is organized as follows: Section 2 reviews the existing literature on climate vulnerability, inequality, and the agricultural channel; Section 3 gives an overview of the situation in the MENA region; Section 4 presents the empirical strategy, data sources, and methodology; Section 5 discusses the econometric results alongside complementary graphical analysis; and Section 6 concludes with a synthesis of the key findings and policy recommendations.

2. LITERATURE REVIEW

Over the past few decades, a growing body of research has highlighted the detrimental effects of global warming on economic output, particularly on Gross Domestic Product (GDP). These studies consistently demonstrate that rising temperatures and climate-related disruptions can significantly hinder economic growth and productivity at the macroeconomic level.

Beyond the direct impact on GDP, researchers have also investigated other consequences of climate change that indirectly affect economic performance. For instance, climate change influences mortality rates (Curriero, F. C. et al., 2002; Deschênes, O. & Greenstone, M., 2011; Deschênes, O. & Moretti, E., 2009), influences crime (Jacob, B. A. et al., 2007), and intensifies the links between drought and conflict

(Miguel, E. et al., 2004). These multifaceted effects contribute to broader social and economic instability, further compounding the challenges posed by climate change.

Despite this extensive research on the macroeconomic and social consequences, comparatively less attention has been devoted to understanding how climate change affects inequality within countries. Importantly, the impacts of climate change are not distributed evenly across populations. Islam and Winkel (2017) offer a theoretical framework explaining how climate change tends to increase within-country inequalities, disproportionately affecting vulnerable groups and deepening existing social and economic divides.

First, disadvantaged groups are often more exposed to climate hazards. Financial constraints frequently force these communities to reside in vulnerable areas such as flood-prone zones. Noted that a significant portion of the population in developing regions lives in "low-elevation coastal zones," making them particularly vulnerable to rising sea levels. These groups are also more likely to be employed in sectors that are highly sensitive to extreme weather, such as outdoor labor.

Second, disadvantaged groups are also more susceptible to the adverse effects of climate change. Bougnoux, N. et al. (2014) found that in some Middle Eastern and North African (MENA) countries, poorer households experienced greater income losses, crop damage, and livestock depletion compared to wealthier households. In addition, they face higher health risks and financial burdens due to climate hazards, which are often worsened by inadequate housing and infrastructure.

Third, these groups often lack the resources necessary to cope with and recover from climate-related disasters. Limited access to insurance, credit, and savings undermines their ability to bounce back, thereby reinforcing and deepening existing socioeconomic inequalities. This limited coping capacity contributes to a widening gap between rich and poor in the face of repeated climate shocks.

To further elucidate this concept, let's consider a scenario where both advantaged and disadvantaged groups face equal exposure and susceptibility to climate hazards. If both groups recover at an identical pace, then the inequality level will remain unchanged from its pre-disaster state.

However, if their recovery rates differ, the inequality gap will widen. There is a lot of evidence indicating disparities in recovery rates (Verner, D., 2010; Kraay & McKenzie, 2014; Ravallion & Jalan, 2001; McDowell & Hess, 2012), with disadvantaged groups typically exhibiting lesser ability to cope.

Islam and Winkel (2017) analyzed two transmission channels through which climate change impacts inequality: the economic and political channels. The economic channel involves the reduction of private resources available to disadvantaged groups. In unequal societies, individuals with lower income and fewer assets are more vulnerable to climate hazards and less capable of recovery afterward.

The political channel, on the other hand, operates through power dynamics. Advantaged groups often wield influence over governmental decisions, potentially delaying climate protection measures and leaving disadvantaged groups more exposed to risks. (See Huynh & Hoang, 2024)

Moreover, adaptation policies may disproportionately benefit advantaged groups. Political economy theories suggest that the concentration of political and economic power increases environmental degradation and hinders pro-environmental actions. For instance, in southern Mali, the increased vulnerability of villagers to catastrophic fires stems from the implementation of state-prescribed fire policies, which were, in turn, responses to international pressure to combat deforestation and desertification. This process involves establishing official links with various international organizations and ultimately leads to conditions of international indebtedness burdening the country. This shows that even mitigation is part of worsening inequality.

The Agricultural Channel

Climate change can impact inequality through several mechanisms, and agriculture is one of the most important channels, especially in developing economies where a significant share of the population depends on farming. As noted by Hallegatte and Rozenberg (2017) and Rao, van Ruijven, Riahi, and Bosetti (2017), the agricultural channel is a primary transmission mechanism through which climate change affects the poor and exacerbates income inequality.

This is largely due to the biological sensitivity of agriculture to climate variables. Dell, Jones, and Olken (2014) explain that, temperature and precipitation are direct inputs into agricultural production, making the sector one of the most immediately affected by climate change. Extreme weather events such as heatwaves and floods can significantly disrupt agricultural activity. Numerous studies have documented the negative impact of rising temperatures on crop yields. For example, Schlenker and Roberts (2009) found that in the U.S., temperatures exceeding 29–32°C sharply reduce crop yields. Similarly, Schlenker and Lobell (2010) found that higher temperatures lower agricultural output for Sub-Saharan Africa.

These findings underscore the importance of the agricultural channel in understanding how climate change contributes to inequality. This is not only because agriculture is climate-sensitive, but also because poor and marginalized populations are disproportionately dependent on this sector. Islam and Winkel (2017) note that these groups are more likely to live in high-risk areas and rely on agriculture, making them highly vulnerable to climate shocks such as droughts, floods, and temperature extremes. Their limited access to insurance, credit, or savings further reduces their ability to cope and recover, reinforcing existing inequalities.

Empirical evidence supports this theoretical link. Paglialunga, Coveri, and Zanfei (2022), using panel data for 150 countries from 2003 to 2017, identified agriculture as a key channel in the climate change–inequality nexus. Their findings show that temperature increases are associated with rising income inequality, especially in countries with large agricultural or rural populations. Specifically, they estimate that a 1% increase in temperature correlates with a 2-point rise in the Gini index in countries with high rural dependency.

Additional studies highlight the income and employment risks for rural populations. Jayachandran (2006) found that low rainfall significantly reduced rural wages in India. Similarly, Yang and Choi (2007) showed that rainfall shocks in the Philippines led to declining rural incomes. As a coping mechanism, households with migrant members received increased remittances, while those without migrants experienced sharp declines in consumption, illustrating the unequal coping capacities across households. In Mexico, Feng, Oppenheimer, and Schlenker (2012) found that declining crop yields due to climate change led to increased emigration, indicating a broader socioeconomic impact beyond agricultural output alone.

Altogether, these studies emphasize that agriculture is a central pathway through which climate change worsens poverty and inequality. Poor households suffer from declining yields, rising food prices, and loss of income, which can push them toward migration or reduce their standard of living. The burden falls disproportionately on rain-fed areas and on communities with limited adaptive capacity.

Beyond direct agricultural effects, climate change also reduces labor productivity, particularly for outdoor agricultural workers, and worsens health outcomes (e.g., child stunting, malaria) further deepening inequality and vulnerability.

Finally, the adverse distributional impacts of climate change through agriculture can be mitigated by inclusive development policies. These include investments in climate-resilient crops, improved irrigation systems, and social protection programs for vulnerable households. Targeted adaptation in the agricultural sector is essential to strengthen resilience and prevent the widening of rural–urban inequality.

3. EMPIRICAL STUDY

This section provides an overview of the data sources, key variables, and sample characteristics used in the empirical analysis

3.1 Sample Description

This study includes annual data from 13 countries in the MENA region (Kuwait, Bahrain, Qatar, Israel, Oman, Saudi Arabia, Libya, Tunisia, Iran, Egypt, Jordan,

Iraq, and Morocco) observed over the period 1996–2021, selected based on two criteria: significant climate vulnerability, as indicated by a ND-GAIN index over 40, and the availability of consistent income inequality data. These countries exhibit considerable structural heterogeneity, particularly in terms of the share of agricultural employment, which allows for the identification of cases where the agricultural sector acts as a key transmission channel between climate shocks and within-country income inequality.

This internal diversity is leveraged as a quasi-experimental source of variation: some countries combine high climate vulnerability with low agricultural dependence, while others face a dual exposure to both climatic stress and agricultural fragility. This configuration enables a differentiated analysis of the mechanisms through which climate change impacts income distribution and offers a relevant empirical ground to assess the moderating role of adaptation policies.

3.2 Data Overview

3.2.1 Vulnerability

The ND-GAIN Index is a composite indicator designed to measure a country's overall vulnerability to climate change as well as its readiness to adapt. It captures exposure to climate hazards, sensitivity of key systems, and adaptive capacity by combining data across multiple dimensions such as infrastructure, health, ecosystems, and socio-institutional preparedness.

The ND-GAIN Index is computed as follows:

$$\text{ND-GAIN Score} = (\text{Readiness score} - \text{Vulnerability score} + 1) \times 50.$$

Therefore, the closer the score is to 100, the better.

However, a key methodological limitation prevents us from using the full ND-GAIN index directly in our empirical model. Its “readiness” component incorporates socioeconomic indicators; including income inequality measures; which creates an endogeneity problem when inequality is the dependent variable of interest. In other words, using an index that partially includes inequality to explain inequality risks contaminating our results with circular reasoning.

To address this issue and ensure a clear conceptual and statistical separation between the explanatory and dependent variables, we isolate and retain only the vulnerability component of the ND-GAIN index collected from the Notre Dame Global Adaptation Initiative database. This sub-index captures a country's structural exposure to climate-related risks across six critical sectors: food, water, health, ecosystem services, human habitat, and infrastructure. Crucially, it does not embed any indicators related to income distribution, making it a more exogenous and theoretically sound measure for assessing the impact of climate vulnerability on within-country income inequality; particularly through the agricultural sector, which is central to our transmission hypothesis.

3.2.2 Inequality

The majority of studies analyzing the effect of climate change on income inequalities use the Gini index due to its simplicity and wide availability.

However, several major limitations justify an alternative approach. As Piketty and Goldhammer (2014) emphasized, it is methodologically reductive to summarize such a complex reality as inequality with a single index. Two countries can have the same Gini index while exhibiting fundamentally different income distributions because of incomparable Lorenz curves. For example, in 2015, Greece and Thailand both had an identical Gini coefficient (0.360), but the ratio of income between the top 10% and the bottom 10% was 13.8 in Greece versus 8.9 in Thailand; a significant difference completely masked by the Gini index.

This observation led us to favor a measure that is more sensitive to the extremes of the distribution: the ratio between the incomes held by the top 10% richest and the bottom 50% poorest. This ratio, collected from the WID database (World Inequality Database), captures two essential segments of the social structure: on the one hand, individuals benefiting from the highest incomes; often better protected against climate, urban, or sector-specific vulnerabilities; and on the other hand, the lower half of the population, which includes the households most exposed and sensitive to climate shocks, especially those dependent on the agricultural sector.

Importantly, this income measure is pre-tax, which aligns with our objective of analyzing the structural impact of climate vulnerability on economic inequality through the lens of agricultural employment. Using pre-tax income makes it possible to isolate the role of market dynamics; such as labor income disparities or productivity shocks; before any redistribution by the state, which may itself vary in capacity or responsiveness across contexts. This allows us to better capture the raw exposure of different income groups to climate-related shocks and their differentiated adaptive capacities, particularly in the agricultural sector.

3.2.3 Agricultural Employment as a Transmission Channel

Beyond climate vulnerability and income inequality, this study also considers agricultural employment as a key mediating variable in the relationship between climate shocks and income distribution. A country's dependence on agriculture can be assessed through various lenses; economic, trade-related, or social. Here, we focus on social dependence, which we measure using the share of agricultural employment in the total labor force. This variable was collected from the World Bank database.

This indicator reflects the proportion of the working population whose livelihoods are directly tied to agricultural activities. We choose this approach based on the hypothesis that households dependent on agriculture are among the most vulnerable to climate shocks, and therefore more likely to experience negative distributional consequences. While this variable does not capture the agricultural sector's contribution to GDP or exports, it highlights the exposure of a structurally sensitive segment of the population, especially in developing economies.

Although our analysis targets national income inequality, we argue that the agricultural sector plays a central role in the transmission of climate effects. In countries where a large share of the labor force depends on agriculture, climate shocks; such as droughts or floods; can significantly reduce rural incomes, thereby widening income gaps with urban or formally employed populations. Thus, even though our inequality measures cover the entire population, agricultural employment serves as a critical transmission channel linking climate vulnerability to distributional outcomes.

3.3 Empirical Evidence

To motivate our empirical strategy, this section presents a set of descriptive patterns based on the key variables of interest: income inequality, climate vulnerability, and agricultural employment. These stylized facts aim to provide preliminary insights into the relationship between structural exposure to climate risks and the distribution of income within countries.

The patterns highlighted below do not constitute causal evidence but serve to illustrate potential channels through which climate-related risks might disproportionately affect income distribution; especially in countries where a large share of the labor force is dependent on agriculture.

3.3.1 Distribution of ND-GAIN Vulnerability Index Across Countries

This figure displays the ND-GAIN vulnerability scores for the countries in the dataset, showing the range and spread of climate vulnerability.

Figure 1. Distribution of ND-GAIN Vulnerability Index Across Countries (1996-2021). This figure illustrates the ND-GAIN index scores, highlighting countries with differing levels of climate risk exposure and resilience, which shape their economic fragility and vulnerability to climate shocks.

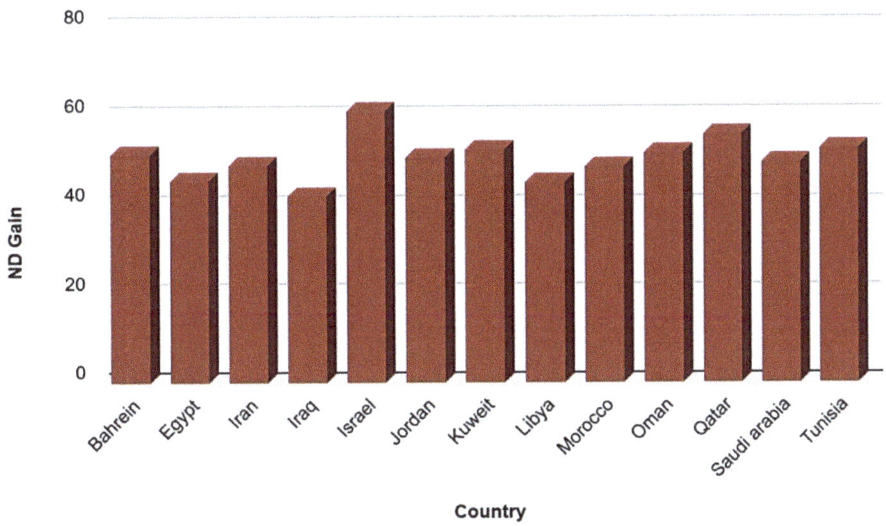

Source: *created by the authors*

As illustrated in Figure 1, the ND-GAIN index highlights significant disparities in climate resilience across countries. A low ND-GAIN score indicates not only a high exposure to risks such as droughts, water stress, and extreme temperatures but also a limited institutional capacity to absorb and manage these shocks, due to insufficient infrastructure, crisis management mechanisms, and social protection systems.

Countries identified as most vulnerable; such as Egypt, Iraq, and Libya; are therefore economically fragile, as climate shocks are poorly mitigated and tend to directly affect household incomes, especially in sensitive sectors like agriculture.

3.3.2 Climate vulnerability and Agricultural Employment: An Economic Risk Factor

The figure below presents the relationship between the climate vulnerability and agriculture employment across countries.

Figure 2. Climate vulnerability and agricultural employment (1996-2021). Classification of countries based on their share of agricultural employment and pre-tax income inequality ratio (top 10% to bottom 50%) using median values for the period 1996–2021.

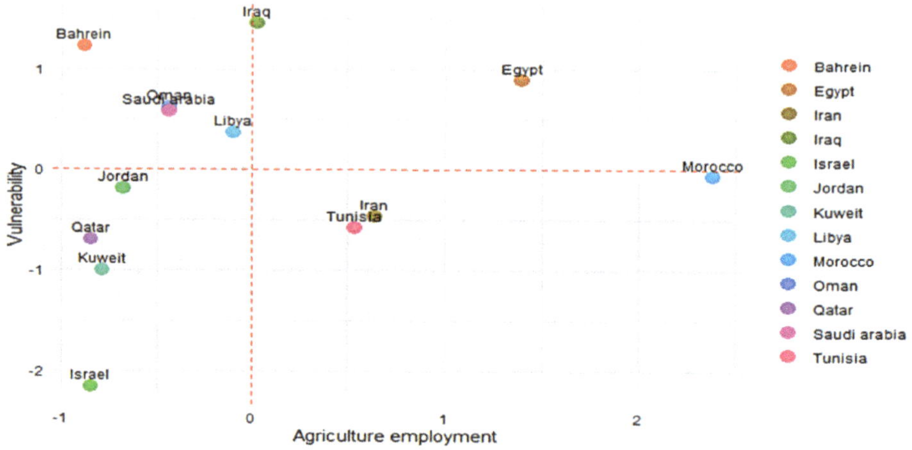

Source: *created by the authors*

Figure 2 indicates that countries are classified according to whether their agricultural employment share and the pre-tax income inequality ratio (top 10% to bottom 50%) are above or below their respective median values.

The Figure shows that in countries like Morocco and Egypt, where agriculture represents a large share of employment, primary market inequalities remain high.

This can be explained by several economic factors:

- The agricultural sector concentrates the least productive jobs, which are often informal and thus poorly paid.
- Agricultural income is highly sensitive to climate variability, increasing its instability.

This implies that in countries with a high reliance on agriculture, the majority of the workforce belongs to the lower half of the income distribution (the bottom 50%) and remains highly exposed to climate shocks without effective compensatory mechanisms.

3.3.3 Cross-Classification of Countries by Agricultural Employment and Income Inequality

Figure 3 divides countries according to the standardized levels of agricultural employment and pre-tax income inequality (top 10%/bottom 50%), using median cutoffs to classify each variable.

Figure 3. Agricultural Employment vs. Income Inequality Ratio (Top 10% / Bottom 50%). Distribution of countries in the MENA region based on agricultural employment share and pre-tax income inequality ratio (top 10% to bottom 50%), highlighting the diverse socio-economic contexts from 1996 to 2021.

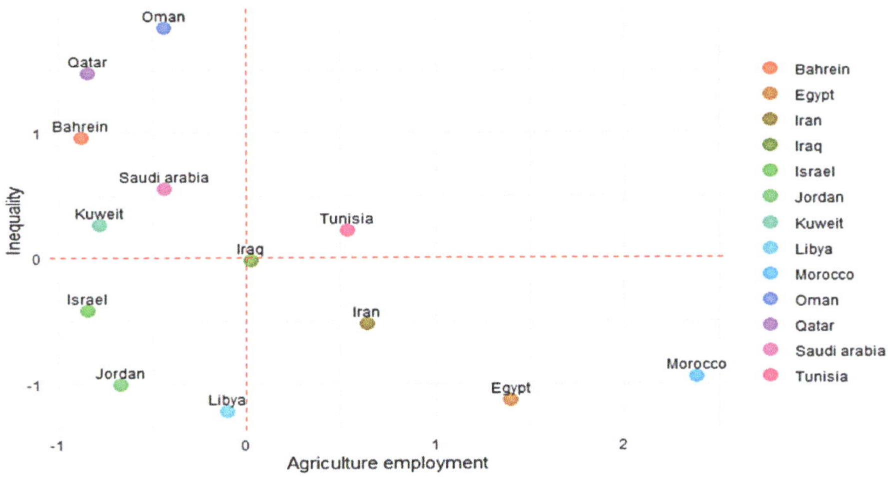

Source: *created by the authors*

However, Figure 3 brings a more nuanced perspective. It shows that the most agriculture-dependent countries are not necessarily the most unequal in terms of pre-tax income distribution. Morocco and Egypt, despite their strong reliance on agriculture, display inequality levels below the regional median. Yet, this does not imply that these societies are egalitarian. On the contrary, both countries face deep-rooted structural inequalities, often not fully reflected in aggregate indicators.

In Morocco, a high share of agricultural employment coexists with persistent rural poverty and marked regional disparities. Similarly, in Egypt, income gaps between rural Upper Egypt and the urbanized Nile Delta remain significant. These territorial and sectoral inequalities; relating to access to education, water, land, and agricultural finance; are rarely captured in standard pre-tax income metrics. Moreover, it is important to note that post-tax inequality in these countries tends to be significantly higher, due to weak redistributive systems and limited social protection.

Tunisia, meanwhile, illustrates a different dynamic. Although it also maintains a relatively high level of agricultural employment, it displays pre-tax inequality levels above the median. This suggests that agricultural dependence alone does not explain income inequality. In Tunisia's case, structural and regional imbalances; particularly between coastal cities and the rural interior; combined with an agricultural sector that is fragmented and unequal, help explain the higher income disparities. Unlike Morocco or Egypt, Tunisia's inequality may reflect not only poverty, but also a greater polarization of opportunity and access to resources within its economy.

These observations invite caution against assuming a mechanical link between vulnerability, agriculture, and high inequality. In reality, it is the economic and institutional structure of a country that determines how vulnerability translates into inequality. In low-diversity rural economies, incomes may be uniformly low, giving an illusion of equality where in fact many are equally vulnerable. Conversely, countries with low vulnerability and minimal agricultural employment, such as Qatar or Oman, exhibit very high pre-tax inequality, often driven by deep socio-economic dualisms; between nationals and migrant workers, or between public and private sectors.

3.3.4 Exploring the Non-Linear Dynamics Between Environmental Exposure and Economic Disparity

The following graph presents a cross-country comparison within the MENA region, using pre-tax inequality measures and overall vulnerability scores.

Figure 4. Climate Vulnerability vs. Pre-Tax Income Inequality in MENA Countries. Relationship between climate vulnerability and pre-tax income inequality across MENA countries, illustrating the varied interactions between environmental risks and economic disparities from 1996 to 2021.

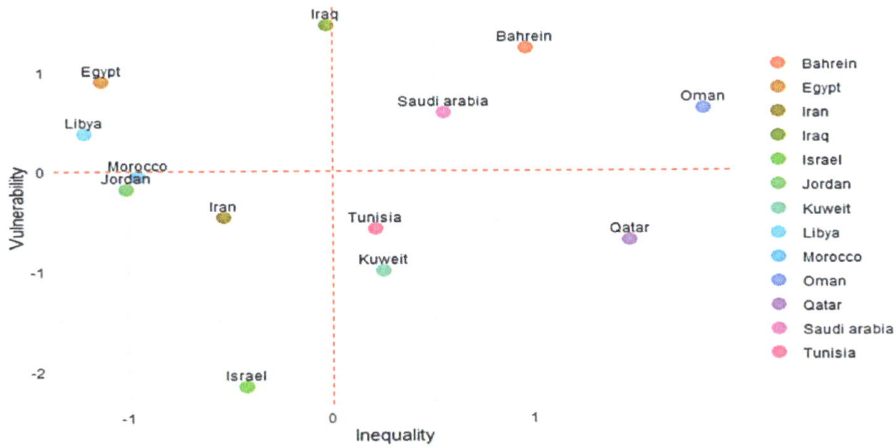

Source: *created by the authors*

The results of Figure 4, which cross climate vulnerability with pre-tax income inequality, confirm that the relationship between these two dimensions is neither linear nor systematic. Some highly vulnerable countries, such as Iraq or Egypt, display moderate levels of inequality, while less vulnerable countries like Qatar or Kuweit exhibit very high inequality levels.

This observation echoes the findings from the previous graphs; particularly those on the share of agricultural employment; which showed that the most agriculture-dependent countries (e.g., Morocco, Egypt) are not necessarily the most unequal. This suggests that a strong reliance on agriculture may lead to income homogeneity, not through equitable distribution, but due to widespread poverty and a lack of economic diversification. Conversely, in rentier or highly urbanized economies, inequality is amplified by dual socio-economic structures.

Thus, climate vulnerability does not mechanically lead to economic inequality: its impact depends on the country's structural context, including its production model, the weight of agriculture, and the quality of its institutions. This reading invites a reconsideration of the initial research question: it is not only a matter of assessing whether vulnerability increases inequality, but of understanding in which contexts it does so, and through which structural channels this transmission occurs.

3.4 Empirical Methodology

The following section presents the methodological framework of this study. It details the empirical strategy employed, including the specification of econometric models and the selection of key variables.

3.4.1 Methodology

This study investigates the impact of climate vulnerability on income inequality in MENA countries, with a particular focus on how this relationship is mediated by the structure of the economy; especially the agricultural sector; and moderated by the quality of institutions. While existing literature suggests a direct link between climate shocks and income disparities, our descriptive analysis highlights a more complex picture: countries with high agricultural employment (such as Morocco and Egypt) may exhibit lower pre-tax inequality. In contrast, Tunisia, with similar agricultural dependence, shows higher inequality, suggesting that structural and institutional factors condition this relationship.

3.4.2 Empirical Model

Building on existing literature, this study explores the empirical relationship between climate vulnerability and income inequality across 13 MENA countries (Kuwait, Bahrain, Qatar, Israel, Oman, Saudi Arabia, Libya, Tunisia, Iran, Egypt, Jordan, Iraq, and Morocco) over the period 1996 to 2021, while controlling for conventional determinants of income distribution.

To ensure the robustness and reliability of our panel data estimation strategy, we undertook a comprehensive set of diagnostic tests related to model specification, error structure, and the appropriateness of our estimation approach. The dependent variable in our model is the logarithm of income inequality, explained by key dimensions of climate vulnerability alongside control variables such as trade openness, governance quality, population density, population size, and real GDP growth. We began by testing for the appropriate panel estimator using a Hausman test, which yielded a chi-squared statistic of 3.47 with a p-value of 0.7481. This result fails to reject the null hypothesis of no systematic difference between fixed and random effects estimators, thereby justifying the use of the random effects approach. To assess potential multicollinearity among regressors, we examined Variance Inflation Factors (VIFs), with results showing a mean VIF of 1.9 and all values well below the commonly accepted threshold of 5, indicating no serious collinearity concerns.

We then conducted a series of residual diagnostics. The Breusch-Pagan test for heteroskedasticity returned a significant result ($p = 0.0326$), suggesting the presence of non-constant error variance. Additionally, the Breusch-Godfrey/Wooldridge test for serial correlation indicated strong autocorrelation in the idiosyncratic errors ($p < 2.2e-16$). The Jarque-Bera test for normality of residuals also yielded a highly significant result ($p < 2.2e-16$), leading to a rejection of the null hypothesis and indicating the presence of non-normality in the residual distribution. However, the Pesaran CD test for cross-sectional dependence did not provide evidence of significant dependence across units ($p = 0.3117$), supporting the assumption of cross-sectional independence. In light of these findings, specifically, the presence of individual effects without significant time effects, the appropriateness of random effects estimation, and the violations of classical assumptions regarding homoskedasticity, independence, and normality, we adopt a random effects panel model with robust standard errors. This specification mitigates the impact of model misspecification and improves the reliability of statistical inference. (Refer to appendix for more details)

The core regression model estimated in this study is as follows:

$$Inequality_{it} = \beta . Vulnerablity_{it} + \gamma . X_{it} + u_t + \varepsilon_{it} \quad (1)$$

Where:

$Inequality_{it}$ is the top 10% / bottom 50% pre-tax income ratio (from the World Inequality Database),

$Vulnerablity_{it}$ is the ND-GAIN climate vulnerability index (Notre Dame University),

X_{it} is a vector of control variables including:

- Log (GDP per capita) (World Bank),
- trade openness (World Bank),
- population density (World Bank),
- institutional quality.

To capture the heterogeneous effect of vulnerability across different structural contexts, we extend the model by including interaction terms:

$$Inequality_{it} = \beta(Vulnerablity_{it} * quality\ of\ institutions_{it}) + \gamma X_{it} + u_t + \varepsilon_{it} \quad (2)$$

$$Inequality_{it} = \alpha_i + \beta\left(Vulnerablity_{it} * employment\ in\ agriculture_{it}\right) + \gamma X_{it} + u_t + \varepsilon_{it} \quad (3)$$

These interactions allow us to test two key hypotheses:

- that vulnerability is more likely to increase inequality in countries with a high share of agricultural employment,

- that strong institutions can mitigate the inequality-enhancing effects of climate vulnerability.

Finally, we disaggregate the ND-GAIN vulnerability index into its six sectoral components (food, water, health, ecosystem services, human habitat, and infrastructure) to identify which specific dimensions of vulnerability are most closely associated with inequality outcomes. This helps clarify whether certain forms of structural exposure (e.g., to water stress or food insecurity) are more inequality-inducing than others.

4. RESULTS

This section presents the empirical results of the analysis exploring the impact of climate vulnerability on pre-tax income inequality in MENA countries from 1996 to 2021.

4.1 Model Presentation

The first model (refer to Table 1), aims to assess the direct effect of climate vulnerability on pre-tax income inequality. It seeks to determine whether increased exposure to environmental risks; such as droughts, water stress, and ecological instability; is significantly associated with higher levels of primary income inequality. This baseline framework captures the structural link between environmental stress and income distribution, independently of other economic or institutional factors.

The second model builds on this relationship by introducing an interaction term between climate vulnerability and the share of agricultural employment in the economy (refer to the second collum of Table 1). This specification allows us to test whether the productive structure of a country; particularly its reliance on agriculture; can mitigate or amplify the impact of climate shocks on income disparities.

The third model explores a different source of heterogeneity: governance. It introduces an interaction between climate vulnerability and institutional quality (refer to the third collum of Table 1). The goal here is to examine the extent to which states with robust administrative capacity and redistributive mechanisms can dampen the inequality-inducing effects of environmental shocks.

Finally, the fourth model disaggregates climate vulnerability into its key components; food security, water availability, health, ecosystems, infrastructure, and housing; to identify which dimensions are most critical in transmitting climate shocks to income inequality. This approach provides a more granular understanding of how different aspects of vulnerability drive inequality. (refer to the fourth column of Table 1)

Table 1. Random Effects Panel Regression Results (Robust Standard Errors, Arellano Method)

Variable	(1) Coef. (SE)	(2) Coef. (SE)	(3) Coef. (SE)	(4) Coef. (SE)
Vulnerability	2.869*** (0.00002)	12.248*** (0.000)	4.223*** (0.00000)	
Empagr			0.063*** (0.0001)	
Water				1.061 (0.064)
Food				-0.807*** (0.0005)
Health				-0.097 (0.615)
Ecosystems				2.433*** (0.000)
Habitat				1.452*** (0.000)
Infrastructure				0.130 (0.587)
Trade openness	0.001*** (0.001)	0.001** (0.001)	0.001** (0.003)	0.001*** (0.00004)
Governance	0.530*** (0.0002)	8.586*** (0.000)	0.436** (0.004)	0.302* (0.032)
Population density	-0.0001* (0.047)	-0.0002** (0.001)	-0.0001 (0.207)	-0.0002*** (0.0003)
Vulnerability × empagr			-0.126** (0.0004)	
GDP per capita	0.084 (0.079)	0.108* (0.043)	0.272*** (0.00002)	-0.058 (0.325)
Vulnerability × governance		-19.783*** (0.000)		
Constant	1.131** (0.010)	-2.811*** (0.001)	-0.243 (0.598)	1.122* (0.020)
Model Statistics:				
	(1)	**(2)**	**(3)**	**(4)**
Observations	338	338	338	338
R^2	0.154	0.239	0.194	0.221
Adjusted R^2	0.142	0.226	0.177	0.197
F Statistic	60.600***	104.141***	79.259***	92.601***

Notes: Robust standard errors estimated using the Arellano method are reported in parentheses.

Significance levels: $*p < 0.05$; $**p < 0.01$; $***p < 0.001$.

Notes: Model 1 estimates the direct effect of climate vulnerability on pre-tax income inequality. Models 2–4 sequentially incorporate interactions with agricultural employment and institutional quality, and disaggregate vulnerability into its main components (food security, water, health, ecosystems, infrastructure, housing) to assess structural and sectoral heterogeneity in the climate–inequality relationship.

Source: authors' own work

4.2 Economic Interpretation of the Empirical Results

The first model highlights a clear and robust link between climate vulnerability and pre-tax income inequality. The positive and highly significant coefficient on the "vulnerability" variable indicates that countries more exposed to climate risks tend to exhibit higher levels of primary income inequality. This result supports the idea of a direct transmission mechanism from environmental shocks to household income, particularly in economies where poor households depend on climate-sensitive sectors such as agriculture or informal rural employment. These populations, lacking social safety nets or institutional buffers, see their incomes fall sharply in the event of climate shocks, while wealthier groups, often employed in protected or tertiary sectors, remain relatively unaffected.

The second model introduces an interaction term between climate vulnerability and the share of agricultural employment. Interestingly, the interaction term is negative and significant, meaning that the impact of climate vulnerability on inequality is attenuated in countries with high agricultural employment. While this may seem counterintuitive, it can be explained by the fact that, in highly agricultural economies, most of the labor force is already concentrated in the lower end of the income distribution. As a result, climate shocks affect a relatively homogeneous group, which limits income dispersion and thereby moderates the rise in measured inequality. In other words, poverty is widespread, but income differences between the top and bottom are compressed due to the lack of a sufficiently differentiated middle class.

The third model examines the role of institutions in moderating this relationship. The interaction between climate vulnerability and governance quality is strongly negative and significant, meaning that strong institutions mitigate the inequality-inducing effects of climate shocks. In practical terms, this means that in countries with effective governance; characterized by legal stability, administrative capacity, and active public policies; the adverse effects of climate vulnerability on income distribution are significantly reduced. This result is economically grounded: well-governed states can implement redistributive policies, social assistance programs, insurance mechanisms, or targeted investments in adaptation (e.g., rural infrastructure, irrigation, early warning systems), thereby buffering vulnerable households from the full brunt of climate shocks. In contrast, in contexts of weak governance, shocks are transmitted directly to incomes without any cushioning, thereby exacerbating disparities.

Finally, the fourth model breaks down climate vulnerability by its specific components. The results show that certain dimensions are particularly relevant in driving inequality. Weaknesses in ecosystems and housing conditions emerge as significant aggravating factors, confirming that physical and environmental exposure; such as living in arid zones or precarious settlements; amplifies income gaps. Notably, the food security variable shows a negative and significant effect, suggesting that countries with stronger food systems tend to experience lower income inequality. This may be attributed to reduced domestic food price volatility and better access to basic goods for poor households. Conversely, the water, health, and infrastructure dimensions show more heterogeneous effects, likely due to the diverse climatic and institutional conditions across countries.

Taken together, these findings confirm the hypothesis of a dual deficit; both economic and institutional; in the most vulnerable countries. The economic deficit stems from heavy dependence on agriculture, which increases direct exposure to climate shocks, while the institutional deficit reflects the inability to compensate or regulate these effects through policy. If unaddressed, this dual structural mechanism risks entrenching long-term social polarization, undermining economic stability and social cohesion.

4.3 Cross-Interpretation with Graphical Analysis

The empirical results from the econometric models are further supported and enriched by the graphical analysis, which provides contextual and visual insight into the effects of climate vulnerability on pre-tax income inequality.

The first figure illustrates the relationship between climate vulnerability and the ND-GAIN index. Countries such as Egypt, Iraq, and Libya; characterized by low ND-GAIN scores; exhibit both high exposure to climate shocks and limited institutional capacity to respond. This observation is consistent with the findings of the first model: in the absence of effective public investment or safety nets, environmental shocks directly translate into income losses, particularly for those in climate-sensitive sectors like agriculture.

The second and third figures examine the relationship between agricultural employment and income inequality, expressed through the top 10% to bottom 50% income ratio. These visuals support the second econometric model's conclusion that vulnerability's impact is dampened in economies with large agricultural sectors. For instance, in Egypt or Morocco, a substantial portion of the labor force occupies the lower segment of the income distribution, leading to limited measured income dispersion despite widespread poverty. This reflects a statistical flattening of inequality in a context of diffuse deprivation.

However, the third figure adds nuance: it shows that countries with high agricultural shares are not necessarily the most unequal according to standard metrics. This suggests that agriculture does not inherently generate inequality but may conceal structural disparities; such as territorial or sectorial inequalities; that are poorly captured by pre-tax statistics.

The third model's insights on governance are further illustrated by regional case studies. Tunisia, despite its agricultural profile, exhibits relatively high inequality, likely due to territorial fragmentation and a dualistic agricultural system not adequately addressed by public policy. In contrast, countries with stronger institutions might better absorb similar shocks, highlighting the moderating role of governance in the inequality-vulnerability nexus.

The final figure combines climate vulnerability and pre-tax inequality, echoing the logic of the fourth model. It confirms that the effect of climate shocks depends not only on overall vulnerability but also on its specific components and national structural characteristics. Highly vulnerable countries like Egypt or Iraq are not always the most unequal, while relatively resilient countries such as Qatar or Kuwait may exhibit significant inequality driven by deep socio-economic cleavages (e.g., dual labor markets, migration status, wealth inequality). This heterogeneity aligns with the component-level analysis, which suggests that factors like ecosystem degradation and food insecurity are more strongly linked to income disparities, whereas the effects of health and infrastructure vary more widely depending on the country.

In sum, the graphical analysis reinforces and nuances the findings from the econometric models. It confirms that the relationship between climate vulnerability and inequality is neither linear nor automatic, but highly contingent on a country's productive structure, agricultural dependency, territorial dynamics, and above all, institutional quality. The combination of empirical data and visual evidence shows that climate vulnerability acts as a magnifier of existing inequalities. However, its actual impact is shaped by the national context; economic, social, and political. Addressing this complex interplay is therefore essential to achieving both climate resilience and social equity.

4.4 Policy Recommendations

This study's findings highlight the urgent need for targeted and context-sensitive policy interventions to mitigate the exacerbating effects of climate vulnerability on income inequality, particularly in countries with structural agricultural dependence and limited institutional capacity.

First, improving governance quality is foundational. In countries such as Iraq and Libya, where institutional fragmentation and low administrative capacity prevail, efforts must focus on building resilient state structures. Enhancing legal stability, administrative effectiveness, and transparency will enable these states to design and enforce redistributive policies and implement adaptation strategies that protect the most vulnerable populations from climate-induced income shocks.

Second, the agricultural sector must be strengthened, particularly in Egypt and Morocco, where a large share of the labor force is engaged in low-productivity and climate-sensitive farming. Investment in climate-smart agricultural practices, such as drought-resistant crops, early warning systems, and sustainable land use, is essential. Equally crucial is the development and rehabilitation of water infrastructure. In these countries, where water scarcity is a pressing issue, upgrading irrigation systems, improving water storage and distribution networks, and implementing equitable water governance mechanisms can reduce both the volatility of rural incomes and regional inequalities. For example, modernizing Nile-dependent irrigation in Egypt or addressing groundwater overuse in Morocco could significantly increase climate resilience and income stability in rural areas.

In Tunisia, where inequality is compounded by dualistic agricultural structures and regional disparities, integrated rural development policies are needed. This includes expanding access to infrastructure, markets, and public services in interior regions, and ensuring that climate adaptation measures do not disproportionately favor capital-intensive or export-oriented farming.

Third, enhancing basic services such as health, education, and housing remains a critical pillar across the region. In fragile contexts like Libya, and in rapidly urbanizing areas across the MENA region, policies must address the vulnerabilities associated with informal settlements, degraded ecosystems, and underdeveloped health infrastructure; all of which can deepen inequality in the wake of climate shocks.

Lastly, international support plays a crucial role, especially for countries facing fiscal and technological constraints. Climate finance, technical assistance, and regional cooperation should be scaled up to help countries implement structural reforms and invest in adaptive infrastructure. Coordinated efforts with international development institutions, including targeted funding for water resource management and climate-resilient agriculture, will be key to reducing both environmental vulnerability and its socioeconomic consequences.

In sum, tackling climate-induced inequality requires not only cross-sectoral national action, but also differentiated strategies adapted to the specific economic and institutional realities of each country. Only through such a comprehensive and localized approach can climate resilience and social equity be jointly achieved.

4.5 Robustness Analysis and Limitations: Toward More Granular Research on Climate Inequality in MENA

As part of the robustness analysis, we employed the GINI Index as an alternative measure of income inequality. The findings consistently confirmed a positive and statistically significant relationship between climate vulnerability and inequality. However, the coefficient differed from the one obtained when using the income share ratio of the top 10% to the bottom 50%, which is expected since the GINI Index captures the entire income distribution rather than just the extremes. Specifically, we found that a one-point increase in a country's vulnerability score increases the GINI Index by 0.04%. This result is in line with the findings of Cevik, S., and Jalles, J. T. (2022) who also used the ND-GAIN vulnerability index.

We also ran the estimations using different panels of countries, and the results remained robust. However, it is important to highlight that these results apply primarily to developing countries, as the coefficient was not statistically significant for developed countries.

In a second robustness check, we replaced the climate vulnerability variable with temperature change (relative to pre-industrial levels), a commonly used indicator in the climate change literature. The results showed that rising temperatures have a positive and significant impact on inequality. Nevertheless, the coefficients obtained with climate vulnerability were both larger and more statistically significant. This suggests that climate vulnerability, being a composite indicator that includes temperature, exposure, and adaptive capacity, captures the impact of climate change on inequality more effectively and earlier than temperature change alone.

Additionally, when analyzing the interaction between temperature change and agricultural employment, the results were consistent with our main findings. However, the interaction between temperature change and institutional quality was not statistically significant, suggesting that institutional resilience alone may not mitigate inequality effects in the short term.

In summary, our findings suggest that climate change consistently worsens within-country inequality, a result that holds under multiple specifications. The stronger and more robust results obtained using the ND-GAIN vulnerability index underscore the multi-dimensional nature of climate risks, which cannot be captured by temperature alone.

Despite these contributions, some limitations remain. The MENA country panel used in this study was relatively small and did not fully capture the region's heterogeneity, due to limited data availability for the period 1995–2021. Although we applied robust standard errors, which are generally reliable for small panels, this restriction limits the generalizability of our findings.

For future research, we recommend using micro-level data where possible, as it could offer more granular insights into how climate change affects inequality at the household or regional level. In the absence of such data, a focused study on the middle class could also be valuable, especially since no existing research has examined the specific impact of climate change on the middle class in the MENA region.

5. CONCLUSION

This research set out to explore the relationship between climate vulnerability and pre-tax income inequality, focusing on the structural and sectoral conditions that shape this dynamic. Through a series of econometric models and a complementary graphical analysis, the findings reveal that the effect of climate vulnerability on inequality is neither automatic nor uniform across countries. Instead, it is deeply conditioned by a country's economic structure, particularly the weight of agriculture in employment, and the quality of its institutional framework.

The analysis shows that countries highly exposed to climate risks tend to exhibit greater income inequality, especially when governance mechanisms are weak and households depend on climate-sensitive sectors. However, the relationship is more nuanced in economies with a large agricultural workforce. In these contexts, climate shocks often affect a relatively homogeneous low-income population, which may compress income disparities while entrenching widespread poverty. On the other hand, countries with stronger institutions appear more capable of buffering the inequality-inducing effects of climate shocks through redistributive policies, infrastructure investment, and social protection systems.

Furthermore, the decomposition of vulnerability into its underlying components highlights the differentiated impact of various climate stressors. Factors such as ecosystem fragility and food insecurity are particularly associated with rising inequalities, whereas others, like access to water or health infrastructure, show more heterogeneous effects depending on national contexts.

However, this study also faces several limitations. First, the analysis relies primarily on pre-tax income inequality, which may underestimate the full extent of disparities, especially in contexts where redistribution is minimal or informal economies are large. Second, the quality and consistency of vulnerability and governance indicators across countries can vary, which may affect the robustness of the estimates. Third, the study captures cross-country patterns but does not account for intra-national inequalities, such as rural-urban divides or regional disparities, which are particularly relevant in countries with dual economies. Lastly, the agricultural sector itself is treated in aggregate terms, whereas a more disaggregated approach, distinguishing between subsistence, export-oriented, and informal farming; could offer deeper insights into the mechanisms linking climate vulnerability to inequality.

In sum, the findings underscore the importance of understanding climate vulnerability not just as an environmental risk, but as a socio-economic amplifier. Its impact on inequality is mediated by structural features, such as reliance on agriculture and regional disparities; as well as by institutional capacity. Policy responses must therefore go beyond climate adaptation and address the underlying economic and institutional deficits that make societies more vulnerable to both environmental and social shocks. Failing to do so risks reinforcing a cycle of exclusion and fragility, with profound implications for long-term development and social cohesion.

REFERENCES

Alboghdady, M., & El-Hendawy, S. E. (2016). Economic impacts of climate change and variability on agricultural production in the Middle East and North Africa region. *International Journal of Climate Change Strategies and Management*, *8*(5), 663–678. DOI: 10.1108/IJCCSM-07-2015-0100

Bougnoux, N., Wodon, Q., Liverani, A., & Joseph, G. (2014). *Climate change and migration: Evidence from the Middle East and North Africa* (World Bank Publications No. 18929). The World Bank.

Cevik, S., & Jalles, J. T. (2022). *For whom the bell tolls: Climate change and inequality* (IMF Working Paper No. 103). International Monetary Fund. https://doi.org/DOI: 10.5089/9798400208126.001

Chancel, L., Bothe, P., & Voituriez, T. (2023). *Climate inequality report 2023: Fair shares in a warming world*. World Inequality Lab.

Curriero, F. C., Heiner, K. S., Samet, J. M., Zeger, S. L., Strug, L., & Patz, J. A. (2002). Temperature and mortality in 11 cities of the eastern United States. *American Journal of Epidemiology*, *155*(1), 80–87. DOI: 10.1093/aje/155.1.80 PMID: 11772788

Del Pozo, A., Brunel-Saldias, N., Engler, A., Ortega-Farias, S., Acevedo-Opazo, C., Lobos, G. A., Jara-Rojas, R., & Molina-Montenegro, M. A. (2019). Climate change impacts and adaptation strategies of agriculture in Mediterranean-climate regions (MCRs). *Sustainability (Basel)*, *11*(10), 2769. Advance online publication. DOI: 10.3390/su11102769

Dell, M., Jones, B. F., & Olken, B. A. (2014). What do we learn from the weather? The new climate–economy literature. *Journal of Economic Literature*, *52*(3), 740–798. DOI: 10.1257/jel.52.3.740

Deschênes, O., & Greenstone, M. (2011). Climate change, mortality, and adaptation: Evidence from annual fluctuations in weather in the United States. *American Economic Journal. Applied Economics*, *3*(4), 152–185. DOI: 10.1257/app.3.4.152

Deschênes, O., & Moretti, E. (2009). Extreme weather events, mortality, and migration. *The Review of Economics and Statistics*, *91*(4), 659–681. DOI: 10.1162/rest.91.4.659

Drine, I. (2011). *Climate change compounding risks in North Africa* (WIDER Working Paper No. 2011/32). United Nations University World Institute for Development Economics Research.

Elkhalfi, O., Chaabita, R., Ghoujdam, M., Zehraoui, K., El Alaoui, H., & Belhaj, I. (2025). The impact of climate change on food security in the Middle East and North Africa: Challenges and adaptation strategies. *Journal of Agriculture and Food Research, 21*, 101963. DOI: 10.1016/j.jafr.2025.101963

Feng, S., Oppenheimer, M., & Schlenker, W. (2012). *Climate change, crop yields, and internal migration in the United States* (NBER Working Paper No. 17734). National Bureau of Economic Research. https://www.nber.org/papers/w17734

Gentle, P., & Maraseni, T. N. (2012). Climate change, poverty and livelihoods: Adaptation practices by rural mountain communities in Nepal. *Environmental Science & Policy, 21*, 24–34. DOI: 10.1016/j.envsci.2012.03.007

Hallegatte, S., & Rozenberg, J. (2017). Climate change through a poverty lens. *Nature Climate Change*, *7*(4), 250–256. DOI: 10.1038/nclimate3253

Huynh, C. M., & Hoang, H. H. (2024). Climate change and income inequality in Asia: How does institutional quality matter? *Journal of the Asia Pacific Economy*. Advance online publication. DOI: 10.1080/13547860.2024.2315700

IEMed. (2021). *Climate change in the MENA region: Environmental risks, socio-economic effects and policy challenges for the future*. IEMed. https://www.iemed.org/wp-content/uploads/2021/01/Climate-Change-in-the-MENA-Region.pdf

IPCC. (2021). *Climate change 2021: The physical science basis. Contribution of Working Group I to the Sixth Assessment Report of the Intergovernmental Panel on Climate Change* (Masson-Delmotte, V., Eds.). Cambridge University Press., https://www.ipcc.ch/report/ar6/wg1/

IPCC. (2022). *Climate change 2022: Impacts, adaptation, and vulnerability. Contribution of Working Group II to the Sixth Assessment Report of the Intergovernmental Panel on Climate Change* (Pörtner, H.-O., Eds.). Cambridge University Press., DOI: 10.1017/9781009325844

IPCC. (2023). *Climate change 2023: Synthesis report. Contribution of Working Groups I, II and III to the Sixth Assessment Report of the Intergovernmental Panel on Climate Change* (Core Writing Team, H. Lee, & J. Romero, Eds.). IPCC. https://www.ipcc.ch/report/ar6/syr/

Islam, S. N., & Winkel, J. (2017). *Climate change and social inequality* (UN DESA Working Paper No. 152). United Nations Department of Economic and Social Affairs. https://www.un.org/esa/desa/papers/2017/wp152_2017.pdf

Jacob, B. A., Lefgren, L., & Moretti, E. (2007). The dynamics of criminal behavior: Evidence from weather shocks. *The Journal of Human Resources*, *42*(3), 489–527. DOI: 10.3368/jhr.XLII.3.489

Jayachandran, S. (2006). Selling labor low: Wage responses to productivity shocks in developing countries. *Journal of Political Economy*, *114*(3), 538–575. DOI: 10.1086/503579

McDowell, C., & de Hess, M. (2012). Social vulnerability and disaster: Understanding pathways to resilience. *Disasters*, *36*(1), 1–18. DOI: 10.1111/j.1467-7717.2011.01255.x PMID: 21702892

Miguel, E., Satyanath, S., & Sergenti, E. (2004). Economic shocks and civil conflict: An instrumental variables approach. *Journal of Political Economy*, *112*(4), 725–753. DOI: 10.1086/421174

OECD/FAO. (2018). *OECD-FAO agricultural outlook 2018–2027*. OECD Publishing., DOI: 10.1787/agr_outlook-2018-

Paglialunga, E., Coveri, A., & Zanfei, A. (2022). Climate change and within-country inequality: New evidence from a global perspective. *World Development*, *159*, 106030. Advance online publication. DOI: 10.1016/j.worlddev.2022.106030

Piketty, T., & Goldhammer, A. (2014). *Capital in the Twenty-First Century*. Harvard University Press. https://www.jstor.org/stable/j.ctt6wpqbc

Rao, N. D., van Ruijven, B. J., Riahi, K., & Bosetti, V. (2017). Improving poverty and inequality modelling in climate research. *Nature Climate Change*, *7*(12), 857–862. DOI: 10.1038/s41558-017-0004-x

Ravallion, M., & Jalan, J. (2001). *Household income dynamics in rural China* (Policy Research Working Paper No. 2612). World Bank. https://openknowledge.worldbank.org/handle/10986/18714

Schlenker, W., & Lobell, D. B. (2010). Robust negative impacts of climate change on African agriculture. *Environmental Research Letters*, *5*(1), 014010. Advance online publication. DOI: 10.1088/1748-9326/5/1/014010

Schlenker, W., & Roberts, M. J. (2009). Nonlinear temperature effects indicate severe damages to U.S. crop yields under climate change. *Proceedings of the National Academy of Sciences of the United States of America*, *106*(37), 15594–15598. DOI: 10.1073/pnas.0906865106 PMID: 19717432

United Nations Economic and Social Commission for Western Asia. United Nations Population Fund Arab States Regional Office, & Save the Children. (2022, November). *Through their eyes: Climate change and economic inequality in the MENA region* (E/ESCWA/CL2.GPID/2022/BOOKLET.1). https://www.unescwa.org/publications/through-their-eyes-climate-change-and-economic-inequality-mena-region

United Nations Economic and Social Commission for Western Asia (ESCWA). (2024). *Arab sustainable development report 2024* (E/ESCWA/CL5.SDGS/2024/1). https://www.unescwa.org/publications/arab-sustainable-development-report-2024

Verner, D. (2010). *Reducing poverty, protecting livelihoods, and building assets in a changing climate: Social implications of climate change in Latin America and the Caribbean*. World Bank., DOI: 10.1596/978-0-8213-8238-7

World Bank. (2018). *Climate change in MENA: Challenges and opportunities for the world's most water-stressed region*. World Bank Group. https://documents.worldbank.org/en/publication/documents-reports/documentdetail/894251519999525186/climate-change-in-mena-challenges-and-opportunities-for-the-world-s-most-water-stressed-region

World Bank. (2020). *Poverty and shared prosperity 2020: Reversals of fortune*. World Bank Group. https://www.worldbank.org/en/publication/poverty-and-shared-prosperity

Yang, D., & Choi, H. (2007). Are remittances insurance? Evidence from rainfall shocks in the Philippines. *The World Bank Economic Review*, *21*(2), 219–248. DOI: 10.1093/wber/lhm003

KEY TERMS AND DEFINITIONS

Adaptive Capacity: The third component of climate vulnerability used by the ND-GAIN index. It captures a country's ability to deal with, adjust to, mitigate potential damage from, or cope with the consequences of climate change (e.g., dam capacity, medical staff—physicians, nurses, and midwives—protected biomes, engagement in international environmental conventions).

Agricultural Employment Share: The percentage of the labor force engaged in agricultural activities, often used to indicate economic structure or rural dependency.

Assessing adaptive capacity: involves evaluating agricultural capabilities, especially in developing nations, where technological advancements play a pivotal role. Agricultural technologies serve as indicators of a country's ability to adapt to climate change. Specific technologies, including irrigation infrastructure, fertilizer and pesticide usage, and mechanization, are indicative of a nation's agricultural resilience. Additionally, child malnutrition serves as a poignant indicator of a society's capacity to meet basic nutritional needs, particularly for its most vulnerable members.

Climate Vulnerability: The first key dimension of the ND-GAIN index. It captures the degree to which a country is likely to experience harm due to exposure to climate hazards, its sensitivity, and its adaptive capacity. The overall vulnerability is assessed across six sectors: food, water, health, ecosystem services, human habitat, and infrastructure. It uses different variables to capture exposure, sensitivity and adaptive capacity of the six sectors mentioned. Consider the first sector, for example. The exposure of the food sector is quantified by the projected changes in cereal yields. Specifically, exposure is assessed through the average impacts on three primary crops: rice, wheat, and maize, which together represent a substantial portion of global food consumption. Additionally, population change is incorporated into this assessment, recognizing that population dynamics and shifting consumption patterns significantly influence food demand. Thus, population growth is utilized as a proxy for estimating future food demand.

Climate–Economic Nexus: The interaction between climate variables and economic systems, highlighting how environmental factors can influence economic growth (e.g., impacts on GDP), poverty, and inequality. This nexus also encompasses the feedback loops between economic activities and climate change—such as emissions from industrial production contributing to global warming, which in turn affects economic systems. Moreover, it includes the differential vulnerabilities of sectors and regions, as well as the role of adaptation and mitigation policies that shape both economic and climate outcomes.

Climate-Inclusive Policy: A policy approach that integrates climate adaptation or mitigation into economic and social development planning.

Climate-Induced Inequality: Disparities that arise or worsen due to the differential effects of climate change across regions, sectors, or/and social groups. It captures the way climate change interacts with existing socioeconomic structures to disproportionately affect those who are already vulnerable. These disparities manifest across multiple dimensions:

Exposure: The first component of climate vulnerability. It shows the extent to which human society and its supporting sectors are stressed by future climate conditions. It captures the physical factors external to the system that contribute to vulnerability, using variables that are projections (e.g., projected change in cereal yields, projected population change).

Institutional Quality: The effectiveness of political, legal, and administrative institutions in delivering services, upholding the rule of law, and responding to citizens' needs. This concept is crucial because it highlights that countries will not experience the impacts of climate change uniformly. Strong institutional quality can help mitigate the impact of climate change on inequality. (Huynh & Hoang, 2024)

ND-GAIN Index: A global index computed by the University of Notre Dame. It measures a country's vulnerability to climate change and its readiness to adapt. The ND-GAIN Index is computed as follows: ND-GAIN score = (Readiness score – Vulnerability score + 1) x 50. The closer the score is to 100, the better.

Sensitivity analysis: considers factors such as the dependence on food import, as nations relying heavily on imported food are vulnerable to price fluctuations in the global market, which may be exacerbated by climate change impacts on agriculture. The significance of rural populations cannot be overstated, as the majority of the world's impoverished reside in rural areas, with agriculture serving as a primary source of income and development. Rural populations, particularly those engaged in subsistence farming, face heightened vulnerability to climate-related shocks.

Sensitivity: The second component of climate vulnerability. It shows the degree to which key sectors of the economy (or people) are dependent on resources that are sensitive to climate change impacts. For example, in the food sector, one variable used is food import dependency. If this dependency is high, the country is highly sensitive because it is more susceptible to food price shocks. As explained by Nelson et al. (2010), climate change and its impacts on the agriculture sector may accentuate price volatility.

Within-Country Inequality: Disparities in income, resources, or opportunities that exist between individuals or groups within a single country. It can be measured by income, wealth, gender (females vs. males), spatial (rural vs. urban), etc.

APPENDIX

Econometric Diagnostic Tests

1. Jarque-Bera Test for Normality of Residuals

Test statistic: X-squared = 145.22
Degrees of freedom: 2
p-value: < 2.2e-16
Interpretation: The very low p-value suggests that residuals are not normally distributed.

2. Breusch-Godfrey/Wooldridge Test for Serial Correlation

Test statistic: chi-squared = 242.15
Degrees of freedom: 25
p-value: < 2.2e-16
Interpretation: Strong evidence of serial correlation in the idiosyncratic errors.

3. Breusch-Pagan Test for Heteroskedasticity

Test statistic: BP = 13.745
Degrees of freedom: 6
p-value: 0.03262
Interpretation: The p-value below 0.05 suggests the presence of heteroskedasticity.

4. Multicollinearity Diagnostics

Results from multicollinearity checks (VIF and Tolerance metrics):

Statistic	N	Mean	St. Dev.	Min	Max
VIF	6	1.904	0.809	1.007	3.006
VIF_CI_low	6	1.668	0.638	1.000	2.548
VIF_CI_high	6	2,824.802	6,913.268	1.521	16,936.450
SE_factor	6	1.354	0.293	1.004	1.734
Tolerance	6	0.613	0.257	0.333	0.993
Tolerance_CI_low	6	0.379	0.241	0.0001	0.657
Tolerance_CI_high	6	0.676	0.245	0.392	1.000

5. Hausman Test (RE vs. FE)

Test statistic: chi-squared = 3.4687
Degrees of freedom: 6
p-value: 0.7481
Interpretation: The high p-value suggests that the random-effects model is consistent and preferred over fixed-effects.

6. Lagrange Multiplier Test for Two-Way Effects

Test statistic: chibar-squared = 1954.2
df0 = 0.00, df1 = 1.00, df2 = 2.00
Weights: w0 = 0.25, w1 = 0.50, w2 = 0.25
p-value: < 2.2e-16
Interpretation: Strong evidence in favor of panel effects being present (significant two-way effects).

Chapter 7
Policy Synergies for Climate-Resilient Agriculture:
Can Algeria and Tunisia Learn From Each Other?

Amine M. Benmehaia
https://orcid.org/0000-0001-8110-1601
University of Biskra, Algeria

Boubaker Dhehibi
ICARDA, Tunisia

Rania Soula
University of Sousse, Tunisia

ABSTRACT

This chapter explores the transformative potential of agricultural policy synergies between Algeria and Tunisia to address shared climate vulnerabilities, including water scarcity, aquifer depletion, and rising food import dependency. By systematically analyzing five key areas of collaboration—water governance exchange, drought-resistant crop development, cross-border aquifer management, agri-tech knowledge transfer, and climate-smart value chains—the study demonstrates how integrated approaches can enhance resilience, optimize resource use, and reduce economic risks. The chapter argues that bilateral cooperation could unlock ecological, economic, and social dividends, such as slower aquifer depletion, lower cereal import costs, and rural job creation. However, realizing these synergies requires

DOI: 10.4018/979-8-3373-3346-5.ch007

overcoming institutional fragmentation and political inertia. The chapter concludes with a systemic framework for Maghreb agricultural resilience, positioning Algeria and Tunisia as potential leaders in regional climate adaptation.

1. INTRODUCTION

The Maghreb region, particularly Algeria and Tunisia, ranks among the world's most water-stressed zones, with climate models projecting a 20–30% decline in renewable water resources by 2050 (World Bank, 2022). Both countries share similar agroecological challenges: erratic rainfall, desertification, and overexploitation of the Northwestern Sahara Aquifer System (NWSAS), a transboundary groundwater reserve critical for agriculture. Algeria's annual wheat imports exceed $9 billion in 2024, while Tunisia spends $1.5 billion on cereal imports (USDA statistics, 2023)—exposing both nations to volatile global markets. These pressures are compounded by rising temperatures, which could reduce arable land by 40% by 2040 (IPCC, 2023). Against this backdrop, agricultural systems must adapt rapidly, yet policy responses remain fragmented and nationally siloed.

Despite parallel climate threats, Algeria and Tunisia have pursued divergent agricultural strategies with mixed results. Algeria prioritizes large-scale infrastructure (e.g., dams, state-led irrigation) but struggles with 60% water loss in distribution networks (MRE, 2023). Tunisia excels in participatory water governance through Agricultural Development Groups (GDAs) but lacks funding to scale innovations (DG-ACTA, 2022). This disconnect is starkly evident in the NWSAS, where uncoordinated extraction has led to annual water table declines of 2–3 meters (OSS, 2021). Such fragmentation undermines resilience, perpetuates import dependency, and squanders opportunities for knowledge exchange. The status quo is untenable: without integration, climate adaptation efforts will remain reactive, inefficient, and underfunded.

This chapter argues that Algeria and Tunisia can achieve far greater resilience by systematically aligning their agricultural policies through targeted synergies. Drawing on Ostrom (1990)'s spirit of principles of polycentric governance and the UN's water-energy-food nexus framework (UNECE, 2018), we demonstrate how collaborative approaches in five key areas—water governance, crop diversification, aquifer management, agri-tech, and value chains—can create multiplicative benefits. Such synergies are not merely theoretical: similar integration in Morocco's Green Generation Plan boosted water productivity by 35% (World Bank, 2021). The central thesis is that shared challenges demand shared solutions, and bilateral cooperation can turn climate vulnerabilities into opportunities for innovation.

This chapter navigates between realism and optimism. It acknowledges political and institutional barriers—from subsidy misalignment to bureaucratic inertia—while mapping actionable pathways to overcome them. It does not shy from hard questions: Can Algeria's centralized system adapt Tunisia's participatory models? Can both nations transcend historical competition over shared resources? The goal is to provoke strategic thinking, showing that the biggest risk is not failure but inaction.

The chapter unfolds in three sections. First, we dissects five priority areas of Synergies for Algeria-Tunisia agricultural collaboration: (1) water governance exchange, (2) drought-resistant crop development, (3) cross-border aquifer management, (4) agri-tech knowledge transfer, and (5) climate-smart value chains. Next, we synthesize these synergies into a coherent framework through a systemic agenda for Maghreb agricultural resilience, visualizing interdependencies and quantifying ecological-economic benefits. Finally, the conclusion issues a call to action, urging policymakers to treat shared climate risks as catalysts for unprecedented cooperation.

2. SELECTED SYNERGIES FOR ALGERIA-TUNISIA AGRICULTURAL COLLABORATION

The Maghreb's agricultural future hinges on transcending national silos. This section dissects five actionable synergies where Algeria and Tunisia's complementary strengths – one's infrastructural scale meeting the other's participatory innovations – can forge climate resilience. Each subsection blends empirical data with policy analysis to reveal how bilateral learning could rewrite the region's adaptation narrative.

2.1. Water Governance Exchange

Water flows beyond borders, yet policies remain dammed by bureaucracy! Algeria's top-down irrigation systems lose 60% of distributed water, while Tunisia's farmer-led GDAs achieve 30% higher efficiency – but lack scaling funds. This subsection explores how marrying Algeria's hydraulic infrastructure with Tunisia's participatory models could quench both nations' thirst for sustainable water management.

Divergent governance models: Infrastructure scale vs. Participatory efficiency

The water management systems of Algeria and Tunisia reflect fundamentally different philosophies in addressing agricultural water scarcity. Algeria has pursued a centralized, infrastructure-heavy approach, investing heavily in large dams and state-led irrigation schemes (Oulmane *et al.*, 2022; Laoubi & Yamao, 2009). Since

2000, the country has constructed over 85 major dams, increasing its water storage capacity to more than 9 billion cubic meters. This ambitious hydraulic mission was driven by the goal of achieving food sovereignty through expanded irrigation. However, the system suffers from significant inefficiencies. Nearly 60% of irrigation water is lost before reaching crops due to aging canals, poor maintenance, and mismanagement. In key agricultural regions like the High Plains, less than 40% of stored water actually benefits farms, contributing to stagnant crop yields. The financial burden is also substantial, with the government covering 90% of irrigation costs, creating a system where farmers have little incentive to conserve water.

Tunisia, by contrast, has developed one of North Africa's most successful decentralized water governance models. Its Agricultural Development Groups (GDAs), established in the late 1980s, place water management in the hands of local farmers. These democratically organized groups oversee 80% of the country's irrigation systems (ONAGRI, 2021), setting water allocations, tariffs, and maintenance schedules through collective decision-making. This participatory approach has led to notable improvements in efficiency, with GDAs achieving 25–30% better water-use outcomes compared to state-managed systems. Fee collection rates exceed 53% (Nawar et al., 2024), and localized governance has reduced water conflicts in regions like Cap Bon (Ferchichi et al., 2022). However, the GDA model faces its own challenges, particularly in financial sustainability and institutional capacity, which can hinder scaling beyond smallholder irrigation perimeters (Mahdhi et al., 2014). Limited state funding means many GDAs rely heavily on member contributions, and in areas like Sfax, excessive and unregulated pumping of groundwater has made it difficult for them to enforce collective rules and manage water sustainably (Elloumi, 2016). A simple comparative analysis reveals stark institutional differences between the two systems. Algeria's top-down governance concentrates decision-making within national ministries, leading to rigid water allocation policies that often fail to adapt to local needs. Tunisia's GDAs, on the other hand, operate at the community level, allowing for more responsive and context-specific management. Financially, Algeria's system is sustained by heavy state subsidies, while Tunisia's GDAs depend on user fees, creating different incentives for conservation. Conflict resolution also differs markedly—Algeria relies on judicial processes, whereas Tunisia's GDAs emphasize local mediation.

These contrasting approaches highlight the strengths and weaknesses of each system. Algeria excels in large-scale infrastructure development but struggles with maintenance and efficiency. Tunisia's participatory model fosters better water stewardship but lacks the resources and coordination to manage broader hydrological challenges. The Table 1 below summarizes these key differences.

Table 1. Key differences of water governance between Algeria and Tunisia

Governance Aspect	*Algeria*	*Tunisia*
Decision-making level	National ministries	Farmer-elected GDAs
Water-use efficiency	40% losses	70–75% efficiency
Financial sustainability	State-subsidized	Member-funded
Conflict resolution	Judicial arbitration	Local mediation

Theoretical perspectives help explain these divergences. Algeria's system aligns with what scholars describe as a "hydraulic bureaucracy," (Evers & Benedikter, 2009; Dasthagir, 2024) where engineering solutions dominate over institutional flexibility. Tunisia's GDAs, meanwhile, reflect principles of polycentric governance, as theorized by Elinor Ostrom, though challenges like groundwater depletion indicate limits to purely local management. Both countries face institutional lock-in—Algeria's reliance on large infrastructure projects and Tunisia's difficulty in integrating broader hydrological planning.

These governance differences are not merely academic; they have real-world consequences. In Algeria, the emphasis on dams and centralized control has led to underutilized infrastructure and inefficient water use (Kherbache, 2020). In Tunisia, the success of GDAs in small-scale irrigation has not translated into effective management of larger water systems, such as regional aquifers (Chebil et al., 2019; Sghaier et al., 2025). The question, then, is not which model is superior, but how each country might learn from the other's successes and failures. Algeria could benefit from Tunisia's participatory ethos to improve maintenance and farmer engagement, while Tunisia might draw lessons from Algeria's experience in securing large-scale investments. The next section explores these potential synergies in greater depth.

Pathways for reciprocal policy learning

The complementary strengths of Algeria's infrastructure scale and Tunisia's participatory governance present concrete opportunities for mutual learning and policy innovation. Rather than advocating for wholesale adoption of either model, the most promising path forward lies in strategic integration of both approaches through targeted collaboration mechanisms.

Hybrid governance pilots in strategic locations: Algeria could initiate experimental projects combining its large-scale irrigation infrastructure with Tunisia's GDA principles. The Mitidja Plain, which supplies 40% of Algeria's horticultural output but suffers from 50% distribution losses, would be an ideal testing ground (Laoubi & Yamao, 2009; Imache et al., 2009). A phased implementation could begin with delegating management of tertiary canals to farmer associations while maintaining state control over primary infrastructure. This mirrors Morocco's partial decentralization success in the Doukkala region, where similar hybrid arrangements led to key achievements, including a reduction of water losses from 50% to about 10%, an increase in cropping intensity from below 100% to 140%, and the development of high value-added crops (AfDB, 2018).. Tunisia could reciprocate by adopting Algeria's technical standards for pressurized irrigation systems in GDA-managed perimeters, potentially reducing energy costs by 22-31%, according to ICARDA estimates from pilot projects in semi-arid regions (ICARDA, 2020).

Institutional knowledge transfer platforms: Establishing a permanent Algeria-Tunisia Water Governance Exchange Program could facilitate systematic learning. This would go beyond one-time study tours to create embedded exchange positions - for instance, Algerian engineers spending six-month rotations working with Tunisian GDAs, while Tunisian extension specialists advise on participatory approaches in Algerian irrigation schemes. The program should include joint training modules on topics like conflict mediation (drawing on Tunisia's experience) and infrastructure financing (leveraging Algeria's expertise). The German-Tunisian partnership project demonstrated that structured institutional exchanges can lead to more effective policy learning and higher adoption rates than conventional technical assistance approaches (GIZ, 2020).

Financial innovation for sustainable water management: Algeria's access to infrastructure financing could be combined with Tunisia's successful cost-recovery models. A joint fund, capitalized through equal contributions from both governments and climate adaptation finance, could support projects that blend large-scale and participatory elements. For instance, Algeria's planned solar-powered desalination plants in coastal areas could incorporate Tunisia's progressive water pricing mechanisms, where tariffs increase with consumption levels. Preliminary modeling suggests this hybrid approach could reduce fiscal burdens on both states by 15-20% while maintaining service quality.

Data integration for transboundary management: Developing shared monitoring systems would create new opportunities for collaborative governance. Algeria's advanced satellite-based water monitoring capabilities (through the Alsat program) could be integrated with Tunisia's dense network of ground sensors. A pilot could focus on the shared Northwestern Sahara Aquifer System, establishing real-time data sharing protocols and joint analysis units. The Swiss-supported Genevois Aquifer management between France and Switzerland offers a proven template, having reduced cross-border conflicts by 75% through similar technical cooperation.

Policy innovation through joint research: Creating a Maghreb Water Policy Lab would institutionalize ongoing collaboration. Co-hosted by Algeria's and Tunisia's respective National Institute of Agronomic Research, this think tank could produce regular policy briefs comparing implementation outcomes. Its first work program will focus on: evaluating Algeria's recent experiments with delegated dam management; assessing Tunisia's efforts to scale GDAs to larger irrigation schemes; and developing common indicators for water governance performance. Such structured learning would represent a significant advance over current ad hoc exchanges.

Overcoming political and institutional barriers: Successful policy learning requires addressing entrenched interests and bureaucratic inertia. Algeria's water sector reforms could draw lessons from Tunisia's gradual approach to decentralization, which started with small pilot areas before expanding nationally. Conversely, Tunisia could benefit from Algeria's experience in building political consensus for major infrastructure projects through high-level inter-ministerial committees. Establishing a joint Algeria-Tunisia Water Council at ministerial level would provide the necessary political backing, while parallel technical working groups could ensure continuity. The Franco-Spanish shared water governance mechanisms demonstrate how such dual-track approaches can sustain cooperation across political cycles.

Measuring success and scaling up: Effective policy learning requires robust monitoring frameworks. Initial hybrid projects should incorporate clear benchmarks: irrigation efficiency gains, farmer satisfaction levels, cost recovery rates, and conflict reduction metrics. Promising results could then inform national policy revisions - for instance, Algeria might amend its Water Code to formally recognize water user associations, while Tunisia could revise its GDA statutes to facilitate larger-scale operations. The key is maintaining flexibility to adapt approaches based on evidence rather than political considerations.

This reciprocal learning approach recognizes that water governance systems are deeply embedded in their national contexts. The goal is not to erase differences but to create channels for continuous exchange and adaptation. By combining Algeria's engineering capabilities with Tunisia's participatory ethos, both countries could develop more resilient, efficient water management systems better equipped to face climate challenges. The proposed mechanisms provide practical pathways to turn this potential into reality, with careful attention to implementation challenges and measurable outcomes.

2.2. Drought-Resistant Crops Development

Two countries, one climate threat, divergent responses. Where Algeria pioneers quinoa in research stations, Tunisia pushes drought-hardy olives through extension networks. Here we map how bridging these approaches through joint seed banks and cross-border field trials could birth a new generation of climate-proof staples for the Maghreb.

Divergent agricultural pathways: Crop adaptation strategies in Algeria and Tunisia

The approaches taken by Algeria and Tunisia to develop drought-resistant crops reveal fundamental differences in agricultural policy and implementation, shaped by distinct institutional frameworks and environmental challenges. These contrasting strategies highlight how each country is attempting to build climate resilience while facing unique constraints and opportunities.

Algeria has pursued a centralized, research-intensive model focused on introducing novel drought-resistant crops through its national agricultural research system. The National Institute for Agronomic Research (INRAA) has led extensive trials of alternative crops like quinoa, sorghum and millet in arid regions (Maamri *et al.*, 2021; Oustani *et al.*, 2023). These crops were selected for their ability to thrive with minimal water and in poor soil conditions. Field experiments have demonstrated quinoa's particular promise, achieving yields of 1.8 tons per hectare while using 30% less water than traditional wheat crops. Similar successes have been documented with sorghum and pearl millet, which show 20-25% better water efficiency than barley, the traditional drought-resistant staple. However, despite these technical achievements, adoption rates among farmers remain disappointingly low, with less than 5% of agricultural producers in trial regions switching to these alternative crops. This implementation gap stems from several systemic challenges including weak extension services that fail to connect research findings with farmers, underdeveloped markets for non-traditional crops, and subsidy policies that continue to favor water-intensive cereals like wheat and barley.

Tunisia has charted a different course by building climate adaptation around existing agricultural systems rather than introducing completely new crops. The country has focused on enhancing the drought resilience of traditional tree crops, particularly olives and dates, through improved varieties and irrigation techniques. Drought-tolerant olive cultivars like Chemlali now cover approximately 1.84 million hectares nationwide (Debbabi et al., 2022), supported by government programs promoting drip irrigation that have boosted yields by 20-30% in central growing regions (OTI, 2012; ElFakih et al., 2023). In southern areas where water salinity is increasing, date palm plantations have successfully adapted to using brackish groundwater while still achieving 12% productivity gains (Dhaouadi et al., 2022). This pragmatic approach benefits from strong farmer engagement through Tunisia's decentralized Agricultural Development Groups (GDAs), which facilitate knowledge sharing and technology adoption at the local level (Ouerghemmi et al., 2024). However, Tunisia's strategy faces limitations of its own, notably the continued dominance of water-intensive cereals which still occupy 14% of irrigated land (FAO, 2018). Other constraints include the prevalence of small landholdings that hinder economies of scale and potential vulnerabilities from over-reliance on olive monocultures in the face of increasingly extreme drought conditions (Mejri et al., 2025).

The institutional and policy differences between the two countries' approaches become clear when examining key aspects side by side. Algeria's strategy emphasizes scientific innovation through its research institutes but struggles with practical implementation, while Tunisia has achieved broader on-the-ground adoption by working within existing farming systems but may face long-term limitations in crop diversity. Algeria maintains a stronger focus on developing entirely new crop options through centralized research programs, whereas Tunisia has concentrated on adapting traditional crops through farmer-participatory methods. These differences extend to subsidy policies, with Algeria continuing to support conventional cereals while Tunisia has created incentives for water-efficient irrigation of perennial crops. The contrast in adoption rates is particularly striking, with Tunisia seeing 45% of farmers take up drought-adapted olive varieties compared to Algeria's minimal uptake of alternative annual crops.

These divergent pathways reflect deeper structural differences in how the two countries approach agricultural innovation. Algeria's model represents a "technology-push" strategy where scientific breakthroughs are expected to drive change, but often encounter barriers in reaching end users. Tunisia exemplifies a "demand-pull" approach that responds to immediate farmer needs but may lack the transformative potential of completely new cropping systems. Both countries have demonstrated partial successes - Algeria in developing innovative crop solutions and Tunisia in achieving wider implementation - but neither approach has yet produced comprehensive climate resilience. This analysis suggests that significant opportunities exist for cross-learning, where Algeria could benefit from Tunisia's effective extension systems and Tunisia might incorporate some of Algeria's promising alternative crops into its diversification efforts.

This comparative analysis sets the stage for exploring how strategic collaboration could help both countries overcome their respective limitations and accelerate progress toward climate-resilient agriculture. The next section will examine specific opportunities for synergizing these divergent approaches through joint research initiatives, knowledge sharing, and policy coordination.

Synergizing research and dissemination: A framework for collaborative innovation

Algeria and Tunisia stand at a critical juncture where collaborative innovation in drought-resistant agriculture could transform their shared struggle against climate change into an opportunity for regional leadership. The path forward requires moving beyond isolated national efforts to create an integrated system that combines Algeria's scientific research strengths with Tunisia's practical dissemination networks, while aligning policies and mobilizing sustained funding. This synergy would work through multiple interconnected channels to accelerate the development and adoption of climate-resilient crops across both countries.

At the core of this collaboration should be a joint research platform that breaks down the current barriers between Algeria's advanced crop science and Tunisia's farmer-tested adaptation strategies. Algeria's National Institute for Agronomic Research has made significant breakthroughs with drought-tolerant quinoa, sorghum and millet varieties that could be game-changers for the region, but these innovations remain largely confined to experimental stations. Meanwhile, Tunisia's agricultural extension system has successfully promoted climate-adapted olive and date cultivation through its decentralized farmer networks, but lacks access to novel crop options that could diversify production systems. A shared research initiative could establish cross-border field testing sites where Algerian-developed crop varieties are evaluated under Tunisia's diverse growing conditions, while Tunisian salt-tolerant cultivars are introduced in Algeria's Saharan oases. This two-way exchange would be strengthened by combining Algeria's biotechnology capabilities with Tunisia's traditional knowledge of dryland farming practices.

The knowledge dissemination component requires building bridges between Algeria's digital agriculture infrastructure and Tunisia's participatory extension model. Tunisia's Agricultural Development Groups have proven effective at delivering climate-smart techniques to farmers through local demonstrations and peer-to-peer learning, an approach that could be adapted to Algeria's context where top-down extension systems have struggled. Conversely, Algeria's investments in satellite monitoring and digital advisory systems could enhance Tunisia's farmer outreach with real-time data on weather patterns and soil conditions. Pilot programs in border regions could test integrated dissemination methods that combine the best of both systems, such as using mobile platforms to amplify the reach of Tunisia's successful farmer field school model.

Policy coordination represents perhaps the most challenging but potentially transformative dimension of this collaboration. Currently, both countries maintain subsidy systems and land tenure policies that inadvertently discourage widespread adoption of drought-resistant crops. Algeria's substantial cereal subsidies continue to incentivize water-intensive wheat and barley production, while Tunisia's land fragmentation limits economies of scale for alternative crops. A bilateral policy task force could develop harmonized approaches such as redirecting a portion of Algeria's cereal subsidies to support farmers transitioning to resilient crops, while adapting Tunisia's innovative land leasing programs to Algeria's context. The creation of shared certification standards for climate-resilient crops could help build market demand and premium pricing for these products across the region.

Financing this transformation will require innovative mechanisms that pool national resources with international climate funding. A dedicated Maghreb Climate-Smart Agriculture Fund could support joint research initiatives, cross-border technology transfer, and farmer training programs over the minimum 10-year timeframe needed to achieve meaningful impact. This should be complemented by institutional structures such as a permanent bilateral commission to oversee implementation and resolve policy conflicts, building on existing joint agricultural committees. Engaging private sector actors from both countries in developing processing and marketing infrastructure for drought-resistant crops will be essential to create sustainable market pull for these innovations.

The potential benefits of this collaborative framework extend far beyond the sum of its parts. Within 3-5 years, targeted pilot programs could demonstrate 20-30% increases in adoption of resilient crops in key regions, while over a decade the approach could transform entire agricultural systems. Challenges such as bureaucratic inertia, unequal distribution of benefits, and the accelerating pace of climate change will require ongoing adaptation, but the alternative of continuing with fragmented national efforts risks leaving both countries dangerously vulnerable to worsening droughts and water scarcity. By combining their complementary strengths through this comprehensive framework, Algeria and Tunisia can position themselves as pioneers in climate-resilient agriculture while addressing their shared food security challenges more effectively than either could achieve alone. The time for parallel but separate efforts has passed - the climate crisis demands nothing less than fully integrated action across research, dissemination, policy and financing.

2.3. Cross-Border Aquifer Management

The Northwestern Sahara Aquifer does not recognize political borders as its levels drop 3 meters yearly. This subsection sounds the alarm on the world's next potential water conflict zone, then charts a cooperative path forward – from shared monitoring wells to crisis-response protocols – before the wells run dry!

The Northwestern Sahara aquifer system: A shared resource in crisis

The Northwestern Sahara Aquifer System, a vast underground water reserve shared by Algeria and Tunisia (and also Libya), faces an existential threat from unsustainable exploitation and climate pressures. This ancient groundwater resource, which took thousands of years to accumulate, is being depleted at rates that far exceed natural replenishment, putting at risk the agricultural livelihoods and water security of both nations. The current extraction patterns reveal a troubling imbalance that threatens the long-term viability of this critical shared resource.

In Algeria, annual groundwater withdrawals have reached approximately 2.5 billion cubic meters, primarily supporting large-scale irrigation projects in the High Plains and growing urban demands in the Sahara region. Tunisia pumps about 1 billion cubic meters yearly, mainly sustaining its precious date palm oases and coastal cities. Together, these withdrawals total 3.5 billion cubic meters annually – more than three times the aquifer's natural recharge capacity of less than 1 billion cubic meters. This massive imbalance has caused water tables to drop precipitously by 2-3 meters each year across the region, with some areas experiencing even more dramatic declines of 10-15 meters over the past decade. These falling water levels represent not just statistics but a looming catastrophe for communities that depend on this vital resource.

The environmental consequences of this overexploitation are already visible and worsening. In Tunisia's Djerid region, the heart of the country's date production, farmers report increasing difficulties accessing groundwater as levels continue to drop. Many traditional irrigation systems, some dating back centuries, are becoming obsolete as the water retreats deeper underground. Along coastal areas, the reduced water pressure has allowed seawater to intrude into aquifers, increasing salinity to levels that threaten agricultural productivity. Similar problems plague Algeria's agricultural zones, where abandoned wells and failing irrigation systems tell the story of a resource under severe stress. These changes are not just environmental indicators but represent real threats to food security and rural livelihoods across the region.

The institutional framework for managing this shared resource remains woefully inadequate to address these challenges. While both countries recognize the problem in principle, their actual cooperation remains limited and ineffective. The 2008 NWSAS Consultation Mechanism, established to foster joint management, lacks any binding authority or enforcement mechanisms. Each country continues to pursue independent water policies without meaningful coordination, monitoring the aquifer using different methods and metrics that make collaborative assessment difficult. Compounding these governance failures, domestic subsidy policies in both nations actively encourage overuse - Algeria through free electricity for agricultural pumping and Tunisia through diesel subsidies that make groundwater extraction artificially cheap.

Climate change acts as an accelerating factor in this crisis, reducing the already minimal recharge rates while increasing agricultural water demands. Scientific projections suggest recharge could decrease by 20-30% by 2040, further straining the system. Without immediate and coordinated action, the aquifer could lose half of its exploitable reserves within two decades - a scenario that would devastate agriculture in both countries and potentially spark conflicts over dwindling resources.

The situation represents a classic tragedy of the commons playing out in real time, where individual rational actions lead to collective disaster without appropriate governance structures.

This growing crisis demands more than technical adjustments or temporary measures. It requires a fundamental rethinking of how Algeria and Tunisia manage their shared water heritage, moving from competition to cooperation, from short-term exploitation to long-term stewardship. The current path leads inevitably to depletion and conflict, while an alternative approach based on joint management could secure water resources for future generations. The solutions exist, but they require political will, institutional innovation, and a shared commitment to sustainability that has so far been lacking.

Toward cooperative governance: Models for Algeria-Tunisia joint management

The escalating crisis of the Northwestern Sahara Aquifer System demands an urgent shift from unilateral water exploitation to collaborative governance. While the challenges are formidable, several concrete pathways exist for Algeria and Tunisia to establish effective joint management of this vital shared resource. These solutions draw on international best practices while remaining grounded in regional realities and the specific hydrological and political contexts of both nations.

A functional cooperative framework must begin with establishing robust technical foundations for joint decision-making. Currently, Algeria and Tunisia monitor the aquifer using different methodologies - Algeria's network of 50 deep observation wells provides different data than Tunisia's 120 shallow sensors. Creating a unified monitoring system with standardized measurement protocols would represent a crucial first step. This could involve installing new jointly-managed monitoring wells along border regions and developing common data standards through a technical working group. The Swiss-French Genevois Aquifer management system demonstrates how such technical coordination can work in practice, having reduced cross-border conflicts by 75% through shared data collection and analysis. Satellite-based monitoring, leveraging Algeria's existing Alsat program, could complement ground measurements to provide comprehensive aquifer health assessments.

The heart of any sustainable solution lies in developing equitable water allocation mechanisms that recognize both countries' needs while respecting ecological limits. A tiered extraction quota system, based on scientifically-determined sustainable yield thresholds for different aquifer zones, could replace the current unrestricted pumping. The 2008 NWSAS Consultation Mechanism failed precisely because it lacked binding limits - a new agreement must include enforceable caps adjusted periodically based on monitoring data. Accompanying this, progressive water pricing that increases with extraction volumes could discourage overuse while generating funds for conservation efforts. Morocco's experience with similar incentive structures in the Souss-Massa basin shows they can reduce groundwater depletion by 15-20% when properly implemented.

Institutional architecture represents perhaps the most critical component of successful cooperation. The current ad hoc consultations should be replaced by a permanent Algeria-Tunisia Aquifer Commission with three key characteristics: legal authority to make binding decisions, technical capacity to analyze aquifer data, and political independence to mediate disputes. This body should include equal representation from both countries' water ministries, agricultural sectors, and scientific communities, with possible international observers to facilitate dialogue. The commission would oversee implementation of extraction quotas, coordinate drought response plans, and develop long-term adaptation strategies. The Senegal River Basin Organization provides a relevant regional model, having successfully managed shared water resources between four West African nations for decades.

Financial sustainability mechanisms must underpin these technical and institutional reforms. A joint aquifer conservation fund, capitalized through initial contributions from both governments and supplemented by international climate adaptation financing, could support critical initiatives like modernizing irrigation systems in high-usage areas or developing alternative water sources for coastal cities. The fund could also compensate farmers who voluntarily reduce groundwater use during drought periods, similar to Australia's successful water buyback programs in the Murray-Darling Basin. Algeria's existing energy subsidies for agricultural pumping could be gradually restructured to promote efficiency rather than waste, while Tunisia might redirect portions of its diesel subsidies toward solar-powered irrigation that reduces groundwater demand.

Grassroots engagement represents a frequently overlooked but essential element of sustainable aquifer governance. Local water user associations, modeled on Tunisia's successful Agricultural Development Groups but adapted to the transboundary context, could help implement conservation measures at the community level. These associations could manage localized quota systems, maintain monitoring equipment, and serve as communication channels between policymakers and end-users. The UNESCO-supported Drâa Valley project in Morocco demonstrates how involving local stakeholders in groundwater management can improve both compliance and outcomes.

The path to cooperative governance will inevitably face significant political and bureaucratic obstacles. Algeria's centralized water management system may resist sharing authority, while Tunisia's complex subsidy regime creates entrenched interests in the status quo. A phased implementation approach - beginning with pilot projects in less contentious border regions and gradually expanding - can help build trust and demonstrate benefits. Confidence-building measures like joint scientific studies or emergency water sharing during droughts could foster goodwill. International partners like the Sahara and Sahel Observatory or the Green Climate Fund could provide neutral technical assistance and financial incentives to maintain momentum.

Climate change adaptation must be integrated throughout the cooperative framework. As recharge rates decline and temperatures rise, the management system must incorporate flexible adjustment mechanisms. This includes establishing clear drought response protocols that automatically trigger reduced extraction limits when monitoring data shows aquifer stress. The system should also promote alternative water sources like treated wastewater reuse and desalination to gradually reduce aquifer dependence in vulnerable areas. Israel's integrated water management approach, which combines strict groundwater controls with large-scale water recycling and desalination, offers valuable lessons in this regard.

The economic case for cooperation grows stronger as aquifer depletion accelerates. Modeling suggests that uncontrolled depletion could cost both countries over 2 billion annually in lost agricultural production and infrastructure damage by 2030. In contrast, the proposed cooperative system would require modest initial investments–perhaps 50-100 million for monitoring infrastructure and institutional capacity building – while potentially saving billions in avoided losses. The Table 2 below compares key aspects of the current situation with the proposed cooperative model.

Table 2. Key aspects of the proposed cooperative model

Governance Aspect	*Current Situation*	*Proposed Cooperative Model*
Decision-making	Unilateral national policies	Joint commission with binding authority
Monitoring	Inconsistent national systems	Unified transboundary monitoring network
Extraction limits	Nonexistent	Zonal quotas based on sustainable yield
Financing	National subsidies for overuse	Joint conservation fund with incentives
Stakeholder involvement	Limited local participation	Transboundary water user associations

Successful implementation would require careful sequencing of reforms. An achievable roadmap might begin with establishing the technical monitoring framework (years 1-2), followed by creating the joint commission and piloting quota systems in selected zones (years 3-5), then gradually expanding to full aquifer-wide management (years 6-10). Regular review mechanisms would allow for adjustments based on experience and changing climate conditions.

The alternative to cooperation - continued competitive overexploitation - promises only mutual depletion and conflict. By contrast, a jointly managed aquifer system could become a regional model for transboundary water cooperation while safeguarding a vital resource for future generations. The technical solutions exist; what's needed now is the political vision and institutional innovation to implement them. For Algeria and Tunisia, the choice is between short-term water nationalism that benefits neither country in the long run, or collaborative stewardship that secures water security for both. The time for decisive action is rapidly diminishing as water tables continue their steady decline.

2.4. Agri-Tech Knowledge Transfer

This subsection plays matchmaker to two disconnected tech ecosystems, revealing how cross-pollinating Tunisia's frugal innovations with Algeria's high-tech capabilities could democratize precision agriculture across the Maghreb.

Technological complementarities: Mapping existing innovations

The current agricultural technology landscapes of Algeria and Tunisia reveal striking complementarities that remain largely untapped due to institutional and operational silos. A systematic examination of each country's distinctive technological strengths exposes clear opportunities for mutually beneficial knowledge exchange that could accelerate climate adaptation across the Maghreb region.

Algeria has developed a relatively significant capacity in capital-intensive precision agriculture technologies through its network of state-supported research institutions. The National Institute for Agronomic Research (INRAA) has pioneered satellite-based yield prediction systems that combine Alsat-2 imagery with machine learning algorithms, achieving 85% accuracy in forecasting cereal harvests three months before maturity. These systems currently serve large-scale agribusinesses in the High Plains, optimizing irrigation and fertilizer use across 200,000 hectares of commercial farms. However, penetration among Algeria's 1.3 million smallholder farmers remains below 5%, constrained by high costs and limited digital literacy. Similarly, Algeria's Center for Renewable Energy Development (CDER) has deployed a significant number of solar-powered irrigation systems in the Sahara region, demonstrating technical success in reducing diesel dependence by 70%, yet maintenance challenges and spare part shortages have left 30% of installations nonfunctional within three years.

Tunisia has charted a different technological path focused on affordable, decentralized solutions tailored to small-scale producers. The Water Research and Technologies Center (CERTE) has developed low-cost soil moisture sensors priced at just $15 per unit, allowing farmers to optimize irrigation timing without expensive infrastructure. Field tests in the Cap Bon region showed 10% water savings among early adopters (Vandôme, 2023), yet nationwide adoption remains constrained by limited production capacity, as well as challenges related to farmer knowledge, financial resources, and institutional support, resulting in coverage of only 8% of irrigated land (Allani et al., 2022). Tunisia's National Agricultural Extension Agency has also innovated in farmer engagement tools, developing interactive voice response (IVR) systems that deliver localized agronomic advice to 300,000 farmers via basic mobile phones - a critical advantage in rural areas with low smartphone penetration. However, these systems lack integration with advanced predictive analytics that could enhance their climate adaptation value.

The technological divergence extends to water management solutions. Algeria's large-scale infrastructure approach has produced advanced automated canal control systems used in major irrigation schemes like the Beni Haroun project, capable of reducing distribution losses to 15% in pilot zones. Tunisia meanwhile has perfected modular drip irrigation kits that smallholders can install without technical expertise, achieving 40% adoption rates in governorates with strong agricultural extension services. This presents a classic case of complementary innovation - Algeria excels at system-level engineering while Tunisia leads in user-appropriate decentralized solutions.

Crop monitoring technologies tell a similar story. Algeria's space agency (ASAL) provides high-resolution (10m) satellite vegetation indices to agricultural cooperatives, while Tunisia's National Observatory of Agriculture combines Sentinel-2 data with ground truthing from its network of 1,200 farmer observers. Neither approach has crossed borders, despite the obvious synergies between Algeria's technical capacity and Tunisia's participatory data collection model.

These technological ecosystems have evolved in response to each country's distinct agricultural structures and policy priorities. Algeria's focus on food sovereignty has driven investments in technologies that enhance productivity of staple crops on large farms, while Tunisia's emphasis on smallholder resilience has fostered innovations in affordable, locally adaptable tools. What emerges clearly is that neither country's technological portfolio alone can fully address the climate challenges facing Maghreb agriculture - but combined, they could form a robust innovation system capable of serving diverse farming systems across the region.

The barriers to knowledge exchange are not technological but institutional. Intellectual property regimes remain nationally siloed, with no framework for joint patenting or licensing. Extension systems operate without cross-border coordination, despite serving similar agroecological zones. Research priorities are set unilaterally, missing opportunities for collaborative R&D that could yield technologies suited to regional rather than just national conditions. Addressing these structural gaps could unlock significant untapped potential in the Algeria-Tunisia agricultural innovation landscape.

Creating a Maghreb Agri-Tech exchange ecosystem

The technological complementarities between Algeria and Tunisia present a compelling case for establishing an integrated agricultural innovation system that transcends national boundaries. Transforming these latent synergies into tangible impacts requires moving beyond ad hoc cooperation to create a structured ecosystem for continuous knowledge exchange and joint technology development tailored to the Maghreb's unique climatic and agricultural conditions.

At the institutional level, a Maghreb Agri-Tech Transfer Alliance could serve as the cornerstone for systematic collaboration. This permanent bilateral body would bring together Algeria's National Agronomic Research Institute (INRAA) and Tunisia's Agricultural Research and Higher Education Institution (IRESA) under a unified governance framework. The alliance would perform three critical functions: coordinating joint research agendas, managing intellectual property rights for shared innovations, and standardizing certification protocols for new technologies. A precedent exists in the ASEAN Agricultural Technology Network, which increased cross-border technology adoption by 40% among member states through similar institutional arrangements. The alliance should establish specialized working groups focusing on priority areas like water-efficient technologies, drought-resistant crop varieties, and renewable energy applications in agriculture.

Financial mechanisms must be redesigned to incentivize collaborative innovation. A Maghreb Agri-Tech Co-Development Fund, capitalized through equal contributions from both governments and supplemented by international partners like the African Development Bank, could provide grants for joint R&D projects. The fund should prioritize technologies that address shared challenges, such as adapting Tunisia's low-cost soil sensors for Algeria's saline conditions or scaling Algeria's solar pump innovations through Tunisia's decentralized maintenance networks. A tiered financing structure could support the entire technology lifecycle – from initial prototyping to pilot testing and full commercialization (repayable loans). This approach mirrors successful elements of India's Agri-Tech Startup Fund while addressing the Maghreb's specific needs.

Farmer-centered dissemination channels are essential to bridge the gap between innovation and adoption. Building on Tunisia's successful Agricultural Development Groups (GDAs) and Algeria's emerging digital extension services, a cross-border "Tech Champions" program could identify and train lead farmers in both countries to demonstrate and adapt new technologies to local conditions. These champions would receive training at joint centers of excellence - likely located in border regions like Tébessa-Kasserine - where they could test technologies in comparable agroecological zones. Digital platforms using Algeria's satellite connectivity could host virtual field days and peer-to-peer learning exchanges, overcoming geographic barriers to knowledge sharing. The German-Tunisian cooperation project demonstrated that such farmer-to-farmer extension can triple adoption rates compared to traditional top-down approaches.

Policy harmonization must underpin these technical and financial interventions. Currently, divergent regulations on agricultural inputs, equipment certification, and data governance create unnecessary barriers to technology transfer. A bilateral task force could develop: standardized protocols for sensor data collection and sharing; mutual recognition agreements for equipment certifications; and aligned subsidy policies that incentivize cross-border tech adoption. Special attention should be given to creating an "innovation passport" system that allows technologies validated in one country to undergo accelerated approval in the other. Morocco's recent reforms to streamline agri-tech regulations offer valuable lessons in this regard.

The private sector's role as an innovation driver should be activated through targeted incentives. Joint innovation challenges could reward Algerian and Tunisian startups that develop solutions for shared problems like groundwater monitoring or early pest detection. Technology parks in border regions could offer tax holidays and streamlined customs procedures for agri-tech firms operating across both markets. The proposed ecosystem should particularly encourage partnerships between Algeria's energy companies (with their expertise in large-scale solar applications) and Tunisia's agricultural cooperatives (with their distributed service models). Israel's success in fostering agri-tech startups through similar public-private partnerships provides a relevant benchmark.

Monitoring and evaluation systems must be built into the ecosystem's architecture from the outset. A shared digital dashboard could track key metrics: number of technologies co-developed, adoption rates across different farm sizes, water/energy savings achieved, and income impacts for smallholders. Regular impact assessments should inform iterative improvements to the exchange mechanisms. The International Center for Agricultural Research in the Dry Areas (ICARDA) could provide independent evaluation using its established protocols for technology adoption studies in arid regions.

Implementation should proceed through phased pilots that build confidence and demonstrate quick wins. Initial focus could be on two or three high-impact areas: perhaps combining Algeria's satellite monitoring with Tunisia's farmer feedback systems for drought early warning, then expanding to joint development of solar-powered desalination units for agricultural use. Success in these focused areas would create momentum for broader collaboration while allowing for lessons learned to inform scaling.

Ultimately, the proposed ecosystem represents more than just technology transfer - it envisions the creation of a shared innovation culture that leverages the complementary strengths of both countries' agricultural research and development systems.

By institutionalizing collaboration across the entire innovation chain - from research prioritization to farmer adoption - Algeria and Tunisia can position the Maghreb as a global leader in climate-resilient agricultural technologies while addressing their shared food security challenges more effectively than either could achieve alone. The time is ripe for such cooperation, as both countries face intensifying climate pressures and recognize the limitations of going it alone in agricultural innovation.

2.5. Climate-Smart Value Chains

From "commodity sellers" to "climate-smart exporters", this could be the ambition. This subsection crunches the numbers on how blending Algeria's quinoa with Tunisia's olive oil expertise could capture premium markets while cutting the region's $10.5 billion annual wheat import bill! The business case for cooperation has never been clearer.

Current gaps and opportunities in cross-border value chains

The agricultural trade relationship between Algeria and Tunisia reveals a paradox of shared vulnerabilities and untapped complementary strengths. Both nations currently operate parallel value chains that remain stubbornly disconnected despite their geographical proximity and similar climate challenges. This fragmentation results in significant inefficiencies and missed opportunities for building collective resilience against worsening environmental stresses.

Algeria's heavy reliance on imported food staples presents a striking case study in climate vulnerability. The country spends approximately 9 billion annually on wheat imports, representing over 701.5 billion in annual cereal imports covering 60% of demand. What makes this situation particularly paradoxical is that both countries simultaneously maintain underdeveloped trade links with each other in agricultural goods that could reduce this import dependency. For instance, Algeria produces surplus quantities of potatoes and tomatoes that frequently rot in storage while Tunisia faces periodic shortages of these staples, yet cross-border trade remains hampered by inconsistent customs procedures and lack of coordinated production planning.

The livestock sector demonstrates similar disconnects. Algeria has invested heavily in modern dairy processing facilities that frequently operate below capacity due to seasonal milk shortages, while Tunisia's smaller-scale dairy farmers often struggle to find stable markets for their production. Climate variability exacerbates these inefficiencies - drought years in Tunisia's northern grazing lands coincide with pasture stress in eastern Algeria, creating simultaneous production drops that could be mitigated through coordinated herd management and feed sharing arrangements. The current lack of integrated early warning systems for pasture conditions means both countries miss opportunities to balance supply through strategic cross-border livestock movements.

Processing and value addition present another layer of unrealized potential. Tunisia has developed sophisticated olive oil and date processing clusters in Sfax and Gabès that could add value to Algeria's expanding plantations of drought-resistant crops. Algeria's growing quinoa production, currently exported as raw grain, could be transformed into premium gluten-free products using Tunisia's food processing expertise. Instead, both countries predominantly export unprocessed agricultural goods to European markets while importing expensive processed foods back from the same destinations - a lose-lose scenario that drains foreign reserves and misses opportunities for job creation in rural processing hubs.

Climate change is amplifying these value chain vulnerabilities in concerning ways. Rising temperatures have already reduced the viability of traditional rainfed cereal systems in both countries' breadbasket regions. The 2021 drought saw simultaneous wheat production shortfalls of 30% in Algeria and 40% in Tunisia, forcing both nations into competitive international buying that drove up global prices. Meanwhile, the countries maintain separate, duplicative systems for grain storage, quality testing, and distribution that could be rationalized through cooperation. Algeria's large grain silos frequently operate below capacity, while Tunisia struggles with insufficient storage that forces premature sales during harvest gluts.

The fisheries sector illustrates how climate impacts could be better managed through coordinated value chains. Warming Mediterranean waters are causing fish stocks to migrate northward, affecting both countries' small-scale fishers. While Tunisia has developed innovative value-added fish processing for export markets, Algeria's catch remains predominantly sold fresh with high spoilage rates. Joint investment in cold chain infrastructure and shared market intelligence could help both nations adapt to changing stock patterns while capturing more value from dwindling resources.

Trade data reveals surprising gaps in what should be natural complementarities. Algeria exports citrus to Europe while importing Tunisian oranges, a pattern driven by non-tariff barriers and lack of coordinated marketing. Both countries maintain separate, competing *Halal* certification systems for meat exports despite sharing identical religious standards, creating unnecessary costs for producers. The tomato sector shows similar inefficiencies - Algeria's greenhouse tomato production peaks in winter when Tunisia has shortages, while Tunisia's summer surpluses could help meet Algeria's Ramadan demand spikes, yet trade flows remain negligible.

These disconnects become even more puzzling when considering the countries' shared climate adaptation goals. Algeria's substantial investments in solar-powered irrigation could be paired with Tunisia's expertise in precision agriculture to create climate-smart horticulture value chains. Tunisia's successful agro-tourism models in Djerba could be adapted to Algeria's Saharan oasis communities facing economic decline from water stress. Instead, both countries continue to develop parallel, competing systems for addressing nearly identical challenges.

The root causes of these missed connections are multifaceted. Outdated trade protocols dating from the 1990s fail to account for climate-induced production pattern changes. Banking systems lack mechanisms for cross-border agricultural financing. Certification systems remain unaligned despite shared food safety standards. Perhaps most critically, there is no permanent institutional platform for joint value chain planning that could anticipate and adapt to shared climate risks.

As temperatures continue rising and water resources dwindle, these fragmented approaches become increasingly unsustainable. The next section explores how strategic integration of climate-smart value chains could transform these vulnerabilities into opportunities for shared resilience and economic benefit. By addressing these gaps systematically, Algeria and Tunisia could build agricultural trade systems that not only withstand climate shocks but actively contribute to food sovereignty and rural development goals for both nations.

Building integrated climate-resilient markets

The transformation of disconnected agricultural systems into integrated climate-resilient markets requires a fundamental reimagining of how Algeria and Tunisia approach cross-border food systems. This shift must move beyond temporary trade agreements to establish permanent structures for collaborative value chain development that actively mitigate shared climate risks while capitalizing on complementary strengths. The foundation for this integration lies in recognizing that the ecological and economic futures of both nations are inextricably linked through their shared climate vulnerabilities and agricultural potentials.

Joint processing infrastructure represents one of the most immediate opportunities for value chain integration. Strategic investments in cross-border agro-processing hubs could transform raw agricultural commodities into higher-value products while reducing post-harvest losses. Algeria's emerging quinoa production in the High Plains region, currently exported as raw grain, could be processed in Tunisia's underutilized food manufacturing facilities in Sfax and Gabès into gluten-free pasta, snacks, and other value-added products. Conversely, Tunisia's olive oil processing expertise could be deployed in eastern Algeria's expanding olive groves, creating new economic opportunities in regions affected by cereal production declines. This approach would mirror Morocco's successful strategy of developing “terroir” processing zones that combine local production with centralized value addition.

Climate-smart logistics networks must underpin these integrated markets. Algeria's extensive grain storage capacity in Oran and Annaba could be strategically linked with Tunisia's port infrastructure in Tunis and Sfax to create a shared buffer system for drought years. A real-time digital platform connecting storage facilities, transportation networks, and market demand indicators could optimize grain flows based on climate-induced production variations. During the 2021 drought, such a system could have reduced import costs by 25-30% through coordinated purchasing and inventory management, according to World Food Programme estimates. The logistics integration should extend to cold chain development for perishables, leveraging Algeria's energy infrastructure and Tunisia's export experience to reduce the 35% post-harvest losses currently experienced in both countries' horticulture sectors.

Branding and market positioning offer another critical avenue for collaboration. A unified "Maghreb Climate-Resilient" certification program could differentiate the region's sustainable agricultural products in international markets. This would build on Tunisia's successful organic date exports and Algeria's nascent solar-grown produce initiatives, creating premium market segments that reward climate-adaptive farming practices. The certification could encompass water-use efficiency metrics, renewable energy integration, and soil conservation standards - all verified through a joint Algeria-Tunisia accreditation body.

Financial innovation must accompany physical market integration. A "Maghreb Climate Adaptation Trade Facility" could provide blended financing for cross-border value chain projects, combining concessional loans for climate-smart infrastructure with technical assistance for market linkages. The facility could prioritize investments that simultaneously reduce emissions and enhance resilience, such as solar-powered cold storage units along border regions or shared testing laboratories for drought-resistant crop varieties. Risk-sharing mechanisms, including climate-indexed insurance products, could protect integrated value chains against increasing weather volatility. The African Development Bank's experience with the Desert to Power initiative provides relevant models for such cross-border climate financing structures.

Policy harmonization is essential to enable these market integrations. A bilateral working group should address non-tariff barriers that currently inhibit climate-smart trade, including: Mutual recognition of organic and sustainability certifications, aligned phytosanitary standards for new drought-resistant crops, coordinated export promotion strategies for target markets, and simplified cross-border movement of perishable goods during climate emergencies. The African Continental Free Trade Area (AfCFTA) framework provides a ready platform for implementing these harmonization measures while maintaining necessary safeguards.

Farmer-centered innovation diffusion will determine the ultimate success of integrated markets. Establishing cross-border producer alliances - such as an Algeria-Tunisia Drought-Resilient Crops Cooperative - could facilitate direct knowledge exchange and collective marketing. Digital platforms using Algeria's satellite connectivity could provide Tunisian farmers with real-time climate adaptation advice, while Tunisia's successful mobile market information systems could be extended to Algerian producers. These peer-to-peer networks would complement formal institutional arrangements with grassroots innovation sharing.

Implementation should begin with pilot value chains that demonstrate quick wins while building trust. Priority candidates include: A cross-border olive oil value chain combining Algeria's expanding organic olive production with Tunisia's processing and export expertise, integrated cereal storage and distribution systems to mitigate drought impacts, and joint halal meat marketing initiatives leveraging both countries' religious certification systems. Monitoring and evaluation frameworks must track both economic and climate resilience indicators, including: Value addition per cubic meter of water used, reduction in post-harvest losses during extreme weather events, income stability for smallholder participants, and the diversification of export markets.

The transformation to integrated climate-resilient markets will require overcoming significant political and institutional inertia. However, the alternative - continuing with fragmented, climate-vulnerable value chains - poses far greater risks to both nations' food security and economic stability. By viewing their agricultural futures as interconnected rather than competitive, Algeria and Tunisia can build food systems that not only withstand climate shocks but create new opportunities for sustainable growth. The proposed integration would position the Maghreb as a global innovator in regional climate adaptation strategies while reducing costly import dependencies that currently drain nearly $11 billion annually from the two economies combined.

3. FROM FRAGMENTATION TO INTEGRATION: AN AGENDA FOR MAGHREB AGRICULTURAL RESILIENCE

The whole proves greater than its parts. Having examined individual synergies, this section weaves them into a unified framework where water savings enable crop diversification, tech optimizes resource use, and value chains fund further innovation. Through system dynamics modeling and case analogs, we reveal how Algeria and Tunisia could pioneer a regional resilience model that outshines their solo efforts.

The interdependence of synergies

The true potential of Algeria and Tunisia's agricultural collaboration lies not in isolated policy improvements, but in how their synergies interconnect to create a self-reinforcing system of resilience. Each of the five proposed synergies—water governance, drought-resistant crops, aquifer management, agri-tech transfer, and climate-smart value chains—does not operate in isolation. Instead, they form a dynamic web of mutual dependencies, where progress in one area amplifies success in others.

Figure 1. Diagram of synergies' interdependence process

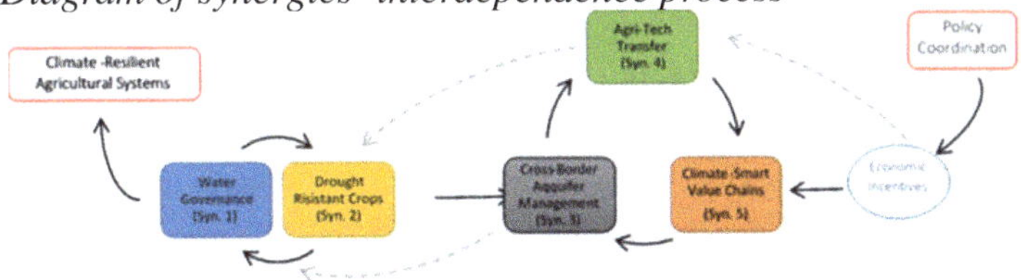

The proposed diagram illustrates how the five key synergies—water governance (1), drought-resistant crops (2), aquifer management (3), agri-tech transfer (4), and climate-smart value chains (5)—interconnect to form a self-reinforcing system for agricultural resilience. At its core, participatory water governance (1) enables the adoption of water-efficient crops (2), which in turn reduces pressure on shared aquifers (3), while agri-tech innovations (4) optimize both water use and crop performance through real-time monitoring and data-driven decisions. Climate-smart value chains (5) complete the cycle by generating economic returns that fund further improvements across all other areas, creating a virtuous feedback loop where market profits are reinvested in water infrastructure, crop research, and technology upgrades.

The system is further supported by two critical enablers—economic incentives that ensure farmer adoption and policy coordination to align bilateral efforts—demonstrating how ecological sustainability, technological innovation, and market forces must work in concert to achieve transformative climate adaptation. Solid arrows show direct operational dependencies between synergies (e.g., water governance enabling crop success), while dotted lines represent reinforcing feedback loops (e.g., value chain profits funding better water management), visually capturing how progress in one dimension amplifies outcomes across the entire system.

Water governance as the foundation: Effective participatory water management (Synergy 1) is the bedrock upon which other interventions depend. Tunisia's decentralized Agricultural Development Groups (GDAs) have proven that farmer-led water allocation improves efficiency, reducing losses by 25–30% compared to centralized systems. If Algeria adopts this model for its irrigation schemes, the saved water directly enables the expansion of drought-resistant crops (Synergy 2), such as quinoa and sorghum, which require precise moisture control. Conversely, without reformed water governance, even the most resilient crops will struggle to scale due to inefficient distribution.

Crop innovation strengthens aquifer sustainability: The adoption of drought-resistant crops (Synergy 2) does more than just improve yields—it alleviates pressure on shared groundwater. Algeria's experiments with quinoa, which uses 30% less water than wheat, could—if paired with Tunisia's extension networks—reduce demand on the overexploited Northwestern Sahara Aquifer System (NWSAS) (Synergy 3). Fewer water-intensive cereals mean slower aquifer depletion, buying time for joint management reforms. In turn, a stabilized aquifer ensures that new crop varieties have a reliable water supply, even in drought years.

Agri-Tech as the nervous system: The exchange of agricultural technologies (Synergy 4) acts as the central nervous system linking water, crops, and markets. Tunisia's low-cost soil moisture sensors, deployed at scale, could help Algerian farmers optimize irrigation under new GDA-led systems. Meanwhile, Algeria's satellite-based yield forecasting could provide Tunisia with early warnings of production shortfalls, allowing proactive adjustments in water allocation. These technologies close the feedback loop between policy (water governance), production (crops), and resource management (aquifers).

Value chains as the economic engine: Finally, climate-smart value chains (Synergy 5) provide the economic incentive to sustain the other four synergies. If Algeria's quinoa and Tunisia's drought-resistant olives are processed and marketed under a shared "Maghreb Climate-Resilient" brand, the higher profits create a funding stream for further water infrastructure improvements, agri-tech R&D, and aquifer monitoring. Without profitable markets, farmer adoption of new crops and technologies remains sluggish.

The feedback loop of resilience: The interdependence of these systems creates a virtuous cycle:

→ Better water governance enables drought-resistant crops, which reduce aquifer depletion.
→ Agri-tech innovations fine-tune both water use and crop performance.
→ Stronger value chains fund further improvements in all four areas.

This is not a linear process but an adaptive system where each component reinforces the others. The failure of one synergy weakens the entire framework—for example, without viable markets (Synergy 5), farmers lack incentives to adopt new crops (Synergy 2), leaving water savings (Synergy 1) unrealized and aquifers (Synergy 3) under continued strain.

Cascading benefits: Ecological, economic, and social dividends

The systemic integration of Algeria and Tunisia's agricultural synergies generates multiplier effects that extend far beyond incremental improvements in water or crop yields. When implemented cohesively, these interconnected interventions produce three categories of cascading benefits—ecological, economic, and social—that collectively transform the Maghreb's climate resilience trajectory.

The combined impact of water governance reforms, aquifer co-management, and drought-resistant crops initiates a positive ecological feedback loop. As participatory irrigation systems (Synergy 1) reduce water waste by 25–30%, and farmers transition to low-water crops (Synergy 2), pressure on the Northwestern Sahara Aquifer (Synergy 3) begins to ease. Simple projections suggest this could slow depletion rates by 40% within a decade, allowing gradual recharge of critically overdrawn groundwater reserves. Simultaneously, agri-tech integration (Synergy 4)—such as soil moisture sensors and satellite-based irrigation advisories—further optimizes water use, reducing the energy footprint of agriculture by minimizing unnecessary pumping. These ecological gains compound over time: healthier aquifers sustain rivers and oases, preserving biodiversity hotspots like Tunisia's Ichkeul Lake and Algeria's Chott Ech Chergui, both UNESCO Biosphere Reserves facing desertification.

The economic benefits operate at three levels:

- Import substitution: Scaling up drought-resistant quinoa and sorghum (Synergy 2) could displace 15–20% of Algeria and Tunisia's wheat imports (currently $10.5B/year combined), conserving foreign reserves.
- Value chain upgrading: Integrating Algeria's solar-powered irrigation with Tunisia's food processing expertise (Synergy 5) creates "Climate-Smart Maghreb" branded products (e.g., gluten-free quinoa pasta, solar-grown olive oil) that command 20–30% price premiums in EU and Gulf markets.
- Job creation: A significant part of qualified and unqualified youth will integrate new rural jobs in agro-processing, tech maintenance, and sustainable farming—critical for stabilizing migration-prone regions like Tunisia's Kasserine and Algeria's High Plains.

The agri-tech transfer (Synergy 4) acts as a catalyst, boosting productivity per drop of water and hectare of land. For instance, adopting Tunisia's 15 soil sensors across Algeria's 1.3 million small farms could highly significant increase in average yields and annual income boost at scale.

The synergies' most profound impact may be their stabilization of vulnerable rural communities. By de-risking agriculture through: Reliable water access (via aquifer management and efficient irrigation), climate-buffered crops (quinoa yields remain stable at 1.8 tons/ha even in drought years) guaranteed markets (through coordinated value chains), the system reduces the boom-bust cycles that currently push smallholders into poverty. Tunisia's Agricultural Development Groups (GDAs) model, if adapted to Algeria, could additionally empower women farmers, who manage 45% of smallholdings but lack decision-making power in water allocation.

The true power of this framework lies in how benefits cascade across domains: Ecological gains (healthier aquifers) enable economic gains (expanded irrigation for high-value crops), which fund social gains (rural jobs). Conversely, social stability (youth employed in agri-tech) sustains ecological stewardship (maintained irrigation systems).

The synergies are not zero-sum—Algeria and Tunisia both gain exponentially more by integrating systems than by pursuing isolated reforms. This interdependence positions the Maghreb as a test case for regional climate adaptation, with lessons for other water-scarce regions worldwide.

A model for integrated climate adaptation

The proposed framework transforms Algeria and Tunisia's fragmented agricultural policies into a cohesive, adaptive system designed to withstand climate shocks while unlocking new economic opportunities. This model is not merely a collection of parallel interventions, but a dynamic architecture where each component reinforces the others, creating resilience greater than the sum of its parts.

The core components of the model are:

1. Resource Optimization Engine
   – Water Governance and Agri-Tech: Tunisia's participatory irrigation groups (GDAs), equipped with Algeria's satellite-based water monitoring, create a closed-loop system where farmers receive real-time data to allocate water precisely—reducing waste while maintaining yields.
   – Aquifer Management and Drought-Resistant Crops: As quinoa and sorghum replace water-intensive wheat, aquifer stress declines, buying time for recharge initiatives like managed aquifer recharge (MAR) using treated wastewater.
2. Productivity Innovation Loop
   – Agri-Tech and Value Chains: Tunisia's low-cost soil sensors help farmers optimize inputs for Algeria's emerging quinoa sector, ensuring premium quality for "Climate-Smart Maghreb" exports. Profits are reinvested in R&D for new drought-tolerant varieties.

– Cross-Border Learning: Algerian engineers and Tunisian extension agents co-develop adaptation packages (e.g., saline-tolerant olive cultivars for coastal zones) through joint innovation hubs.

3. Risk Distribution Mechanism

   – Shared Early Warning Systems: Integrated drought alerts (leveraging Algeria's satellites and Tunisia's farmer networks) trigger pre-agreed responses: Farmers shift to short-cycle crops during dry spells, and cross-border grain reserves (Algeria's storage and Tunisia's logistics) stabilize prices.

   – Diversified Production Basins: Coordinated crop zoning ensures that climate shocks in one region (Tunisia's cereal-focused north) are offset by surpluses in another (Algeria's drought-resilient High Plains).

The model's design allows gradual expansion:

→ Phase 1 (2025–2030): Pilot in border regions (e.g., Tébessa-Kasserine) focusing on water-tech-crop linkages.
→ Phase 2 (2030–2035): Scale aquifer and market integration.
→ Phase 3 (2035–2040): Institutionalize as a Maghreb Climate Adaptation Platform, inviting Morocco and Libya to join.

Unlike siloed projects, this model embeds adaptation into economic systems, ensuring sustainability through Market incentives (premium prices for resilient crops), community ownership (GDAs as frontline implementers), and policy interoperability (harmonized subsidies, data standards).

From theory to implementation: Overcoming barriers to systemic change

The transition from a conceptual framework to tangible on-the-ground impact faces significant political, institutional, and financial hurdles. While the synergies between Algeria and Tunisia present a compelling vision for integrated climate resilience, their implementation requires handling complex realities that have historically stifled cross-border cooperation.

Political economy challenges pose the most formidable barrier. In Algeria, centralized decision-making and entrenched interests in water-intensive agriculture create resistance to decentralizing water governance or reallocating subsidies toward drought-resistant crops. Tunisia's frequent policy shifts, driven by short-term economic pressures, risk undermining long-term commitments to aquifer management or agri-tech investments. Both countries must confront these inertial forces by building coalitions of reform-minded actors—progressive ministries, farmers' unions, and private sector partners—who can champion the synergies as national priorities rather than experimental projects. The 2015 Algeria-Tunisia Joint Agricultural Commission offers a cautionary tale; despite ambitious declarations, its work stalled due to competing domestic agendas. Learning from this, future efforts must anchor cooperation in high-level political agreements with clear accountability mechanisms, such as biennial progress reviews chaired by both countries' agriculture and water ministers.

Institutional mismatches between the two systems further complicate implementation. Algeria's top-down agricultural extension services lack the local engagement channels that make Tunisia's Agricultural Development Groups (GDAs) effective, while Tunisia's fragmented research institutes struggle to absorb Algeria's large-scale technological innovations. Bridging these gaps demands adaptive governance structures—perhaps a joint Algeria-Tunisia Resilience Office with rotating leadership and blended technical teams. This entity could standardize protocols for data sharing, synchronize subsidy reforms, and mediate disputes over shared resources like the Northwestern Sahara Aquifer. The Senegal River Basin Organization provides a regional precedent, having overcome similar institutional divergences among Mali, Senegal, and Mauritania through neutral technical secretariats and consensus-based water allocation rules.

Financing constraints threaten to derail even the most well-designed synergies. Algeria's hydrocarbon-dependent budget cycles and Tunisia's fiscal austerity limit domestic funding for long-term adaptation. Creative solutions are needed, such as a Maghreb Climate Resilience Bond co-guaranteed by both governments to attract diaspora investments, or a Debt-for-Adaptation Swap that redirects portions of Tunisia's external debt toward aquifer recharge projects. International climate funds like the Green Climate Fund could play a catalytic role by offering outcome-based financing—for instance, releasing tranches of funding only when verifiable reductions in aquifer depletion are achieved. The Great Green Wall Initiative's experience in the Sahel demonstrates how blended finance can sustain cross-border environmental projects despite political volatility.

Cultural and knowledge barriers also impede progress. Algerian and Tunisian agronomists often work in isolation, publishing in different languages (French, English, and Arabic) and rarely collaborating on joint research. Establishing bilingual innovation platforms—digital hubs where scientists, farmers, and policymakers from both countries co-develop solutions—could foster shared ownership. Farmer-to-farmer exchange programs, like those that enabled Morocco's successful adoption of drip irrigation, would build trust at the grassroots level.

The path forward requires phased experimentation rather than wholesale transformation. Pilot zones along the Algeria-Tunisia border—where climate risks and interdependencies are most acute—could test the synergies in practice. For example, a Tébessa-Kasserine Resilience Corridor might combine Algeria's solar pumping with Tunisia's water-user associations, monitored by shared sensors. Successes here could then inform national scaling, while failures would provide low-stakes learning opportunities. This approach mirrors China's "ecological civilization" pilot cities, which trialed integrated environmental policies before broader rollout.

Ultimately, overcoming these barriers hinges on reframing the synergies from optional cooperation to essential survival strategies in a climate-disrupted future. As desertification advances and aquifers dwindle, the cost of inaction—measured in lost livelihoods, forced migration, and geopolitical instability—will far exceed the political and financial investments required today. Systemic change is possible, but only if implementation addresses power dynamics, institutional legacies, and financing gaps with the same rigor as the technical synergies themselves.

4. CONCLUSION

Algeria and Tunisia face converging climate threats—deepening droughts, aquifer depletion, and rising food import dependency—that demand urgent, collaborative solutions. Despite sharing similar ecological vulnerabilities, their agricultural policies have evolved in isolation, leading to redundant investments and missed opportunities for resilience-building. This study aims to break this cycle by systematically identifying policy synergies that leverage each country's strengths, transforming parallel struggles into a unified framework for climate adaptation. By assessing complementary innovations in water governance, crop diversification, and technology exchange, the chapter provides a roadmap for bilateral cooperation that could reduce import costs, stabilize rural livelihoods, and safeguard shared resources.

Five critical synergies emerge as pillars of Algeria-Tunisia agricultural collaboration: participatory water governance (combining Tunisia's decentralized irrigation groups with Algeria's infrastructure scale), drought-resistant crops (linking Algeria's quinoa research to Tunisia's extension networks), cross-border aquifer management (joint monitoring and quotas for the Northwestern Sahara Aquifer System), agri-tech transfer (integrating Tunisia's low-cost sensors with Algeria's satellite analytics), and climate-smart value chains (co-branding processed goods like gluten-free quinoa and solar-grown olive oil). Together, these interventions form a self-reinforcing system where water savings enable crop diversification, technology optimizes resource use, and value chains fund further innovation—creating a multiplier effect that neither country could achieve alone. The synergies' power lies in their interdependence: each strengthens the others, building a foundation for regional food sovereignty in an era of climate disruption.

The climate crisis leaves Algeria and Tunisia with a stark choice: compete for dwindling resources or collaborate to reinvent them. The synergies outlined in this chapter are not mere policy options—they are survival strategies for two nations bound by geography, ecology, and now, climate destiny. By integrating water governance, drought-resistant crops, aquifer management, agri-tech, and value chains, Algeria and Tunisia can transform their agricultural systems from fragile and import-dependent to resilient and self-sustaining.

Yet time is the scarcest resource of all. With the Northwestern Sahara Aquifer nearing irreversible depletion and temperatures rising faster than global averages, incrementalism is a luxury neither country can afford. The real question is not whether these synergies will work—evidence from pilot projects and global analogs confirms they can—but whether political leaders will act before the window of opportunity closes.

If Algeria and Tunisia fail to seize this moment of collaboration, history will remember them not as climate victims, but as architects of their own food insecurity. The tools for resilience exist; the missing ingredients are urgency and collective will. The Maghreb's future hinges on answering one question: Will these nations remain prisoners of borders drawn by colonial powers, or will they redraw the map of climate adaptation through unity? The land, the water, and the next generation await their reply.

REFERENCES

African Development Bank (AfDB). (2018). *Morocco: National Irrigation Water Saving Programme Support Project (PAPNEEI) – Project Completion Report. Regional Department North (RDGN)*. African Development Bank.

Allani, M., Frija, A., Nemer, R., Ribbe, L., & Sahli, A. (2022). Farmers' perceptions on an irrigation advisory service: Evidence from Tunisia. *Water (Basel), 14*(22), 3638. DOI: 10.3390/w14223638

Chebil, A., Souissi, A., Frija, A., & Stambouli, T. (2019). Estimation of the economic loss due to irrigation water use inefficiency in Tunisia. *Environmental Science and Pollution Research International, 26*(11), 11261–11268. DOI: 10.1007/s11356-019-04566-8

Dasthagir, K. G. (2024). Hydraulic bureaucracy and irrigation management transfer: Analysing the efficacy of two models in South-India. *International Journal of Water Resources Development, 40*(5), 885–905. DOI: 10.1080/07900627.2024.2341992

Debbabi, O. S., Amar, F. B., Rahmani, S. M., Taranto, F., Montemurro, C., & Miazzi, M. M. (2022). The status of genetic resources and olive breeding in Tunisia. *Plants, 11*(13), 1759. DOI: 10.3390/plants11131759

Dhaouadi, L., Ben Maachia, S., Namsi, A., Abouthief, F., Nagaz, K., & Masmoudi, M. M. (2022). Literature review on water productivity of date palm trees in Tunisian agro systems. *Journal of Oasis Agriculture and Sustainable Development, 2*(4), 30–44. DOI: 10.56027/JOASD.222022

ElFakih, S., Ben Abdallah, S., & Elkadri, A. (2023). Water accounting for food security: The case of rainfed and irrigated olive growing in arid regions in Tunisia. *Journal of Arid Arboriculture and Olive Growing, 2*(1), 12–26. https://iosfax.agrinet.tn/annonce/revue/Volume2-N1/Article%20saida.pdf

Elloumi, M. (2016). La gouvernance des eaux souterraines en Tunisie (Report No. 7). International Water Management Institute (IWMI). https://gw-mena.iwmi.org/wp-content/uploads/sites/3/2017/04/Rep.7-Groundwater-governance-in-Tunisia_final_cover.pdf

Evers, H. D., & Benedikter, S. (2009). Hydraulic bureaucracy in a modern hydraulic society-strategic group formation in the Mekong delta, Vietnam. *Water Alternatives, 2*(3), 416.

FAO. (2018). Water efficiency, productivity and sustainability in the NENA regions (WEPS-NENA). https://www.fao.org/in-action/water-efficiency-nena/countries/tunisia/background-and-sites/ar/?

Ferchichi, I., Zairi, A., & Marlet, S. (2022). Water scarcity conflicts in a community-managed irrigation system in Northern Tunisia: Supporting dialogue and negotiation. IAHS-AISH Scientific Assembly, Montpellier, France, 385, 253–258. DOI: 10.5194/piahs-385-253-2024

GIZ. (2020). Regional cooperation in the water sector in the Maghreb (CREM II). Deutsche Gesellschaft für Internationale Zusammenarbeit (GIZ). https://www.giz.de/en/worldwide/24690.html

ICARDA. (2020). Ultra-low-energy drip irrigation helps small farmers save water, energy, and money! https://icarda.org/impact/impact-stories/ultra-low-energy-drip-irrigation-helps-small-farmers-save-water-energy-and

Imache, A., Bouarfa, S., Kuper, M., Hartani, T., & Dionnet, M. (2009). Integrating "invisible" farmers in a regional debate on water productivity: The case of informal water and land markets in the Algerian Mitidja plain. *Irrigation and Drainage*, *58*(3), 264–272. DOI: 10.1002/ird.523

IPCC. (2023). [Impacts, Adaptation and Vulnerability.]. *Climatic Change*, •••, 2023.

Kherbache, N. (2020). Water policy in Algeria: Limits of supply model and perspectives of water demand management (WDM). *Desalination and Water Treatment*, *180*, 141–155. DOI: 10.5004/dwt.2020.25009

Laoubi, K., & Yamao, M. J. W. T. (2009). Management of irrigation schemes in Algeria: An assessment of water policy impact and perspectives on development. *WIT Transactions on Ecology and the Environment*, *125*, 503–514. DOI: 10.2495/WRM090451

Maamri, K., Djerroudi-Zidane, O., Chaabena, A., & Bazile, D. (2021). Preliminary evaluation of Quinoa genotypes introduced in the Saharan zone of Algeria, case of Ouargla. In *Euro-Mediterranean Conference for Environmental Integration* (pp. 283–286). Springer Nature Switzerland.

Mahdhi, N., Sghaier, M., & Smida, Z. (2014). Efficiency of the irrigation water user association in the Zeuss-Koutine region, south-eastern Tunisia. *New Medit*, *13*(2), 47–55.

Mejri, R., Dhraief, M. Z., Souissi, A., Dhehibi, B., Oueslati, M., Charry, A. C., Frija, A., Ouerghemmi, H., Oumer, A. M., Fendri, M., & Larbi, A. (2025). Empowering smallholder olive growers in northwest Tunisia through an agroecological business model. *Frontiers in Sustainable Food Systems*, *9*, 1587318. Advance online publication. DOI: 10.3389/fsufs.2025.1587318

Nawar, B., Scardigno, A., & Sagardoy, J. A. (2024). Treatment and reuse of water: Economic feasibility and assessment of water pricing policies in Ouardanine irrigation district (Tunisia). *New Medit*, *23*(1), 39–74. DOI: 10.30682/nm2401e

Olive Tree Institute (OTI). (2012). Olive tree irrigation manual: Techniques. Ministry of Agriculture, Institution of Agricultural Research and Higher Education. http://www.iosfax.agrinet.tn/annonce/Manuel%20d'irrigation%20agencee.pdf

ONAGRI (National Observatory of Agriculture). (2021). National report of the water sector. Ministry of Agriculture, Water Resources and Fisheries, Office of Planning and Hydraulic Balancing, Republic of Tunisia. http://www.onagri.tn/uploads/secteur-eau/RNE_2021_V-Anglaise.pdf

OSS. (2021). *Northwestern Sahara Aquifer System: Diagnostic Report*. Sahara and Sahel Observatory.

Ostrom, E. (1990). *Governing the Commons*. Cambridge University Press. DOI: 10.1017/CBO9780511807763

Ouerghemmi, H., Frija, A., Souissi, A., Carpentier, I., Shiri, Z., Dhehibi, B., & Rejeb, H. (2024). Are farmer organizations effective intermediaries and facilitators of agricultural innovations processes? Evidence from Tunisia. *Frontiers in Sustainable Food Systems*, *8*, 1440086. DOI: 10.3389/fsufs.2024.1440086

Oulmane, A., Kechar, A., Benmihoub, A., & Benmehaia, A. M. (2022). Assessment of surface water management institutions: A case of public irrigation schemes in northern Algeria. *Water Policy*, *24*(2), 229–241. DOI: 10.2166/wp.2022.025

Oustani, M., Mehda, S., Halilat, M. T., & Chenchouni, H. (2023). Yield, growth development and grain characteristics of seven Quinoa (Chenopodium quinoa Willd.) genotypes grown in open-field production systems under hot-arid climatic conditions. *Scientific Reports*, *13*(1), 1991. DOI: 10.1038/s41598-023-29039-4

Sghaier, M., Frija, A., Postigo, J., Speelman, S., Alary, V., & Sghaier, M. (2025). Assessing pastoral reforms through the performance of agro-pastoral community-based organizations in South Tunisia. *Rangeland Ecology and Management*, *98*, 49–62. DOI: 10.1016/j.rama.2024.07.008

Vandôme, P. (2023). Rendre l'innovation technologique accessible aux systèmes irrigués: Co-conception, évaluation et implications de la low-tech numérique pour accompagner la gestion de l'eau en agriculture. Sciences agricoles, Institut national d'enseignement supérieur pour l'agriculture, l'alimentation et l'environnement. https://theses.hal.science/tel-04747653v1/file/23-0025_Vandome.pdf

World Bank. (2022). Beyond Scarcity: Water Security in the Middle East and North Africa.

Chapter 8

Abiotic Stress and Crop Production in a Changing Global Environment

Muhammad Zain ul Abideen
Faculty of Agricultural Sciences, University of the Punjab, Lahore, Pakistan

Mian Muhammad Ahmed
https://orcid.org/0009-0001-4496-048X
College of Life Science and Technology, Tarim University, Alar, China

Muhammad Asim
https://orcid.org/0009-0009-2768-0687
College of Horticulture and Forestry Sciences, Huazhong Agricultural University, Wuhan, China

Pan Zhiyong
College of Horticulture and Forestry Sciences, Huazhong Agricultural University, Wuhan, China

Zhang Rui
College of Horticulture and Forestry, Tarim University, Alar, China

Kashif Razzaq
MNS-University of Agriculture, Multan, Pakistan

Yerzat Yerhazi
College of Horticulture and Forestry, Tarim University, Alar, China

Muhammad Atiq Ashraf
https://orcid.org/0009-0006-2675-8604
College of Horticulture and Forestry Sciences, Huazhong Agricultural University, Wuhan, China

Syeda Maira Hamid
https://orcid.org/0009-0005-2065-1999
College of Life Science and Technology, Tarim University, Alar, China

Ayaz Ahmad
College of Life Science and Technology, Tarim University, Alar, China

DOI: 10.4018/979-8-3373-3346-5.ch008

ABSTRACT

Drought, salt, high temperatures, pH imbalances in the soil, heavy metal toxicity, and floods are examples of abiotic stresses that pose serious challenges to agricultural sustainability and crop output worldwide. Plant physiological, biochemical, and molecular processes are negatively impacted by these environmental stresses, which are made worse by the rapid rate of climate change. This results in stunted development, decreased photosynthetic efficiency, disturbed nutrient balance, and eventually significant yield losses. The fundamental processes via which abiotic stressors affect plant systems are thoroughly examined in this chapter, with special attention paid to oxidative stress, hormone signaling, osmotic and ionic imbalances, and stress-induced metabolic reprogramming. It also clarifies crop-specific susceptibilities and geographical differences in stress exposure, emphasizing the disproportionate impacts on rainfed agricultural systems and staple food crops in areas like the Middle East, South Asia, and sub-Saharan Africa. The chapter offers an integrative framework of mitigation strategies in addition to evaluating the environmental and economic effects of prolonged abiotic stress, such as soil degradation and biodiversity decline. These consist of microbial inoculants, sophisticated biotechnology treatments, traditional agronomic techniques, and sustainable land management strategies. In the context of creating resilient agroecosystems, the function of precision technology, climate-smart agriculture, and international policy frameworks is also covered. The chapter emphasizes the critical need for flexible, science-based approaches to protect global food security in the face of growing environmental stressors via multidisciplinary insights and case studies tailored to certain regions.

1. OVERVIEW

The detrimental effects of non-living environmental factors on plant growth, development, and productivity are referred to as abiotic stress. Abiotic stress results from physical and chemical factors in a plant's environment, as opposed to biotic stress, which is brought on by living things like insects, fungus, bacteria, and weeds (Umar et al., 2021). These stresses interfere with vital physiological and biochemical functions, which often result in lower crop yields, worse-quality crops, and, in the worst situations, complete crop failure (Figure 1).

The basic kinds of abiotic stress include:

Drought: Water scarcity leads to reduced turgor pressure, stomatal closure, and impaired photosynthesis, ultimately stunting growth. *Salinity:* High salt concentrations in soil or irrigation water cause osmotic stress, ion toxicity (Na^+, Cl^-), and nutrient imbalances, damaging roots and leaves (Vurukonda et al., 2016). *Extreme Temperatures:* Both heat and cold stress disrupt enzyme activity, membrane stability, and metabolic functions. Heat can denature proteins, while frost causes ice crystal formation in cells. *Soil pH Imbalances:* Acidic or alkaline soils alter nutrient availability (e.g., aluminium toxicity in acidic soils, iron deficiency in alkaline soils) and harm microbial communities critical for plant health. *Heavy Metals:* Contaminants like arsenic (As), cadmium (Cd), and lead (Pb) induce oxidative stress, inhibit root growth, and accumulate in edible plant parts, posing health risks. *Flooding:* Waterlogged soils deprive roots of oxygen, resulting in anaerobic conditions that destroy cells and encourage root rot. Seldom do these challenges happen alone; for instance, drought and heat stress often coexist, and floods may make soil salinity worse. Abiotic stress is a major problem in agriculture as its combined effects increase crop losses.

Globally, abiotic stressors are becoming more frequent and severe due to climate change. Droughts and heatwaves have become more frequent due to rising global temperatures, especially in arid and semi-arid areas like South Asia and the African Sahel. At the same time, coastal farmlands like those in Bangladesh and Vietnam have increased soil salinity due to melting glaciers and increasing sea levels. These problems are exacerbated by unpredictable rainfall patterns, which lead to alternating droughts and floods that disrupt agricultural seasons.

There are significant implications for food security. Millions of people are fed by staple crops like maize, rice, and wheat, which are very vulnerable to abiotic stresses. For instance, dryness during blooming may reduce wheat yields by 50%, while salt leaves millions of hectares uncultivable yearly. Smallholder farmers, who depend on rainfed agriculture, are disproportionately impacted, risking economic ruin and food insecurity. Beyond crop losses, abiotic pressures erode soil health, diminish biodiversity, and strain water supplies, producing feedback loops that undermine agricultural sustainability.

Addressing these difficulties involves a dual approach: modifying crops to tolerate harsher circumstances and employing land-management strategies that prevent environmental deterioration. Innovations in breeding, biotechnology, and precision agriculture provide promise, but their success rests on global collaboration and legislative support to enable fair access for disadvantaged agricultural communities.

2. PURPOSE AND SCOPE

A thorough examination of abiotic stress and its effects on crop productivity worldwide is given in this chapter. It starts by outlining the physiological and molecular processes by which the main stressors, such as salt, drought, and very high or low temperatures, affect plant development. Expanding upon this basis, the conversation delves into regional susceptibilities, emphasizing how climate change intensifies stress in crucial agricultural areas.

Additionally, the chapter assesses new and existing approaches to improve crop resilience.

These consist of:

- Genetic solutions: Using genetic engineering (such as CRISPR-Cas9) and conventional breeding to create crop types that can withstand stress.
- Agronomic techniques: Improving crop rotation, soil amendments, and irrigation to lessen the effects of stress.
- Technological interventions: Making effective use of precision farming, AI-driven forecasting, and remote sensing.

To protect food systems, this chapter emphasizes the critical need to treat abiotic stress by combining findings from agronomy, climate science, and biotechnology. To ensure sustainable agriculture in a time of environmental uncertainty, it urges coordinated, interdisciplinary efforts to close the gap between scientific innovation and practical application.

3. TYPES OF ABIOTIC STRESS

3.1. The Physiological Impact and Yield Consequences of Drought Stress

When the amount of water in the soil is insufficient to fulfill a plant's transpiration and metabolic demands, drought stress results. Drought is acknowledged as one of the most damaging abiotic stressors in agriculture, causing a series of physiological disruptions that significantly reduce plant output and health (Khan et al., 2025; Zhu, 2016). Stomatal closure, a protective mechanism meant to reduce water loss via transpiration, is one of the first and most important physiological reactions to drought (Chaves et al., 2009; Farooq et al., 2009). Although this aids in preserving internal moisture, it also limits the intake of carbon dioxide (CO_2), which significantly lowers photosynthetic activity. As a result, the generation of carbohydrates, which is essential for plant development and energy supply, drastically decreases. Enzymatic processes, especially those linked to the Calvin cycle as RuBisCO, are compromised under extended drought conditions, which further reduces photosynthetic efficiency. In addition to metabolic difficulties, drought stress results in cellular dehydration, which lowers turgor pressure and produces outward signs like wilting, leaf curling, and early senescence. Water deprivation raises the production of reactive oxygen species (ROS) at the molecular level, which, if left unchecked by antioxidant defense mechanisms, may harm essential cellular constituents including proteins, lipids, and DNA. Plants respond by activating intricate stress-signaling networks, which primarily include the production of abscisic acid (ABA). This hormone stimulates the production of genes that respond to drought and improves stomatal closure. These survival strategies, however, often divert energy from development and reproduction, exacerbating yield losses.

When drought stress occurs during a plant's vital developmental phases, its effects are very severe. The plant's ability to absorb light and perform photosynthesis is weakened during the vegetative stage when a lack of water limits cell proliferation, resulting in stunted growth, smaller leaves, and poor canopy formation. The overall vigor of the plant is adversely affected by this decrease in biomass accumulation. Drought may be particularly harmful during the reproductive stage. Water shortage during blooming and grain filling in cereals like maize and wheat may lead to pollen sterility, poor fertilization, and the formation of shriveled grains. Depending on the intensity and timing of the drought, output decreases can range from 30% to 90%. Additionally, drought affects root architecture, often encouraging the development of shallow roots that reach moisture close to the soil's surface. Although this could be advantageous in the short term, it hinders the plant's capacity to access deeper water supplies during protracted dry spells.

The susceptibility of various crops to drought stress varies (Mittler, 2006). During the tillering and grain-filling phases, wheat is very sensitive, and stress causes a decrease in the weight and quantity of kernels. Intermittent drought causes considerable output losses for rice, which grows well under flooded environments, especially because of spikelet sterility. During tasseling and silking, maize is particularly vulnerable; water stress during these phases interferes with pollination and leads to inadequate grain production. Long-term, recurrent drought occurrences lead to microbial activity reduction, organic matter content fall, and soil structure degradation, all of which lower soil fertility and resilience. These modifications start a negative feedback loop in which future droughts are made worse by diminished soil health. This is especially troubling for areas like sub-Saharan Africa and portions of South Asia that rely significantly on rainfed agriculture. The threat to food security increases as aridity in these regions is exacerbated by climate change.

3.2. Mechanisms of Salinity Stress and Its Effects on Crop Production

When too many soluble salts, mostly sodium (Na^+) and chloride (Cl^-) ions, build up in the soil or irrigation water, it may cause salinity stress, which makes it difficult for plants to thrive (Mohanavelu et al., 2021). Plants are impacted by this kind of abiotic stress via a number of interconnected physiological and biochemical processes. Osmotic stress, ion toxicity, and nutritional imbalance are the main ways that salinity has a negative impact on agriculture. These factors all contribute to stunted development, decreased production, and long-term problems. The first reaction brought on by elevated salt levels in the soil is osmotic stress. The osmotic potential of the soil solution decreases with increased salt content, making it more challenging for plant roots to absorb water. This results in a state called physiological drought, in which plants suffer from water stress even while the soil is wet. Plants respond by closing their stomata to reduce transpirational water loss. But this also limits the intake of carbon dioxide, which lowers the rate of photosynthesis. Carbohydrate synthesis slows down when photosynthesis decreases, which eventually leads to less biomass accumulation and plant development.

When these ions build up to dangerous quantities in plant tissues, ion toxicity, mainly from Na^+ and Cl^-, becomes a serious problem. By competing with potassium (K^+), a vital nutrient involved in stomatal control and enzyme activation, sodium ions disrupt vital cellular processes. Functional K^+ deficit may result from sodium's competitive absorption, even in soils that are rich in potassium. Chloride ions, on the other hand, disrupt photosynthetic processes and may result in leaf chlorosis, which is characterized by yellowing brought on by the breakdown of chlorophyll. Cellular integrity is further jeopardised by high amounts of Na^+, which further impair protein synthesis and damage membrane stability.

Another effect of salinity stress is nutrient imbalance. Other vital cations like calcium (Ca^{2+}), magnesium (Mg^{2+}), and potassium (K^+) compete with excess sodium in the soil for binding sites and absorption in plant roots. For example, Na^+ undermines the stability of the cell wall and the integrity of the membrane by competing with Ca^{2+} at the root surface. The plant's capacity to sustain essential metabolic functions is weakened by this disturbance in nutrient uptake and transport, which intensifies the consequences of salt stress.

Both outward signs and physiological dysfunctions are indicative of the cumulative effects of salt stress on crops, which eventually result in lower yields and worse crop quality. Root inhibition is among the most noticeable results right away. Excessive salt limits a plant's ability to efficiently absorb water and nutrients by stunting root growth. Symptoms of affected roots include poor branching, necrosis, and browning. Because it prevents appropriate crop establishment, this is particularly detrimental in the early phases of plant development.

Leaf burn and chlorosis are two more typical symptoms. Damage from the build-up of Na+ and Cl+ ions in the leaves manifests as premature senescence, necrosis, and tip burn. Chlorophyll deterioration results in chlorotic patches, which lower the photosynthetic efficiency of the plant. The plant's energy supplies may be depleted and growth may be further hindered if complete leaves dry out and fall off.

One of the main effects of salt stress is a decrease in yield, especially during the reproductive phases of crop growth. Unbalances in water and ions affect the growth of fruit or grains, fertilization, and blossom production. While salt causes issues like blossom-end rot and deformed fruits in tomatoes, it also decreases tiller production and panicle emergence in rice. Yield losses might vary from 20% to 50%, depending on the crop's susceptibility and the intensity of the stress.

Salinity stress causes soil deterioration in addition to impacts on individual plants, which poses long-term problems for sustainable agriculture. By scattering soil particles and decreasing porosity, aeration, water penetration, and salt buildup may degrade soil structure. Additionally, this reduces microbial activity, a crucial component of soil fertility. Plant development is further hampered by the soil's rising degradation, which feeds a vicious cycle of decreased production and ecological collapse. Different crops react differently to salt. Even at modest saline levels (4–6 dS/m), salt-sensitive crops, such as wheat, rice, and common beans, have significant production reductions. Conversely, halophytes, crops that can withstand salt, such as barley, quinoa, and certain millets, have developed adaptations to live in salty conditions. These processes include ion compartmentalization inside vacuoles, salt exclusion at the root level, and the buildup of osmoprotectants, which support metabolic activity and cell turgor.

To sum up, salinity stress poses a serious risk to the world's food security, especially in areas that are irrigated and along the coast where salt buildup is common. Developing successful methods requires a thorough knowledge of physiological processes and crop responses to salt. Developing salt-tolerant cultivars, enhancing irrigation methods, and implementing sustainable soil management strategies are a few examples. In order to provide resilient agricultural systems in sensitive areas and lessen the negative consequences of salt, such actions are essential.

3.3. Extremes in Temperature

Due to their great sensitivity to temperature changes, plants may suffer significant physiological and biochemical disruptions from exposure to extremes of heat or cold (Chaudhry and Sidhu, 2022). Climate change has led to a rise in temperature stress, with unseasonal frost occurrences and more frequent heatwaves endangering global agricultural production. When temperatures climb over a plant's ideal range, which is normally between 10 and 15°C above its regular growth circumstances, heat stress results. Protein denaturation, which occurs when structural proteins and vital enzymes lose their functional structure due to prolonged heat exposure, hinders metabolic processes, including respiration and photosynthesis. Furthermore, high temperatures hasten the generation of reactive oxygen species (ROS), which oxidatively damage proteins, lipids, and DNA in cells. Early senescence, withering, and burning of the leaves are visible indicators. In crops like wheat, maize, and tomatoes, heat stress during crucial development phases, such as blooming, may cause pollen sterility and poor fruit set, significantly affecting yields.

Cold stress encompasses both freezing stress (below 0°C) and chilling stress (temperatures just above freezing, 0–15°C). Low temperatures cause membranes to become hard, which decreases fluidity and interferes with the movement of nutrients across cell membranes. Frost damage is the term for the mechanical harm that occurs to tissues when ice crystals grow inside and outside of cells under freezing temperatures. Additionally, enzyme activity is slowed down by cold stress, especially in tropical and subtropical crops that aren't naturally tolerant to cold, like soybeans and rice. Growth retardation, yellowing (chlorosis), and necrotic lesions are among the symptoms. Most crops are still susceptible to abrupt temperature reductions, although certain temperate plants may adapt to cold by physiological changes, including building up antifreeze proteins or changing the lipid content of their membranes.

A major obstacle to global agriculture is the growing unpredictability of temperature extremes brought on by climate change. To reduce these pressures and preserve food security in a warming world, breeding crop types that can withstand heat and cold is essential, as is using better agronomic techniques like mulching and modifying planting times.

3.4. Acidity/Alkalinity of Soils

An important abiotic stressor that has a major effect on crop output is an imbalance in the pH of the soil, either toward acidity or alkalinity (Msimbira and Smith, 2020). Through a variety of processes, excessively alkaline soils (pH >8.0) or too acidic (pH <5.5) make it difficult for plants to thrive. Aluminium (Al) and manganese (Mn) toxicity, as well as deficits in vital elements such as phosphorus (P), calcium (Ca), and magnesium (Mg), are the main problems in acidic soils. At low pH values, aluminium becomes soluble and damages root systems by preventing cell elongation and division, which significantly limits the intake of nutrients and water. Manganese concurrently accumulates to hazardous levels that disrupt photosynthesis and other metabolic functions. Additionally, the acidic environment decreases phosphorus availability, resulting in insoluble compounds containing iron and aluminium that are insoluble to plants.

Alkaline soils, on the other hand, provides distinct difficulties for crop cultivation. Important micronutrients including iron (Fe), zinc (Zn), and copper (Cu) precipitate due to high pH levels, rendering them inaccessible to plants. This leads to typical indications of deficiencies, such as stunted development and chlorosis (leaf yellowing). Moreover, sodium (Na) buildup brought on by alkaline circumstances might worsen salinity stress. Both acidic and alkaline soils have reduced nutrient availability, which impairs photosynthetic efficiency, interferes with plant metabolism, and eventually lowers crop yields. Extremes in soil pH have a substantial impact on soil biology in addition to their direct impacts on plants. Beneficial soil microorganisms, such as nitrogen-fixing bacteria and mycorrhizal fungi that develop symbiotic associations with plant roots, are less active and diverse in acidic soils. Crops become increasingly reliant on synthetic fertilisers as a result of this disturbance of microbial populations, which reduces natural nutrient cycling mechanisms. In a similar vein, alkaline environments prevent microorganisms from breaking down organic matter, which delays the release of nutrients in forms that plants can use. A difficult environment for crop production is created by the combined effects of nutrient imbalances, toxic elements, and compromised soil biology. To maintain agricultural productivity, this environment must be carefully managed through the use of pH-tolerant crop varieties, sulfur applications for alkaline soils, and liming for acidic soils.

3.5. Toxicity of Heavy Metals

One important abiotic stress that has far-reaching effects on crop productivity and food safety is heavy metal contamination in agricultural soils (Rahim et al., 2024). Both natural and man-made processes, such as mining, industrial emissions, inappropriate waste disposal, and the overuse of fertilisers and pesticides containing metals, can cause metals like arsenic (As), cadmium (Cd), and lead (Pb) to build up in soils. These harmful metals, when present in bioavailable forms, are easily absorbed by plant roots and disrupt vital physiological functions, resulting in significant growth inhibition and yield losses. Heavy metals have phytotoxic effects that interfere with normal plant function through a variety of mechanisms. These metals cause oxidative stress at the cellular level by producing reactive oxygen species (ROS), which harm proteins, nucleic acids, and lipids. The plant's antioxidant defences are overpowered by this oxidative burst, which causes DNA strand breaks, membrane peroxidation, and enzyme inactivation. By competing with necessary components for transport channels, heavy metals also hinder the uptake of nutrients. For example, As interferes with phosphorus metabolism, while Cd disturbs calcium and zinc homeostasis. Chlorosis, reduced root growth, leaf necrosis, and early senescence are typical toxicity symptoms brought on by these disturbances. The fact that heavy metals bioaccumulate in the food chain and pose serious health risks to humans is perhaps the most concerning. Leafy vegetables and cereal grains are especially vulnerable to metal uptake, and crops cultivated in contaminated soils can accumulate these harmful elements in edible portions. Long-term exposure to food tainted with metals has been associated with several illnesses, such as cancers (As), kidney problems (Cd), and neurological damage (Pb). These metals' trophic-level biomagnification and environmental persistence pose long-term ecological and public health problems that go well beyond agricultural systems.

Integrated strategies that address the causes and effects of contamination are necessary to reduce the toxicity of heavy metals. Agronomic techniques like pH adjustment and the addition of organic matter can lower metal bioavailability, while phytoremediation techniques employing plants that accumulate metals show promise for cleaning up soil. A long-term way to protect food production in impacted areas is to breed crops with limited metal uptake or improved detoxification systems. These efforts must be complemented by stricter regulations on industrial emissions and waste management to prevent further contamination of agricultural lands.

4. IMPACT ON CROP PHYSIOLOGY, YIELD, GROWTH AND DEVELOPMENT

4.1. Growth and development

Abiotic stressors profoundly disrupt fundamental plant growth processes, leading to significant reductions in crop productivity. One of the most immediate effects is the inhibition of cell division, particularly in meristematic tissues responsible for plant growth. Drought, salinity, and extreme temperatures all impair the enzymatic activities required for DNA replication and cell cycle progression, resulting in stunted growth. For example, in maize, water deficit during vegetative stages can reduce cell expansion by up to 70%, leading to shorter plants with fewer leaves.

Leaf area expansion is similarly compromised under abiotic stress conditions. Drought and salinity reduce turgor pressure, limiting cell expansion and causing leaves to remain small and thick. Heat stress accelerates leaf senescence, while cold stress inhibits chlorophyll synthesis, both of which diminish the photosynthetic capacity of crops. In wheat, every 1°C increase above optimal temperatures during growth can reduce leaf area by 5-10%, directly impacting grain yield potential.

Root systems are particularly vulnerable to abiotic stress. Drought and soil compaction physically restrict root elongation, while salinity and heavy metals chemically inhibit root tip growth. Aluminium toxicity in acidic soils causes severe root tip swelling and necrosis, reducing water and nutrient uptake efficiency by up to 50% in sensitive crops like barley. These root impairments create a cascade effect, limiting the plant's ability to access soil resources even when they are theoretically available.

The reproductive phase suffers dramatically under stress conditions. Many crops exhibit accelerated or premature senescence when subjected to abiotic stress, effectively shortening their growth cycle. Rice plants exposed to drought during flowering often show panicle sterility rates exceeding 60%, while heat stress during pollen development in tomatoes can cause complete flower abortion. Cold stress similarly damages reproductive structures, with frost events during flowering causing up to 90% yield loss in sensitive crops like citrus. These developmental disruptions not only reduce current yields but can also have carryover effects on subsequent growing seasons through impacts on seed quality and viability.

The cumulative effect of these physiological disruptions is substantial yield reduction across all major crop species. For instance, drought stress at critical growth stages can decrease cereal yields by 40-60%, while salinity may cause complete crop failure in severely affected areas. The economic implications are staggering, with global annual losses from abiotic stresses estimated to exceed $100 billion. These figures underscore the urgent need to understand stress physiology and develop mitigation strategies to protect global food production systems.

The complex interplay between different stress factors often leads to synergistic negative effects. For example, drought frequently coincides with high temperatures, creating compound stress that is more damaging than either factor alone. Similarly, flooding often leads to secondary stresses like nutrient leaching and root hypoxia. Understanding these interactions is crucial for developing comprehensive strategies to enhance crop resilience in the face of multiple, simultaneous abiotic challenges.

4.2. Metabolic Disruptions

Abiotic stressors trigger profound metabolic disruptions in plants that directly compromise crop productivity. One of the most critical consequences is the accumulation of reactive oxygen species (ROS), including superoxide radicals, hydrogen peroxide, and hydroxyl radicals. Under normal conditions, plants maintain ROS at beneficial levels for signaling and defense. However, abiotic stresses like drought, salinity, and extreme temperatures overwhelm the plant's antioxidant systems, leading to oxidative stress. This results in widespread cellular damage, including lipid peroxidation of membranes, protein denaturation, and DNA strand breaks. The photosynthetic apparatus is particularly vulnerable, with ROS attacking chloroplast membranes and degrading chlorophyll, thereby reducing the plant's energy production capacity. In cereal crops like wheat and rice, oxidative damage during grain filling can reduce starch accumulation by up to 40%, significantly impacting yield and quality.

Concurrently, abiotic stresses induce significant hormonal reprogramming that alters plant growth patterns. The stress hormone abscisic acid (ABA) shows rapid accumulation, triggering stomatal closure to conserve water during drought but simultaneously limiting CO_2 uptake and photosynthesis. Ethylene production surges under multiple stresses, including flooding and heavy metal toxicity, often accelerating leaf senescence and abscission. These hormonal changes create a physiological dilemma for crops - while the responses enhance short-term survival, they frequently occur at the expense of growth and reproduction. For example, in tomatoes, heat stress-induced ethylene promotes flower abortion, while in maize, ABA-mediated drought responses can inhibit kernel development. The hormonal crosstalk becomes particularly complex when plants face multiple simultaneous stresses, as occurs in many field conditions where drought and heat stress frequently co-occur.

The metabolic consequences extend beyond oxidative damage and hormonal changes. Stress conditions force plants to divert energy from growth to survival mechanisms, altering primary and secondary metabolite profiles. Osmotic adjustment through the accumulation of compatible solutes like proline and glycine betaine becomes energy-intensive, while secondary metabolites with protective roles (such as flavonoids and phenolic compounds) increase at the expense of growth-related compounds. In legumes like soybean, drought stress reduces nitrogen fixation activity as carbohydrate allocation to nodules decreases, creating a cascade effect on protein content in seeds. The metabolic shifts also affect post-harvest quality - wheat grown under heat often shows reduced gluten strength, while grapes exposed to water deficit may accumulate higher sugars but lose aromatic compounds. These physiological responses collectively determine the ultimate impact of abiotic stress on crop yield and quality, with the severity depending on stress intensity, duration, and crop growth stage at exposure.

4.3. Reproductive Failure

Abiotic stress exerts profound effects on crop reproductive processes, often leading to severe yield reductions even when vegetative growth appears unaffected. During the critical reproductive phase, plants become particularly vulnerable to environmental stressors, with even brief exposures capable of causing irreversible damage to floral structures and seed development. High temperatures during flowering, for instance, can induce pollen sterility in cereals like wheat and rice by disrupting microsporogenesis—the process of pollen formation. Heat stress damages the tapetal layer of the anther, which normally provides nutrients to developing pollen grains, resulting in shriveled, non-viable pollen. Similarly, in rice, daytime temperatures above 35°C during anthesis cause spikelet sterility, where pollen fails to germinate on the stigma, leading to empty grains.

Drought stress compounds these reproductive challenges by interfering with multiple physiological processes. Water deficit during flowering causes poor anther dehiscence (pollen release) and inhibits pollen tube growth, preventing successful fertilization. In maize, drought-induced delays in silk emergence create asynchrony with pollen shed, dramatically reducing kernel set. Salinity stress similarly disrupts reproduction by causing ionic imbalances that affect sugar metabolism in developing flowers; in tomatoes, high soil salinity reduces pollen viability and fruit set by up to 60%. Even short-term flooding during reproductive stages can be catastrophic—submergence of rice panicles for just 4–5 days during flowering results in complete pollen abortion due to oxygen deprivation.

The consequences of these reproductive failures manifest in both yield quantity and quality. Cereals exhibit reduced grain number and shriveled seeds, while fruit crops like tomatoes and strawberries produce malformed or undersized fruits with poor market value. In wheat, heat stress during grain filling accelerates senescence, shortening the period for starch accumulation and resulting in lighter test weights—a critical quality parameter for milling. For cross-pollinated crops like maize and sunflower, abiotic stress not only reduces seed set but also compromises hybrid vigour in subsequent generations. These reproductive impairments disproportionately affect smallholder farmers who lack irrigation buffers, as rainfed systems cannot compensate for stress timing coinciding with critical flowering windows. Understanding these physiological disruptions is essential for developing targeted interventions, whether through breeding heat-tolerant pollen traits or implementing precision irrigation to protect reproductive phases.

4.4. Crop-Specific Responses

Different species and development stages react differently to abiotic stressors, which cause unique physiological disturbances in crops. Global food production is at risk because these stress-induced alterations often result in significant yield reductions. When plants devote their greatest resources to reproductive structures, such as during blooming and grain filling, the effects are very severe.

Two of the most significant cereal crops in the world, maize and wheat, are particularly vulnerable to drought stress during blooming. In wheat, a lack of water at this point impairs ovule fertilization and pollen viability, resulting in shriveled kernels and poor grain set. Similar to this, partial silk emergence and asynchronous pollen shed under drought cause barren ears and yield decreases of 40–60% in maize. Heat stress also affects both crops, shortening the grain-filling time and decreasing starch accumulation by speeding up phenological development. For instance, maize yields drastically drop when exposed to extended heat over 35°C, whereas temperatures above 30°C during wheat grain filling might reduce kernel weight by 20–30%.

The main grain of more than half of the world's population, rice, is particularly susceptible to salt stress in irrigated and coastal environments. High salt levels disturb the delivery of sucrose to growing panicles, which hinders grain filling and prevents tiller development, a crucial factor in yield. In poorly drained soils, flood-affected rice paddies experience hypoxic stress, where oxygen deprivation slows root respiration and nutrient uptake. Even moderate salinity (4–6 dS/m) can reduce rice yields by 50%, with sodium and chloride ions causing leaf chlorosis and premature senescence. Furthermore, pollution of groundwater-irrigated rice fields with arsenic presents two risks: it inhibits root development and builds up in grains, posing a risk to food safety.

In addition to cereals, horticulture crops exhibit unique stress responses, such as fruit trees and tomatoes. Because of disturbances in calcium transport, tomato plants exposed to saline or drought often show decreased fruit size and blossom-end rot. In fruit trees like citrus and apple trees, heat stress during blooming results in fruit loss and uneven ripening, whereas cold stress damages vascular tissues, causing bark splitting and delayed budbreak. These physiological problems affect marketability in addition to yield because stressed crops often exhibit poor taste or cosmetic flaws like sunburn or cracking.

These crop-specific reactions add up to a global yield gap, in which actual production is less than genetic potential. Abiotic stressors, for instance, are thought to lower rice and wheat yields by 30–70% in areas that are prone to stress, which exacerbates food insecurity. To address these losses, agronomic interventions like deficit irrigation or soil amendments must be combined with customized solutions, such as breeding rice lines that are resistant to salt or wheat varieties that are resistant to drought. It is essential to comprehend these physiological subtleties in order to create focused strategies that maintain productivity in the face of increasing environmental pressures.

5. WORLDWIDE EFFECTS: REGIONAL DIFFERENCES AND AT-RISK REGIONS CONNECTIONS TO CLIMATE CHANGE

Abiotic stress' effects on agriculture are not evenly distributed throughout the world; in some places, because of their distinct climatic and geographical circumstances, problems are more severe and frequent. In agricultural systems, climate change acts as a force multiplier, intensifying preexisting stressors and generating new vulnerabilities. Due to increasingly unpredictable rainfall patterns in the Sahel area, extended and worsening droughts have emerged as a serious danger in Africa. Crop failures are common in nations like Niger, Chad, and Mali because of drought, which lowers soil moisture, depletes groundwater supplies, and inhibits the development of key crops like sorghum and millet. The problem is made worse by high temperatures, which raise the demand for water and speed up evaporation, straining already marginal agricultural systems.

As sea levels rise and saltwater seeps into freshwater systems, coastal portions of Asia, including a sizable portion of Bangladesh, Vietnam, and Indonesia, are struggling with increasing soil salinity. By flooding agricultural regions with salty water, cyclones and storm surges, which are occurring more often and with greater intensity as a result of climate change, exacerbate the issue. The foundation of food security in these areas is rice paddies, which are especially under risk since salt stress lowers germination rates, stunts plant development, and significantly lowers harvests. In Bangladesh alone, approximately 30% of arable land is affected by salinity, forcing farmers to either abandon their fields or shift to less productive salt-tolerant crop varieties. The economic and nutritional ramifications are significant, since diminishing rice output affects both livelihoods and food supply for millions of people.

South Asia, home to some of the world's most densely inhabited and agriculturally reliant countries, confronts growing heatwaves that interrupt crop cycles and diminish harvests. In India and Pakistan, temperatures regularly surpass 45°C during the summer months, creating heat stress in crops like wheat, which is extremely susceptible to high temperatures during its grain-filling stage. Studies suggest that each 1°C rise over ideal growing conditions may lower wheat yields by up to 10%, a troubling trend considering that South Asia accounts for a large share of world wheat output. Furthermore, the effects of already scarce irrigation resources are exacerbated by heat stress, which speeds up water loss from plants and soils. Smallholder farmers, who lack access to innovative cooling technology or drought-resistant seedlings, are struck hardest, with many experiencing dwindling earnings and food insecurity.

Beyond these regional hotspots, additional locations experience compounding stresses that further destabilize food production. In the Middle East and North Africa (MENA), water shortages and soil degradation are important challenges, with over 80% of the region's agriculture classed as dry or semi-arid. Countries like Egypt and Iraq, which depend significantly on riverine agriculture, are suffering lower water flows owing to upstream damming and climate-induced decreases in precipitation. El Niño and La Niña-related variable rainfall patterns in Latin America, on the other hand, cause alternating droughts and floods that interfere with the production of important export commodities like maize, soybeans, and coffee.

The economic and social ramifications of these geographical vulnerabilities are severe. Crop losses lead into higher food costs, lower farmer earnings, and increasing dependence on food imports, which may strain national economies. Rural populations, where farming is the main source of income, are more vulnerable to poverty and relocation as farming becomes more and more unfeasible due to recurrent abiotic pressures. In addition, environmental deterioration that affects biodiversity, water security, and even urban migration patterns, such as desertification in Africa or dropping water levels in South Asia, causes long-term problems that go beyond agriculture.

Solutions tailored to the local climate, agricultural systems, and socioeconomic circumstances are needed to address these discrepancies. African drylands would benefit from greater irrigation and drought-resistant seeds, whereas coastal Asia might need to invest in salt-tolerant rice cultivars and enhanced drainage systems. Similarly, modified planting schedules and shade-producing agroforestry might be used as heat mitigation techniques in South Asia. However, governments, academics, and agricultural communities must work together in concert for these interventions to be successful. International assistance is also required to finance and scale adaptive solutions. The uneven weight of abiotic stress will keep escalating worldwide disparities in food security and agricultural sustainability unless immediate action is taken.

Particularly in areas where farming is the main source of income, the socioeconomic effects of abiotic stress-induced agricultural disturbances are profound. Increased hunger, higher market costs, and less food availability result from crop losses of 20% to 50% for basic crops including wheat, rice, and maize. Smallholder farmers are most affected since they lack the funds to purchase adaptable technology, which often results in debt or forced migration in pursuit of other forms of income.

Increasing food costs put further burden on family finances, especially in low-income nations where households spend a significant amount of their income on necessities. As seen by previous food price crises, social unrest may result from this economic strain, which can further exacerbate already-existing disparities between rural and urban people. Furthermore, the decline in agricultural output leads to rural unemployment, which in turn forces more people into congested cities and increases the strain on already scarce infrastructure and services.

Targeted solutions, such as enhanced water management, climate-resilient crop varieties, and policy assistance for vulnerable agricultural communities, are desperately needed, as shown by the combined effects of these issues. The cycle of poverty and food insecurity in these areas would only worsen in the absence of such measures, which will have an impact on commerce, migration, and geopolitical stability globally.

6. ENVIRONMENTAL AND ECONOMIC REPERCUSSIONS OF ABIOTIC STRESS

Abiotic stress cause significant financial losses to agriculture worldwide by reducing production and upsetting food systems. Case studies from important crop-producing areas provide a clear illustration of these effects. Between 2017 and 2020, Australia had protracted droughts that severely damaged wheat production, with yields in certain regions falling by as much as 50%. This caused price instability in worldwide markets by lowering both the local supply and Australia's wheat exports, which make up 10% of global commerce. Similarly, frequent heat waves have harmed rice harvests in India during crucial growth phases. Due to crop sterility brought on by a heatwave in Punjab and Haryana in 2022, rice yields were reduced by 20–30%, and local rice prices increased by 15%. Because they often lack irrigation infrastructure and financial buffers to recover from crop failures, small-scale farmers are disproportionately affected by such losses. Many are compelled to take out loans or give up farming completely, which exacerbates poverty and migration in rural areas.

Abiotic stressors cause disruptions to whole agricultural value chains in addition to direct production losses. Harvests are reduced by heat waves and droughts, which causes shortages and raises food costs. Food-importing countries in Africa and the Middle East were disproportionately affected by the 2010 Russian heatwave, which caused a 33% reduction in wheat output and a 60–80% surge in world wheat prices. These price increases put a burden on family finances and make millions of people more food insecure. Furthermore, farmers must pay more for inputs when they make investments in soil amendments, stress-tolerant crops, or irrigation to deal with harsher circumstances. These expenditures are sometimes unaffordable for producers with little resources.

The environmental effects are just as bad. Ecosystems deteriorate under persistent abiotic stressors, generating feedback loops that reduce agricultural resilience even further. Drought speeds up desertification and soil erosion, especially in areas like the African Sahel, where more than 80% of agriculture is already destroyed. An area the size of France is lost each year to salt-affected lands worldwide. Salinization, which is often brought on by inadequate irrigation or saltwater intrusion, makes soils infertile. Heavy metal pollution from mining or industrial runoff (such as cadmium in Chinese rice fields) may linger for decades, building up in crops and making its way into food chains. While flooding removes nutrients and topsoil, wet conditions encourage the release of methane, a powerful greenhouse gas that contributes to climate change.

These environmental effects are not limited to farmers. As pollinators and other important soil creatures become unfriendly in stressed agroecosystems, biodiversity decreases. Drought and excessive water extraction for irrigation exacerbate water shortages, endangering aquatic environments and conflicting human demands. For example, in India, 60% of wells are expected to run dry by 2030 due to groundwater depletion brought on by unsustainable irrigation for crops that need a lot of water, like rice. Degradation like this reduces agricultural output over the long run, forcing farmers to increase their use of finite resources in order to sustain yields.

It need systemic measures to address these environmental and economic costs. Government safety nets and insurance programs may protect farmers from income shocks, and diverse agricultural techniques lessen reliance on staples that are sensitive to climate change. Conservation tillage and agroforestry are examples of on-farm techniques that improve soil health and water retention. Global cooperation is necessary to break the cycle of deterioration and guarantee food security for future generations, however, and this includes prioritizing research into stress-resilient crops, tougher pollution restrictions, and climate funding to promote adaptation.

7. MITIGATION STRATEGIES

7.1. Traditional Methods

Traditional mitigation techniques provide workable, proven ways to counteract abiotic challenges in agriculture. Water-saving measures like drip irrigation and rainwater harvesting are frequently used to alleviate drought stress. Especially in arid areas, drip irrigation increases water-use efficiency by delivering water straight to plant roots, reducing evaporation losses. Ponds and check dams are examples of rainwater collecting structures that collect and hold rainfall for use during dry spells, acting as a buffer against water shortages. These techniques can greatly improve crop resilience and yield stability, making them particularly beneficial for smallholder farmers in drought-prone regions.

Traditional methods like gypsum application and leaching are frequently used to address salinity stress. By substituting calcium ions for sodium ions, gypsum (calcium sulfate) helps sodic soils regain their permeability and structure. Applying too much water to remove salts from the root zone is known as leaching, but it needs proper drainage to avoid waterlogging. Farmers also use salt-tolerant crop varieties and raised bed cultivation to reduce root exposure to saline groundwater in coastal areas where seawater intrusion makes salinity worse. These techniques may lessen the effects of salt and preserve agricultural output in impacted regions, even if they might not completely eradicate it.

These traditional methods are feasible for farming communities with limited resources because they are frequently accessible and affordable. However, appropriate implementation and localization are necessary for their efficacy. Long-term sustainability in stressful environments can be ensured by combining these practices with contemporary technologies, such as soil moisture sensors for irrigation scheduling, to maximize their benefits.

7.2. Biotechnological Innovations

Cutting-edge technologies to improve crop tolerance against abiotic challenges are provided by modern biotechnology. A precise and effective technique for creating stress-tolerant crops without adding extraneous DNA is CRISPR-Cas9 genome editing. For example, scientists have altered the genes in maize that control stomatal closure and root architecture using CRISPR, creating cultivars that continue to produce even in drought-prone environments. By altering genes related to sodium ion transport, salt-resistant rice has also been created, allowing plants to survive in salty soils and keep excess salt out of their tissues. Unlike traditional breeding, CRISPR speeds the production of better crops while avoiding the long timescales associated with conventional approaches.

Another significant biotechnological strategy includes transgenic crops, where genes from other species are inserted to enhance stress tolerance. A famous example is Bt cotton, which has been modified not just for insect resistance but also for better heat tolerance. By adding genes that create heat-shock proteins, these cotton types can retain physiological functioning under significant temperature swings. Similarly, transgenic drought-tolerant soybean expressing bacterial cold-shock proteins has showed enhanced growth under water-limited circumstances. While transgenic crops have legal and public acceptability issues in certain countries, their potential to treat abiotic stress is clear, especially when paired with other sustainable approaches.

Beyond single-gene alterations, modern biotechnological tactics include transcription factor engineering and RNA interference (RNAi) technology. By regulating many stress-responsive genes at once, these methods enable the development of crops with broad-spectrum resilience. In wheat and barley, for instance, overexpression of DREB (Dehydration-Responsive Element-Binding) transcription factors has improved resistance to cold stress and drought by triggering protective gene cascades. To suppress negative regulators of stress responses, such those that cause premature senescence under abiotic stress, RNA interference (RNAi) approaches are being investigated.

These methods are further improved by combining biotechnology with machine learning and high-throughput phenotyping. Researchers can find the best genomic targets for editing or transformation by examining enormous databases on how plants react to stress. However, the deployment of transgenic crops must be supported by thorough risk evaluations, ethical concerns, and farmer education to assure equal benefits. When deployed responsibly, these advances provide a potent armament in the battle against abiotic stress, combining conventional breeding and agroecological methods to ensure global food supply under increasingly severe environmental circumstances.

7.3. Sustainable Practices

By increasing crop flexibility and ecosystem resilience, sustainable agriculture methods are essential for reducing abiotic stress. In order to create agricultural systems that are more resilient to environmental difficulties, these techniques emphasize working with natural processes rather than against them. Agroecological practices and soil health enhancement approaches are two of the most successful sustainable tactics; they serve as the cornerstone of stress-resilient agriculture.

A comprehensive foundation for creating agricultural systems that resemble natural ecosystems is provided by agroecology. Growing two or more crops next to one another, or intercropping, improves biodiversity and lessens stress susceptibility. For instance, when drought-tolerant millet is planted with legumes, a microclimate is created that retains soil moisture while the legumes absorb nitrogen from the atmosphere, lowering the requirement for fertilizer. Similarly, cover crops like vetch or clover control weeds that compete for water, enhance soil structure by adding organic matter, and prevent bare soil from eroding during dry times. These methods are especially helpful in areas where monoculture systems often fail due to drought or irregular rainfall. Agroecology creates redundancy in the system by diversifying plant species; if one crop experiences stress, other crops may compensate, stabilizing total production.

Management of soil health is equally important for reducing abiotic stress. In pressured situations, the addition of biochar—a carbon-rich substance made from organic waste, has shown impressive advantages. Biochar increases nutrient availability in damaged soils, improves soil water retention in dry regions, and even has the potential to bind heavy metals, lowering their toxicity to plants. Compost and manure are examples of organic amendments that assist plants more effectively access water and nutrients by reestablishing microbial populations. Applying gypsum together with organic matter to salty soils may aid in the leaching of excess salts and restore soil structure. In smallholder agricultural systems with limited access to synthetic inputs, these treatments are especially crucial. Because they retain more water during droughts, discharge more effectively during floods, and maintain constant pH levels despite variations, healthy soils serve as stress-reduction tools.

A positive feedback loop is produced when these sustainable techniques are combined: stronger plant growth is supported by healthier soil, which in turn improves ecosystem services like carbon sequestration and water filtering. Importantly, farmers of all income levels may use these techniques since they are often knowledge-intensive rather than capital-intensive. Localized adaptation is necessary for their effective use, however, since what works in one area's soil and climate may need to be modified in another. Thus, extension services and information exchange amongst farmers become essential elements of long-term stress reduction plans.

These natural remedies provide a method to preserve production while restoring the agricultural landscapes that are essential to food security, as abiotic pressures are exacerbated by climate change on a worldwide scale.

The multifunctionality of sustainable practices, which concurrently address many stresses while offering co-benefits like enhanced biodiversity, decreased greenhouse gas emissions, and better farm profitability, is what really makes them powerful. They are thus crucial elements of climate-smart agriculture for an uncertain future, not only adaptation tactics. Building food systems that can resist the growing abiotic pressures of the twenty-first century will depend on their broad acceptance.

7.4. Microbiological Interventions

The employment of beneficial microbes, which develop symbiotic relationships with plants to increase stress tolerance, is one of the most promising strategies to treat abiotic stress in crops. Plant growth-promoting rhizobacteria (PGPR) and mycorrhizal fungi are two important microbial methods that increase plant resilience via a variety of physiological and biochemical processes.

Pseudomonas and Bacillus species are examples of Plant Growth-Promoting Rhizobacteria (PGPR), which invade the rhizosphere, or root zone, and directly improve nutrient availability and absorption. Under stress, these bacteria increase the availability of these vital nutrients to plants by fixing atmospheric nitrogen, solubilizing phosphorus, and producing siderophores that chelate iron. Furthermore, PGPR strains release phytohormones that promote root development and enhance plant vigor, such as auxins and cytokinins. Certain PGPR species generate exopolysaccharides in response to drought or salt stress, which serve to mitigate osmotic stress by reducing sodium intake and retaining soil moisture. Through enhanced antioxidant activity and stress-responsive gene expression, some strains also cause plants to develop systemic resistance, preparing them to endure future stressors.

The formation of vast hyphal networks by mycorrhizal fungi, especially arbuscular mycorrhizae (AM fungi), functionally extends the root system and significantly enhances a plant's capacity to absorb nutrients and water. In nutrient-poor or drought-prone soils, these symbiotic fungi are particularly beneficial because they improve phosphorus absorption and move water more effectively than roots alone. Mycorrhizal relationships aid in preventing harmful sodium ions from entering the body while preserving potassium homeostasis, which is essential for cellular processes under salt stress. By generating glomalin, a glycoprotein that binds soil particles, mycorrhizae also help stabilize soil structure by lowering erosion and enhancing water retention. In stressed agroecosystems, they are also beneficial for long-term soil health due to their function in carbon sequestration.

Mycorrhizae and PGPR have been proven to work in concert to mitigate a variety of abiotic stressors. For example, co-inoculation of salt-tolerant PGPR and AM fungi has increased crop yields in salty soils by improving osmotic adjustment and lowering ion toxicity at the same time. Similar to this, these microbial consortia support plant hydration and metabolic activity in drought-affected areas, even in the face of water scarcity. However, microbial treatments' efficacy varies depending on environmental circumstances, crop species, and soil type, necessitating customized applications for best outcomes.

By lowering the demand for synthetic fertilizers and encouraging sustainable soil management, microbial solutions provide ecological benefits over chemical inputs in addition to their stress-reduction capabilities. The goal of ongoing research is to find better microbial strains, improve delivery systems (such as seed coatings and soil inoculants), and combine them with other mitigation techniques like precision irrigation and stress-tolerant crop cultivars. By providing an economical and sustainable route to resilient crop production, using the full potential of these tiny companions has the potential to transform agriculture in areas that are prone to stress.

7.5. Technological Solutions

By combining cutting-edge technologies, precision agriculture has become a game-changing method for reducing abiotic stress in crops. Farmers may make data-driven choices to maximize resource utilization and reduce stress consequences by using drones, Internet of Things (IoT) sensors, and artificial intelligence (AI)-driven climate models to monitor field conditions in real time. IoT sensors placed throughout fields gather vital information on temperature, salinity, nutrient levels, and soil moisture, allowing for precise fertilization and irrigation based on crop requirements. Both water waste and nutrient leaching, which are frequent under conventional farming practices, are avoided with this focused approach. For instance, during drought conditions, soil moisture sensors can activate automated irrigation systems, guaranteeing crops receive enough water without being overused vital adaptation in areas with limited water resources.

High-resolution aerial assessments of crop health are made possible by drones fitted with thermal and multispectral imaging cameras, which can detect stress symptoms before they are apparent to the human eye. These instruments identify regions impacted by heat stress, salt, or drought by detecting changes in plant temperature, chlorophyll content, and canopy structure. In order to reduce stress before yields are jeopardized, farmers may then take action with targeted treatments, such as modifying irrigation schedules or adding growth-promoting biostimulants. Drone technology has been crucial in helping almond orchards in California's Central Valley manage water stress, lowering water use by up to 25% without sacrificing yield.

Precision agriculture is further improved by AI-driven climate modeling, which forecasts stress events and suggests preventative measures. In order to predict droughts, heat waves, or floods weeks in advance, machine learning algorithms examine past meteorological data, crop performance, and soil conditions. Farmers may use these models to choose crop types that can withstand stress, change the dates of planting, or use protective measures like shade netting or mulching. To assist smallholder farmers in India deal with unpredictable monsoon patterns, AI-based tools such as "KisanAI" provide hyper-local weather forecasts and stress management guidance. Precision agriculture ensures sustainable crop production in a climate that is becoming more unpredictable by integrating various technologies to reduce abiotic stress and increase resilience. The scalability of precision agriculture, however, depends on overcoming barriers such as high initial costs, digital literacy gaps among farmers, and infrastructure limitations in rural areas. Governments and private sector partnerships are critical to subsidizing technology adoption and providing training to bridge these gaps. As these techniques become more available, their potential to transform stress reduction in global agriculture rises, giving a road to food security in the face of climate change.

8. CASE STUDIES OF MENA REGIONS

8.1. Drought Mitigation in the Sahel Region

The Sahel area of Africa, a semi-arid transition zone between the Sahara Desert and the Sudanian savanna, experiences chronic water shortages aggravated by climate change. Prolonged droughts, irregular rainfall patterns, and soil degradation have significantly reduced agricultural output, endangering food security for millions. To overcome these issues, farmers and researchers have used a variety of novel and conventional approaches aimed at boosting water efficiency and crop resilience.

One commonly accepted strategy is the employment of water-harvesting techniques, such as the creation of half-moon basins and zai pits. By capturing and holding onto limited rainfall, these structures enable it to seep into the soil instead of escaping as runoff. These pits improve soil fertility and moisture retention when mixed with organic materials like compost or manure, which greatly increases crop yields. Agroforestry, which involves intercropping staple crops like millet and sorghum with drought-resistant plants like acacias or baobabs, is another effective tactic. The trees' deep root systems allow them to reach groundwater that annual crops cannot, while their leaf litter improves soil structure, lowers soil evaporation, and provides shade.

The advent of drought-tolerant, short-cycle agricultural cultivars has also had a revolutionary impact. For instance, farmers have been able to sustain output even in years with little rainfall because to better millet and sorghum cultivars that are engineered to develop fast and tolerate prolonged dry seasons. The creation and spread of these hardy cultivars have been greatly aided by institutions such as the International Crops Research Institute for the Semi-Arid Tropics (ICRISAT).

Initiatives rooted in the community have also proven essential. Through the preservation and trimming of native tree stumps, farmer-managed natural regeneration (FMNR) has recovered degraded areas in Niger, increasing groundwater recharge and stabilizing the microclimate. These initiatives, together with training in sustainable land management funded by the government and non-governmental organizations, have helped Sahelian populations adjust to the region's increasingly drier climate.

Notwithstanding these achievements, difficulties still exist. Infrastructure, education, and policy support investments are necessary to scale up these methods. But the Sahel's experience shows that combining traditional wisdom with contemporary inventions might increase drought resistance, providing important insights for other dry areas dealing with comparable risks.

8.2. Controlling Salinity in Coastal Farming Systems

Due to storm surges, rising sea levels, and excessive groundwater usage for irrigation, coastal agricultural systems across the globe are increasingly threatened by soil salinity. Millions of people's food security is at risk due to saltwater intrusion, which contaminates freshwater supplies and makes fertile regions unproductive. Salinity has lowered rice harvests by 30 to 50% in areas like the Mekong Delta in Vietnam and the Ganges-Brahmaputra Delta in Bangladesh, compelling farmers to give up traditional crops. Nevertheless, new strategies are being developed to address this issue.

Growing halophyte crops, which are salt-tolerant plants that flourish in salty environments, is one viable remedy. In high-salinity soils, species such as quinoa, saltwort, and sea asparagus (Salicornia) not only thrive but also generate nutrient-dense crops. Farmers in the Indus Delta region of Pakistan have embraced salt-tolerant wheat cultivars that have been created via breeding initiatives and sustain yields on soils that are moderately salinized. Similar to this, scientists are experimenting with bio-saline agriculture in Egypt's Nile Delta, which involves combining halophytes with conventional crops to restore damaged areas.

In addition to replacing crops, engineering solutions are working well. To maintain water tables and avoid salt buildup, farms in the Netherlands use controlled drainage systems. To reduce salinity during dry seasons, Bangladesh has instituted community-based rainwater collection, storing freshwater throughout the monsoon season. Furthermore, soil additions such as organic matter and gypsum aid in the displacement of sodium ions, enhancing the structure of the soil.

These case studies show how biological resilience (halophytes), technological innovation (water management), and farmer education can turn endangered coastal agriculture into profitable systems, even when salt presents significant hurdles. In order to empower disadvantaged agricultural communities dealing with salt-induced land degradation, scaling up these solutions calls for governmental support, research funding, and exchange of information.

8.3. Adaptation to Temperature Stress in Tropical Areas

Rising temperatures are posing more and more problems for tropical countries, where heat stress is increasingly limiting agricultural output. Staple crops like rice, maize, and beans are already grown close to their limits of heat tolerance in many tropical nations. Extended exposure to high temperatures lowers pollen viability, speeds up water loss, and interferes with photosynthesis, all of which result in large production losses. For instance, heatwaves during the blooming period pose a hazard to rice production in South and Southeast Asia, where they may result in spikelet sterility and up to 30% reductions in grain output.

Researchers and farmers are using a variety of tactics to address these issues, with genetic adaptability being a key tactic. By choosing for characteristics including better stomatal control, increased antioxidant capacity, and stable pollen formation in hot conditions, breeding efforts are creating crop types that can withstand heat. For example, in India, farmers have been able to sustain production in spite of increasing spring temperatures because to the creation of heat-tolerant wheat cultivars like HD 3385. Heat-resilient maize hybrids, like those from the Drought Tolerant Maize for Africa (DTMA) program, have also shown increased yields in sub-Saharan Africa when subjected to heat stress and drought.

To improve heat resistance, biotechnological methods are being used in addition to traditional breeding. Heat shock proteins (HSPs) and other stress-responsive genes are being introduced or altered via the use of genetic engineering and CRISPR-Cas9 gene editing. Transgenic rice lines that overexpress OsHSP101, for instance, have shown increased survival rates in very hot conditions. Additionally, by identifying important genetic loci linked to thermotolerance, marker-assisted selection is hastening the production of heat-tolerant cultivars.

To lessen the effects of temperature stress, agronomic techniques are also being modified. In tropical agriculture, effective options include modifying planting dates to avoid times of high heat, using mulching to lower soil temperature, and utilizing shade netting in horticultural systems. In order to protect themselves from heat stress, producers in Central America are moving their coffee production to higher elevations and using shade-grown systems.

These initiatives demonstrate how crucial it is to combine agronomic, genetic, and biotechnological approaches in order to improve crop resilience in tropical areas. However, owing to the need for farmer education, financial limitations, and restricted access to better seeds, broad adoption continues to be difficult. To make sure that tropical agriculture can adapt to a warming climate, it will be essential to strengthen extension services, advance climate-smart legislation, and promote international cooperation.

8.4. Integrated Methods in the Indian Subcontinent

Extreme heat, soil salinity, and drought are only a few of the complex interactions of abiotic stressors that endanger agricultural output in a variety of agroecological zones in the Indian subcontinent. In response, academics and farmers have created integrated strategies that boost resilience by fusing traditional knowledge with contemporary technology. Water shortage has been lessened in drought-prone areas like Rajasthan by reviving traditional rainwater gathering techniques like johads, which are little earthen check dams, as well as by using drip irrigation and drought-tolerant crop types like pearl millet (Pennisetum glaucum). These techniques show how conventional approaches may be expanded with modern advances while simultaneously improving groundwater recharge and conserving water.

Farmers have embraced salt-tolerant rice varieties such Luna Sankhi and CSR 43, which were produced by traditional and marker-assisted selection, in coastal regions of West Bengal and Odisha, where storms and increasing sea levels worsen soil salinity. These types are used in conjunction with agroecological techniques as pokkali farming, a custom in Kerala where rice is grown on salty floodplains during the rainy season and shrimp are grown during the dry season. This comprehensive strategy maintains soil health, generates a variety of revenue streams, and optimizes land usage.

Farmers have switched to early maturing wheat varieties like HD 3086 and modified planting dates based on predictive climate models in order to prevent heat stress in the Indo-Gangetic Plains, where wheat yields drastically decrease when temperatures above 30°C during grain filling. Furthermore, conventional soil conservation techniques like residue mulching and zero-tillage are decreasing canopy temperatures and minimizing soil moisture loss. These efforts are further supported by government programs such as the National Innovations in Climate Resilient Agriculture (NICRA) program, which trains farmers in stress management and distributes climate-smart technology.

These case studies demonstrate the effectiveness of combining scientific advancements (such as stress-tolerant seeds and precise equipment) with traditional wisdom (such as crop diversity and organic additives). There are still issues, however, such as uneven resource availability and the need for legislative frameworks that provide incentives for scalable solutions. The experience of the Indian subcontinent emphasizes the use of interdisciplinary, context-specific approaches to increase adaptive capacity in smallholder agricultural systems that are subject to increasing abiotic pressures.

9. FUTURE PROSPECTS: MOVING TOWARD SUSTAINABLE AND RESILIENT AGRICULTURE

9.1. Climate-Smart Agriculture

Climate-Smart Agriculture (CSA), an integrated strategy that concurrently increases production, improves adaptability, and reduces greenhouse gas emissions, is the way of the future for stress-resilient farming. Agroforestry, conservation tillage, and precision irrigation are examples of CSA techniques that enhance soil carbon storage and help crops resist heat waves and droughts. In East Africa, for instance, intercropping maize with nitrogen-fixing plants such as Gliricidia has enhanced soil fertility and decreased water loss, resulting in production increases of 30% even during dry seasons. Similar to this, solar-powered drip irrigation systems have reduced water use by 40% while preserving agricultural yields in water-constrained areas like the Middle East. CSA provides a route to sustainable intensification, or increasing food production per unit of land without harming ecosystems, by coordinating agricultural techniques with climate adaptation objectives.

9.2. Needs for Research and Development

Accelerated research in two crucial areas, genetic resources and local adaptive capacity, is necessary to advance crop resilience. First, genes for drought, salinity, and heat tolerance may be obtained by using genomics to uncover the potential of crop wild relatives and neglected landraces. For example, to create wheat varieties with deeper root systems for water uptake, researchers are investigating wild wheat species from arid regions. Second, to guarantee that solutions are adapted to local conditions, farmers test and choose stress-tolerant cultivars in their own fields through participatory breeding programs. Short-duration maize varieties that mature before seasonal droughts worsen have been successfully introduced in sub-Saharan Africa thanks to such programs. Furthermore, by bridging the gap between research and practical implementation, digital tools such as climate-smart advisories distributed through mobile apps are assisting smallholders in anticipating and managing stress events.

9.3. Global Cooperation and Policy

International collaboration and more robust policy frameworks are necessary for scaling up these solutions. Funding for research on stress-resilient crops and subsidies for smallholders to adopt CSA technologies must be given top priority by national governments. For instance, millions of farmers benefit from India's National Mission on Sustainable Agriculture, which advocates for micro irrigation and drought-tolerant seeds. In order to maximize crop management under stress, global programs such as the CGIAR's Excellence in Agronomy program encourage cross-border knowledge sharing. Dispersed initiatives and unequal access to technology, however, continue to be obstacles. A cohesive worldwide approach supported by organizations such as the UNFCCC and FAO could finance innovation, harmonize policies, and guarantee fair technology transfer to areas that are at risk. The agricultural sector can only feed a growing population and adapt to climate change by working together.

10. CONCLUSION

The production of crops worldwide is severely hampered by abiotic factors such as drought, salt, high temperatures, soil imbalances, heavy metals, and floods. Particularly in vulnerable areas already battling climate change, these environmental stresses disturb plant physiology, hinder growth and production, and jeopardize food security. In some regions, drought and salt alone may cut agricultural output by as much as 50%. Temperature extremes and soil deterioration exacerbate these losses. Beyond farms, the ripple effects worsen hunger and poverty, destabilize rural economies, and raise food prices. Given how serious these effects are, quick action is required to protect agricultural systems and guarantee sustainable food supply for the world's expanding population.

Integrated solutions that include sustainable habits, scientific innovation, and legislative support are needed to address these issues. Promising means to improve resilience are provided by genetic developments, such as stress-tolerant crop types created by CRISPR and marker-assisted breeding. Agroecological techniques, such as precision farming, better water management, and soil conservation, may lessen the effects of stress at the field level. However, without systemic changes, agronomic and technological solutions alone are inadequate. To exchange knowledge, finance research, and put adaptive strategies into practice, especially in developing countries where farmers are most at risk, international cooperation is crucial.

Collaboration and urgency are essential for the future. Building resilience in agriculture is a top priority because climate change will continue to exacerbate abiotic stresses. To protect against future shocks, investments in infrastructure, farmer education, and climate-smart practices are required. The role of policymakers in developing frameworks that support smallholder farmers and encourage sustainable practices is equally important. The world community can create agricultural systems that can withstand environmental stresses and guarantee food security for future generations by combining science, policy, and practical action. Now is the moment to act because reducing abiotic stress is not only necessary for agriculture but also for humanitarian reasons.

REFERENCES

Chaudhry, S., & Sidhu, G. P. S. (2022). Climate change regulated abiotic stress mechanisms in plants: A comprehensive review. *Plant Cell Reports*, *41*(1), 1–31. DOI: 10.1007/s00299-021-02759-5 PMID: 34351488

Chaves, M. M., Flexas, J., & Pinheiro, C. (2009). Photosynthesis under drought and salt stress: Regulation mechanisms from whole plant to cell. *Annals of Botany*, *103*(4), 551–560. DOI: 10.1093/aob/mcn125 PMID: 18662937

Farooq, M., Wahid, A., Kobayashi, N. S. M. A., Fujita, D. B. S. M. A., & Basra, S. M. (2009). Plant drought stress: effects, mechanisms and management. In *Sustainable agriculture* (pp. 153–188). Springer Netherlands. DOI: 10.1007/978-90-481-2666-8_12

Mittler, R. (2006). Abiotic stress, the field environment and stress combination. *Trends in Plant Science*, *11*(1), 15–19. DOI: 10.1016/j.tplants.2005.11.002 PMID: 16359910

Mohanavelu, A., Naganna, S. R., & Al-Ansari, N. (2021). Irrigation induced salinity and sodicity hazards on soil and groundwater: An overview of its causes, impacts and mitigation strategies. *Agriculture*, *11*(10), 983. DOI: 10.3390/agriculture11100983

Msimbira, L. A., & Smith, D. L. (2020). The roles of plant growth promoting microbes in enhancing plant tolerance to acidity and alkalinity stresses. *Frontiers in Sustainable Food Systems*, *4*, 106. DOI: 10.3389/fsufs.2020.00106

Rahim, H. U., Ali, W., Uddin, M., Ahmad, S., Khan, K., Bibi, H., & Alatalo, J. M. (2024). Abiotic stresses in soils, their effects on plants, and mitigation strategies: A literature review. *Chemistry and Ecology*, •••, 1–34.

Umar, O. B., Ranti, L. A., Abdulbaki, A. S., Bola, A. U. L., Abdulhamid, A. K., Biola, M. R., & Victor, K. O. (2021). Stresses in plants: Biotic and abiotic. In *Current trends in wheat research*. IntechOpen. Khan, A. A., Wang, Y. F., Akbar, R., & Alhoqail, W. A. (2025). Mechanistic insights and future perspectives of drought stress management in staple crops. *Frontiers in Plant Science*, *16*, 1547452.

Vurukonda, S. S. K. P., Vardharajula, S., Shrivastava, M., & SkZ, A. (2016). Enhancement of drought stress tolerance in crops by plant growth promoting rhizobacteria. *Microbiological Research*, *184*, 13–24. DOI: 10.1016/j.micres.2015.12.003 PMID: 26856449

Zhu, J. K. (2016). Abiotic stress signaling and responses in plants. *Cell*, *167*(2), 313–324. DOI: 10.1016/j.cell.2016.08.029 PMID: 27716505

Chapter 9
Role of Microbial Biotechnology in Mitigating Climate-Induced Agricultural Stress

Mian Muhammad Ahmed
https://orcid.org/0009-0001-4496-048X
College of Life Science and Technology, Tarim University, China

Syeda Maira Hamid
https://orcid.org/0009-0005-2065-1999
College of Life Science and Technology, Tarim University, China

Zhang Rui
Department of Horticulture and Forestry, Tarim University, China

Pan Zhiyong
College of Horticulture and Forestry Sciences, Huazhong Agricultural University, Wuhan, China

Guo Xingyu
College of Horticulture and Forestry, Tarim University, China

Ye Qian
College of Life Science and Technology, Tarim University, China

Shangqi Yu
College of Life Science and Technology, Tarim University, China

Muhammad Asim
https://orcid.org/0009-0009-2768-0687
College of Horticulture and Forestry Sciences, Huazhong Agricultural University, Wuhan, China

Xia Qiuhao
College of Horticulture and Forestry, Tarim University, China

Ayaz Ahmad
College of Life Science and Technology, Tarim University, China

DOI: 10.4018/979-8-3373-3346-5.ch009

Yerzat Yerhazi
College of Horticulture and Forestry, Tarim University, China

Yue Qinyang
College of Horticulture and Forestry, Tarim University, China

ABSTRACT

Climate change is worsening drought, salinity and extreme heat across MENA, threatening food security. Microbial biotechnology offers a sustainable solution by harnessing beneficial soil microbes like PGPR, AMF and endophytes that enhance crop resilience through nutrient mobilization, stress hormone regulation and anti-oxidant protection. Field trials in key crops (wheat in Egypt, olives in Morocco, date palms in Saudi Arabia) show 15-28% yield increases, though challenges remain in strain optimization, farmer adoption and policy support. Scaling up requires region-specific microbial consortia, advanced delivery systems like nano-encapsulation, and integration with digital agriculture. By overcoming these barriers, microbial solutions can become vital for climate-smart agriculture in MENA, supporting both food security (SDG 2) and climate action (SDG 13) amid growing environmental stresses.

1. INTRODUCTION

1.1. Climate Change and Agriculture in the MENA Area

The Middle East and North Africa (MENA) region is one of the most climate susceptible regions in the world, with increasing environmental problems that may be a threat to its agriculture sustainability. MENA itself, which is dominated by arid and semi-arid climates and suffers from scarce fresh water resources- reliant and heavily irrigate d crop production (HIZ), has already witnessed considerable changes in hydro climatology associated with climate warming. The region is seeing temperatures rise faster than the average, changing patterns of rainfall, more frequent and serious droughts and increasing sea levels. These climate changes are exacerbating food security and agriculture productivity, further threatening the economy and people of the region. The average annual temperature in the MENA region has been observed to increase at a considerable rate over the last four decades with the trend of 0.36°C for each decade between 1981 and 2020. Average summer temperatures have increased even more quickly, at a rate of 0.45°C per decade (Francis & Fonseca, 2024). This warming has resulted in increased rates of evaporation and greater demand for water for crop production, as well as in extreme events (i.e., heatwaves), which are likely to become more common in the next decades. Models predict that, in a business as usual scenario, a high proportion of the population in MENA will be exposed to annual extreme heatwaves of up to 56°C by the second half of the 21st century (Hamed et al., 2025). It would play havoc with our agriculture, especially with crops already susceptible to heat stress. Such conditions would also put at risk livestock farming, increase water stress and stress food production systems in an already water-scare region. As per report of the Intergovernmental Panel on Climate Change (IPCC) there are already countries, like Algeria, Tunisia or Jordan, where wheat is one of the essential crops and that daily temperatures are high and have to endure a pronounced heat stress and have reported lower yields already, with crop models currently projecting yield reductions of up to 50% for wheat, maize and barley in around 30 years. The dangers posed by climate change are not purely ecological, but also socioeconomic, given that in MENA, agriculture is a major employer to a large segment of the population and plays an important role in the region's GDP. Another important challenge aggravated by climate change in the MENA area is the water shortage. Already one of the world's most water-stressed regions, MENA is consuming its freshwater resources faster than they can be replenished. MENA was the first region listed by the UN to be at risk of becoming one of the first in the world to "Run out of water". The combination of over-extraction, bad water management and the impacts of Climate change will leave MENA with no drinkable water. The area is dangerously dependent on fossil groundwater and international water sources including the Nile, for its agriculture. Nevertheless, these sources are dwindling at an alarming pace. In 2025, it is projected that up to 80-90 million people in the MENA will be subject to different levels of water stress, most of which will be related to agricultural water use (Bhattarai & Yousef 2025). For instance, Egypt, which relies on the Nile for some 95% of its irrigation water, has much at stake in climate change, especially as the upstream water demands from countries such as Ethiopia increase. Moreover, enhanced evaporation faster rates and the blocked river flow as a result for high temperatures would decrease water availability for farming consumption.

Soil salinization also poses a major threat to agricultural productivity in the MENA region, as warming temperatures and over-irrigation have led to an increased accumulation of salt in the soil. According to the United Nations Food and Agriculture Organization (FAO), about 1.4 billion hectares of land worldwide are affected by salinity, while a further 1 billion hectares are at risk of becoming saline in the next few decades. In MENA, where mass irrigation is the key factor behind crop growth, soil salinity is rendering boundless hectares of arable land infertile. Salt is deposited in the soil by the evaporation of water from the soil surface which created mineral salts making it difficult for plants to absorb water and nutrients from soil. This factor along with the effect of sea-level rise and temperature elevation, has caused extensive soil erosion. In countries like Egypt, Iraq and Syria, soil salinization is already having an impact on large swaths of cropland, leading to lower agricultural yields and an additional cost of crop production. The damage caused by soil salinization in the MENA area is immense as it is assumed that high soil salinity can lead to 70% reduction in crop yields in highly salinized zone (Bhattarai & Yousef, 2025). In Egypt, for instance, one of the countries' most prodigious agricultural lands the Nile Delta, has been increasingly impacted by the creeping intrusion of saltwater as result of higher sea levels as well as over irrigation. As a result of these shortages, yields of rice, a key food staple in Egypt, have dropped, intensifying the country's food insecurity. Other MENA countries, like Tunisia and Morocco, have also problems of salinity in their irrigated areas which become more severe due to high evaporation rates and wrong practices of water management. Heat stress, water scarcity and soil salinization are a deadly combination for agricultural systems in the region, compromising yields and thus threatening food security in a region where farming is not only a massive contributor to the economy, but also the livelihood of millions.

In light of these constraints, the MENA region urgently requires technology to enable sustainable agriculture and climate adaptation. The mitigation of impacts of climate change on agriculture will require a multi-pronged approach that involves better water management, development of salinity- and drought-resistant varieties, and the employing of high-tech methods to improve productivity of agriculture. Microbial biotechnology is a very powerful tool with regard to elicitation of stress by climate change in agriculture (Priyadarshini et al., 2024) (Figure 1). Beneficial microbes, such as plant growth-promoting rhizobacteria (PGPR) and mycorrhizal fungi, may induce plant tolerance to abiotic stresses, i.e.plant drought stress and salinity stress, increased nutrient uptake, and soil fertility. In the subsequent pages, we will examine how using microorganism-based products provide sustainable solutions to climate-change induced challenges that affect the agricultural systems of MENA.

1.2. Specific Abiotic Stresses in the Region (Drought, Salinity, Heat)

The Middle East and North Africa region is characterized by high exposure to a variety of abiotic stresses affecting agricultural production. These stressors, many of them include drought and salinity and heat are driven by and intensified by climate change, are largely shaped and influenced by the region's natural conditions. Although these stresses' manifestations are interconnected, they commonly present in several categories, namely impacts on plant performance, soil quality, and overall productivity. The summative impacts of the combination of these several factors increasingly driven by the warming climate present a critical challenge to the region's agricultural systems, compromising food security, farmers' income, and economic stability at large. While drought is a particularly pressing and extensive issue, salinity and heat also represent critical stressors with links to climate and environmental changes. Drought: Widespread and Severe Abiotic Stress Drought is a major abiotic stressor and directly affects many agricultural systems and practices in the MENA region. As a naturally arid area, many MENA countries receive less than 250 mm of rainfall annually, which is the minimum level for arid climate classification. However, the definition of drought in this case is not: "precipitation is below normal," but instead: "drought includes the entirety of deficient water as a result of water shortage and excess demand, compared to a long-term of average" (Mishra & Singh, 2010). Drought is a natural phenomenon and results from deficits in precipitation, long-term low soil moisture, and inadequate irrigation and water supply. The current drought conditions are also exacerbated by changing precipitation patterns, and global climate change is increasing drought severity.

The impacts of drought on agriculture are diverse. First, the water scarcity exerts a constraining impact on the level of irrigation indispensable for crop growth in MENA. Numerous crops in the region, such as wheat, maize, and rice, are high water-requiring crops, and in case of drought, their growth and yield capacity are significantly reduced. The lack of water strain water sources which are already overused and less efficient for irrigation water. As per report of IPCC, crop yields have already significantly decreased in certain years in countries such as Morocco, Tunisia and Algeria, where wheat production has decreased by as much as 50% compared to normal years due to lack of rainfall and water scarcity. Additionally, long-term droughts result in soil deterioration, such as erosion and depletion of organic matter, which can reduce the productive capacity of the land. Drought also negatively affects livestock production, specifically food and water sources for animals, because drinking water and forage become limited. It has brought massive loss of livestock, deepened food insecurity and threatened the survival of well over 90 per cent of hundreds of thousands of rural households in the affected areas, where reliance on goats, cows and sheep is critical – Sudan and Somalia, for instance (Bogale & Erena, 2022). As water and pasture is reduced, farmers must deal with lower crop and livestock productivity and the economic devastation of climate extremes, having a double blow.

The Silent Killer of Soils: Salinity

Another important abiotic stress affecting agricultural production in MENA is soil salinity. Soil had become saline as a result of both natural and anthropogenic influences such as increasing sea levels, over irrigation, and inadequate water management. Widespread irrigation for agriculture use (notably in countries such as Egypt and Iraq and Saudi Arabia) in the region has resulted in the build-up of salts in the soil. The region's rapid evaporation only compounds this process and in turn creates higher concentrations of salts in soils where crops are growing and being drawn up by the crop, or being evaporated at the surface.

The impact of soil salinization is severe, in particular for salt sensitive crops. Wheat, Barley and Rice crops become very susceptible to high soil salinity levels, which affect plant uptake of water and nutrients. This not only causes poor growth and yield reduction, but also causes soil structure and fertility loss. These are known as saline soil, or when in high concentrations throughout the surrounding area, as soil salination in terms of soil salinity. In countries that depend on irrigation – like Egypt, where the majority of the country's farmland is located along the Nile Delta – rising sea levels have resulted in saltwater seepage, which has worsened soil salinity and reduced the amount of arable land. Salinity affects 30% of Egypt's irrigated land, with comparable but slightly lower trends in Tunisia (20%) and Morocco (25%) because of common climatic and irrigation issues (Choukr-Allah et al., 2023). Salinity stress not only affects crop yield but it also alters soil quality and entire soil microbiota. Excess salt can harm the microorganisms in soil and disturb the soil microbial community that plays a key role in nutrient cycling, organic matter decomposition, and soil fertility. This disturbance may cause a less healthy soil, decreasing agriculture productivity. Particularly, increases in the costs of water treatment, irrigation systems, and soil reclamation creates economic pressure on farmers, especially smallholders, that are less equipped to pay for such. Under the pressure of salinity impact on both soil and crops, the vicious circle of production deterioration, food shortage, and economic loss has been formed.

Intensifying Heat Stress on Crops

Another expanding threat to agriculture in MENA is one of extreme heat. The area is already a hot area in summer and climate change has exacerbated this situation; annually mean temperature is increasing at a rate of 0.45 °C per decade (Francis & Fonseca, 2024). Heatwaves, which are occurring more regularly, are an added pressure on crops and livestock. Extreme temperatures, in particular above 40 °C, create a physiological stress on plants that results in decreased crop yield and quality. Heat stress might also induce premature maturation, decrease grain filling, and increase susceptibility to crop diseases, and consequently, pose a significant threat to food supply. For crops such as wheat, barley and maize that are integral components of regional staples, spike flowering or grain filling during hot periods can have a dramatic impact on the potential yield. For instance, in some North African region, wheat production reduces up to 20% when exposed to high temperature during grain filling phase (Farhad et al., 2023). High temperature not only affects crops but also has negative impact on the health of livestock with loss of reproductive efficiency, milk yield, and growth rate.

In addition, high heat means animals and crops need more water, increasing demand on scarce water supplies. Not only does this diminish the water set aside for irrigation or human consumption, it also raises the potential for water conflicts, especially in regions where rivers that span more than one country – like the Nile or Tigris-Euphrates – run through territory where needs are at variance with each other.

Interrelations between Abiotic Stresses

Both drought and salinity not only have their own effects, but can also interact with each other. For instance, drought can create high salinity levels by increasing the rate of evaporation, while irrigation water becomes scarce, and heatwaves can exacerbate the impact of drought by increasing the rate of evaporation. The combined impacts of these stresses present a complex problem for farming systems in MENA and demand integrated mitigation and adaptation solutions. While the occurrence of these ever-increasing abiotic stresses is widespread in the region, it is no longer an open secret as the coping mechanisms in conventional farming would be inadequate to secure food production. Agriculture requires sustainable, climate-resilient solutions that can help withstand these challenges. Microbial biotechnology, being a source of smart solutions for drought, salinity and heat tolerance in crops, might be an indispensable tool in this aspect. Here we discuss how microorganisms can be used as tools to increase the resilience of agriculture in the MENA area.

1.3. Why Microbial Biotechnology? A Sustainable Solution

Increasing climate change challenges such as water scarcity, salinity, temperature, and drought stress require the push for sustainable and new solutions to sustain food productivity in the MENA region. Such conventional agricultural methods that rely on application of chemicals for example synthetic fertilizers, pesticides, and high irrigation requirements are found to be unsustainable in the long run (Harsonowati et al., 2024). These approaches are actually worsening further the destruction of the environment, depletion of the soil and shortage of water and more so in an already water-stressed region as MENA. In the face of ever intensified abiotic stresses on the agricultural systems of the region, environmentally sound and sustainable solutions must be found as a matter of urgency. Microbial biotechnology has been considered one of the most environmentally friendly and sustainable technologies for overcoming these obstacles, as it could maintain the health of soil, improve the plants resistance to stress, and use water and nutrient resources more significantly (Malkawi et al., 2024).

Microbial biotechnology is defined as the use of microorganisms such as bacteria, fungi, algae, and viruses in agriculture. The idea is to use the natural abilities of these microbes to stimulate plant growth, protect plants from pathogens, optimize nutrient perception and tolerance to stress. Microbial biotechnology has received more and more attention in recent years because of its potential to reduce adverse impacts of climate change on agriculture. The MENA region can achieve substantial increases in agricultural productivity with reduced environmental impact if it would adopt these biotechnological approaches.

Improve the Tolerance for Stress in Crops

One of the most important advantages of microbial biotechnology is to improve plant resistance to abiotic stresses including drought, salinity, and heat, which are common in the MENA countries. Some microorganisms, such as plant growth promoting rhizobacteria (PGPR), have been found to enhance the resistance of their host plants to stresses of diverse origins. This ability of PGPR can help in drought tolerance by the production of compounds that can increase root length and boost the ability of the plant to take up water in water-deficient conditions. Furthermore, some PGPR species has the ability to synthesize the phytohormone auxin, and regulate plant growth during stress condition to enable energy and nutrient saving for plant under stress (Chieb & Gachomo, 2023). Salinity stress, one of the major constraints in MENA, because of high evaporation rates of the region and use of extensive irrigation, can also be reduced by these microorganisms. These isolates can also reduce saline content in saline soil or encourage some of the salt tolerant bacteria to protect the plant by converting very harmful salt ions to low active forms or to produce osmoprotectants to help the plants to regulate osmotic balance and to accept salt stress of the environment (Yadav et al., 2023). These bacteria can also stimulate nutrient uptake in plants which become more resistant to saline habitats.

Another field where microbial biotechnology can have an impact is the situation of heat stress, which is becoming more and more prevalent (and severe) in MENA countries as the temperature rises. Microorganisms like (e.g.) some strains of arbuscular mycorrhizal fungi (AMF) have been reported to aid plants to adjust to thermal stress by an enhancement of nutrient and water availability, induction of antioxidant systems and stabilization of cell membranes (Wahab et al., 2023). These micro-organisms can aid in ameliorating the adverse effects of heat stress on cellular structures of plants, thus, resulting in plant growth and productivity maintenance at high-temperature stress.

Improvement in Soil Health and Fertility

Soil quality is a critical factor for sustainability of farming systems, particularly in regions such as MENA where it has been degraded by extended use and salinization. Microbial biotechnology is of prime importance for good soil health since it contributes for the growth of beneficial organisms in the soil. These microorganisms can contribute to soil aggregation, nitrogen cycling and organic matter decomposition, resulting in increased soil fertility and plant growth (Anas et al., 2025).

Nitrogen fixing bacteria like *Rhizobium* and *Azotobacter* is one of the most important biological agent employed in agricultural biotechnology. These bacteria have the capacity to turn atmospheric nitrogen into a form that plants can use, decreasing reliance on synthetic nitrogen fertilizers (expensive and ecologically harmful). In areas such as MENA where the access to organic fertilizers is restricted, and the nitrogen fertilizers tend to be over-applied, the application of nitrogen-fixing bacteria would contribute towards enhancing nitrogen utilization efficiencies, and mitigating fertilizers based pollutants.

Apart from nitrogen fixation, organic matter decomposition by beneficial soil microbes can also stimulate the soil organic carbon content. These microbes are beneficial for soil structure and aggregations, thus increasing water retention capacity. And this is especially crucial in the MENA area, where soil erosion and water holding problems are a common occurrence. For instance, arbuscular mycorrhizal fungi (AMF) and other microbes could improve root development and water retention and, in turn, make crops more resistant to both drought and salinity stress (Boorboori & Lackóová, 2025)

Reduced Chemical Inputs and Environmental Footprint

Another important benefit of microbial biotechnology in MENA is that it can decrease dependence on chemical fertilizers, pesticides, and herbicides. For decades, agricultural chemicals have been critical to conventional agriculture, yet they carry numerous risks to health and the environment. Over-application of chemical fertilizers results in soil acidification, nutrient imbalances, and water pollution in the form of runoff. Pesticides and herbicides may be harmful to non-target organisms, through causing loss of biodiversity, or causing the development of pesticide resistant pests. Microbial biotechnology allows alternative sustainable practices in farming by substitution of or reduction in use of chemical inputs. Microbial products like biofertilizers and biopesticides can enhance the growth of plants and inhibit harmful pathogens growth with minimal environmental degradation. For example, some fungi, bacteria and nematodes can be employed as biocontrol agents to control plant pests and diseases. These so-called "natural bio-based" products can also be included as part of integrated pest management (IPM) systems, and such an integration can reduce chemical usage (Abd-Elgawad, 2025).

Utilisation of microbially based products decreased chemical residues in foods and environment pollution due to agricultural effluent. This is especially true in the MENA region, where water is a scarce resource and the focus often involves water resources that run through several countries, thus creating the importance of maintaining water quality. In addition, to lowering cost, and improving nutrient care utilization and environmental protection in aspect to sustainable farming, microbial biotechnology can help reduce the negative impact of environmental degradation while commercializing farming.

Enhanced Water Utilization Efficiency

Water shortage is considered one of the most limiting factors for increasing production in MENA region where water scarcity together with aridity and limited freshwater sources are further aggravated by climate change and increasing pressure from the expanding population. Optimizing water use efficiency (WUE) has become an urgent priority for sustainable agriculture in a world where more than 80% of all water withdrawals accounts for irrigation (FAO, 2020). This challenge could potentially be met by microbial biotechnology, exploiting the natural functions of drought resistant microorganisms to improve water use efficiency during crop production.

Studies have proven that some plant-growth-promoting microorganisms like such as Pseudomonas, Bacillus, and arbuscular mycorrhizal fungi (AMF) can notably improve crop performance under water stress. These beneficial microorganisms exert multiple modes of action on WUE by promoting root growth and branching for enhanced water uptake, generating exopolysaccharides for improved soil structure and water holding capacity, and/or by curbing water loss through transpiration by affecting stomatal movement. Furthermore, drought-resistant microbes also produce osmoprotectants, which act as carriers of water in plants, including proline and glycine betaine, and stress-relieving enzymes, like ACC deaminase, which together enable plants to maintain their cellular function under limiting water conditions.

Field uses of such microbial inoculants have demonstrated remarkable results with reported WUE gains of 20-40% for various crops (Rashid et al., 2022). This means farmers can get similar or even higher yields using much less water - important in water-stressed areas. There are particularly important implications for the MENA region where groundwater over-exploitation and pollution exacerbate water scarcity issues.

In addition to the immediate water savings, microbial interventions contribute to long-term soil health and ecosystem resilience. Unlike chemical inputs, these biologic interventions are sustainable, and can be tailored to local specifics. With climate predictions showing more aridification in MENA in the future, it will be crucial to include microbial biotechnology in agricultural water management to safeguard food security. Strengthened policies and farmer training schemes will be instrumental in aiding dissemination of innovative practices and ensuring the region's agricultural systems can be sustainable even under increasing water scarcity. By using nature's own solutions, creating microbial biotechnology as an opportunity for climate-resilient agriculture in water scarce areas is possible and becomes an indispensable part of future sustainable food production systems.

A Roadmap to Sustainable Agriculture in MENA

Facing increasing climate-driven stresses (drought, salinity, and heat), the MENA region needs solutions that are efficient, sustainable, scalable, and affordable. Of all these methods, microbial biotechnology has emerged as the most promising technique having several unique advantages over the traditional ones. By promoting crop resilience to abiotic stress, soil health, minimizing usage of chemical inputs and water use, it tackles the underlying indicators of agricultural decline with respect to the constraints posed by the region's resources. Unlike resource concentrated interventions (e.g., transgenics crops), microbial solutions are cost effective, environmentally friendly and also work across various agroecological conditions. Importantly, these biological tools are inclusive, benefiting smallholder farmers that drive MENA's agriculture, and not requiring expensive infrastructure or energy for operation. Other approaches can only give short-term respite, but microbial biotechnology offers a holistic approach, harnessing environmentally friendly bioprocesses to establish a more permanent state of resistance. As the region experiences further climate-related stresses, embracing these inventions will be critical to provide sustainable agricultural production, enhance food security and reduce environmental degradation. Hence, incorporating such microbial-based practices into regional policies and farming practices is crucial for not only safeguarding the livelihoods but also for sustaining the ecological equilibrium in MENA. This shift in paradigm towards nature-based solutions, is the best way open to the region for climate-resilient agriculture.

2. MICROBIAL PLAYERS AND THEIR FUNCTIONAL ROLES

The agricultural systems of the MENA region face escalating threats from climate-induced abiotic stresses, including drought, salinity, and extreme temperatures. In response, microbial biotechnology has emerged as a sustainable solution, leveraging the natural symbiotic relationships between plants and beneficial soil microorganisms. These microbes enhance crop resilience through a variety of biochemical and physiological mechanisms, offering a cost-effective and eco-friendly alternative to conventional agricultural inputs.

Among the most impactful microbial groups are Plant Growth-Promoting Rhizobacteria (PGPR), which colonize the rhizosphere and stimulate plant growth through direct and indirect pathways. Strains of *Pseudomonas* and *Bacillus* are particularly notable for their ability to produce siderophores, solubilize phosphates, and reduce ethylene stress via ACC deaminase activity. Similarly, *Azospirillum* species contribute to nitrogen fixation and secrete phytohormones like auxins, which promote root development and improve water uptake under drought conditions. Another critical group is Arbuscular Mycorrhizal Fungi (AMF), which form symbiotic associations with plant roots, extending their hyphal networks to enhance nutrient and water absorption. AMF also secrete glomalin, a glycoprotein that improves soil structure and water retention, making them invaluable in arid environments.

Beyond rhizobacteria and mycorrhizae, endophytic fungi and actinomycetes play vital roles in stress mitigation. Endophytes, such as *Trichoderma* and *Fusarium*, reside within plant tissues without causing harm, where they produce antioxidants and enzymes that combat oxidative stress. Actinomycetes, including *Streptomyces*, contribute to soil fertility by decomposing organic matter and suppressing pathogenic microbes, further supporting plant health.

The mechanisms by which these microbes alleviate stress are multifaceted. Nutrient solubilization, particularly of phosphorus, potassium, and zinc, is a key function, with microbes like *Bacillus megaterium* secreting organic acids to release bound nutrients in the soil. Phytohormone modulation is another critical pathway; for example, PGPR that produce ACC deaminase lower ethylene levels, preventing stress-induced growth inhibition. Additionally, microbes induce osmotic adjustment in plants by stimulating the accumulation of osmolytes like proline and glycine betaine, which help maintain cellular turgor under drought or salinity. To combat oxidative damage, microbes upregulate antioxidant enzymes such as superoxide dismutase (SOD) and catalase (CAT), scavenging reactive oxygen species (ROS) that accumulate under stress. Finally, certain microbes prime plants for induced systemic resistance (ISR), enabling faster and more effective responses to environmental challenges through hormonal signaling pathways.

Collectively, these microbial functions provide a robust toolkit for enhancing crop resilience in the MENA region. However, their efficacy depends on selecting the right microbial strains for specific crops and stresses, as well as optimizing delivery methods to ensure survival and activity in harsh field conditions.

3. INNOVATIONS IN MICROBIAL FORMULATIONS AND DELIVERY

The effectiveness of microbial biotechnology in agricultural application heavily relies on the efficient formulation and delivery of the microorganisms that can maintain their viability, stability and physiologically activity in the field. Advances in bioformulation manufacturing methods, encapsulation technologies, and application procedures have greatly improved the practicability of microbial inoculants. Additionally, smartly applied microbial consortia, utilising one microbe to combat more than one abiotic stress, are on the rise and there is a rising affordability of profit for unfortunate farmers. These advances will be reviewed here, based on available literature, with a focus on recent results that show their relevance for climate resilient agriculture.

The preparation of formulations with high cell viability at the time of storage and the application is still a challenging task. Microorganisms must withstand various environmental stress factors, such as UV radiation, temperature change, atmosphere, atmospheric, oxidative stress, and desiccation, while still remaining biologically active in the rhizosphere or phyllosphere. Traditional liquid and solid-based formulations sometimes show poor stability, field performance and are not suitable for commercial agriculture applications. In order to overcome these shortcomings, recent years have seen the introduction of sophisticated encapsulation technologies based on biodegradable polymers, such as alginate, chitosan and starch, and innovative carriers, such as silica nanoparticles, liposomes and carbon nanotubes. For example, alginate microbeads were used to protect Azospirillum brasilense and Pseudomonas fluorescens which effectively improved their viability when exposed to drought and high salt (Saad et al., 2020). The B. subtilis nanoparticle formulation with chitosan nanoparticles has also shown to enhance both heat tolerance and microbial viability in drier soils (Yarullina et al., 2023). More recently, hybrid encapsulation were reported, where inorganic nano-materials were combined with polymer to impart dual layer encapsulation. Humidity responsive release systems designed using silica-based nanocarriers for controlled activation of microbes by root exudates. Moreover, a lipid-based formulation like liposome-encapsulated beneficial rhizobacteria profoundly improve microbial adhesion to plant surfaces and colonization efficiency.

In addition to the physical encapsulation, synthetic biology provides great potential for improving microbial tolerance. For instance, Rhizobium strains genetically manipulated with heat shock and osmoprotective protein-encoding genes have shown increased tolerance to high temperature and salinity (Sharma et al., 2023). Further, the generation of "smart" bioformulations that respond to defined environmental cues (e.g., changes in pH, moisture or specific root exudates) is likely to be an emerging and exciting area of enhancing controlled microbial colonization and community promotion in the future.

The addition of cryoprotectants and stress adapted additives during the composition preparation has proved to yield good results. Such as trehalose, glycerol and other osmolytes are added in to protect cells such as microorganism cells during drying and storage (e.g. lyophilization and spray-drying). This method has proven highly successful for maintaining the infectivity and viability of *Bradyrhizobium* and *Bacillus* strains over 6 months at room temperature (Vorobey et al., 2023).

Electrospinning methods are being studied for fibrous mat materials containing microbial inoculants. These mats life the the hazards of desiccation and UV irradiation but also ensure the slow release of microorganisms near the vicinity of which in contact with the root area, promoting plant growth under stress. Moreover, the development of microfluidic technology has made it possible to prepare uniform microcapsules with controllable release behavior and provide prospects for precision agriculture. Liquid microbial products (i.e., aqueous suspensions and emulsions) are especially popular because of their ease of use and compatibility with existing irrigation systems. However, the organisms in these products generally need to be stored at low temperature to retain viability, and they have a short shelf life, which restricts their use in resource-poor settings. Temperature variations during storage and transportation can also negatively impact microbial performance.

On the contrary, with solid preparations, such as granules, fine powders and tablets, the storage stability is improved, although there is a drawback that the release of microorganisms is slow, to such an extent that it takes a small amount of time until some favorable effect can be brought about when, for example, grown plants have entered upon growth stages immediately after sowing. Recent developments in freeze-drying have overcome some of these restrictions. For instance, the microbial survival was greatly enhanced by supplementation with cryoprotectants (e.g., trehalose, skim milk) during freeze-drying. Lyophilized Rhizobium species have demonstrated up to 90% viability after 12 months of ambient storage, and rapid reactivation upon soil application provides a promising alternative to conventional cold-chain dependent products (Berninger et al., 2018).

Carriers based on biochar are another potentially novel approach. Biochar serves as a shelter against such environmental stresses as UV radiation and desiccation of the microbial inoculum and as a porous medium suitable for microbial colonization and reproduction. Biochar—based carrier bioinoculants have been demonstrated to stimulate plant growth in degraded and saline soil, and contribute to the bioremediation of polluted sites, by activating useful microbial populations.

Furthermore, nano-facilitated delivery vehicles reshape the microbial inoculation methods. For example, wheat root colonization and drought tolerance can be improved using silica nanoparticles encapsulating Trichoderma harzianum (Aljeddani et al.,

2024). Similarly, liposome-encapsulated plant growth-promoting rhizobacteria on plants led to higher in salt tolerance when compared with regular inoculants treated plants. The nanoformulations adhere far better to plant surfaces and are released over a longer duration despite harsh environmental conditions. Recent advances in 3D printing and other additive manufacturing approaches are attracting increasing interest due to their ability to provide a spatial distribution of microbial inoculants with high resolution across the porous soil matrix, which can be used to facilitate better root–microbe synergies in promoting nutrient supply and stress tolerance. Concurrently, artificial intelligence (AI) is becoming more prominent in the design and optimization of microbial products. Microbial consortia as well as high-performing microbes with superior agronomic traits are being discovered through AI techniques.

However, despite these encouraging developments there are also challenges ahead. The high cost of production, limitations due to scale-up effects and uncertain regulatory climate still represent some of the major obstacles for the marketability of advanced microbial consortia. Future research should focus on designing cheap and scalable technologies and perform systematic evaluations of the long-term ecological effects of modern materials and genetically modified inoculants on soil microbiomes. The integration of nanobiotechnology, synthetic biology, and smart materials has greatly improved the performance of microbial formulation and delivery systems. These technologies have enormous potential to revolutionize agricultural practices and improve resilience, sustainability and productivity in the face of increasing climate challenges.

3.1. Application Methods: Seed Coating, Soil Amendment, and Foliar Sprays

Application of microbial inoculums in agriculture has undergone a tremendous change with application in seed coating, soil application, foliar spray to improve plant resilience and productivity under diverse conditions of stress.

Seed Coating: Seed treatment with microbial inoculants: gaining strength and cutting the cost of harmless microbes for plants' natural defenses Seed dressing with microbial inoculants to ensure rapid root colonization, protect plants from stress right from the start. New developments include polymer-encapsulated and Bradyrhizobium japonicum-coated seeds that have shown to improve soybean nodulation under drought (Jarecki, 2022). The electrostatic coating methods have also increased the adhesion of microbes to the seeds, resulting in less wash-off during irrigation and higher inoculation efficiency. Recent developments in microencapsulation have also resulted in the production of seed coatings that shield beneficial microorganisms from the stresses in the environment. For example, wheat germination and seedling growth were improved in the presence of salt stress when Pseudomonas fluorescens was entrapped in alginate beads, which ensured that the microorganisms remained alive and could also provide controlled release. Microencapsulated microbial seed coatings, were also found to enhance soil environment and maize grain yield, in saline soil indicating their promising role in improving crop growth when subjected to stress conditions. Seed coatings with plant growth-promoting bacteria (PGPB) have also been shown to enhance seed germination, plant growth, and tolerance to abiotic stresses (such as salinity and drought), with a decrease in the use of agrochemicals.

Soil Amendment: Another promising carrier of biofertilizers can be granular forms using soil amendments or microbial compositions. Biochar is a carbonaceous material produced through thermochemical decomposition of biomass and it has attracted attention as a promising carrier of microbial inoculum because of its high porosity and nutrient content. Some studies have reported that biochar-based microbial inoculants can improve plant growth in severe environment by stimulating microbial colonization and activity. For instance, co-application of biochar and phosphorus has been shown to increase soil microbial biomass, mycorrhizal colonization and growth of subterranean clover, suggesting enhanced uptake of nutrients and health of the plants (Solaiman et al., 2025). Biochar application could enhance maize growth and shift the bacterial community structure in phenanthrene polluted soils, indicating that biochar could be used to remediate soils and improve soil fertility. In addition, biochar immobilized with plant growth-promoting rhizobacteria (PGPR) improve the physicochemical properties in soil, agronomic traits, and microbial community structures in lettuce seedling development, which support the synergistic effects of biochar and PGPR for sustainable agriculture.

Foliar Sprays: Foliar sprays are one alternative used to targeted deploy of microbes on the phyllosphere, and make for an easy and effective way to help plants cope with heat and UV stress. Foliar application of Bacillus amyloliquefaciens has been reported to reduce the heat-induced yield loss in rice up to 25% (Hossain et al., 2024). Nanoemulsion based sprays have also enhanced microbial adhesion to leaves, which is beneficial even at increased evaporation rates, to be highly applicable for use in arid and semi-arid areas. Biochar based foliar application has also been researched lately. For example, the positive effect of biochar in terms of enhancing beneficial microbial populations in the soil, indirectly beneficial for foliar application through overall plant health, is also enhanced by a synergism with PGPR. Arbuscular mycorrhizal fungi (AMF) together with biochar have also been described to be effective soil amendments that improve salinity tolerance, growth and lipid metabolism on common wheat, highlighting the promise of an integration to enhance plant stress tolerance.

3.2. Use of Consortia for Multi-Stress Tolerance

Single strain microbial inoculants often fail to offer complete protection against the range of abiotic stresses that crops encounter, including drought, salinity, and nutrient-deficiency stresses. In contrast, microbial aggregates with synergistic strains such as plant growth-promoting rhizobacteria (PGPR), arbuscular mycorrhizal fungi (AMF), and the nitrogen fixer can provide greater effectiveness. Such consortia take advantage of synergistic interactions to improve plant resistance and yield under various stress conditions.

Recent research highlights the success of such consortia. For example, consortium of *Azotobacter, Pseudomonas*, and *Trichoderma* strain increased the wheat yield by 35% in response to combined drought and salinity stress (Khan et al., 2022). Likewise, co-inoculation of AMF with Bacillus strains enhanced water use efficiency of maize more than single strain application. The ability of microbial consortia to deliver synergistic advantages therefore is a viable approach for multi-stress tolerance. The mechanisms by which these favourable effects of activity occur are complex. PGPR can synthesize phytohormones, solubilize phosphates and fix atmospheric N thus increasing nutrient contents. AMF expand the capacity for water and nutrient acquisition of the root system. Nitrogen-fixing bacteria help to add to the nitrogen pool, which is important under nutrient-poor conditions. In combination, these microorganisms regulate plant hormones, reinforce antioxidant protection mechanisms, and ameliorate osmotic adjustment, which play a role in enhancing plant tolerance to abiotic tensions.

Moreover, microbial communities positively contribute to health of the soil. They improve the soil structure, diversity of microorganisms and nutrient cycling, which result in a more favorable environment to grow plants. Biochar and phosphorus applied jointly with a microbial seed inoculation, observed to enhance soil microbial biomass, mycorrhizal colonization, and plant growth implying better nutrient uptake and plant health.

However, despite these encouraging outcomes, there are still many practical obstacles in the use of microbial consortia. Efficacy of consortia also depends on factors like compatibility of microbial strains, environmental conditions and crop specificity. Thus, tailored formulations and strategic application are needed to gain maximum potential advantages. Future studies will be directed toward the fine tuning of these consortia for various crops and stresses as a basis toward the wider implementation of such bioinoculants in sustainable agriculture. Overall, microbial consortia offer a great potential to improve plant tolerance to several abiotic stresses. Such consortia can enhance plant health, productivity, and crop sustainability by exploiting synergistic relationships among the various beneficial microbes in response to environmental challenges.

While there has been substantial advancement in microbial formulations and delivery technologies, there are still several challenges. Scaling these technologies for broad adoption in agriculture demands that cost and regulatory barriers and farmer adoption are addressed. Other long-term field experiments are required to evaluate the ecological risks of microbial inoculants to soil diversity. In order to improve microbial consortia design, future studies should consider multi-omics approaches (e.g., metagenomics, metabolomics) to be integrated. Rapid development of nanotechnology and synthetic biology could also offer new opportunities to engineer better microbial products in future. Successful development of microbial-formulations and -delivery systems is pivotal to translate research from the laboratory to the field. The technology of encapsulation, nano-carriers and application methods (seed coating, soil amendment, spray) play a major role in improving the efficacy and utility of the microbials. In addition, application of microbial consortia allows a systemic solution for coping with more than one-factor stress simultaneously. Future work must concentrate on the large scale integration of these technologies particularly in the regions that are at high risk of climate change (e.g., Middle East and North Africa (MENA)) where there is an urgent need for sustainable farming practices.

4. APPLICATIONS IN MAJOR MENA CROPS

The diverse agroecological zones of the MENA region support a variety of staple and cash crops, each facing unique challenges under climate change. Microbial biotechnology has demonstrated significant potential to address these challenges, with tailored applications yielding measurable improvements in crop productivity and stress tolerance.

In wheat and barley, which are critical for food security in North Africa and the Levant, drought and salinity are major constraints. In Morocco, inoculation with *Bacillus aryabhattai* has been shown to increase wheat yields by 22% under water-deficient conditions, while in Egypt, co-inoculation of *Rhizobium* and AMF improved barley growth in saline soils with electrical conductivity levels as high as 8 dS/m. These microbes enhance nutrient uptake and root development, enabling crops to thrive despite limited water availability and high salt concentrations.

Date palms and citrus fruits, vital to the economies of the Arabian Peninsula and North Africa, are particularly vulnerable to salinity and water scarcity. In Saudi Arabia, applications of *Streptomyces* spp. have reduced salt stress in date palms, resulting in a 15% increase in fruit yield. Similarly, in Tunisia, *Pseudomonas fluorescens* has been used to boost citrus root growth under drought conditions by producing auxins that stimulate root elongation. These interventions not only improve yield but also enhance fruit quality, which is crucial for export markets.

Olive cultivation, a cornerstone of Mediterranean agriculture, is increasingly threatened by heatwaves and erratic rainfall. In Jordan, the introduction of AMF (*Glomus intraradices*) has improved water-use efficiency in olive trees by 35%, ensuring stable oil production even under water stress. Meanwhile, in Algeria, *Sinorhizobium meliloti* has been employed to enhance nitrogen fixation in chickpeas, a key legume crop, leading to an 18% increase in nodulation and overall yield under drought conditions.

Despite these successes, the widespread adoption of microbial solutions faces several barriers. One major challenge is the lack of region-specific microbial strains, as many commercial biofertilizers are developed for temperate climates and may not perform optimally in MENA's harsh environments. Additionally, farmer awareness of microbial inoculants remains limited, and extension services are often inadequate to promote their use. Practical issues such as storage and shelf-life also pose obstacles, as high temperatures can degrade microbial viability before application.

To overcome these barriers, targeted research is needed to identify and optimize microbial strains for local conditions. Policymakers and agricultural stakeholders must also invest in farmer education and infrastructure to facilitate the large-scale adoption of microbial technologies. By addressing these challenges, microbial biotechnology can play a transformative role in securing the future of agriculture in the MENA region, ensuring resilience in the face of escalating climate pressures.

5. FIELD CASE STUDIES IN THE MENA REGION

The practical application of microbial biotechnology in the MENA region has yielded promising results, as demonstrated by several field case studies. These real-world examples highlight the potential of microbial solutions to address region-specific agricultural challenges while also revealing key lessons for scalability.

In Egypt, where salinity intrusion in the Nile Delta threatens rice and wheat production, farmers have trialed plant growth-promoting rhizobacteria (PGPR) to rehabilitate degraded soils. A 2022 study in the Sharkia Governorate showed that inoculating wheat with *Pseudomonas fluorescens* and *Bacillus subtilis* increased yields by 28% in saline-affected fields compared to untreated controls. The microbes enhanced nutrient availability and reduced sodium uptake, allowing crops to thrive despite soil salinity levels exceeding 10 dS/m. However, farmers noted challenges in consistent application due to fluctuating water availability, underscoring the need for drought-resistant microbial formulations.

Morocco has pioneered the use of arbuscular mycorrhizal fungi (AMF) in drought-prone olive orchards. A five-year project in the Marrakech-Safi region demonstrated that AMF-inoculated olive trees retained 40% more soil moisture during dry spells, leading to a 15% increase in oil yield. The fungi's extensive hyphal networks improved water and phosphorus absorption, reducing the need for irrigation. Despite these benefits, adoption remains limited by the high upfront cost of fungal inoculants and a lack of local production facilities, pointing to the necessity of government subsidies and public-private partnerships for wider implementation.

In Saudi Arabia, where date palm cultivation faces extreme aridity, researchers have explored endophyte-assisted farming. Trials in the Al-Ahsa Oasis revealed that date palms treated with salt-tolerant endophytic bacteria (*Enterobacter cloacae*) exhibited 20% higher fruit set under saline irrigation (EC 12 dS/m). The microbes produced osmoprotectants that mitigated salt stress, preserving fruit quality. However, scaling this technology has been hindered by the region's reliance on imported microbial products, emphasizing the need for localized biofertilizer production to ensure affordability and accessibility.

These case studies illustrate that while microbial biotechnology holds immense promise for climate-resilient agriculture, its success depends on addressing logistical, economic, and educational barriers. Tailoring solutions to local conditions and fostering stakeholder collaboration will be critical for broader adoption.

6. CHALLENGES AND FUTURE DIRECTIONS

Despite the demonstrated efficacy of microbial biotechnology, several challenges impede its widespread adoption in the MENA region. One major limitation is the lack of region-specific microbial strains. Many commercial bioinoculants are developed for temperate climates and may not perform optimally under MENA's extreme heat, salinity, and water scarcity. For instance, some PGPR strains exhibit reduced viability at soil temperatures above 40°C, which are common in Arabian Peninsula summers. This highlights the urgent need for research programs to isolate and test native microbial strains adapted to local agroecological conditions.

Another critical barrier is farmer awareness and accessibility. Smallholder farmers, who dominate MENA's agricultural sector, often lack knowledge about microbial inoculants' benefits and proper application techniques. In Tunisia, a 2023 survey revealed that only 12% of farmers in arid regions were familiar with biofertilizers, citing mistrust of alternatives to chemical inputs. To overcome this, extension services must prioritize hands-on training and demonstration plots to build confidence in microbial solutions. Additionally, the high cost and short shelf-life of bioformulations deter adoption. Unlike chemical fertilizers, microbial products often require refrigeration and have a viability window of 3–6 months, posing storage challenges in rural areas with unreliable electricity.

To address these hurdles, future efforts should focus on:

1. Strain Optimization: Leveraging metagenomics and synthetic biology to engineer microbes with enhanced stress tolerance (e.g., heat-stable *Azotobacter* strains).
2. Innovative Delivery Systems: Developing nano-encapsulated or biochar-based carriers to prolong microbial survival in harsh environments.
3. Policy Support: Governments should establish biofertilizer subsidies, certification frameworks, and local production hubs to reduce costs.
4. Digital Integration: AI-driven tools could predict optimal microbial combinations for specific crops and soils, enabling precision agriculture.

Public-private partnerships will be essential to scale these innovations. For example, Morocco's "Green Generation" initiative could integrate microbial solutions into its climate-smart agriculture targets, while Saudi Arabia's Vision 2030 might invest in domestic biofertilizer production. By aligning research, policy, and education, microbial biotechnology can transition from niche trials to mainstream practice.

7. CONCLUSION

The escalating impacts of climate change on MENA's agriculture demand urgent, sustainable solutions. Microbial biotechnology offers a transformative approach to enhance crop resilience, improve soil health, and reduce dependency on synthetic inputs. As demonstrated by field successes in Egypt, Morocco, and Saudi Arabia, tailored microbial applications can mitigate salinity, drought, and heat stress in key crops like wheat, olives, and date palms. These innovations align with global sustainability goals by reducing agriculture's environmental footprint while boosting productivity.

However, realizing this potential requires concerted action. Researchers must prioritize the development of locally adapted microbial strains and user-friendly formulations. Policymakers need to foster enabling environments through funding, infrastructure, and farmer education programs. Meanwhile, the private sector can drive commercialization by making bioinoculants affordable and accessible.

The path forward is clear: integrate microbial biotechnology into national agricultural strategies as a pillar of climate adaptation. By uniting science, policy, and community engagement, the MENA region can harness the power of microbes to build a more resilient and food-secure future. The time to act is now—before climate stresses outpace conventional farming's capacity to adapt.

REFERENCES

Abd-Elgawad, M. M. (2025). Integrated Nematode Management Strategies: Optimization of Combined Nematicidal and Multi-Functional Inputs. *Plants*, *14*(7), 1004.

Aljeddani, G. S., Hamouda, R. A., Abdelsattar, A. M., & Heikal, Y. M. (2024). Stress-Responsive Gene Expression, Metabolic, Physiological, and Agronomic Responses by Consortium Nano-Silica with Trichoderma against Drought Stress in Bread Wheat. *International Journal of Molecular Sciences*, *25*(20), 10954.

Anas, M., Khalid, A., Saleem, M. H., Ali Khan, K., Ahmed Khattak, W., & Fahad, S. (2025). Symbiotic synergy: Unveiling Plant-Microbe interactions in stress adaptation. *Journal of Crop Health*, *77*(1), 18.

Berninger, T., González López, Ó., Bejarano, A., Preininger, C., & Sessitsch, A. (2018). Maintenance and assessment of cell viability in formulation of non-sporulating bacterial inoculants. *Microbial Biotechnology*, *11*(2), 277–301.

Bhattarai, K., & Yousef, M. (2025). Water Scarcity and Climate Change in MENA: Challenges, Innovations, and Geopolitical Impacts. In *The Middle East: Past, Present, and Future* (pp. 105–136). Springer Nature Switzerland.

Bogale, G. A., & Erena, Z. B. (2022). Drought vulnerability and impacts of climate change on livestock production and productivity in different agro-Ecological zones of Ethiopia. *Journal of Applied Animal Research*, *50*(1), 471–489.

Boorboori, M. R., & Lackóová, L. (2025). Arbuscular mycorrhizal fungi and salinity stress mitigation in plants. *Frontiers in Plant Science*, *15*, 1504970.

Chieb, M., & Gachomo, E. W. (2023). The role of plant growth promoting rhizobacteria in plant drought stress responses. *BMC Plant Biology*, *23*(1), 407.

Choukr-Allah, R., Mouridi, Z. E., Benbessis, Y., & Shahid, S. A. (2023). Salt-affected soils and their management in the Middle East and North Africa (MENA) region: a holistic approach. In *Biosaline agriculture as a climate change adaptation for food security* (pp. 13–45). Springer International Publishing.

Farhad, M., Kumar, U., Tomar, V., Bhati, P. K., & Krishnan, J, N., Barek, V., ... & Hossain, A. (2023). Heat stress in wheat: A global challenge to feed billions in the current era of the changing climate. *Frontiers in Sustainable Food Systems*, *7*, 1203721.

Food and Agriculture Organization of the United Nations. (2020). AQUASTAT database. https://www.fao.org/aquastat

Francis, D., & Fonseca, R. (2024). Recent and projected changes in climate patterns in the Middle East and North Africa (MENA) region. *Scientific Reports*, *14*(1), 10279.

Hamed, M. M., Ullah, S., Jasim Al-Hasani, A. A., Nashwan, M. S., & Shahid, S. (2025). Rising Threats of Compound Droughts and Heatwaves in MENA: Implications for Population and Agriculture under Paris Agreement Scenarios. *Earth Systems and Environment*, •••, 1–20.

Harsonowati, W., Manohara, D., Dwiastuti, M. E., Widawati, S., Sodiq, A. H., Khastini, R. O., & Purwani, J. (2024). Microbial-based stimulants on plant adaptation to climate change. In *Biostimulants in Plant Protection and Performance* (pp. 369–383). Elsevier.

Hossain, M. M., Ahmed, S., Alam, M. S., & Hossain, A. (2024). Adverse effects of heat shock in rice (Oryza sativa L.) and approaches to mitigate it for sustainable rice production under the changing climate: A comprehensive review. Heliyon.

Jarecki, W. (2022). Physiological response of soybean plants to seed coating and inoculation under pot experiment conditions. *Agronomy (Basel)*, *12*(5), 1095.

Khan, M. Y., Nadeem, S. M., Sohaib, M., Waqas, M. R., Alotaibi, F., Ali, L., & Al-Barakah, F. N. (2022). Potential of plant growth promoting bacterial consortium for improving the growth and yield of wheat under saline conditions. *Frontiers in Microbiology*, *13*, 958522.

Malkawi, H. I., & Kapiel, T. Y. S. (2024). Microbial biotechnology: A key tool for addressing climate change and food insecurity. *European Journal of Biology and Biotechnology*, *5*(2), 1–15.

Mishra, A. K., & Singh, V. P. (2010). A review of drought concepts. *Journal of Hydrology (Amsterdam)*, *391*(1-2), 202–216.

Priyadarshini, S., Mulage, P., Conchita, S., Pavani, P., & Wasule, D., KB, C.K., Balaji, M.P. and Ashwini, K.S. (2024). Microbial Biotechnology: Innovations for Crop Protection and Improvement. *Journal of Advances in Biology & Biotechnology*, *27*(7), 238–256.

Rashid, U., Yasmin, H., Hassan, M. N., Naz, R., Nosheen, A., Sajjad, M., & Ahmad, P. (2022). Drought-tolerant Bacillus megaterium isolated from semi-arid conditions induces systemic tolerance of wheat under drought conditions. *Plant Cell Reports*, •••, 1–21.

Saad, M., Abokoura, H., Abdellatif, K., El-Latif, M., & Aly, M. (2020). Application of Alginate Bead Encapsulated N 2-Fixing Bacteriais Improvingwheat Yield under Drought Stress. *Plant Archives*, *20*, 1735–1747.

Sharma, M. P., Grover, M., Chourasiya, D., Bharti, A., Agnihotri, R., Maheshwari, H. S., & Bagyaraj, D. J. (2020). Deciphering the role of trehalose in tripartite symbiosis among rhizobia, arbuscular mycorrhizal fungi, and legumes for enhancing abiotic stress tolerance in crop plants. *Frontiers in Microbiology*, *11*, 509919.

Solaiman, Z. M., Blackwell, P., Shafi, M. I., Salman, N. D., Storer, P., & Babur, E. (2025). Co-application of biochar and phosphorus increases soil microbial biomass, mycorrhizal colonization, growth, and nutrition of subterranean clover. *AIMS Microbiology*, *11*(1), 59–73.

Vorobey, N. A., Kukol, K. P., Kots, S. Y., Pukhtaievych, P. P., & Patyka, V. P. (2023). Viability of nodule bacteria Bradyrhizobium japonicum on soybean seeds by treatment with fungicides during extended storage.

Wahab, A., Muhammad, M., Munir, A., Abdi, G., Zaman, W., Ayaz, A., & Reddy, S. P. P. (2023). Role of arbuscular mycorrhizal fungi in regulating growth, enhancing productivity, and potentially influencing ecosystems under abiotic and biotic stresses. *Plants*, *12*(17), 3102.

Yarullina, L. G., Kalatskaja, J. N., Cherepanova, E. A., Yalouskaya, N. A., Tsvetkov, V. O., Ovchinnikov, I. A., & Mardanshin, I. S. (2023). Approaches to improving the biological activity of agricultural formulations based on bacteria of the genus Bacillus and chitosan nanocomposites. *Applied Biochemistry and Microbiology*, *59*(5), 549–560.

Chapter 10

Using Multispectral and Radar Imagery to Estimate Citrus Water Requirements in the Context of Climate Change in the MENA Region:

Case From Morocco

Yasmine Bouchibti
Institute Agronomic and Veterinary Hassan II, Rabat, Morocco

Noureddine Belkadi
International Academy of Agricultural Technologies and Innovations, Casablanca, Morocco

Nour-Eddine Bassa
International Academy of Agricultural Technologies and Innovations, Casablanca, Morocco

Allal Hamouda
Institute Agronomic and Veterinary Hassan II, Rabat, Morocco

Yassine Mouniane
https://orcid.org/0000-0002-6037-5301
Faculty of Sciences, Ibn Tofail University, Kenitra, Morocco

DOI: 10.4018/979-8-3373-3346-5.ch010

ABSTRACT

Citrus farming in the MENA region faces increasing challenges due to climate-induced water scarcity. Ensuring precise and efficient irrigation is essential to sustain production under semi-arid conditions. This chapter presents a remote sensing-based method that combines multispectral (Sentinel-2) and radar (Sentinel-1) imagery to estimate the water requirements (ETc) of citrus trees in southern Morocco over a five-year period. Vegetation indices such as NDVI and radar-based indices like VH-VV were used to compute crop coefficients (Kc), which were then coupled with meteorological data to estimate ETc. Radar imagery, less affected by weather conditions, provided high accuracy (RMSE = 0.050) and continuous monitoring. Spatial and temporal variability in water demand was effectively captured, enabling targeted irrigation strategies. This approach offers a scalable and cost-effective tool to improve water management and resilience in agriculture, contributing to key climate adaptation priorities in the MENA region.

1. INTRODUCTION

Climate change is currently one of the main threats facing agriculture in the MENA (Middle East and North Africa) region (Lange, 2019; Le Mouël et al., 2023). The MENA region is largely arid to semi-arid and climate-related environmental stresses, including increasingly warmer temperatures, unexpected shocks associated with rainfall reduction generally, and spikes in the frequency of extreme weather events, are compounded by pressures placed on already constrained water resources, thereby sacrificing agricultural systems and regional food security (Salimi & Al-Ghamdi, 2020; Namdar et al., 2021; Alrteimei et al., 2022; Al-Jawaldeh et al., 2022).

Irrigation is a key component of economic development, supporting agricultural growth, improving rural livelihoods, generating export revenues and building resilience to the impacts of climate change (Aleksandrova et al., 2014; Gebeyhu et al., 2024). In arid and semi-arid regions, where water is a precious and limited resource, optimized and sustainable irrigation management becomes essential to maximize agricultural yields and preserve water resources (Mouniane et al., 2024), particularly in a context of climate change and increasing food demand (Li et al., 2024).

Citrus, one of the main fruit categories grown worldwide, is particularly sensitive to fluctuations in water supply, making it a strategic crop in the Mediterranean region (Cerdà et al., 2021). Their production requires precise management to guarantee a quality harvest (Zhang et al., 2022). *Citrus* is also cold-sensitive plants, and exposure to extreme temperatures can lead to environmental stress (Zabihi et al., 2016). Various studies have highlighted the importance of examining the impacts of climate change on farmers and understanding the strategies adopted by them, as well as by the organizations involved, to strengthen the resilience of farms and rural communities in the face of these challenges (Ezrari et al., 2022). Although citrus orchards are among the most widely grown crops in Mediterranean basins, they stand out for their high-water requirements (Rallo et al., 2017), in contrast to other crops that have a high capacity to withstand arid environments (Volschenk, 2020).

Traditionally, the assessment of crop water requirements has relied on classical agronomic methods, such as soil moisture measurements and visual observations. Although commonly used, these methods are often costly, time-consuming and lack precision when it comes to monitoring large agricultural areas. The emergence of geospatial technologies, in particular remote sensing, now offers innovative approaches, characterized by high spatial resolution and continuous monitoring capability, paving the way for more efficient and sustainable irrigation management.

In this context, advances in technology for remote sensing offer new perspectives for targeted and sustainable water management in agriculture. In particular, the Sentinel-1 (radar imagery) and Sentinel-2 (multispectral imagery) satellite missions of the Copernicus program provide free, regular, and high-resolution spatial and temporal information, well-positioned to monitor the development of vegetation, assess its physiological status, and indirectly estimate crop water needs based on various indices (NDVI, VH-VV, etc.) and ET estimation models. Radar imagery, in particular, stands out for its ability to penetrate clouds and provide accurate data, whatever the weather conditions. The satellites of the Sentinel-1 mission regularly provide detailed radar images, enabling us to monitor water requirements over large geographical areas. At the same time, multispectral imagery obtained via the Sentinel-2 satellite is used to calculate vegetation indices, such as NDVI, which are essential for assessing the water status of crops and estimating their water requirements.

In Morocco, agriculture consumes approximately 85% of total water withdrawals, and citrus fruit production is arguably one of the most water-dependent sectors, especially in regions like Souss-Massa. The past decade of recurrent droughts has had a detrimental impact on available water reserves and agricultural production in Morocco. National programs such as the Plan Maroc Vert, Generation Green 2020-2030, and the National Water Plan 2020-2050 are intended to foster sustainable irrigation and climate-resilient agriculture. Still, one of the biggest bottlenecks to successful irrigation planning in Morocco is the limited availability of reliable, near real-time data related to crop water status.

Aside from the technical scope of remote sensing, this research also engages with larger questions about climate adaptation in the Mediterranean. The combination of radar and multispectral data presents an opportunity to provide a scalable way to assess crop water requirements under climatic uncertainties, supporting both scientific aims and decisions in water-scarce contexts at farm and institutional levels.

2. MATERIALS AND METHODS

2.1. Description of the study area

The study area is a plot of land located in the Souss Massa region, in the province of Taroudant, near the rural commune of Oulad Aissa. The plot is atypical in its irregular trapezoidal polygonal shape, with eight points delimiting it geographically (30°34'42.60 "N 8°36'6.48 'W; 30°34'31.44 'N 8°36'13. 0 "W; 30°34'26.04 'N 8°36'14.04 'W; 30°34'26.04 'N 8°36'11.88 'W; 30°34'25.68 'N 8°36'5.40 'W; 30°34'25.68 'N 8°36'0.72 'W; 30°34'26.40 'N 8°35'57.48 'W; 30°34'43.32 'N 8°36'2.16 'W). This plot is planted with Navel orange trees, whose dense foliage largely covers the ground, limiting their visibility from an aerial view. Moreover, the trees are close together, covering the ground uniformly. These are mature trees, with foliage covering 70% or more of the ground.

Figure 1. Map of the experimental site (google maps)

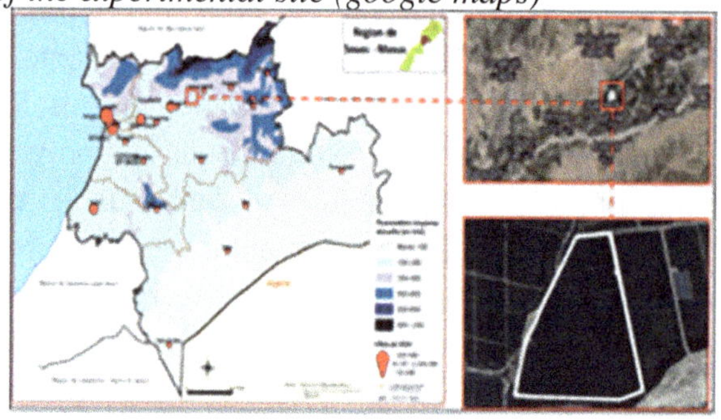

2.2. Meteorological Conditions

The climate of the study area is semi-arid to sub-desert, with low and irregular annual rainfall, varying between 140 and 200 mm. The average annual temperature is 19°C, with peaks of over 27°C in summer and lows of around 11°C in winter. The hot, dry "Chergui" wind can influence temperatures in spring and summer. The region enjoys over 3,000 hours of sunshine a year.

Figure 2. Trends in average daily temperatures in the study area from 2018 to the present day

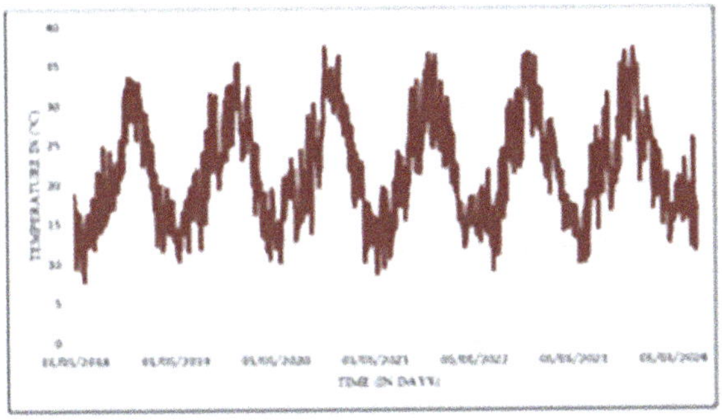

Figure 3. Daily rainfall trends in the study area from 2018 to the present day

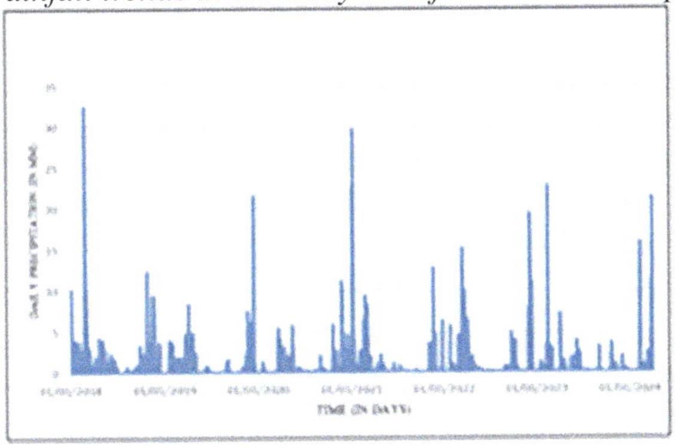

2.3. Data and equipment used

2.3.1. Multispectral images

Multispectral images were collected from the Sentinel-2 mission (Sentinel 2A and 2B), freely accessible via the Copernicus program. These images, with a spatial resolution of 10 m, cover the visible and near infrared. The study period runs from January 2019 to December 2023, with a total of 195 images, selected for their low cloud cover (less than 5%).

2.3.2. Radar imagery

Radar images are from the Sentinel-1 mission (Sentinel 1A and 1B), collected via Google Earth Engine. Ground Range Detected (GRD) images in Interferometric Wide Swath (IW) mode were used, as they are already pre-processed for radiometric and geometric corrections. The collection period is the same as for multispectral images.

2.3.3. Meteorological data

Meteorological data for calculating reference evapotranspiration were provided by the CGMS-Morocco system, with a resolution of 4.5 km.

2.3.4. Reference Crop Coefficients data

Reference crop coefficients, based on phenological stages and ground cover of adult orange trees, were used to calibrate and validate the model.

2.4. Software and tools used

The software and tools used in this study include Google Earth Engine, Jupyter Notebook, and ArcGIS Pro. Google Earth Engine was used to access and manipulate multispectral and radar image collections, enabling efficient large-scale data management. Jupyter Notebook was used for data processing, providing an interactive environment for analysis and process automation. Finally, ArcGIS Pro was used for spatial analysis and visualization of results, facilitating interpretation of geographic data and presentation in the form of detailed maps and graphs.

2.5. Data pre-processing

Multispectral and radar images were subjected to several corrections to ensure data accuracy. Multispectral images underwent atmospheric, radiometric and geometric corrections to mitigate distortions, using the 'COPERNICUS/S2_SR' collection, which includes these corrections by default. A filter was applied to select images with less than 5% cloud cover. Radar images were corrected for radiometric calibration, terrain correction, decibel conversion and thermal noise suppression. The 'COPERNICUS/S1_GRD' collection was used, with images in VH and VV polarization to minimize distortions.

2.6. Estimation of crop coefficients

The Normalized Difference Vegetation Index (NDVI) was calculated from multispectral images to assess the chlorophyll activity of orange trees. This NDVI was then used to estimate crop coefficients, by applying a linear regression based on the Tasumi et al. (2005) equation, adapted specifically to the citrus trees in our plot. In addition, radar images were used to estimate crop coefficients using indices specific to radar features, although the details remain to be further worked out to better match citrus peculiarities.

3. RESULTS

3.1. NDVI results from multispectral imagery

The Normalized Difference Vegetation Index (NDVI) enables us to monitor the health and strength of growth, and we can clearly see that NDVI values are particularly high at the beginning of the year, almost reaching values of 0.9 and even exceeding it in January 2022. On the other hand, we observe a set of typical seasonal variations that are repeated for the years 2019 to 2021. This seasonality is no longer respected for the two successive years 2022 and 2023, when we observe a severe drop in NDVI followed by a period of fluctuating values stabilizing around 0.6-0.7.

Figure 4. NDVI visualization from 2019 to 2023

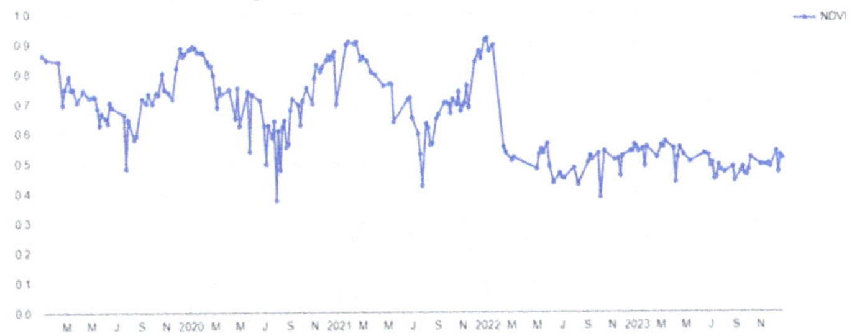

3.2. Radar imagery index

As we can see, radar images offer more values than multispectral ones, but their variation is much greater due to noise. For this reason, we applied local weighted regression to smooth the data, making it less noisy but also taking local variation into account.

Figure 5. Visualization of the VH index from 2019 to 2023

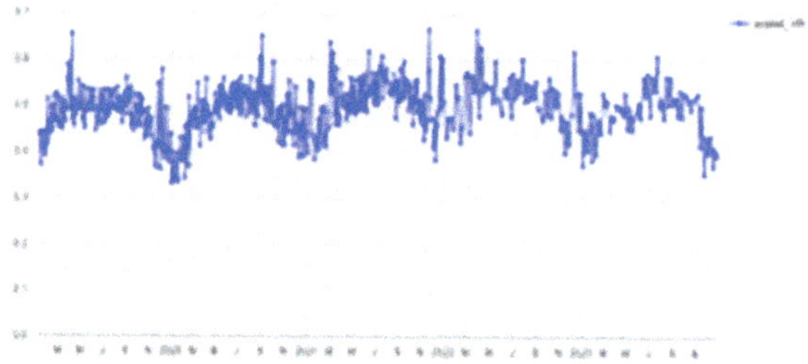

Figure 6. Visualization of the VH-VV index from 2019 to 2023

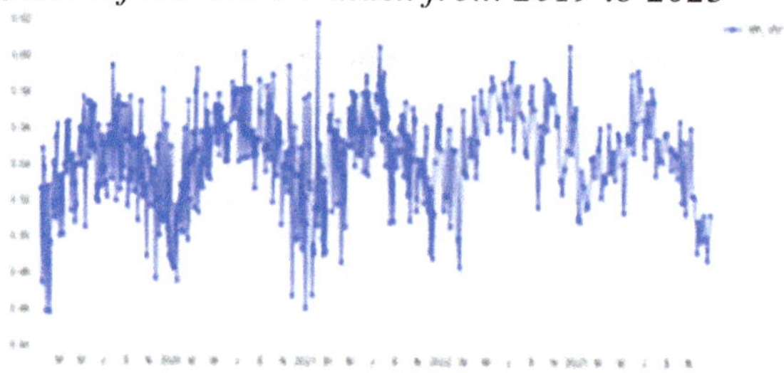

Figure 7. Visualisation de l'indice du SNI de 2019 à 2023

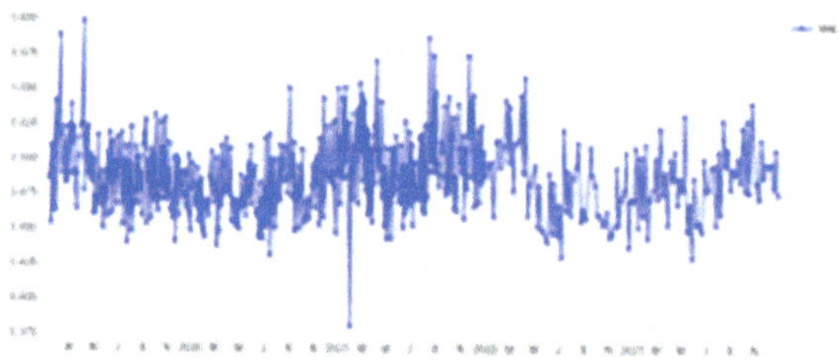

Figure 8. Visualization of the VH+VV index from 2019 to 2023

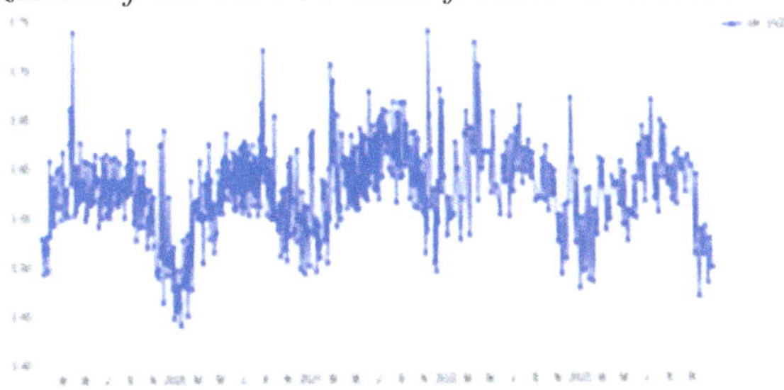

3.3. Estimation of crop coefficients

3.3.1. Crop coefficients estimated using multispectral imagery

By applying the regression equation of Beeri et al (2020) to our NDVI values, we obtain our series of crop coefficients estimated at the level of the period concerned. We will then use the metrics to assess the performance and accuracy of our model. Finally, we'll determine a correction coefficient to make our model specific to citrus, which we'll integrate into the general equation before estimating water requirements.

Figure 9. Visualization of Kc crop coefficients from 2019 to 2023

As we can see, we obtain a time series with variable fluctuations according to phenological stages, we note that despite the presence of the cloud filter up to 5%, the series obtained offers a certain regularity and image capture i.e. an estimate of the crop coefficient throughout the period presented. We note the presence of seasonality, with peaks in crop coefficients reaching values of up to 1.1 and lower and minimum values around 0.5 and 0.6, reflecting the respective phenological stages where vegetative growth is at its peak and others where vegetative growth is weaker.

3.3.2. Crop coefficients estimated using multispectral imagery

Having obtained our estimated crop coefficients we compare them with the reference data based on the bias, RMSE and nRMSE metrics. We obtain the following values:

Table 1. Tables of error metrics for estimating water requirements using multispectral imagery

Biais	**0.2713**
RMSE	0.3305
nRMSE	3.3047

The positive bias above, with a value of 0.2713, informs us that a systematic overestimation of crop coefficients is present, and that the overestimation averages 0.2713 at reference values for citrus crop coefficients, making it a high average value. Turning to the other metrics, the RMSE value also indicates a high average error value of 0.3305, making predictions less close to actual values. The nRMSE confirms the lack of fit, with a value of 3.3047. For a good fit, and therefore a good estimate, this must be less than 1, so the model does not predict crop coefficients well.

We opted for the determination of a correction coefficient that we established for each stage, then by calculating its average on a global scale, in order to no longer be impacted by overestimation we found a value of -0.3065 for this correction coefficient, which we added to the global equation for calculating the crop coefficient. After adding the correction coefficient, we obtained the following table of metrics.

Table 2. Table of error metrics for plot water requirement estimates after model calibration

Biais	**-0.0502**
RMSE	0.1938
nRMSE	1.9385

3.3.3. Crop coefficients estimated by radar imagery

Before applying local weighted regression, we need to homogenize the data and avoid the influence of outliers. We can see some of these in the graphs above.

To this end, we have proceeded to normalize the values in order to apply the next analysis steps.

As indicated in the previous section, we will apply local weighted regression to the four collected indices, varying the smoothing factor. We have opted for robust smoothing factors in the same way as the Berri et al study, with smoothing factors having values of 10, 20, 30, 40 and 50 respectively.

Thus, for each of the radar indices presented above, we smoothed its values by first determining the weight matrix and the regression coefficients specific to each point in order to obtain the smoothed values.

Figure 10. ScaledVH values from 2019 to 2023 based on five smoothing factors

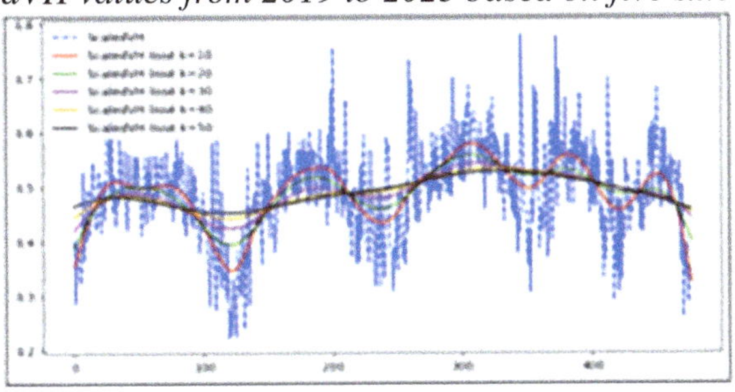

Figure 11. Scaled VH-VV index values from 2019 to 2023 based on five smoothing factors

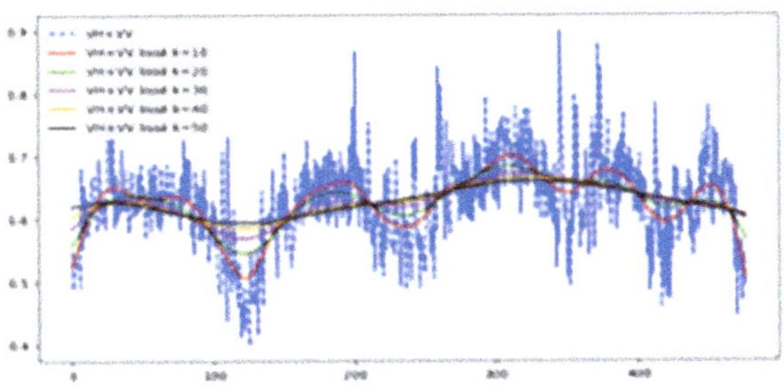

Figure 12. Smoothed SNI index values (ScaledSNI) from 2019 to 2023 based on five smoothing factors

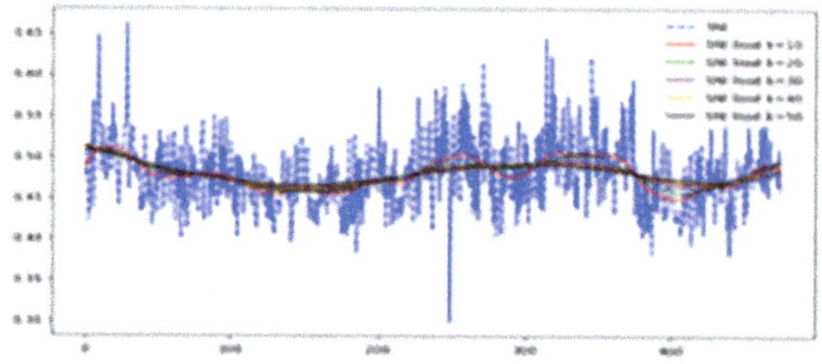

For each point in the series, a neighborhood of nearby points was selected according to the smoothing parameter k. We can clearly see that as the smoothing factor increases, the model captures local trends in order to reduce noise and obtain less variation. In contrast to the multispectral images, the raw series data show rapid and noisy variations, making it difficult to clearly distinguish phenological stages.

We note that the red smoothed curve relating to the smoothing factor k = 10, is the closest to the raw data series, indicating a very close sensitivity to local variations. The green curve minimizes certain variations while following the general curvature of the data. This smoothing, with a smoothing factor equal to 20, preserves the general curvature of the data while smoothing well and reducing noise. Smoothing with k = 30, which is much more pronounced, shows us that short-term variations are weakened, retaining only the long-term trends in the data. As for the last curves with smoothing factors of k = 40 and k = 50, here the smoothing is very pronounced, we can no longer distinguish short-term nuances and see that the curve is approaching a general constant. With the help of error metrics, we'll see which smoothing factor is closest to the reference data, and which radar index minimizes errors and estimates the cultural coefficient as accurately as possible.

3.3.4. Crop coefficients estimated using radar imagery

Having smoothed our data, we now need to obtain the Kc crop coefficients from the smoothed data for each combination of index and factor and K smoothing. So, we obtain our crop coefficients from the equation in the Beeri et al (2020) study.

Figure 13. Crop coefficients estimated from the SAR ScaledVH index

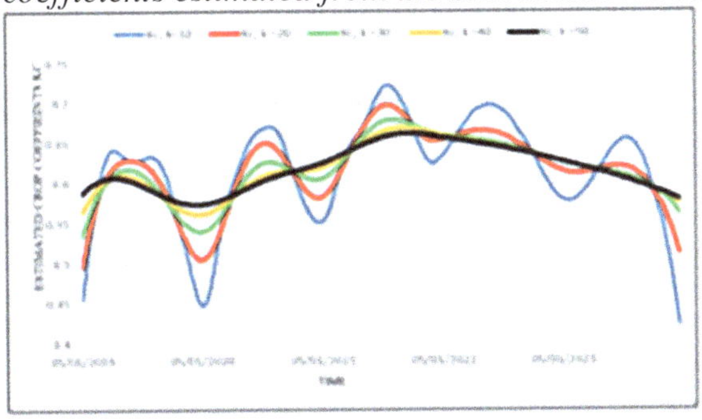

Figure 14. Crop coefficients estimated from the SAR index ScaledVH-VV

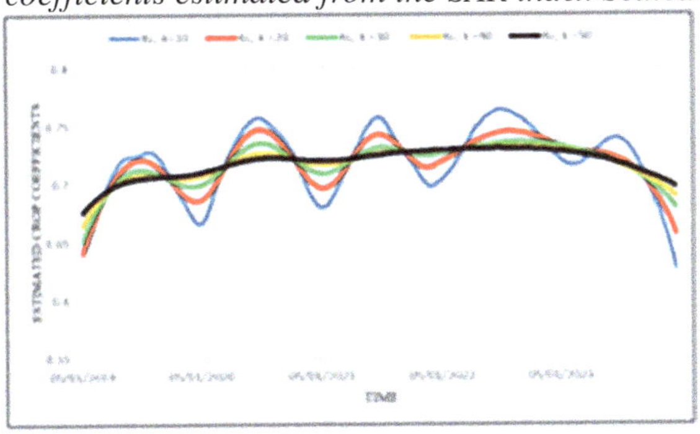

Figure 15. Crop coefficients estimated from the SAR Scaled VH+VV index

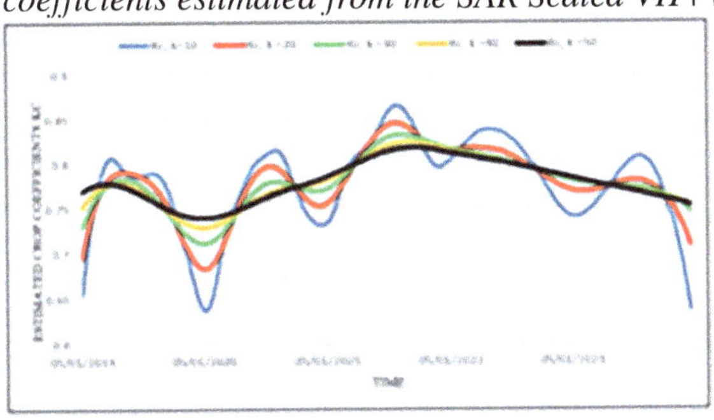

Figure 16. Crop coefficients estimated from the SAR ScaledSNI index

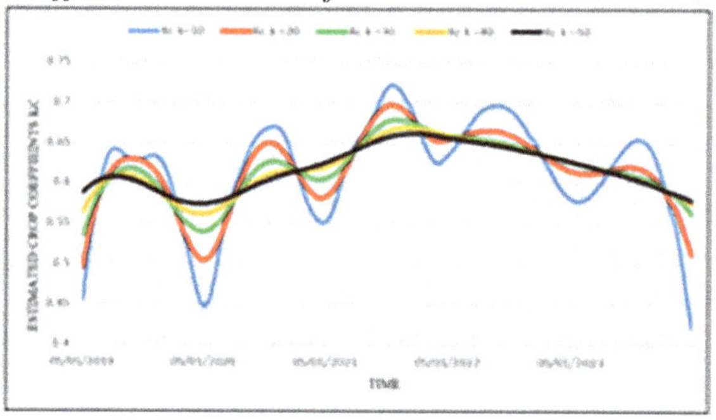

Having estimated the crop coefficients from each of the smoothed indices presented above, we note similar patterns in the crop coefficients despite the differing degrees of variation, and we also see that these patterns occur at the same periods in each of the indices: the December and January periods are those characterized with the lowest crop coefficients, while the periods with the highest crop coefficients are in the summer months of July and August. As for values, we can see that they are all between 0.6 and 0.7 in general, with values varying according to growth stage and vegetative growth, except for crop coefficients determined by the SNI index, which show lower values around 4.5 for minimum peaks; on the other hand, the VH+VV index shows us higher values, with values exceeding 0.8. Using the metrics, we shall see which index and which smoothing parameter are selected and preferred over the others for comparison and subsequent calibration with the reference crop coefficients for citrus.

3.3.4. Choice of model for estimating the cultural coefficient in radar imagery

In order to determine crop water requirements, it is essential to determine which index and which smoothing factor to use. To do this, we compared our estimated values, i.e. for each combination of index and smoothing factor, we calculated error metrics in relation to the reference values.

Table 3. Tables of error metrics for each combination of index and smoothing factor

		KcVH	KcVH-VV	KcVH+VV	KcSNI
K =10	Biais	0,0524	0,0253	0,0858	-0,0766
	RMSE	0,0742	0,0624	0,1145	0,1171
	nRMSE	0,7418	0,6235	1,1451	1,1715
K = 20	Biais	0,0523	0,0247	0,0860	-0,0767
	RMSE	0,0686	0,0553	0,1065	0,1049
	nRMSE	0,6861	0,5528	1,0649	1,0492
K =30	Biais	0,0528	0,0245	0,0866	-0,0762
	RMSE	0,0677	0,0500	0,1016	0,0964
	nRMSE	0,6768	0,5002	1,0160	0,9638
K = 40	Biais	0,0532	0,0246	0,0870	-0,0758
	RMSE	0,0683	0,0470	0,0988	0,0913
	nRMSE	0,6831	0,4700	0,9880	0,9134
K = 50	Biais	0,0536	0,0248	0,0872	-0,0754
	RMSE	0,0688	0,0455	0,0972	0,0884
	nRMSE	0,6878	0,4549	0,9719	0,8835

We're looking for the index associated with the smoothing factor that offers the lowest error and the highest accuracy. To do this, we've based ourselves on three metrics: the bias, which indicates the average difference between predicted and actual values. The average error between predictions and actual values (RMSE) and finally the nRMSE, which is in fact a normalization of the RMSE with respect to the data range. The lowest value of these three metrics is desired.

We first note that for the VH index used to estimate crop coefficients, the bias is almost constant for the different values of the smoothing factor. The RMSE and nRMSE values are minimal for the smoothing factor K =30.

With regard to the contribution of the VH-VV index, we note that the bias observed is about half that of the VH index, which is very small, making the estimation of crop coefficients by the latter effective, and characterized by very small fluctuations as a function of the smoothing factor. The parameters RMSE and nRMSE decrease with increasing smoothing factor, reaching respective minimum values of 0.0430 and 0.4303. We can consider the VH-VV index as an interesting index for estimating crop coefficients.

The VH+VV index offers us the highest bias values of all the indices, and is almost constant apart from very slight fluctuations. The other two indices are also particularly high compared to those observed for the other indices, with nRMSE values starting at 1.0831 with K =10 and decreasing slightly with increasing values of smoothing factors, stabilizing at around 0.9. We can therefore consider other indices than VH+VV for accurate estimation of water requirements.

The RMSE and nRMSE values are also high and decrease with smoothing, while the observed values are slightly lower than those of VH+VV.

As we have seen with the metrics, whatever the value of the smoothing factor, the VH-VV index offers the lowest values and is therefore optimal. As far as the K value is concerned, based on the values of the metrics we must by nature select the VH-VV index with a K =50. However, taking into account the notion of smoothing, which refers to a form of adjustment or correction to reduce the effect of noise, we must consider the effect of oversmoothing that this method also offers. Oversmoothing refers to the fact that the model smoothes the data too much, resulting in a significant loss of detail: in our case, a loss of the variability needed to properly distinguish each stage. We saw this phenomenon clearly in the crop coefficient curves estimated by radar imagery earlier.

Based on these analyses, the best index to opt for is the VH-VV index with a smoothing factor of K =30. Indeed, RMSE values decrease progressively from K=10 to K =50, but as we can see, the differences become smaller and smaller as the smoothing factor K increases. The nRMSE values also follow a similar trend to RMSE, but the errors decrease from K=10 to K=40 and then improve only slightly from K=40 to K=50. We didn't choose K = 40 because it offers only a slight improvement and may lead to a risk of over-smoothing. So the VH-VV index with K=30 appears to be an optimal choice, giving us good performance and minimizing the risk of over-smoothing.

3.4. Water requirement results (ETc)

3.4.1. ETo values determined using meteorological data

By applying the Hargreaves-Samani method to the meteorological data received, we were able to determine the reference evapotranspiration from the year 2018 to the month of March in the year 2024. We note the presence of a marked and well-paced seasonality over the five years, with low evapotranspiration values of between 1-2 mm in the months at the end of the year from October onwards, which can be explained in particular by low temperature data. We also observe significantly high values during the summer period, with maximum values of 8-8.5 mm around mid-July, when the climate is particularly hot.

Figure 17. Reference potential evapotranspiration ETo estimated using the Hargreaves-Samani equation from 2018 to mid-2024

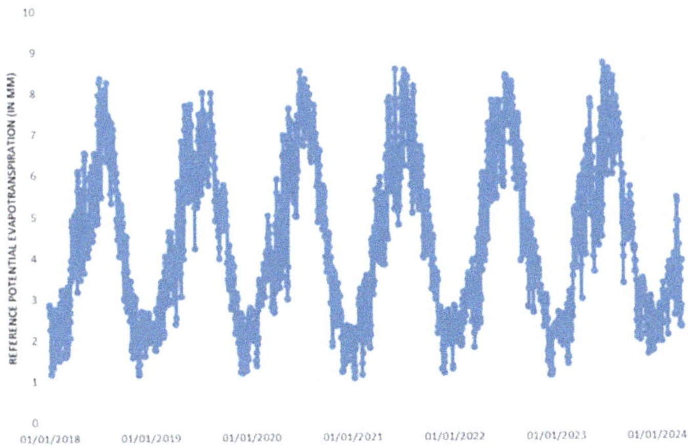

3.5. ETc results using multispectral imagery

3.5.1. Results of plot water requirements as a function of time

Having multiplied the estimated and calibrated crop coefficients by the reference evapotranspiration estimated using the Hargreaves-Samani equation, we obtain the water requirements throughout the five-year period for the plot. We note that the seasonality effect is always respected and of the same type as that of the reference evapotranspiration presented above, with extremes observed at the same periods. However, from the year 2022 onwards, we observe a drop in minimum and maximum values.

The lowest values obtained correspond to tree dormancy periods, which generally begin in December and end in January, when requirements are particularly low. The period of flowering followed by fruit set depends of course on the variety, according to its earliness and characteristics, but also on the environment. Trees generally flower in spring, from March to May, during which period our model shows values of between 3 and 5 mm. These values are in line with the water requirements for this stage, even if within this same period we find some overestimates or underestimates, as the values depend on the varieties with which we compared them. As for the period of active fruit growth, i.e. from April to August, when water requirements are naturally at their highest, we find values ranging from 5 to 7 mm, which are also those theoretically found for this period. Finally, for the period of fruit ripening, which depends on the targeted harvest period, we find values of between 5 and 3 mm, which are very close to the water requirements for this period.

We can therefore assume that our model estimates water requirements close to reality, based on our approach with multispectral imagery.

Figure 18. Citrus water requirements estimated using multispectral images for the plot from2019 to 2023

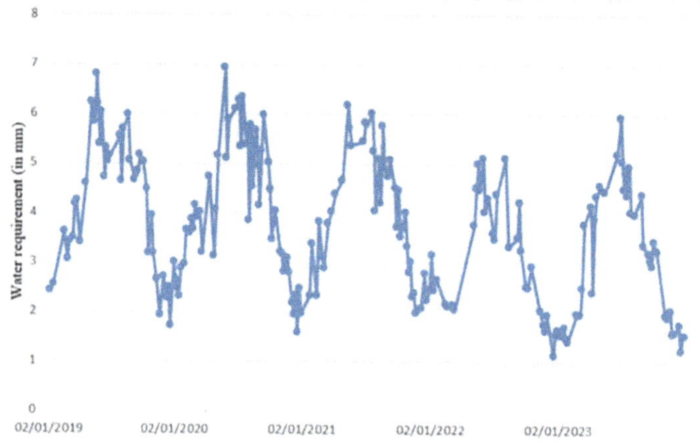

3.5.2. Results of plot water requirements as a function of space

In the same way as our determination of the plot's water requirements on a temporal scale, we can visualize the water requirements of the citrus trees in the plot spatially, taking into account the variability within the plot. This will enable us to know which areas of the plot on a 10 m scale have higher water requirements than others, and thus to favor more precise and efficient irrigation.

We can thus visualize the variation in water requirements at plot level over a given period.

We will visualize water requirements from July to September, the period of the citrus fruit-growing season, and show us the zone affected and the zones most in need of water, in order to maximize and standardize the supply of irrigation water to obtain water-rich fruit of the right size. This is the wettest period for citrus, so we expect evapotranspiration to be at its highest

We'll also be visualizing water requirements from December to March, a so-called dry phase for citrus, as it requires the least water throughout the cycle, It is therefore interesting to visualize spatial variability, and to see which areas require the most water, and which areas of the plot require the least.

Figure 19. Multispectral imagery map of the plot's water requirements for the period December 1, 2022, to March 1

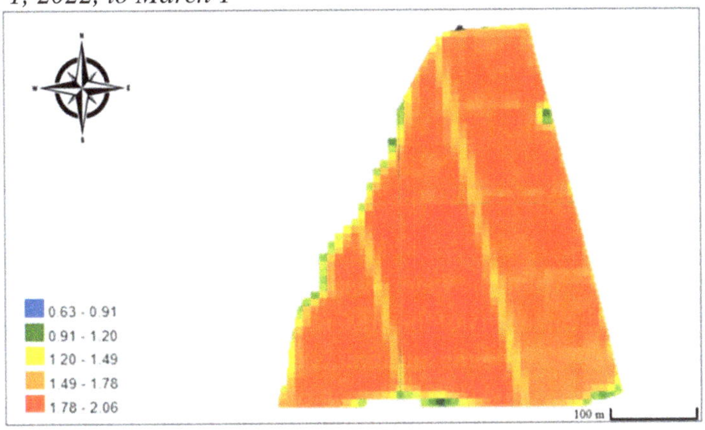

We note that the image produced by multispectral imagery for the period when citrus fruits require the least water, the values estimated on a spatial scale show us that the maximum areas requiring water are at the level of 2 mm and the others the lowest are close to 0. These results are close to the values estimated on a temporal scale where we see in the graph that for this period the values were mainly between 1 and 2. We note that the multispectral image also enabled us to distinguish between the sub-plots of the plot, and to distinguish them with much lower requirements. The bare soil around the plot was determined to be very low, since it has no vegetation.

Figure 20. Multispectral imagery map of the plot's water requirements from May 2023 to September 2023

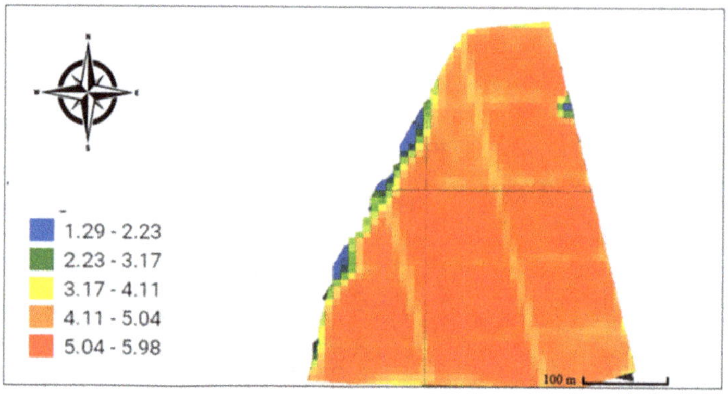

As with the previous plot, the water requirements determined during the summer period show us that the majority of the present plot requires high water requirements during the period when evapotranspiration is at its highest and the fruit swells and soaks up water to be of a good size. The map confirms the high water requirements of citrus during this period to keep the plants healthy.

Based on the maps, this can help plot managers to optimize irrigation and avoid giving too much or too little water to areas that need respectively more or less water, thus improving plant health and water use efficiency.

3.6. ETc results using radar imagery

3.6.1. Results of plot water requirements as a function of time

As with the previous graph of estimated water requirements for multispectral imagery, the graph of estimated water requirements for radar imagery reflects the same seasonal variations, with peaks in water requirements of up to 6 mm per day in the hottest month of July, dropping to 1 mm in the coldest months of December and January. We note that for radar imagery, the estimated extremes are lower, reaching 6 mm for summer peaks and between 0.8 and 1 mm in winter. As far as seasonality is concerned, the annual cycles are clear and repetitive over time. As for the requirement values for the different phenological stages, the estimated values are in line with the requirement values for each phenological stage, but at times we can see some higher or lower values.

Figure 21. Water requirements of citrus trees estimated using radar images for the plot

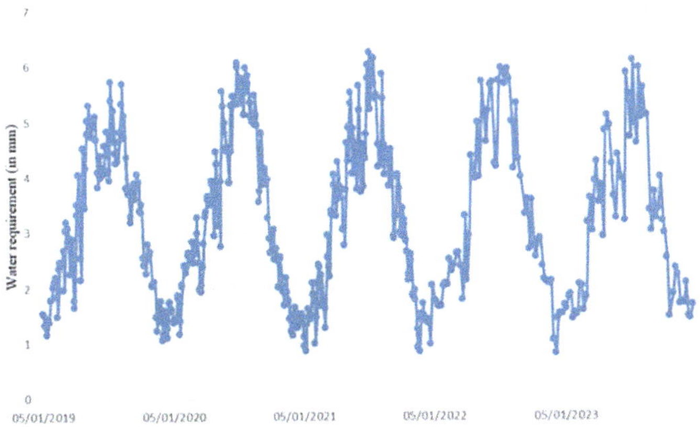

3.6.2. Results of plot water requirements as a function of space

This map (Figure 22) shows that, spatially, the distribution is globally between the second and third class during this period, showing values in the same range as those proposed by the multispectral images. However, the spatial distribution is different: we note that for the multispectral images, the estimated needs in the subplots are uniform and belong to the same class, unlike those provided by the radar imagery, which show great variation on a very small scale.

We can interpret this difference by the characteristics of radar images, which are noisy and variable for the same area, and very sensitive to soil characteristics such as roughness and soil moisture.

Figure 22. Radar imagery map of water requirements for the period December 1, 2022 to March 1, 2023

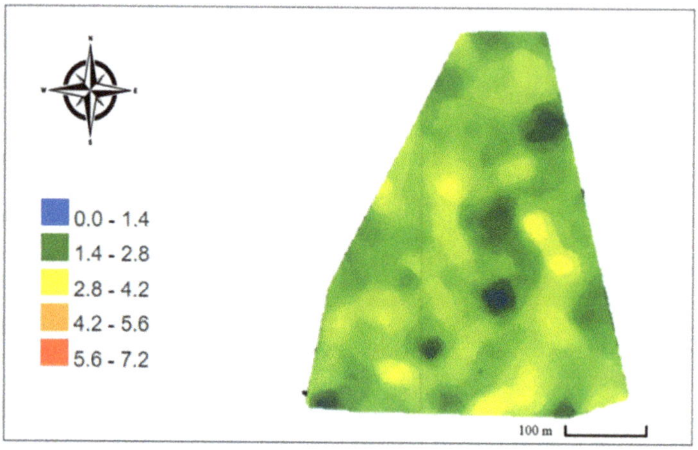

This map (Figure 23) shows that, spatially, the distribution is globally between the second and third class during this period, showing values in the same range as those proposed by the multispectral images. However, the spatial distribution is different: we note that for the multispectral images, the estimated needs in the subplots are uniform and belong to the same class, unlike those provided by the radar imagery, which show great variation on a very small scale.

We can interpret this difference by the characteristics of radar images, which are noisy and variable for the same area, and very sensitive to soil characteristics such as roughness and soil moisture.

Figure 23. Map of water requirements for the plot from May 2023 to September 2023 using radar imagery

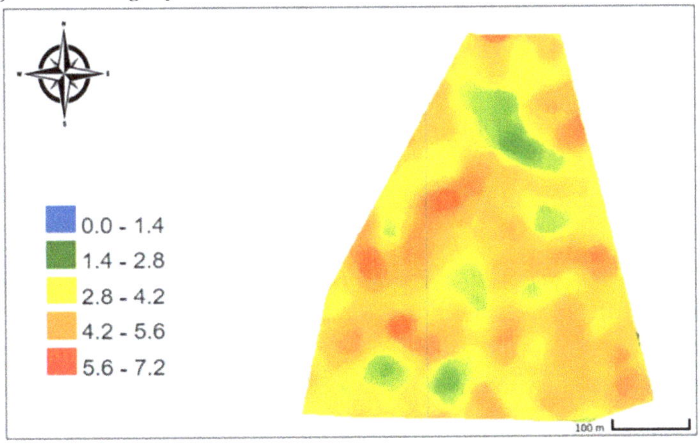

4. DISCUSSION

The technical results previously described have broader implications beyond the experimental paradigm. The periodicity of evapotranspiration (ETc) and its interannual variability of ETc is an important indicator of changing patterns of water availability due to climate change. The reduction in NDVI and ETc for the years 2022 and 2023 can be connected to extreme drought over the Souss-Massa basin, which triggered emergency protocols for the allocation of irigation supplies. These results provide a clear validation of the sensitivity of citrus crops to climate induced stress and show how satellite monitoring can be used to assess these risks before they occur. The present study highlights the effectiveness of integrating radar and multispectral imagery for estimating the water requirements of citrus crops in a semi-arid context. The results obtained demonstrate that this approach is not only technically feasible, but also scientifically robust for large-scale application. The use of Sentinel-2-derived indices, such as NDVI, provided relatively reliable estimates of crop coefficients (Kc) after adjustment. However, multispectral images proved to be sensitive to meteorological conditions, particularly cloud cover. Conversely, radar data from Sentinel-1 had the major advantage of providing continuous monitoring, irrespective of weather conditions or brightness. Processing using weighted local regression (LOESS) reduced the noise inherent in radar data, while maintaining good sensitivity to phenological variations. The VH-VV radar index, in particular with a smoothing factor of K=30, gave the best results in terms of Kc estimation, with an RMSE of 0.050, significantly lower than that obtained with multispectral images (0.1938).

The spatio-temporal maps of water requirements generated in this study reveal not only well-marked seasonal cycles, but also spatial disparities within plots. This ability to visualize intra-plot variations represents a major asset for localized irrigation management, optimizing both yields and water consumption (Sofia et al., 2024; Miltiadou et al., 2025; Silva et al., 2025). This work thus offers an operational method for implementing precision irrigation, based on a "data-driven" approach, adaptable to farms of different sizes. The effectiveness of this method paves the way for intelligent irrigation systems directly integrating satellite data into agronomic decision-making (Fuentes-Peñailillo et al., 2024).

The methodological approach developed in this study - which combines radar and multispectral imagery to estimate crop water requirements, is a concrete response to the climatic, agricultural and technological challenges facing MENA countries. This region, one of the most exposed to water stress and global aridification, is seeing its agricultural production jeopardized by decreasing availability of natural resources, accelerated urbanization, and often obsolete water management models. In this context, the tool proposed by this research goes beyond the strictly local framework to form part of a broader reflection on the modernization of agricultural practices, adaptation to climate change and regional food sovereignty. Several of the main themes of the book Impacts of Climate Change on Agriculture in the MENA Region are thus directly reflected in this study.

4.1. Achievements and challenges of agriculture and irrigation in MENA countries

One of the most pressing structural challenges facing MENA countries is water management for agriculture. Irrigation accounts for up to 85% of total water consumption, making any improvement in its efficiency absolutely strategic (Hameed et al., 2019; Al-Addous et al., 2024). However, the majority of farms still rely on traditional techniques, often gravitational, which result in considerable losses through evaporation, run-off or percolation (Levidow et al., 2014). What's more, water allocation is rarely based on actual crop needs, due to a lack of reliable, up-to-date data. The approach presented in this chapter responds to this problem by proposing a management system based on satellite observation, capable of providing continuous, localized and scientifically-based information on plant water status.

By exploiting NDVI indices for multispectral images and VH/VV indices for radar images, it becomes possible to assess the water requirements of each plot according to its phenological stage, microclimatic conditions and actual plant cover (Ouaadi et al., 2023;). This information paves the way for differentiated, targeted and therefore much more efficient irrigation (Nikolaou et al., 2020; Kang et al., 2023). Applied on a regional scale, such an approach could transform the way irrigated perimeters are managed, enabling better resource planning, reduced wastage and improved yield per unit of water used. This is a major step forward in the face of the structural challenges facing irrigated agriculture in the region.

4.2. Impact of technology and science on agricultural yields and food sovereignty

This study also illustrates how the introduction of new technologies, even those that are easy to access, can have a considerable impact on agricultural yields and, more broadly, on food security (Kang et al., 2009; Vermeulen et al., 2012). The use of free satellite images, coupled with open-source tools such as Google Earth Engine or Python, makes it possible to set up a data processing chain that is robust, reproducible and potentially accessible to cooperatives, the technical departments of ministries of agriculture, or even farmers trained in digital agriculture (Kassie et al., 2011).

The benefits are twofold: on the one hand, to enable finer control of irrigation by reducing periods of water deficit or surplus, which has a direct impact on crop quality and quantity; on the other, to provide decision-making tools that no longer rely solely on experience or intuition, but on objective, visualizable and interpretable agronomic indicators (Liu et al., 2024). This transition to data-driven agriculture would increase producers' autonomy, while strengthening their ability to cope with climatic hazards. Ultimately, this approach could make a significant contribution to strengthening food sovereignty in MENA countries, by stabilizing local production and limiting dependence on imports (Alomari-Mheidat et al., 2024).

4.3. Technological levers for adapting to climate change

The study is also clearly in line with the logic of Climate-Smart Agriculture, integrating the use of technology as a lever for adaptation. In a context where seasons are becoming increasingly unpredictable, and heat waves and prolonged droughts are disrupting crop cycles, flexibility is becoming an imperative (Raihan et al., 2024). One of the main benefits of remote sensing, as exploited here, lies in its ability to detect weak signals - incipient water stress, a drop in vegetative vigor, a thermal anomaly - before their effects are visible to the naked eye.

This capacity for early detection enables growers to react more quickly, readjust their practices and, ultimately, minimize losses. It also represents an opportunity to develop intelligent irrigation systems, capable of automatically adjusting volumes according to satellite data, via connected management platforms. These innovations are particularly well-suited to the realities of the MENA region, where inter- and intra-annual climate variability requires constant adaptation of agricultural practices. Making these tools available to the rural world, via public institutions or public-private partnerships, could considerably accelerate the region's agro-technological transition.

4.4. Analysis and action frameworks for adapting agriculture to climate change

Finally, beyond the technical aspects, this study calls for an evolution in agricultural governance frameworks. The transition to a more reactive agriculture, with more refined water management, calls for a revision of the institutional instruments currently in place. This means not only training producers in the use of these tools, but also adapting land-use planning policies, water distribution rules and technical support methods. For example, agricultural extension services could integrate remote sensing data into their field diagnostics; irrigation plans for large-scale irrigation schemes could be based on weekly updated maps; and irrigation subsidies could be made conditional on rational water use based on accurate estimates (Filchev & Chanev, 2025).

Furthermore, this approach could contribute to the creation of regional observatories for adaptation to climate change, capable of cross-referencing agricultural, hydrological and climatic data to guide political decisions. By integrating these tools into territorial planning, MENA countries could gain in collective resilience, by anticipating crises rather than enduring them (Gerald et al., 2025).

This study illustrates that remote sensing, and specifically with radar imagery, can be an excellent resource for policy making on a national and local level with directed water governance. Public agricultural agencies can build on satellite crop coefficient maps and provide irrigation advisories for demanding plots, aligning with some degree of equity in water distribution. Further, extension agents and farmers can be trained on the use of open platforms like Google Earth Engine that will help broaden equitable access to climate-smart technology. Future institutional structures should incorporate satellite monitoring in monitoring systems, subsidy targeting, and adaptive irrigation schedules.

5. CONCLUSION

Remote sensing data contribute in different but complementary ways when estimating the water requirements of crops in environments that have a desert and semi-desert(s). Consolidating the temporal and spatial nature of the information, we observed that the highest accuracy maps of water requirements for crops came from mult-spectral data while the radar imagery was less impacted by the effect of weather systems and therefore had more cloud-free representation in time. The combined use of remote sensing data allows the option of geographically differentiated irrigation and alike might provide a legitimate option to monitor temporal crop stress. Similar to the goals of agricultural adaptation to climate change in the Mena region, if agricultural policies or agricultural services are to take advantage of remote sensing for irrigation management, they need to a) incorporate remote sensing into their irrigation management policy frameworks, b) establish prototyping platforms for automated irrigation water scheduling, and c) provide stakeholders with resources where satellite based open-access data is used in operational contexts. Looking ahead, agriculture has specific and serious challenges if we are to sustain food production as climate-change driven events increase, therefore developing capacity for using data as control mechanisms for managing variations in their production system must be pursued. Radar and multispectral imagery offer advantages and disadvantages for estimating water requirements, depending on the objectives and characteristics sought. Radar images offer the advantage of being independent of weather conditions, penetrating clouds and rain, and can be captured day and night, providing continuous, weather-independent monitoring. These images can be used to detect variations in soil characteristics such as soil moisture, surface roughness and texture, but we have seen that they can also be used to estimate water requirements. However, we have also seen that they present limitations due to their complexity of use and processing, and we have had to smooth the data according to various smoothing factors in order to obtain values ready for use. Spatial variability also impacts the visualization of needs, displaying heterogeneity within a small area. Multispectral images stand out for their ability to capture information in a wide range of bands of the electromagnetic spectrum, enabling precise detection of plant characteristics, health status and water requirements. They are widely used to estimate water requirements, unlike radar images, where only a few studies have been carried out. On the other hand, unlike radar images, multispectral images are sensitive to climatic vagaries and are affected by clouds, smoke and other atmospheric conditions, which can limit their continuous acquisition. What's more, multispectral images are dependent on solar radiation, and can only be captured during daylight hours. All these characteristics mean that data acquisition can be much less frequent and available than with radar imagery. We obtained interesting results at the temporal scale, where the requirements estimated by the two approaches are different, but come very close to the requirements theoretically provided for citrus. We found that multispectral imagery enabled us to map the spatial distribution of water requirements in a coherent manner, with a clear distribution at plot level, unlike radar imagery, which showed interesting values at temporal scale, but much more variable at spatial scale, due in particular to the specific characteristics of radar images, which are sensitive to variations in soil characteristics.

REFERENCES

Al-Addous, M., Bdour, M., Rabaiah, S., Boubakri, A., Schweimanns, N., Barbana, N., & Wellmann, J. (2024). Innovations in solar-powered desalination: A comprehensive review of sustainable solutions for water scarcity in the Middle East and North Africa (MENA) region. *Water (Basel), 16*(13), 1877. DOI: 10.3390/w16131877

Al-Jawaldeh, A., Nabhani, M., Taktouk, M., & Nasreddine, L. (2022). Climate change and nutrition: Implications for the Eastern Mediterranean Region. *International Journal of Environmental Research and Public Health, 19*(24), 17086. DOI: 10.3390/ijerph192417086

Aleksandrova, M., Lamers, J. P. A., Martius, C., & Tischbein, B. (2014). Rural vulnerability to environmental change in the irrigated lowlands of Central Asia and options for policy-makers: A review. *Environmental Science & Policy, 41*, 77–88. DOI: 10.1016/j.envsci.2014.03.001

Alomari-Mheidat, M., Corell, M., Martín-Palomo, M. J., Castro-Valdecantos, P., Medina-Zurita, N., de Sosa, L. L., & Moriana, A. (2024). Moderate water stress impact on yield components of greenhouse tomatoes in relation to plant water status. *Plants, 13*(1), 128. DOI: 10.3390/plants13010128

Alrteimei, H. A., Ash'aari, Z. H., & Muharram, F. M. (2022). Last decade assessment of the impacts of regional climate change on crop yield variations in the Mediterranean region. *Agriculture, 12*(11), 1787. DOI: 10.3390/agriculture12111787

Cerdà, A., Novara, A., & Moradi, E. (2021). Long-term non-sustainable soil erosion rates and soil compaction in drip-irrigated citrus plantation in Eastern Iberian Peninsula. *The Science of the Total Environment, 787*, 147549. DOI: 10.1016/j.scitotenv.2021.147549

Ezrari, S., Radouane, N., Tahiri, A., El Housni, Z., Mokrini, F., Özer, G., Lazraq, A., Belabess, Z., Amiri, S., & Lahlali, R. (2022). Dry root rot disease, an emerging threat to citrus industry worldwide under climate change: A review. *Physiological and Molecular Plant Pathology, 117*, 101753. DOI: 10.1016/j.pmpp.2021.101753

Filchev, L., & Chanev, M. (2025). Remote Sensing in Climate Change Research. In Dobrinkova, N., & Fidanova, S. (Eds.), *Environmental Protection and Disaster Risks (EnviroRisks 2024). EnviroRISKs 2024. Lecture Notes in Networks and Systems* (Vol. 883). Springer., DOI: 10.1007/978-3-031-74707-6_17

Fuentes-Peñailillo, F., Gutter, K., Vega, R., & Silva, G. C. (2024). Transformative technologies in digital agriculture: Leveraging Internet of Things, remote sensing, and artificial intelligence for smart crop management. *Journal of Sensor and Actuator Networks.*, *13*(4), 39. DOI: 10.3390/jsan13040039

Gebeyhu, B., Dagalo, S., & Muluneh, M. (2024). Soil moisture-based irrigation interval and irrigation performance evaluation: In the case of lower kulfo catchment, Ethiopia. *Heliyon*, *10*(16), e36089. DOI: 10.1016/j.heliyon.2024.e36089

Gerald, A. B., Vibhute, A. D., Aitouhanni, I., Mouniane, Y., El-Khadir, I., Chriqui, A., Mehdioui, S., & Hmouni, D. (2025). Precision agriculture and artificial intelligence: Towards sustainable resource management in the face of climate challenges. In Vibhute, A. D. (Eds.), *Generative adversarial networks for remote sensing* (pp. 227–256). IGI Global., DOI: 10.4018/979-8-3693-6900-5.ch009

Hameed, M., Moradkhani, H., Ahmadalipour, A., Moftakhari, H., Abbaszadeh, P., & Alipour, A. (2019). A review of the 21st century challenges in the food-energy-water security in the Middle East. *Water (Basel)*, *11*(4), 682. DOI: 10.3390/w11040682

Kang, Y., Chen, Z., Li, L., & Zhang, Q. (2023). Construction of multidimensional features to identify tea plantations using multisource remote sensing data: A case study of Hangzhou city, China. *Ecological Informatics*, *77*, 102185. DOI: 10.1016/j.ecoinf.2023.102185

Kang, Y., Khan, S., & Ma, X. (2009). Climate change impacts on crop yield, crop water productivity and food security – A review. *Progress in Natural Science*, *19*(12), 1665–1674. DOI: 10.1016/j.pnsc.2009.08.001

Kassie, M., Shiferaw, B., & Muricho, G. (2011). Agricultural technology, crop income, and poverty alleviation in Uganda. *World Development*, *39*(10), 1784–1795. DOI: 10.1016/j.worlddev.2011.04.023

Lange, M. A. (2019). Impacts of climate change on the Eastern Mediterranean and the Middle East and North Africa region and the water–energy nexus. *Atmosphere (Basel)*, *10*(8), 455. DOI: 10.3390/atmos10080455

Le Mouël, C., Forslund, A., Marty, P., Manceron, S., Marajo-Petitzon, E., Caillaud, M.-A., Dumas, P., & Schmitt, B. (2023). Can the Middle East–North Africa region mitigate the rise of its food import dependency under climate change? *Regional Environmental Change*, *23*(2), 52. DOI: 10.1007/s10113-023-02045-y

Levidow, L., Zaccaria, D., Maia, R., Vivas, E., Todorovic, M., & Scardigno, A. (2014). Improving water-efficient irrigation: Prospects and difficulties of innovative practices. *Agricultural Water Management, 146*, 84–94. DOI: 10.1016/j.agwat.2014.07.012

Li, M., Zhou, S., Shen, S., Wang, J., Yang, Y., Wu, Y., Chen, F., & Lei, Y. (2024). Climate-smart irrigation strategy can mitigate agricultural water consumption while ensuring food security under a changing climate. *Agricultural Water Management, 292*, 108663. DOI: 10.1016/j.agwat.2023.108663

Liu, X., Yin, B., Bao, X., Hou, X., Wang, T., Shang, C., Yang, M., & Zhen, W. (2024). Optimization of irrigation period improves wheat yield by regulating source-sink relationship under water deficit. *European Journal of Agronomy, 156*, 127164. DOI: 10.1016/j.eja.2024.127164

Miltiadou, M., Grieve, S., Ruiz-Benito, P., Astigarraga, J., Cruz-Alonso, V., Tijerín Triviño, J., & Lines, E. R. (2025). PlotToSat: A tool for generating time-series signatures from Sentinel-1 and Sentinel-2 at field-based plots for machine learning applications. *Environmental Modelling & Software, 188*, 106395. DOI: 10.1016/j.envsoft.2025.106395

Mouniane, Y., Chriqui, A., El-Khadir, I., Hbyaj, K., El-Moustaqim, K., Mezouara, M., Zirari, M., Mabrouki, J., & Hmouni, D. (2024). Impact of remote sensing on irrigation management in precision agriculture. In Mabrouki, J., & Mourade, A. (Eds.), *Technical and technological solutions towards a sustainable society and circular economy* (pp. 237–245). Springer., DOI: 10.1007/978-3-031-56292-1_19

Namdar, R., Karami, E., & Keshavarz, M. (2021). Climate change and vulnerability: The case of MENA countries. *ISPRS International Journal of Geo-Information, 10*(11), 794. DOI: 10.3390/ijgi10110794

Nikolaou, G., Neocleous, D., Christou, A., Kitta, E., & Katsoulas, N. (2020). Implementing sustainable irrigation in water-scarce regions under the impact of climate change. *Agronomy (Basel), 10*(8), 1120. DOI: 10.3390/agronomy10081120

Ouaadi, N., Jarlan, L., Khabba, S., Le Page, M., Chakir, A., Er-Raki, S., & Frison, P.-L. (2023). Are the C-band backscattering coefficient and interferometric coherence suitable substitutes of NDVI for the monitoring of the FAO-56 crop coefficient? *Agricultural Water Management, 282*, 108276. DOI: 10.1016/j.agwat.2023.108276

Raihan, A., Ridwan, M., & Rahman, M. S. (2024). An exploration of the latest developments, obstacles, and potential future pathways for climate-smart agriculture. *Climate Smart Agriculture, 1*(2), 100020. DOI: 10.1016/j.csag.2024.100020

Rallo, G., González-Altozano, P., Manzano-Juárez, J., & Provenzano, G. (2017). Using field measurements and FAO-56 model to assess the eco-physiological response of citrus orchards under regulated deficit irrigation. Agricultural Water Management, 180(Part A), 136–147. https://doi.org/DOI: 10.1016/j.agwat.2016.11.011

Salimi, M., & Al-Ghamdi, S. G. (2020). Climate change impacts on critical urban infrastructure and urban resiliency strategies for the Middle East. *Sustainable Cities and Society*, *54*, 101948. DOI: 10.1016/j.scs.2019.101948

Silva, E. R. O., Silva, T. L., Wei, M. C. F., Souza, R. A., & Molin, J. P. (2025). Spatial and temporal variability management for all farmers: A cell-size approach to enhance coffee yields and optimize inputs. *Plants*, *14*(2), 169. DOI: 10.3390/plants14020169

Sofia, G., Zaccone, C., & Tarolli, P. (2024). Agricultural drought severity in NE Italy: Variability, bias, and future scenarios. *International Soil and Water Conservation Research*, *12*(2), 403–418. DOI: 10.1016/j.iswcr.2023.07.003

Tasumi, M., Trezza, R., Allen, R. G., & Wright, J. L. (2005). Operational aspects of satellite-based energy balance models for irrigated crops in the semi-arid U.S. *Irrigation and Drainage Systems*, *19*(4), 355–376. DOI: 10.1007/s10795-005-8138-9

Vermeulen, S. J., Aggarwal, P. K., Ainslie, A., Angelone, C., Campbell, B. M., Challinor, A. J., Hansen, J. W., Ingram, J. S. I., Jarvis, A., Kristjanson, P., Lau, C., Nelson, G. C., Thornton, P. K., & Wollenberg, E. (2012). Options for support to agriculture and food security under climate change. *Environmental Science & Policy*, *15*(1), 136–144. DOI: 10.1016/j.envsci.2011.09.003

Volschenk, T. (2020). Water use and irrigation management of pomegranate trees - a review. *Agricultural Water Management*, *241*, 106375. DOI: 10.1016/j.agwat.2020.106375

Zabihi, H., Vogeler, I., Mat Amin, Z., & Ramezani Gourabi, B. (2016). Mapping the sensitivity of citrus crops to freeze stress using a geographical information system in Ramsar, Iran. *Weather and Climate Extremes*, *14*, 17–23. DOI: 10.1016/j.wace.2016.10.002

Zhang, M., Riaz, M., Xia, H., Li, Y., Wang, X., & Jiang, C. (2022). Four-year biochar study: Positive response of acidic soil microenvironment and citrus growth to biochar under potassium deficiency conditions. *The Science of the Total Environment*, *318*, 152515. DOI: 10.1016/j.scitotenv.2021.152515

Chapter 11
Technoscientific Impact of Agriculture on Farmers' Seed Sovereignty

Ananya Pandey
https://orcid.org/0000-0002-2419-6314
Christ University, Bangalore, India

Achyuta Nanda Mishra
https://orcid.org/0000-0002-4113-0974
Christ University, Bangalore, India

ABSTRACT

The technoscientific transformation in the agriculture sector has significantly altered the traditional seed system that has profoundly affected the rights of the farmers. The evolution of biotechnology including genetically modified crops, gene editing, hybrid seeds protected under the Intellectual Property regime has reshaped the farmers right to seed sovereignty, sharing, saving and exchanging practices. In the Global South, there exist a wide spectrum of conflict between the technological advancement and the rights of farmers. In light of the above perspective, by tracing the historical shift from community based traditional seed knowledge to corporate based seed and patent technologies, this chapter critically analyzes the innovation perspective that undermined the indigenous practices and farmer led initiatives concerning seed sovereignty. Therefore, this chapter calls for a rights based and inclusive approach towards the agricultural innovation that safeguards seed diversity and traditional knowledge.

DOI: 10.4018/979-8-3373-3346-5.ch011

INTRODUCTION

The agricultural sector is considered to be the most important and foundational sector that has continued to evolve with the waves of technological and scientific advancement and intervention. Beginning from the Neolithic revolution, humans transitioned widely from the foraging to the farming perspective and from farming to industrial farming which included the mechanized plows and later which took the shape of the Green Revolution in the mid-20th century. In present times, agriculture is widely shaped by science and innovation technology (Mazoyer & Roudart, 2006). This technological innovation has led to increased productivity, food security and modernization of the farming system. Additionally, there are certain unintended consequences of technological development that include the marginalization of the traditional seed system, erosion of farmers rights, specifically the small holder farmers. In the current scenario, the evolution of the new technoscientific paradigms such as the biotechnological regime, use of the Artificial Intelligence (AI), and precision farming has transformed the agricultural landscape.

The advent of the technoscientific reshaping refers to the integration of scientific research and technological system that process and input the agricultural production of goods, services and products. It also includes the development in the biotechnological sector i.e., the incorporation of the genetically modified crops, CSISPR gene technologies, smart farming technologies such as irrigation system, satellite imaging and AI- supportive tools (Wolfert et al., 2017). The organization and industries that utilize the new technologies such as the blockchain, data analytics, machine learning are capable of overcoming the challenges such as outbreak of the pests, soil and nutrition deficiency and certain climatic variations. Such kind of innovations are often used as tools for empowering farmers, increasing their resilience, cutting down the labor costs and enhancing the sustainable development perspective. The ownership, control and access to these resources and technologies shape the economic, legal and political perspective the further increases the complexities. For instance, the protection of the seed varieties and plants under the intellectual property regime limit the traditional rights and practices of farmers, to save, exchange and reuse the seeds (Kloppenburg, 2010. In a similar way, the incorporation of the AI regime in the agriculture sector raises several critical concerns about the data protection, seed sovereignty and the commodification of the farers rights held back by the private agricultural companies and multinational corporations.

Furthermore, such advent of technological and scientific interventions in the sector, raises critical question involving the issue that does such innovation empowers or marginalizes the farmers (specifically the small holder farmers). The central point of discussion that revolves around this question is based on the aspect of seed sovereignty and innovation. On one hand the supporters of the agricultural innovation and technologies are of the opinion that they are capable of increasing and improving the efficiency, yield of crops along with the adaptation to the changing climatic conditions, on the other side the critics of these technologies highlight the dependency over the inventions and warns about the new forms of the upcoming environmental oversights. In the Global South, the small holder farmers experience the contradictions over the universalistic propagation of the progress. Majority of these farmers face the pressure from the markets, states and corporations to adopt these technologies which is often involuntary. Rather than resolving the issues such as lack of infrastructure, high cost of the usage of these advancements, transparent nature of the commercial seed system enhances the existing inequalities in the sector (Patel, 2013). In light of this, the rights of the farmers and their empowerment needs to be critically explored i.e., not just in terms of the agricultural productivity but through the lens of the seed sovereignty, economic autonomy, preservation of their cultural rights and tradition along with the ecological sustainability. . The global seed systems that are often controlled by the multinational agribusinesses, several institutions and several international treaties and conventions such as the Agreement on the Trade-Related Aspects of Intellectual Property Rights (TRIPS) and the Convention on the International Union for the Protection of New Varieties of Plants (UPOV), gives priority to foster innovation thereby limiting the rights of farmers and transferring the control of seeds, plants and soil data to the corporate entities (Brush, 2005; Shiva, 2016). This has adverse impact not only on the local and indigenous knowledge system but also severely impacts the food security and biodiversity in the environment.

Keeping in view the above, this chapter explores the integration of technological advancements in the agricultural sector that not just reshapes the farmers practicing but also the identity and empowerment of the farmers rights. It also analyzes the intersection of Intellectual Property Rights (IPRs) with the technoscientific intervention that often governs environment issues and socio-economic perspective and justice to the farmers rights. Therefore, in order to address the challenges and issues mentioned above, this chapter firstly discusses the historical trajectory of the agricultural innovation inclined from the traditional seed saving practices towards the technoscientific advancements and interventions. This section of the chapter also analyzes the role of Green Revolution in laying down the ground framework for the incorporation of technology in the agricultural sector. The next section of the chapter analyzes the evolution and involvement of technoscientific interventions in reshaping the agricultural section in terms of genetically modified crops, biotechnology and their challenges and benefits in the present scenario. Furthermore, the chapter widely explores the sovereignty of farmers over the seeds that is not only legal but also economic involving the dignity, autonomy and rights of the farmers. Finally, the chapter concludes by stating that the rights and knowledge of the farmers who are responsible for feeding the world should be protected.

THE HISTORICAL TRAJECTORY OF AGRICULTURAL INNOVATION

The dynamic interplay between the food security, technological advancements with the nature and farmers rights represents the evolution of agricultural systems. The community resilience and sustainability of agriculture depends on the system of traditional knowledge that is deeply grounded in the indigenous environmental practices. For the preservation of the traditional seed systems and the transmission of the traditional knowledge is often conducted orally and embeds back in its cultural and traditional practices. It also prioritizes the preservation of biodiversity and balances the ecology (Altieri, 2004). Historically, the food security and agroecological balance was ensured by the communities through the practice of crop rotation, seed selection that aligns with the climate adaptability and intercropping (Shiva, 1991). With the integration and rapid transformation in the scientific and political imperatives in the agricultural regime, the Green Revolution in the mid-20th century is considered to be a period of significant transition. The Green Revolution involved the support of the Western philanthropic and government institutions that promoted the elimination of poverty and hunger in the Global South (specifically in Asia and Latin America by improvising yield varieties, inducing chemical fertilizers, pesticides and elevating the irrigation infrastructure (Evenson & Gollin, 2003).

Around 10,000 years BC, humans gathered their sustenance from the natural biological diversity in their environment and eventually began to domesticate plants and animals. As domestication progressed, humans began to choose superior plant materials for propagation and animals for breeding—at first without realizing it, but eventually with the aim of cultivating enhanced food crops and livestock (Pingali, 2012). For millennia, farmers have chosen crops based on advantageous characteristics, thereby enhancing the plants' suitability for agriculture (Mazoyer & Raudart, 2006). Traits that were sought after included crop varieties (or cultivars, short for "cultivated varieties") with shorter growing seasons, enhanced resistance to diseases and pests, larger seeds and fruits, improved nutritional content and shelf life, and greater adaptability to the various ecological conditions in which crops were cultivated (Tilman et al., 2002; Joseph & Pandey, 2025a, Joseph & Pandey, 2025b)). Throughout the centuries, agricultural technology has advanced to create a wide range of options for producing food, feed, and fibre. Technology reduces the time we devote to fundamental activities like producing food, making our lives easier and more enjoyable in many respects (Shiva, 1991). Everyone knows how transportation has evolved over time to become more efficient and safer. The agricultural sector has experienced tremendous changes, many of which have increased the efficiency and safety of food and fibre production (Howard, 2009). Throughout history, agriculture has undergone major revolutions that changed farming methods. The "fourth agricultural revolution" or "agriculture 4.0" is already progressing. Every earlier agricultural revolution was considered radical in its time (Lacy, 2006; Fuglie et al., 2011). The initial revolution was the changeover from hunting and gathering to settled farming. The second concerns the agricultural revolution of the 18th century, during which a shift to commercial farming occurred. The third revolution pertains to the rise in productivity after the war due to mechanization and the Green Revolution in developing countries (Shiva, 2016). Every age saw the emergence of important technological progress. With the shift from traditional to modern methods, increased productivity and scalability became possible. Israel transformed agriculture in the modern agricultural world through the invention of drip irrigation. This was just the start of the AgTech revolution.

The inclination of the technology in agriculture has several implications concerning the farming community across the world. The traditional practices that are subtenant to agriculture are devalued in the modern times and displaced by the modern models of expert agronomy. Historically, what emerged from the experimentation and environment feedback by the farmers is not replaced by the prescriptions from the laboratories given by the corporate breeders (Cleveland & Soleri, 2002). The monoculture of the cash crops emerging after the experimentation of the breeders has significantly contributed to the nutritional deficiencies, environment vulnerabilities and loss of the cultural heritage (Patel, 2013). Therefore, the technoscientific reshaping of the agriculture is subject to raise several critical concerns about the control and production of the goods and services for a sustainable future to be cultivated. Where the scientific production allows the enhancement of food security, it also introduces several disturbances in the form of ecological risk to biodiversity, equity concerns and the like (Altieri & Nicholls, 2017). The rapid transformation from the traditional knowledge to the biotechnological phase has aided to deepen the inequalities between the farmers and the corporations.

DIGITAL AND BIOTECHNOLOGICAL TOOLS: THE NEW ERA

Globally the integration of the digital technologies and biotechnological innovations has redefined the food production system. An interesting new realm in farming has been opened up by biotechnology. The new methods offered by biotechnology are comparatively rapid, highly targeted, and efficient in their use of resources. A significant benefit is that a shared set of techniques, such as gene identification and cloning, can be applied widely. At the centre lies the introduction of genetically modified crops and plants and advanced gene editing technologies (CRISPR-Cas9 for example), which has genuinely modulated the opportunities for enhancing the traits. The term "GMO" (genetically modified organism) is now commonly used by consumers and in popular media to refer to foods produced through genetic engineering (Qaim, 2020). However, the precision offered by gene editing, particularly CRISPR, marks a more refined approach, allowing for targeted changes in plant genomes without introducing foreign DNA. This shift is significant as it may mitigate some public and regulatory resistance previously associated with transgenic crops. Gene editing has enabled the development of drought-resistant maize, rice with improved photosynthetic efficiency, and tomatoes with extended shelf life (Zhang et al., 2018). Despite the technological promise, access to and control over these innovations remain unequally distributed, often dominated by multinational agribusinesses that dictate the terms of use through intellectual property rights and patents.

Alongside biotechnological interventions, digital agriculture has emerged as a critical pillar in the technoscientific reshaping of farming. Digital tools—ranging from Internet of Things (IoT) sensors to artificial intelligence (AI) platforms—are enabling precision agriculture practices that optimize input use, improve yields, and reduce environmental impact. For example, AI-driven platforms can analyze satellite imagery and real-time field data to forecast crop performance and detect pests or diseases early (Kamilaris et al., 2018). IoT-enabled sensors provide granular insights into soil moisture, nutrient levels, and weather conditions, allowing for timely interventions. Furthermore, big data analytics aggregates information across farms, regions, and seasons to identify patterns and inform decision-making. This data-intensive model holds potential for climate-resilient agriculture by facilitating adaptive strategies that respond to shifting environmental conditions. Yet, its success depends on infrastructure availability, digital literacy, and data accessibility, which are often lacking in many rural farming communities, particularly in the Global South.

Despite the promise of digital and biotechnological tools, their proprietary nature raises significant concerns about ownership, licensing, and accessibility. Most biotechnological innovations, including genetically engineered seeds and CRISPR-edited crops, are protected under intellectual property regimes such as patents or plant variety protection (PVP). These legal frameworks grant exclusive rights to developers—typically large multinational corporations—over the use, reproduction, and sale of innovations (Howard, 2015). Consequently, farmers are often prohibited from saving or replanting patented seeds, thereby eroding traditional practices of seed sovereignty. Licensing agreements can be restrictive, legally binding farmers to annual seed purchases and compliance with rigid technology-use guidelines. For instance, Monsanto's "Technology Use Agreement" forbade the saving of genetically modified seeds, with violations leading to litigation and financial penalties for farmers (Kloppenburg, 2010). These dynamics create a dependency loop, where farmers increasingly rely on corporate-controlled technologies for continued productivity.

In the digital domain, similar patterns of proprietary control are evident. Many digital agriculture platforms operate as closed systems, where data generated by farmers is collected, stored, and analyzed by private companies with limited transparency or accountability. Questions of data ownership, consent, and benefit-sharing are often unresolved, leading to concerns over digital colonization in agriculture (Carbonell, 2016). Farmers may unknowingly cede rights to their farm data, which is then monetized by corporations for commercial gain. This asymmetry undermines farmers' autonomy and can lead to exclusion from the value generated through data analytics. Additionally, the entry barriers to such technologies—including high costs, language incompatibility, and the need for constant connectivity—further marginalize smallholder farmers, reinforcing structural inequalities in agriculture.

Moreover, these technologies are often developed without adequate participation from farmers or consideration of local ecological knowledge. The top-down nature of innovation dissemination risks imposing solutions that may be technologically sound but socio-culturally inappropriate. For instance, precision agriculture technologies optimized for large-scale monocultures in developed countries may not translate effectively into the diversified, resource-constrained farming systems of the Global South. The exclusion of local actors from the innovation ecosystem not only undermines the sustainability of these interventions but also perpetuates a technocratic model of agricultural development that prioritizes efficiency over equity (Sumberg et al., 2021).

While public-sector research institutions and open-source initiatives have begun to challenge corporate dominance in agricultural innovation, they remain underfunded and fragmented. Organizations like the CGIAR are developing gene-edited and digital solutions tailored to smallholder needs, yet their global impact is often limited by regulatory barriers and intellectual property entanglements (Cohen & Paarlberg, 2004). The growing movement for open-access biotechnology and farmer-led innovation offers a promising counter-narrative, emphasizing collaborative knowledge production, transparency, and rights-based approaches. These include community seed banks, participatory breeding programs, and digital platforms co-designed with farmers to ensure relevance and inclusivity.

FARMERS' SEED SOVEREIGNTY: CONCEPT AND CONTEMPORARY CHALLENGES

Farmers' seed sovereignty is an evolving concept that encapsulates the rights, autonomy, and control that farmers, especially smallholders and indigenous cultivators, possess over their agricultural practices, resources, and decision-making processes. It extends beyond food sovereignty by embedding the farmer at the core of agricultural policy, innovation, and cultural sustainability. The term is often used to denote the degree of economic independence, ecological stewardship, and cultural continuity that farmers exercise over their land, seeds, knowledge systems, and livelihoods. Economically, seed sovereignty implies the farmer's ability to maintain sustainable income, access markets fairly, and choose the means of production without being subordinated to corporate or state control (Patel, 2009). Culturally, it involves the preservation of traditional farming knowledge, rituals, seed exchange practices, and the intergenerational transfer of indigenous wisdom. Ecologically, seed sovereignty includes the right to steward and regenerate natural resources, preserve agrobiodiversity, and practice environmentally sustainable agriculture aligned with local conditions and ecosystems (Altieri, 2004).

However, the advancement of technoscientific innovations—particularly genetically modified organisms, digital agriculture, and data-driven decision-making tools—has introduced complex tensions between innovation and farmers' autonomy. While these technologies promise improved productivity, climate resilience, and resource efficiency, they often come embedded with corporate control mechanisms that limit the farmer's agency. For example, genetically modified seeds are typically patented and sold under licensing agreements that prohibit seed saving and reuse, effectively making farmers dependent on annual purchases from multinational corporations (Kloppenburg, 2010). This undermines the traditional practice of seed sovereignty, whereby farmers produce, save, and exchange seeds adapted to local environmental conditions. Moreover, digital platforms used for precision agriculture frequently operate under proprietary models where data collected from farms is owned by companies, not the farmers themselves. This data asymmetry leads to an erosion of informational seed sovereignty, where farmers lose control over crucial knowledge regarding soil, climate, and crop performance (Carbonell, 2016).

The imposition of standardized technological packages has also created a disconnect between scientific knowledge and local agrarian realities. Innovations developed in laboratories often overlook the socio-ecological diversity of rural farming systems. As a result, traditional farming systems rooted in agroecological principles and local cultural identities are either marginalized or actively displaced. For instance, in many parts of the Global South, the introduction of high-yield varieties during the Green Revolution led to the abandonment of indigenous seeds that were not only more resilient to local conditions but also central to the communities' cultural identity (Shiva, 2016). The loss of such seeds represents not just genetic erosion but also the disintegration of cultural memory, rituals, and ecological relationships that have been maintained through generations. These seeds are carriers of community heritage and resilience strategies—attributes that modern agricultural technologies rarely account for.

Furthermore, farmers' seed sovereignty is constrained by the global regime of intellectual property rights, particularly the Trade-Related Aspects of Intellectual Property Rights (TRIPS) agreement and the International Union for the Protection of New Varieties of Plants (UPOV) conventions. These legal frameworks prioritize breeders' and corporations' rights over those of farmers and have been widely criticized for imposing Eurocentric standards of innovation that fail to recognize collective, informal, and non-commercial forms of seed innovation (De Schutter, 2009). Such legal regimes transform seeds—once a shared and sacred commons—into privatized commodities, thereby restricting the farmer's autonomy to engage in traditional practices like seed saving and sharing. Consequently, farmers are increasingly subjected to legal risks and economic dependency, undermining the very basis of their seed sovereignty.

Moreover, this technoscientific restructuring exacerbates existing inequalities within rural communities. Wealthier farmers with access to credit and literacy in digital technologies are more likely to benefit from innovations, while smallholder and marginalized farmers are often left behind. This technological stratification leads to a further concentration of power and resources, reducing the collective bargaining capacity of the agrarian community. Women farmers, who play a vital role in seed preservation and ecological stewardship, are especially vulnerable, as they are often excluded from technology dissemination channels and decision-making structures (Bezner Kerr, 2010). As such, the reshaping of agriculture through technology tends to reinforce patriarchal and socio-economic hierarchies, diluting the democratic and inclusive character of farmers' seed sovereignty.

Addressing these challenges requires a reconfiguration of innovation systems to recognize and prioritize the needs, rights, and knowledge of farmers. Farmer-led innovation, participatory plant breeding, open-source seed systems, and community-controlled data initiatives are critical pathways toward restoring seed sovereignty. Policy frameworks must shift from a top-down, corporate-led model to a pluralistic and rights-based approach that upholds the cultural and ecological dimensions of farming (Pandey & Mishra, 2025a). The Food and Agriculture Organization (FAO) and the United Nations Declaration on the Rights of Peasants (UNDROP) advocate for the recognition of farmers' rights to seeds, land, and traditional knowledge, affirming that seed sovereignty is inseparable from justice and sustainability (FAO, 2018; Pandey & Mishra, 2025b)). Ultimately, a technoscientific future that empowers rather than displaces farmers must be grounded in ethical reflexivity, legal pluralism, and a deep respect for agrarian cultures and ecosystems.

REGULATORY PERSPECTIVE OF AGRICULTURAL TECHNOLOGIES

The political economy of agricultural technologies reveals a power-laden landscape where corporate actors dominate critical inputs such as seeds, agrochemicals, and increasingly, farm data. This consolidation of control reshapes agricultural practices and farmer autonomy, raising significant concerns about equity, access, and sustainability. At the core of this transformation lies the growing influence of multinational agribusiness corporations—such as Bayer-Monsanto, Corteva, and Syngenta—which have gained monopolistic control over the seed and pesticide markets through vertical integration and intellectual property rights (IPRs). These firms leverage patents, plant breeders' rights, and licensing arrangements to restrict seed saving, exchange, and reuse—practices that have traditionally sustained farming communities across generations (Kloppenburg, 2010; Howard, 2015). The introduction of genetically modified (GM) seeds and hybrid varieties, protected under international regimes like the Agreement on Trade-Related Aspects of Intellectual Property Rights (TRIPS) and the International Union for the Protection of New Varieties of Plants (UPOV), has effectively privatized the genetic commons, thereby subordinating farmers' rights to corporate interests (Shiva, 2016).

TRIPS, administered by the World Trade Organization, obligates member states to adopt IPR regimes that protect plant varieties either by patents or by an effective sui generis system, or a combination of both (WTO, 1994, Art. 27.3(b)). Many countries, particularly in the Global South, have implemented UPOV-compliant plant variety protection laws that align with the 1991 Act of the UPOV Convention. This act sharply curtails farmers' traditional practices of saving, exchanging, or selling farm-saved seeds, especially for commercial purposes (Gibson, 2009). India's Protection of Plant Varieties and Farmers' Rights Act (PPV&FR), 2001, is a notable exception, as it attempts to balance breeders' and farmers' rights by legally recognizing the latter's right to save, use, sow, resow, exchange, and even sell farm produce, provided it is not branded as a registered variety (Prasad & Reddy, 2020). However, in practice, legal ambiguities and enforcement challenges often result in de facto corporate control over seed systems, undermining the spirit of such progressive legislation (Rangnekar, 2004).

The ramifications of these IPR regimes are evident in landmark legal disputes. The case of *Monsanto Canada Inc. v. Schmeiser* (2004) epitomizes the tension between farmers' autonomy and corporate patent enforcement. Percy Schmeiser, a Canadian farmer, was sued by Monsanto for allegedly planting genetically modified canola without a license, even though he claimed the seeds blew onto his fields without his knowledge. The Canadian Supreme Court ruled in favor of Monsanto, emphasizing that possession of patented genes—even if unintentional—constitutes infringement (Monsanto Canada Inc. v. Schmeiser, [2004] 1 S.C.R. 902). This precedent has had global repercussions, reinforced the legal muscle of biotech firms and placing an onerous burden on farmers to monitor and regulate the genetic content of their crops.

In India, the commodification of seeds and the aggressive marketing of Bt cotton—a genetically modified pest-resistant variety—has been linked to a crisis in agrarian livelihoods. Despite initial promises of higher yields and reduced pesticide use, farmers faced high input costs, fluctuating yields, and mounting debt (Stone, 2011). The loss of traditional seed diversity, combined with inadequate access to institutional credit and insurance, contributed to widespread distress. Numerous studies and advocacy groups have connected these structural factors to the tragic phenomenon of farmer suicides, particularly in Maharashtra, Andhra Pradesh, and Telangana (Sainath, 2013; Gruère et al., 2008). While technology alone cannot be blamed, the broader political economy—including restrictive IPRs, market volatility, and corporate monopolies—has deepened vulnerabilities for smallholder farmers.

Beyond seeds, the digitalization of agriculture is introducing new arenas of control. Smart farming tools—such as precision agriculture systems, remote sensors, and AI-driven analytics—collect massive volumes of farm-level data, often stored and owned by agritech firms rather than farmers (Carbonell, 2016). This raises profound questions about data sovereignty, transparency, and informed consent. As agricultural value chains become increasingly data-driven, the asymmetry of information and power between technology providers and farmers is widening, potentially replicating the same dependency structures seen in the seed industry. Scholars warn of an emerging "digital enclosure" in agriculture, where farmers are not just consumers of proprietary seeds or inputs but become data sources for platform-based agribusiness models (Rotz et al., 2019). Without robust data governance frameworks and farmer-led digital infrastructure, this could lead to new forms of techno-feudalism in agriculture.

TECHNOSCIENTIFIC TOOLS IN AGRICULTURE: A BANE OR BOON?

The rapid infusion of technoscientific advancements into agriculture presents a complex interplay between empowerment and dependency for farmers. On one hand, technologies such as genetically modified crops, precision farming, AI, and remote sensing tools have significantly enhanced productivity, improved pest and nutrient management, and facilitated real-time decision-making. On the other hand, these advancements have entrenched new forms of dependency—on proprietary knowledge systems, digital platforms, and external inputs—thereby challenging farmers' autonomy and long-term sustainability. This dual nature raises a critical question: as to consider that do the technoscientific revolution in agriculture liberate farmers by enhancing their capabilities, or does it deepen systemic vulnerabilities and reduce sovereignty.

An Empowerment Narrative

At the core of the empowerment narrative is the undeniable improvement in yield efficiency and resource optimization. Technologies such as AI-powered irrigation systems, satellite-based crop monitoring, and automated machinery have revolutionized traditional farming by allowing farmers to make data-informed decisions. These systems reduce input costs, minimize environmental damage, and increase profitability. For instance, precision agriculture allows site-specific application of fertilizers and pesticides, leading to higher efficiency and reduced ecological footprints (Wolfert et al., 2017). Similarly, GM crops engineered for resistance to pests or tolerance to drought have enabled higher productivity in regions facing climatic stress (Qaim, 2020). Such tools are often presented as critical to feeding a growing global population and combating climate change-induced agricultural risks. Despite these concerns, there are growing instances where technoscientific tools have genuinely empowered farmers, especially through inclusive and open models. Smart farming cooperatives—such as digital farmer producer organizations (FPOs) in India—pool resources to access technologies collectively, negotiate better market prices, and enhance bargaining power (Mittal & Mehar, 2016). Open-source platforms like FarmOS or OpenAg promote transparency, data ownership, and adaptability by providing farmers with free, customizable tools for monitoring and managing farm operations (Pandey, 2025a). These initiatives shift the narrative from dependency to participatory innovation, enabling farmers to tailor technology to local needs and contexts. Agroecological innovations also present a compelling model of empowerment by integrating traditional knowledge with scientific research in a way that enhances ecological resilience and community sovereignty (Pandey & Joseph, 2025). Practices such as intercropping, seed saving, and community seed banks empower farmers to resist corporate seed monopolies and maintain biodiversity (Altieri & Nicholls, 2020). By emphasizing ecological balance and local autonomy over industrial inputs, agroecology challenges the notion that modern science and technology must always be proprietary and top-down.

A Dependency Perspective

However, this technological progress comes with a growing reliance on external actors—corporations, digital platforms, and biotech firms—that control access to these innovations. Farmers often face restricted access to seeds, data, and knowledge that are protected by intellectual property rights (IPRs), licensing agreements, and proprietary algorithms. Companies like Monsanto (now part of Bayer) have aggressively pursued patent protections on GM seeds, leading to legal battles over seed saving and replanting practices, as seen in the landmark case of *Monsanto Canada Inc. v. Schmeiser* (2004). Such legal frameworks limit farmers' control over their own means of production, turning them into consumers of corporate technologies rather than stewards of agricultural knowledge.

Moreover, the digitization of agriculture introduces a new layer of dependency—data colonialism. Many digital platforms collect vast amounts of data from farmers but do not always ensure equitable access to or control over that data. While such platforms promise personalized recommendations and optimization, they also raise concerns about transparency, privacy, and monopolization of agricultural intelligence (Carbonell, 2016). This creates an asymmetrical relationship where farmers generate value through data but lack negotiating power or ownership over its use, reinforcing precarity rather than empowerment. This technoscientific dependency is particularly acute in the Global South, where infrastructural disparities, financial barriers, and digital illiteracy further marginalize smallholder farmers. For example, subscription-based agricultural services or expensive patented seeds are often out of reach for many rural farmers in Africa and South Asia (Fuglie et al., 2019). The push for high-tech solutions, without adequate regulatory support or capacity-building, risks exacerbating existing inequalities and creating a "technological treadmill," where farmers must continuously invest in new technologies just to maintain competitiveness (Stone, 2011).

The technoscientific reshaping of agriculture is neither inherently liberating nor wholly exploitative—it is a landscape of contradictions shaped by who owns, governs, and distributes technological resources. Empowerment is achievable when farmers are placed at the center of innovation processes, when technologies are adapted to local contexts, and when legal and institutional frameworks protect their rights to knowledge, data, and seeds (Pandey, 2025b). Conversely, unchecked corporate control, restrictive IPRs, and data asymmetries risk transforming farmers into dependent actors in a high-tech agrarian ecosystem. A just and sustainable future for agriculture must therefore strike a balance—leveraging the potential of science while safeguarding the seed sovereignty of those who feed the world.

TRENDS IN AGRO-BIOTECHNOLOGY: A GLOBAL POLICY APPROACH

Global policy responses to the technoscientific reshaping of agriculture increasingly reflect the need to reconcile innovation with sustainability, equity, and farmers' sovereignty over the seeds. International organizations such as the Food and Agriculture Organization (FAO), the World Intellectual Property Organization (WIPO), and frameworks like the United Nations Declaration of Indigenous Peoples (UNDRIP) play a pivotal role in guiding this transition. The FAO has been instrumental in promoting agroecological practices as a counterbalance to industrial agriculture, emphasizing the importance of biodiversity, local knowledge systems, and participatory approaches. Its "Scaling up Agroecology Initiative" presents a framework for integrating ecological principles with food system transformation, aiming to empower smallholders and indigenous communities through localized, knowledge-intensive methods rather than input-dependent technologies (FAO, 2018). In parallel, WIPO's work on traditional knowledge seeks to protect farmers' and indigenous communities' intellectual contributions from misappropriation, advancing mechanisms such as sui generis systems to accommodate community-based innovations outside conventional patent frameworks (WIPO, 2020). These efforts align with UNDRIP, which affirms indigenous peoples' rights to maintain, control, and develop their cultural heritage, seeds, traditional knowledge, and agricultural practices (United Nations, 2007). Such global frameworks collectively signal a shift toward more inclusive governance in agricultural innovation.

At the heart of this shift is the global food sovereignty movement, which critiques the dominant neoliberal agricultural paradigm and calls for the democratization of food systems. Coined by the international peasant movement La Via Campesina in 1996, food sovereignty asserts the right of people to define their own food and agriculture systems, prioritize local production, and protect natural resources (La Via Campesina, 2007). This paradigm resists the corporate concentration of agri-food chains and challenges the commodification of seeds and land. It promotes agroecology not only as a set of ecological practices but also as a political and cultural framework that restores agency to farmers and local communities (Pandey & Mishra, 2025). Agroecology, in this context, becomes both a scientific and social movement, advancing alternatives that counter the dependency created by proprietary technologies, such as genetically modified organisms and synthetic inputs. Numerous empirical studies underscore how agroecological systems increase resilience, improve biodiversity, and offer socio-economic benefits for marginalized farming communities (Altieri & Nicholls, 2017; IPES-Food, 2016). Furthermore, agroecology facilitates farmer-to-farmer knowledge sharing, enhancing horizontal learning and community solidarity, which are often disrupted in top-down techno-scientific interventions.

Participatory technology development (PTD) and community seed banks represent grounded alternatives to the linear model of innovation in agriculture. PTD challenges the notion of farmers as passive recipients of scientific knowledge and repositions them as co-creators of technologies. This approach involves iterative collaboration between farmers, researchers, and extension agents to adapt and develop context-specific innovations that align with local needs and values (Chambers et al., 1989). PTD has proven effective in fostering trust, ensuring relevance, and reinforcing the cultural dimensions of agricultural practices. Similarly, community seed banks have emerged as critical tools for preserving agrobiodiversity, especially in the face of genetic erosion caused by commercial seed markets. These banks operate on principles of reciprocity, conservation, and local control, offering an alternative to seed monopolies and patent restrictions (Vernooy et al., 2015). Supported by national and international networks, they contribute to seed sovereignty by enabling farmers to save, exchange, and improve local varieties without dependency on external suppliers. Countries like Nepal, Ethiopia, and India have demonstrated how seed banks—coupled with legal recognition and institutional support—can enhance both resilience and innovation in rural settings.

Together, these policy responses and grassroots alternatives illustrate a growing consensus on the need for more equitable and sustainable agricultural futures. The convergence of global rights-based instruments and local agroecological practices signals a move beyond the binary of tradition versus technology. Instead, it recognizes that agricultural innovation must be co-designed, democratically governed, and ecologically grounded to be genuinely transformative. While international organizations must continue to reform global rules—particularly in intellectual property, trade, and investment—to favor public over private interest, local initiatives must be supported to ensure the agency and rights of farming communities remain central to agricultural development. By integrating global policy frameworks with bottom-up innovations like PTD and community seed banks, a pluralistic model of agricultural transformation becomes not only possible but necessary in navigating the challenges of food security, climate change, and social justice.

CONCLUSION AND RECOMMENDATIONS

The technoscientific transformation of agriculture stands at the crossroads of progress and precarity. This chapter has explored how innovations such as genetically modified organisms (GMOs), precision agriculture, artificial intelligence, and proprietary agri-biotechnologies are redefining the agrarian landscape. While these developments have brought significant improvements in yield optimization, pest control, and climate adaptability, they have also raised urgent questions regarding farmers' seed sovereignty, autonomy, and long-term sustainability.

One of the key arguments advanced in this chapter is that the benefits of techno-scientific progress are not uniformly distributed. Instead, they are often mediated by complex networks of corporate ownership, intellectual property regimes, and data monopolies that disproportionately favor large agribusinesses over smallholder and indigenous farmers. The imposition of seed patents, licensing models, and digital agriculture platforms has led to a shift from open, community-based systems of knowledge exchange to a model of closed, proprietary control. This undermines farmers' historical rights to save, exchange, and improve seeds—practices that are essential not just for agricultural resilience but also for cultural and ecological continuity.

Moreover, the discourse around agricultural innovation has often been framed in a technocentric manner, sidelining the socio-cultural and ethical dimensions of farming. Farmers are increasingly positioned as passive recipients of technology rather than active co-creators. This not only diminishes their agency but also fosters a dependency on external inputs, knowledge systems, and capital-intensive technologies. The digital divide further exacerbates this dependency, particularly in the Global South, where infrastructure gaps, low digital literacy, and unequal access to financing limit the uptake of these innovations.

Paradoxically, the progress of such advancement lies in the dual approach of the technology i.e., the empowerment of the farmers on one side and marginalization of the farmers on the other end. Where some of the tools and techniques offer a progressive means of combating poverty and hunger, some of them are supposed to only govern the market logics and elites that push the farmers economic cycle in debt, dependency and dispossession The pluralistic, human rights-based approach towards the innovation in agriculture must be recognized. Farmers should be awarded the status of owners on their own field data rather than being only considered as beneficiaries. This shall ensure that farmers are counted under the broader domain of scientific knowledge and research innovations. The incorporation of the effective sui-generis system under the intellectual property protection related to farmers rights should be diversified. Therefore, the future of agriculture is supposed to be crafted by the inclusive and balanced policy approach that does not come at the cost of farmers rights.

Policy Action And Recommendations for a Balanced Technoscientific Future

As there is witnessed the acceleration in the growth of the usage of technoscientific approach that safeguards the rights of farmers that aligns with the fostering of innovation, it is essential to adopt the balanced approach. The rights of the farmers such as access to seeds, its usage and exchange is subject to unequal dynamics that is protected by the intellectual property regime under the broader realm of ITPGRFA, UPOV Convention and the TRIPS Agreement (FAO, 2009; UNGA, 2018). Exemptions from the national law initiatives should be granted to the farmers that enhance global food security and farmers seed sovereignty (De Schutter, 2011). Furthermore, in order to overcome the monopolization of seeds by the breeders and corporate entities incentives should be in direct reach of the farmers led research initiatives. The local relevance to the open-source seed sharing platforms must ensure agroecological knowledge exchange and participatory plant breeding. Community focused initiatives such as Open-Source Seed System and the open access community seed banks must uphold the rights of farmers and back up the resilience (Kloppenburg, 2010). The infrastructural and financial support system that is directed towards the institutions that prioritize the rights of farmers, enhance sustainability and preserve the traditional knowledge of the indigenous people are required to be strengthened. In the Global South, the focus on the inclusive research paradigms that focus on the productivity measures rather than the focusing on the productivity driven model (Chambers, 1994). Such kind of models not only produce the inclusive measures but also establishes the trust between the scientific and the local community.

In the present time, there is a dire need to strengthen the legal, regulatory and policy framework concerning the access to technology and data in the agricultural sector. With the integration of AI and digital tools the legal architecture with the equitable access to data and technology is should be protected from being excluded. Along with this step transparency in data collection, licensing practices and benefit sharing should be mechanized (Pandey, 2026). For instance, the data gathered, integrated and used in any forms should be disclosed to the farmers as a means of securing their ownership and co-ownership over the shared data (Carbonell, 2016). The formulation of the policies and mechanism for regulating the price of the developed and protected varieties in the agricultural markets should be encouraged that also protects small holder farmers from any exploitive practices (Howard, 2016). In addition to these legal reforms, the capacity building programs such as educating and making aware the farmers of their technological capacities is essential. This should also be accompanied with the assurance that farmers have the skills to access and utilize the innovative technologies effectively. Also, the integration of the national agricultural extension that serves as a two-way knowledge exchange platform (where traditional knowledge along with the scientific expertise can be shared at the same time). These inclusive efforts should also be backed by the international cooperation with special focus laid on the marginalized and vulnerable section of farmers. A balanced technoscientific approach including the principles of equity, sustainability and justice should be incorporated within the agricultural sector. The inclusive technoscientific approach should be moved forward as a holistic vision where all the aspects such as culture, ecology and social justice are embedded. Furthermore, the alignment with the Sustainable Development Goals i.e., Goal 2 (end hunger and poverty) and Goal 12 (responsible production and consumption) provides an inclusive framework for incorporating such kind of approach (UN, 2015). Therefore, in order to ensuring fair protection to the farmers and protecting their rights is imperative to establish a sustainable and equitable food security system.

REFERENCES

Altieri, M. A. (2004). *Agroecology: The science of sustainable agriculture*. Westview Press.

Altieri, M. A., & Nicholls, C. I. (2017). Agroecology: A brief account of its origins and currents of thought in Latin America. *Agroecology and Sustainable Food Systems*, *41*(3-4), 231–237. DOI: 10.1080/21683565.2017.1287147

Altieri, M. A., & Nicholls, C. I. (2017). The adaptation and mitigation potential of traditional agriculture in a changing climate. *Climatic Change*, *140*(1), 33–45. DOI: 10.1007/s10584-013-0909-y

Bezner Kerr, R. (2010). Reconceptualizing food sovereignty: Peasant women, agroecology and feminist politics. *Development*, *53*(1), 112–117. DOI: 10.1057/dev.2009.47

Carbonell, I. M. (2016). The ethics of big data in agriculture. *Internet Policy Review*, *5*(1), 1–13. DOI: 10.14763/2016.1.405

Chambers, R., Pacey, A., & Thrupp, L. A. (1989). *Farmer First: Farmer Innovation and Agricultural Research*. Intermediate Technology Publications. DOI: 10.3362/9781780440149

Cleveland, D. A., & Soleri, D. (2002). *Farmers, scientists and plant breeding: Integrating knowledge and practice*. CABI Publishing. DOI: 10.1079/9780851995854.0000

Cohen, J. I., & Paarlberg, R. (2004). Unlocking crop biotechnology in developing countries—A report from the field. *World Development*, *32*(9), 1563–1577. DOI: 10.1016/j.worlddev.2004.05.003

De Schutter, O. (2009). Seed policies and the right to food: Enhancing agrobiodiversity and encouraging innovation. Report of the UN Special Rapporteur on the Right to Food, A/64/170.

Evenson, R. E., & Gollin, D. (2003). Assessing the impact of the Green Revolution, 1960 to 2000. *Science*, *300*(5620), 758–762. DOI: 10.1126/science.1078710 PMID: 12730592

FAO. (2018). *The right to food and the rights of peasants and other people working in rural areas*. Food and Agriculture Organization of the United Nations.

FAO. (2018). *Scaling up agroecology to achieve the sustainable development goals: Summary for policy makers*. Food and Agriculture Organization of the United Nations. https://www.fao.org/documents/card/en/c/I9049EN

Fuglie, K., Gautam, M., Goyal, A., & Maloney, W. F. (2019). *Harvesting Prosperity: Technology and Productivity Growth in Agriculture*. World Bank Group.

Fuglie, K. O., Heisey, P. W., King, J. L., & Schimmelpfennig, D. (2011). *Research investments and market structure in the food processing, agricultural input, and biofuel industries worldwide*. USDA Economic Research Service. DOI: 10.2139/ssrn.2027051

Gibson, J. (2009). *Intellectual property, traditional knowledge and the UN*. Cambridge University Press.

Gruère, G. P., Mehta-Bhatt, P., & Sengupta, D. (2008). Bt cotton and farmer suicides in India: Reviewing the evidence. *IFPRI Discussion Paper No. 00808*.

Howard, P. H. (2009). Visualizing consolidation in the global seed industry: 1996–2008. *Sustainability (Basel)*, *1*(4), 1266–1287. DOI: 10.3390/su1041266

Howard, P. H. (2015). Intellectual property and consolidation in the seed industry. *Crop Science*, *55*(6), 2489–2495. DOI: 10.2135/cropsci2014.09.0669

Joseph, J., & Pandey, A. (2025a). Green Industrialization for the Sustainable Future of Developing Countries: A Critical Study. In *Adaptive Strategies for Green Economy and Sustainability Policies* (pp. 227-244). IGI Global Scientific Publishing.

Joseph, J., & Pandey, A. (2025b). Gender Equality: A Catalyst for Global Sustainable Future. In *Social System Reforms to Achieve Global Sustainability* (pp. 101-126). IGI Global Scientific Publishing. Brush, S. B. (2005). *Farmers' Bounty: Locating Crop Diversity in the Contemporary World*. Yale University Press. DOI: 10.4018/979-8-3373-1280-4.ch005

Kamilaris, A., Kartakoullis, A., & Prenafeta-Boldú, F. X. (2018). A review on the practice of big data analysis in agriculture. *Computers and Electronics in Agriculture*, *143*, 23–37. DOI: 10.1016/j.compag.2017.09.037

Kloppenburg, J. (2010). Impeding dispossession, enabling repossession: Biological open source and the recovery of seed sovereignty. *Journal of Agrarian Change*, *10*(3), 367–388. DOI: 10.1111/j.1471-0366.2010.00275.x

Kloppenburg, J. (2010). Impeding dispossession, enabling repossession: Biological open source and the recovery of seed sovereignty. *Journal of Agrarian Change*, *10*(3), 367–388. DOI: 10.1111/j.1471-0366.2010.00275.x

Lacy, W. B. (2006). Agricultural biotechnology, societal implications, and the need for public dialogue. *Agriculture and Human Values*, *23*(3), 273–280.

Mazoyer, M., & Roudart, L. (2006). *A History of World Agriculture: From the Neolithic Age to the Current Crisis*. Monthly Review Press.

Mittal, S., & Mehar, M. (2016). Socio-economic factors affecting adoption of modern information and communication technology by farmers in India: Analysis using multivariate probit model. *Journal of Agricultural Education and Extension, 22*(2), 199–212. DOI: 10.1080/1389224X.2014.997255

Monsanto Canada Inc. v. Schmeiser, [2004] 1 S.C.R. 902

Pandey, A. (2025a). The Transformative Potential of Artificial Intelligence in Building Sustainable Agricultural Systems: Innovations, Challenges, and Future Directions. In *AI and Ecological Change for Sustainable Development* (pp. 161-182). IGI Global Scientific Publishing.

Pandey, A. (2025b). Agroecological Transformation Through Farmers' Empowerment and IP Reform: A Human Rights-Based Approach. In *Social System Reforms to Achieve Global Sustainability* (pp. 1-22). IGI Global Scientific Publishing.

Pandey, A. (2026). Technological Innovation and Intellectual Property in Agriculture: A Critical Study With Reference to Farmers' Rights. In *Precision and Intelligence in Agriculture: Advanced Technologies for Sustainable Farming* (pp. 29-50). IGI Global Scientific Publishing.

Pandey, A., & Joseph, J. (2025). Leveraging Artificial Intelligence for Sustainable Development in Agriculture: An Exploratory Study. In *Artificial Intelligence and Data Science for Sustainability: Applications and Methods* (pp. 187-212). IGI Global Scientific Publishing.

Pandey, A., & Mishra, A. (2025). Farmers' Rights in the Age of Agricultural Intellectual Property: From Soil to Statute. In *Geostatistical Insights on Mapping Flood Hazards and Wetland Dynamics* (pp. 123-150). IGI Global Scientific Publishing.

Pandey, A., & Mishra, A. (2025a). Intellectual Property and Human Rights of Farmers: Striking a Balance in the Era of Globalization. In *Globalization, Global Security, and New International Realities for Modern Democracies* (pp. 221-246). IGI Global Scientific Publishing.

Pandey, A., & Mishra, A. (2025b). Conflict and Coexistence of Human Rights: An Exploratory Study with Reference to Intellectual Property Rights. *Journal of Human Rights and Social Work, 10*(1), 1–12. DOI: 10.1007/s41134-024-00361-9

Patel, R. (2009). Food sovereignty. *The Journal of Peasant Studies, 36*(3), 663–706. DOI: 10.1080/03066150903143079

Patel, R. (2013). *Stuffed and Starved: The Hidden Battle for the World Food System*. Melville House.

Pingali, P. (2012). Green Revolution: Impacts, limits, and the path ahead. *Proceedings of the National Academy of Sciences of the United States of America, 109*(31), 12302–12308. DOI: 10.1073/pnas.0912953109 PMID: 22826253

Prasad, R. S., & Reddy, D. N. (2020). Farmers' rights in India: A policy perspective. *Journal of Intellectual Property Rights, 25*(4), 152–158.

Qaim, M. (2020). Role of new plant breeding technologies for food security and sustainable agricultural development. *Applied Economic Perspectives and Policy, 42*(2), 129–150. DOI: 10.1002/aepp.13044

Rangnekar, D. (2004). The socio-economics of geographical indications: A review of empirical evidence from Europe. *UNCTAD-ICTSD Project on IPRs and Sustainable Development.*

Rotz, S., & Duncan, E.. (2019). The politics of digital agricultural platforms: Markets, institutions and technology. *Canadian Food Studies, 6*(3), 48–86.

Sainath, P. (2013). *Everybody loves a good drought: Stories from India's poorest districts*. Penguin Books.

Shiva, V. (1991). *The violence of the Green Revolution: Third World agriculture, ecology, and politics*. Zed Books.

Shiva, V. (2016). *Who Really Feeds the World? The Failures of Agribusiness and the Promise of Agroecology*. North Atlantic Books.

Stone, G. D. (2011). Field versus farm in Warangal: Bt cotton, higher yields, and larger questions. *World Development, 39*(3), 387–398. DOI: 10.1016/j.worlddev.2010.09.008

Sumberg, J., Thompson, J., & Woodhouse, P. (2021). Science, technology and innovation for agricultural development: Reframing the discourse. *Outlook on Agriculture, 50*(2), 91–97. DOI: 10.1177/0030727021995701

Tilman, D., Cassman, K. G., Matson, P. A., Naylor, R., & Polasky, S. (2002). Agricultural sustainability and intensive production practices. *Nature, 418*(6898), 671–677. DOI: 10.1038/nature01014 PMID: 12167873

Wolfert, S., Ge, L., Verdouw, C., & Bogaardt, M. J. (2017). Big Data in Smart Farming – A review. *Agricultural Systems, 153*, 69–80. DOI: 10.1016/j.agsy.2017.01.023

WTO. (1994). *Agreement on Trade-Related Aspects of Intellectual Property Rights*. Marrakesh Agreement Establishing the World Trade Organization.

Zhang, H., Zhang, J., Lang, Z., Botella, J. R., & Zhu, J.-K. (2018). Genome editing—Principles and applications for functional genomics research and crop improvement. *Critical Reviews in Plant Sciences*, *37*(4), 215–234.

Chapter 12

Digital Transformation, Climate Change, and Agriculture: Towards a Better Sustainability in the MENA Region

Houda Mahboub
https://orcid.org/0000-0003-3903-2387
SUPMTI, Morocco

Hicham Sadok
https://orcid.org/0000-0003-4402-0954
Mohammed V University in Rabat, Morocco

ABSTRACT

With the increasing food demand, water scarcity, inefficient resource management, but also climate change, agriculture is facing serious problems. The situation gets even worse as time passes by due to several reasons, including the actual context of wars. As the key basis for several populations and Economies, opting for concrete actions to ensure a sustainable future is a must. The inclusion of digital technologies in agriculture was proposed by several researchers and practitioners. However, there is no clear response on the link between digital transformation and agriculture. Moreover, this research era is still scarce, with more limitations regarding the MENA region. Considered among the most affected regions by climate change, but also with a high dependence rate on agriculture, exploring how digital technologies can help to achieve better agriculture and more sustainability in MENA countries remains important. This chapter provides an integrative approach to the existing literature on this topic.

DOI: 10.4018/979-8-3373-3346-5.ch012

INTRODUCTION

Climate change remains one of the serious issues that the World suffers from. Several states, institutions, and researchers are calling for serious consideration of this challenge. Indeed, one of its serious threats is its impact on agriculture. As one of the basics to ensure sustainability and human continuity, but also its key contribution to GDP and the economy of several countries, agriculture surely needs interesting support (Manucharyan, 2025). As one of the proposed solutions to limit the destruction of agriculture by climate change issues, digital transformation imposes itself on several aspects of our lives.

In the African context, several digital agriculture applications are used for supply-chain management, business intelligence, advisory services, financial services and so on (Goedde et al., 2021). Despite this advancement, these solutions do not really create value for their users and face a very high risk of failure. This is due to the lack of retention since the registration rate reaches 80% but only 30% of the farmers ensure active usage (30%) (Goedde et al., 2021). In fact, several elements can explain this limitation. First, the lack of attractiveness of the added value to gain in case of the use of these digital solutions is not sufficient which pushes the users to have this appetite to discover what is new in these digital options but then do not get convinced or cannot see the plus (crop yields, cost reduction, better profits) provided by these options (Farah et al., 2025). In parallel with this point being addressed, policymakers need to make interesting investments in digital infrastructure, platforms, and regulation. This will lead to the reduction of the digital divide, enhance the inclusion regarding access to digital solutions, but also propose an acceptable fee for farmers, reduce the interesting difference between urban and rural areas, and have warehouses for better data storage and analysis (Román-Vázquez et al., 2025). Furthermore, better assistance (physical and digital) remains important to retain more clients but also to explain as best as possible to farmers in order to get the best results from the digital technologies, therefore satisfying the existing clients and seducing new ones with the support of the positive word of mouth (Abebaw, 2025). Regarding the reality that shows that many countries in Africa do not have access to electricity, the limited rural areas network coverage, the interesting cost of entry-level second or third-generation wireless (up to 70% of the farmer's monthly income), and the limited public investment in the digital era due to the existence of other priorities or high debt, the flourishment of digital technologies in Africa or the MENA region seems to be not a close reality to live.

However, several authors believe that this dream can be true, despite the interesting limitations, if the motivation to move further is combined with the correct investment / public initiatives. Indeed, the inclusion of digital transformation as a possible solution to climate change issues and enhancing agriculture remains an interesting research area that has seduced many researchers in recent years. This chapter synthesises the existing literature by applying an integrative approach. It remains general and synthetic, which can perfectly serve as the introduction to this book.

In summary, this chapter will be dedicated to the link between climate change, digital transformation, and agriculture. It will underline how these three elements can be linked, and the impact of the latter on sustainability and real life.

Digital transformation is nowadays a reality and no longer a myth. Whereas interesting gaps are remarkable between developed and developing countries, some policy-makers are doing their best to compensate for these differences despite the interesting roadblocks faced (digital divide, inertia, …). As an era that continuously attracts the interest of several policy-makers, researchers, and practitioners, and despite the interesting advancement in this topic, it still represents a fuzzy word. Indeed, the literature and reports are using digitalization and digital transformation as if they were synonyms. However, its key basis, which is getting a clear understanding of its definition to limit any misuse, remains a challenge. This statement was confirmed by several authors who recognised the vagueness of the existing definitions, a mismatch of the conceptualisation of the basic DT keywords, and the immaturity of the DT literature landscape and understanding (Besson & Rowe, 2012; Gong & Ribiere, 2021). Therefore, the first step in implementing and understanding digital transformation (DT) requires conceptual harmonisation and its contours to address the existing semantic and conceptual difficulties (Besson & Rowe, 2012; Cha & Lee, 2013, 2013; Goerzig & Bauernhansl, 2018; Gong & Ribiere, 2021, 2021; Gray & Rumpe, 2017; Haffke et al., 2016; Morakanyane et al., 2017; Westerman et al., 2014).

In previous research, the term digital transformation is used to refer to various ideas, stages and perceptions of digitalization implementation. It refers to all digital implementations and related business challenges. Indeed, the research conducted on digital transformation is still in a nascent stage despite the interesting evolution of research in the digital transformation era and has several shortcomings (Gong & Ribiere, 2021; Ji & Li, 2022). To remedy these limitations, we will provide, through this section, the main definitions (Table 1) regarding digital transformation and its basic concepts before assessing its direction of evolution in the Moroccan post-COVID context.

Table 1. Digital transformation definitions

Definition	Author(s)
DT is the use of technology to radically improve an organization's performance or reach.	Gemini (2011)
"The use of technology to radically improve performance or reach of enterprises"	Westerman et al. (2014)
DT is the use of information technology tools and ideas to transform business structures and workflows.	Zhou Zhiming et al. (2014)
"DT involves leveraging digital technologies to enable major business improvements, such as enhancing customer experience or creating new business models".	Piccinini et al. (2015)
Use of digital technologies to radically improve the company's performance.	Bekkhus (2016)
A generic term for converting information, processes, products, or services into a form that can be processed or supported by information technology.	Zimmermann et al. (2016)
"DT is concerned with the changes digital technologies can bring about in a company's business model, which result in changed products or organizational structures or in the automation of processes. These changes can be observed in the rising demand for Internet-based media, which has led to changes in entire business models (for example in the music industry).	Matt et al. (2016)
"Changes and transformations that are driven and built on a foundation of digital technologies [...]Whereas organizations are constantly transforming and evolving in response to the changing business landscape, digital transformation is the changes built on the foundation of digital technologies, ushering unique changes in business operations, business processes, and value creation".	Nwankpa & Roumani (2016)
DT "reflects the pervasiveness of changes induced by digital technologies throughout an organization"	Chanias & Hess (2016)
"Extended use of advanced IT, such as analytics, mobile computing, social media, or smart embedded devices, and the improved use of traditional technologies, such as enterprise resource planning (ERP), to enable major business improvements".	Chanias (2017)
The best understanding of digital transformation is adopting business processes and practices to help the organization compete effectively in an increasingly digital world.	Kane et al. (2017)
"Fundamental alterations in existing and the creation of new business models [...] in response to the diffusion of digital technologies such as cloud computing, mobile Internet, social media, and big data".	Remane et al. (2017)
"Digital Transformation (DX) is about adopting disruptive technologies to increase productivity, value creation, and social welfare."	Ebert & Duarte (2018)
"process through which companies converge multiple new digital technologies, enhanced with ubiquitous connectivity, with the intention of reaching superior performance and sustained competitive advantage".	Ismail et al. (2018)
"DT is the integration of digital technologies into a company's activities to keep pace with the changing competitive environment, radically improve performance, and enhance the customer experience by creating new value chains".	Güler & Büyüközkan (2019)
"a process that aims to improve an entity by triggering significant changes to its properties through combinations of information, computing, communication, and connectivity technologies"	Vial (2019)
DT is the ongoing process of using digital technologies in daily organizational life to create value for the enterprise.	Jinghua (2020)

continued on following page

Table 1. Continued

Definition	Author(s)
"DT is the process of combining digital technologies with sound business models to generate great value for enterprises."	Lee et al. (2021)
"A change in how a firm employs digital technologies, to develop a new digital business model that helps to create and appropriate more value for the firm".	Verhoef et al. (2021)
"DT is a comprehensive and all-around strategic transformation process that involves innovations in information technology, human resources, business models, and operating models, from management to operations, from production to sales, and from leaders to employees, through the use of digital technology".	Ji & Li (2022)

Source: Made by the author based on literature (Gong & Ribiere, 2021; Ji & Li, 2022; Vial, 2019)

Through the analysis of the existing definitions outlined in Table 1, we can detect several limitations. First, as Ji and Li (2022) outlined, the digital transformation existing definitions are fragmented and are usually more focused on a single perspective (technical or organizational). From the organizational perspective, they are either oriented toward the strategy perspective or the business model update. Moreover, there are not enough review papers to clarify the digital transformation concept, and the existing literature dealing with this topic is fragmented. On the other hand, all these definitions agreed on the use of digital technologies driven by a determinant to get an output, mainly increasing economic or organisational performance (strategy perspective) as well as looking at value through business models. This is the perspective with which most of the work on digital transformation has been oriented.

In general, the digital journey involves three key stages: digitization, digitalization, and digital transformation. The notion of digitization consists of exploiting and converting data to the digital form (Dougherty & Dunne, 2012; Loebbecke & Picot, 2015; Verhoef et al., 2021); then digitalization referring to the transforming the business process into digital (Dougherty & Dunne, 2012; Li et al., 2016) and finally the digital transformation which is a kind of implementation of a series of technological and human changes to restructure the existing managerial models and culture (Mahboub & Sadok, 2022).

In the literature, we globally use digital transformation for the organizational perspective and the concept of digital economy for the macro level, or digitalization for both the micro and macro levels. The digital economy can be defined as an "*amalgamation of several general-purpose technologies and the range of economic and social activities carried out by people over the Internet and related technologies*" (Alibekova et al., 2020; Dahlman et al., 2016; Mahboub & Sadok, 2024). For Deloitte, the digital economy represents "*the economic activity that results from billions of everyday online connections among people, businesses, devices, data, and processes. The backbone of the digital economy is hyperconnectivity, which means growing interconnectedness of people, organizations, and machines that re-*

sults from the Internet, mobile technology, and the Internet of Things" (Benazzouz & Sadok, 2025; Deloitte, 2020; Mahboub & Sadok, 2024).

However, digital transformation is not the only challenge faced nowadays. Among the interesting issues faced nowadays is climate change, given its repercussions on agriculture, making sustainability harder to achieve. Climate change is defined by the Intergovernmental Panel on Climate Change (IPCC) as " *a change in the state of the climate that can be identified ... by changes in the mean and/or the variability of its properties and that persists for an extended period*" (IPCC, 2022; Kabir et al., 2023; Santos & Bakhshoodeh, 2021). On the other hand, sustainability refers to "*the equitable, ethical and efficient use of natural resources for fulfilling the needs of current and future generations of human beings and enhancing their well-being*" (Ruggerio, 2021; Sakalasooriya, 2021).

To attain sustainability through addressing climate change problems, the digital transformation paradox was included in agriculture based on the recommendation of many institutions and researchers. The agriculture digital transformation was suggested as a solution, but also a productivity driver to enhance environmental and economic performance (Sargani et al., 2025). Through the inclusion of digital technologies into agriculture, the innovation could compensate for the active continuous destruction derived from climate change but also social issues, leading to better data analysis, resource allocation optimization, more resilience, higher sustainability, and better efficiency in agriculture (Galanakis, 2020; Sargani et al., 2025; Wu et al., 2023; Zhong et al., 2024). We verify the consistency of this statement in the coming section through the analysis of the current research.

Sustainable Agriculture: Can Digital Transformation Help?[1]

With this interesting increase in food demand, but also the increased diseases requiring some specific lifestyles and meals, and the serious issues related to climate change, agriculture is no longer only interesting due to its contribution to economic development, but rather a determinant of sustainability and human well-being in the future. Indeed, more than 27% of the world's population's livelihoods rely directly on agriculture (Rudrakar & Rughani, 2024). However, agriculture is facing serious problems such as water scarcity, soil erosion, resource depletion, political wars (such as Ukraine-Russia), world-power disturbance, famine, resource depletion and so on (Aguilera et al., 2020; Hou et al., 2024). With 33% of soils moderately to severely degraded and water pollution due to fertilisers and pesticide overuse (FAO, 2021; Parra-López et al., 2024; Tripathi et al., 2020), taking agriculture seriously is no longer a game! Accordingly, the agriculture sector requires a digital revolution to improve food value chains and apply innovative practices with less waste and a greener future, such as precision agriculture, gene editing, traceability technology and so on to support the agribusiness and rural economy (Hemamalini et al., 2024; Hong & Xiao, 2024; Hou et al., 2024; OECD & Food and Agriculture Organisation of the United Nations, 2023; Sargani et al., 2025; Usigbe et al., 2023). Therefore, digital transformation will contribute to the amelioration of the supply chain resilience in agriculture, defined as "*the capacity to maintain the availability and integrity of food supply during various disturbances*" (Liu et al., 2022; Sargani et al., 2025). The research of Alkhatib and Momani (2023) and Ivanov et al. (2021) showed that digital transformation significantly impacts supply chain resilience (Sargani et al., 2025).

Indeed, the introduction of digital transformation into agriculture has passed through several stages (Figure 1). The inclusion of digital into agriculture has been visible since the introduction of Agriculture 2.0 after the Industrial Revolution, in which agriculture workers started getting support from machines to enhance their performance. After that, the Agriculture 3.0 stage was more marked by the introduction of computers and programming and a lower contribution of humans. However, this evolution was not satisfying given the digital divide, context, state limitations and cybersecurity issues. Agriculture 4.0, or smart agriculture, was introduced through the support of AI and IoT technologies. Given the Agriculture 4.0 limitations, an updated version (Agriculture 5.0) appears to fit within the new orientations but also to induce humans again by relying on a human-machine perspective, contributing also to social challenges faced in agriculture.

Figure 1. Agriculture digital transformation evolution

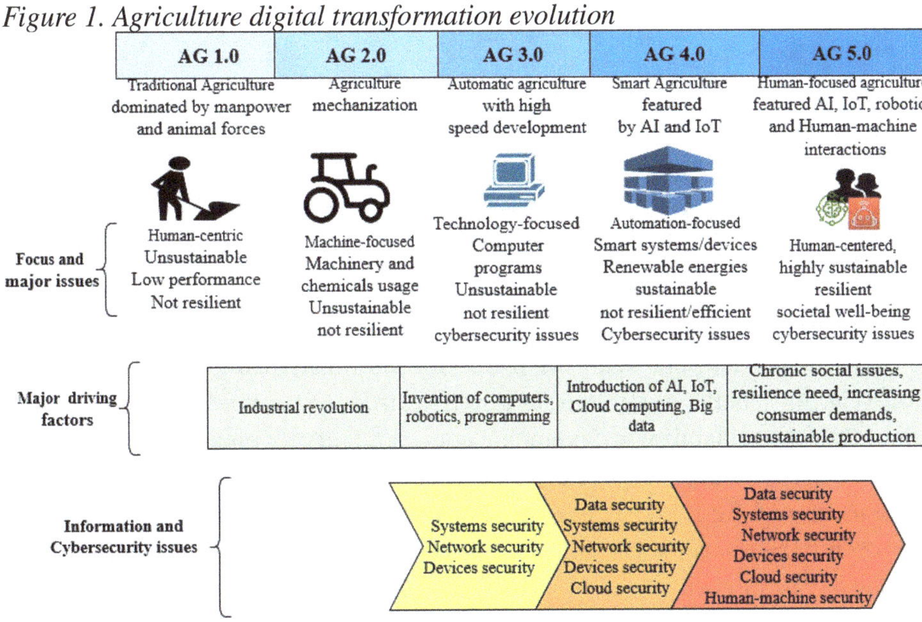

Source: Bissadu et al. (2025)

To verify if digital transformation is really a supportive solution to agricultural issues, Sargani et al. (2025) conducted a study on the role of digital technologies in enhancing agricultural productivity and supply chain resilience. The digital transformation was translated by many variables related to the Digital Technology Adoption Rates (DTAR), such as the adoption of IoT devices, the utilisation of data analytics, and the use of digital platforms for farm management (Figure 2). Their SEM analysis revealed that the adoption of digital technologies in agriculture boosts productivity and resilience. When combined with a proper agriculture infrastructure, the digital transformation explains 68% of the variance of agriculture productivity and 62% of the supply chain resilience. The digital transformation positively and significantly impacts both directly and indirectly agriculture.

Figure 2. Conceptual model of the link between digital transformation and agriculture

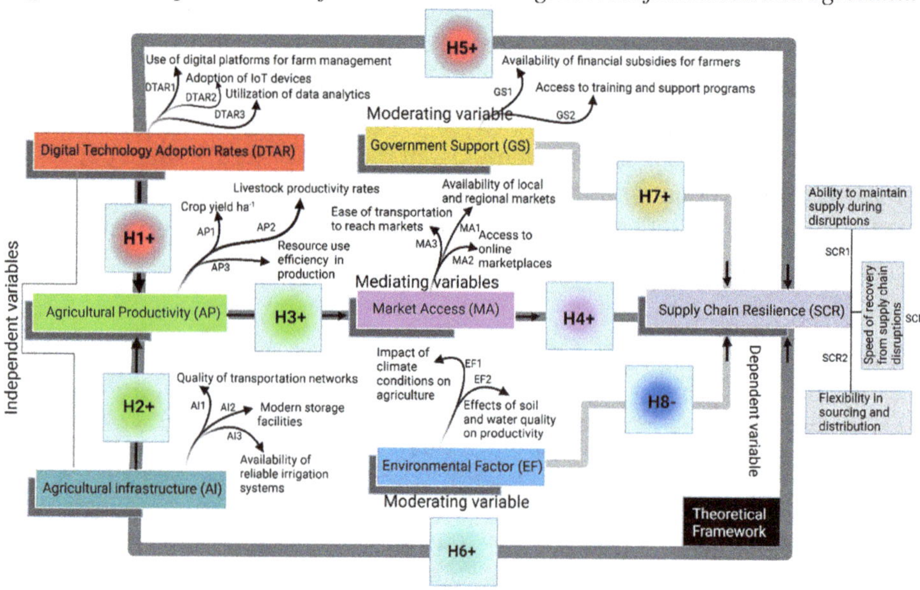

Source: Sargani et al. (2025)

On the other hand, Costa and al.'s (2023) study reveals that digital transformation positively impacts the agriculture industry, optimises resource management, and enhances its sustainability in developing countries that rely on agriculture.

In a context of great digital development and interesting technologies with a fabulous growth rate, digital transformation remains surely a great opportunity that needs to be seized. However, this may be seen as a far reality given the digital divide, interesting costs that cannot always be supported by traditional farmers, lack of trust in traditional agriculture leading to financial exclusion, complexity of use and technology-implementation requirements (Parra-López et al., 2024) ... On the other hand, including digital transformation in agriculture can lead to serious challenges. For example, the introduction of digital technologies (such as Artificial Intelligence, Internet of Things, Big Data, and Remote sensing) in agriculture led not only to new concepts such as digital agriculture / Agriculture 4.0 and smart farming but also to new practices such as crop and pest forecasting, smart harvesting, smart irrigation, robotic pest control (Hassoun et al., 2023; Parra-López et al., 2021, 2024). This advancement led to more efficient use of water, automation of agriculture practices, sustainable production, cost reduction, technology-driven procedures, renewable energy use, and a more resilient agriculture (Mourtzis et al., 2022). However, this digitalization of processes triggered the human contribution and work (Contreras-Medina et al., 2022). Moreover, these new practices require specific competencies and skills, which results in the exclusion of traditional farmers on whom several developing countries rely. As a consequence, the poverty situation becomes even more serious. To retain their core missions, these farmers find themselves in the obligation of either looking for external support (like introducing new stakeholders) or investing in the education of their children, who must opt for an agriculture specialisation to take the lead.

Indeed, the digital economy contributes to a more sustainable agriculture through low-carbon technological innovation, reshaping therefore business models and moving towards environment-friendly practices. However, the impact of the digital economy on agricultural sustainability transition depends on the region's capacity to promote low-carbon innovation. In intense low-carbon innovation, digital and environment-friendly technologies allow farmers to adopt more efficient and cleaner methods (Hou et al., 2024; Shen et al., 2022). This leads to more effective utilisation of resources, restructuring of the agricultural sector, and generally improved ecological outcomes.

Where low-carbon innovation is weak, the digital economy has little effect. These regions usually lack the technological foundation to support green transformation, have high decarbonization costs, and have insufficient capability for the development and adoption of carbon-reducing technology (Hou et al., 2024). As a result, agricultural digital transformation cannot attain its full optimization potential. Furthermore, in low-capacity innovation areas, governmental efforts to invest in green technology are challenged. These consist of extended development times, low production, and inefficient use of resources. It is increasingly challenging to unite the cooperation of research centres, universities, and businesses, and to recruit skilled workers. Insufficient digital infrastructure and a lack of regulation also complicate data sharing and digital development. The authors synthesised the impact of digital transformation on agriculture, based on the change theory, in the following conceptual model (Figure 3).

Figure 3. The impact of the introduction of digital technologies in agriculture

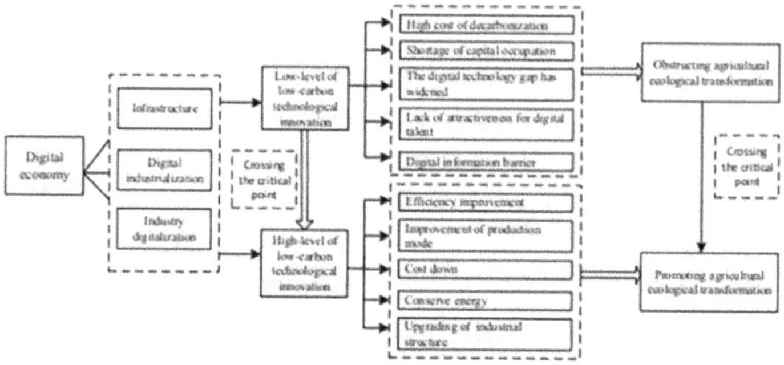

Source: Hou et al. (2024)

The study of Hou et al. (2024) showed that some conditions need to be respected in order to get results from the digital transformation of agriculture for a sustainable future. First, low-carbon innovation technology needs to reach a certain level in order to positively influence agriculture and resource optimisation, leading to better agriculture. Moreover, their study underlines that the link between digital economy and agriculture follows a U-shaped relationship (Hou et al., 2024). Additionally, green innovation capacity needs to be sufficient to seize a return on investment (ROI) from government investment. Otherwise, these investments are going to take much time and will be further considered as a waste of money and time, given the invisibility of results (Hou et al., 2024).

In order to gather the maximum possible from the agriculture digital transformation and achieve better sustainability, Wepner et al. (2025) proposed several public-policy initiative recommendations by themes (Table 2). These recommendations can serve as a guide for different countries with similar issues. The government needs only to align these actions through contextualization based on its current situation.

Table 2. Proposed public policies for better sustainability through agriculture digital transformation

Thematic area	Measures
Technology, research & innovation	· Promote the synergy of digital technologies with traditional agriculture to enhance sustainability and resource-efficient agriculture. · Motivate digital technologies and AI programming to favour agricultural diversity and regional variety adaptation. · Accelerate research in critical areas, including solutions to pressing environmental issues, such as drought-tolerant crops and pollutant-binding plants. · Focus on broader system-wide improvements rather than individual digital tools or applications, ensuring measures align with long-term sustainable agriculture goals. · Foster affordable, low-cost, and open-source digital solutions to enhance accessibility and usability across agri-food systems. · Develop reliable sensors for biological parameters and AI monitoring tools to support precision agriculture and resource-efficient farming. · Ensure transparent, explainable, intuitive and comprehensible open-source AI systems. · Address digital security risks and reliance on AI in agriculture while ensuring its transparency to increase innovation potential and support informed decision-making. · Establish continuous evaluation frameworks for AI tools to monitor and optimize resource consumption.
Market & business models	· Encourage collaborations and cooperatives to reduce investment costs for automation and digitalisation at the local and regional levels. · Expand concepts and cooperatives for shared access to machinery, personnel and e-platforms for direct marketing, · Incite collaborations of farmers to adopt alternative energy sources and decentralized energy production for additional income. · Develop new models, such as "Food Cooperatives", to enable direct marketing from multiple suppliers. · Facilitate information sharing on digital tools and their applications at local and regional levels through cooperative networks. · Utilize AI for real-time trade analysis and predictions (e.g., yields, demand, prices), helping align agricultural production with market trends.
Financial support	· Establish legislative measures and funding priorities to incentivize digital technology investments. · Reallocate agricultural funding to prioritize innovative projects while introducing new funding instruments involving diverse actors such as national banks, research funding organisations, and provincial governments. · Explore new funding mechanisms, including partnerships with, e.g., insurance companies or crowdfunding opportunities. · Appoint a single institution to oversee all funding instruments related to the agri-food system, encompassing basic and applied research at regional, national, and European levels. · Provide additional financial support for strengthening infrastructure in agriculture, developing decentralized server structures and expanding alternative energy production.

continued on following page

Table 2. Continued

Thematic area	Measures
Culture & social values	· Raise awareness about sustainable agriculture at various levels to foster sustainable and appreciative relationships between consumers, the public, and farmers. · Utilize digital tools, such as apps and gamification, to increase public awareness. · Provide education on sustainable agriculture, digitalisation and healthy nutrition through all educational levels from schools to lifelong learning programs. · Address the ethical considerations of AI and data analysis in agriculture, ensuring a balance between technological advancement and moral principles, aligned with EU AI legislation. · Strengthen lobbying efforts for agricultural interests, incorporating digitalisation across regional, national, and European levels, fostering collaboration between ministries (e.g., agriculture and education) to organize awareness campaigns.
Networks	· Develop systems to disseminate relevant, actionable information targeted across the network, ensuring agriculture can sustainably benefit from advances in digitalisation. · Establish educational networks with universities, technical colleges, and associations to enhance training and knowledge sharing. · Facilitate knowledge-sharing and technology transfer into practical applications for farmers. · Use digital platforms to connect stakeholders at all levels for efficient communication and faster information to strengthen common interests and share.
Competencies & knowledge	· Foster holistic agricultural education, integrating sustainability, biodiversity, and environmental protection alongside digitalisation competencies. · Emphasize interdisciplinary approaches between agriculture and digitalisation. · Promote adult education programs on digitalisation and AI in agriculture. · Establish training courses in data science and AI in agricultural schools, universities and agricultural institutions.
Infrastructure	· Expand critical infrastructure, including fibre-optic networks and mobile data transmission for robust digitalisation and automation adoption. · Implement energy-efficient technologies. · Develop decentralized server structures to enhance data security and minimize risks. · Prioritize alternative and self-sufficient energy sources and non-fossil energy generation.
Policy framework	· Adapt policies and develop new legislation to integrate sustainability aspects and ensure data security and environmental protection. · Ensure the trustworthiness of AI systems through clear regulations, in line with the EU's AI law. · Develop AI-driven forecasting for environmental policies. · Introduce tax incentives to encourage investment in digital tools for agriculture. · Revise building regulations to enable urban indoor farming and renewable energy projects in urban and rural areas.

Source: Wepner et al. (2025)

The shift towards the inclusion of digital technologies into agriculture is relying on the interesting potential of digital transformation to promote agricultural diversity and resilience. This includes, for example, the development and dissemination of new crop varieties that are more resource-efficient and better suited to evolving environmental conditions (Wepner et al., 2025). Digital tools can also support transparency, accessibility, and affordability in farming technologies, provided they are implemented with attention to digital equity and data security risks (Abbate et al., 2023; Dörr & Nachtmann, 2022; Finger, 2023).

On the other hand, introducing artificial intelligence (AI) and customized data analytics into agri-food systems further enhances decision-making capacity. In the context of global agricultural trade, such innovations enable more accurate forecasting of crop yields and market demand, thereby improving supply chain coordination and resource allocation (Dayioğlu & Turker, 2021; McFadden et al., 2022). However, the effects of digital transformation do not operate in isolation. Agricultural systems are composed of interlinked thematic areas, meaning that the impact of any single technological measure is often mediated or amplified by others. This complexity necessitates a nuanced framework for digital transformation, differentiating between systemic-level changes and operational-level initiatives.

Two primary levels of intervention emerge. System-level measures involve macro-level factors such as digital infrastructure, financing, and policy, and require the proactive engagement of governments and institutions. In contrast, operational-level measures are focused on farm-level activities such as training, education, and cooperative strategies that directly influence agricultural practice (Wepner et al., 2025). Both levels are essential for realizing the full benefits of digitalization in the agri-food sector.

At the system level, robust and resilient digital infrastructure is fundamental. Reliable internet connectivity, adequate energy supply (including renewables), and access to digital devices are prerequisites for the successful integration of automation and data technologies into agriculture (FAO, 2021; McFadden et al., 2022). These infrastructural elements must be strategically developed at both national and regional levels, guided by long-term planning and inclusive policy design. Infrastructure development is not just a technical task—it reflects broader socioeconomic priorities and requires investment frameworks that are sustainable and equitable.

Equally important is the question of financing. While financial support is essential for digital transformation, monetary investment alone is insufficient. A variety of financial strategies are required to catalyze the widespread adoption of new technologies. This includes tailored subsidies, innovation grants, tax incentives, and private-sector investment models that encourage experimentation without disproportionately burdening smallholders (Nishimoto et al., 2021; Wepner et al., 2025). Importantly, financial interventions must be embedded within larger systems of support, including market access and legal protection.

Transformation of infrastructure and financing must be underpinned by a forward-looking policy framework. Policymakers must integrate environmental considerations with digital agendas, aligning taxation, data governance, and sustainability goals into a coherent structure. For instance, policy incentives can promote sustainable farming practices while also facilitating the uptake of digital tools. At the same time, ethical and legal frameworks are necessary to ensure data protection, fair access, and responsible technology use, especially given the growing role of AI in shaping agricultural decision-making (Annosi et al., 2023; Finger, 2023).

Another vital operational mechanism is **cooperation**, particularly in the form of collective business models. Cooperative approaches (shared ownership of machinery, labour pooling, or joint marketing strategies) can reduce financial burdens and support access to digital innovations.

Emerging models like community-supported agriculture (CSA), food cooperatives, and collective farm shops enable small and medium producers to directly engage with consumers while enhancing economic resilience. These strategies not only improve efficiency but also foster community ties and decentralize power within the food system (Lucas et al., 2019).

Beyond systemic reforms, transformation at the operational level plays a crucial role in embedding sustainability within the daily practices of farmers. Education and training are central to this process. As digital competencies become essential to modern farming, it is critical to provide accessible learning opportunities for farmers, advisors, students, and technology developers. Such training should not only cover technical skills but also engage with broader topics like biodiversity, climate change, and socio-ecological systems (FAO, 2021). In rural regions in particular, where access to formal education and digital infrastructure may be limited, these efforts are especially urgent.

Operational transformation also involves addressing the role of **public engagement**. Farmers alone cannot shoulder the responsibility for agri-food system sustainability. Raising public awareness about agricultural challenges through targeted campaigns and educational programs can foster greater empathy and support from consumers. Such engagement can improve consumer-farmer relationships and generate societal support for sustainable policies and practices (Goller et al., 2021). For example, **gamification strategies** (mobile applications that combine education with entertainment) offer innovative ways to increase public understanding of agriculture and sustainability in a digital age (Efremova, 2023).

Ultimately, the transformation toward a sustainable digital agriculture system requires a **multi-level, systemic, and participatory approach.** It must simultaneously address environmental, economic, and social dimensions, and involve diverse stakeholders from across the agri-food spectrum. This includes not only farmers and policymakers, but also researchers, industry leaders, educators, and consumers. Only through coordinated collaboration can the full potential of digital agriculture be realized (UNDP, 2025).

For instance, partnerships between agricultural chambers, professional associations, and academic institutions can play a key role in **building capacity** and

improving data literacy among farmers and rural communities. These collaborations should be supported by national strategies and funding mechanisms that recognise the value of local knowledge and bottom-up innovation (Walz et al., 2007). Meanwhile, integrating societal perspectives and collaborations into policymaking (through deliberative forums, citizen assemblies, or community consultations) can help ensure that technological transformation remains inclusive and equitable (De Bernardi, 2020).

In general, the digital transformation of agriculture represents a profound opportunity to reshape food systems for greater sustainability and resilience. However, this transition cannot be achieved through technological solutions alone. It requires thoughtful planning, inclusive governance, and a systems-oriented vision that connects macro-level infrastructure and policy with on-the-ground practices and social dynamics. By aligning these dimensions, stakeholders can build a future-proof agri-food system capable of meeting both current challenges and future demands.

From an economic and sustainability perspective, agriculture has the opportunity to address not only famine and food assurance but also to help with social problems such as the unemployment rate. In fact, several countries, including developed ones, suffer from serious steep labour supply gaps reaching more than 100.000 people (Rotz et al., 2019). In fact, agriculture represented a sector with the lowest requirement rate of a high educational level and contributed significantly to the employment rate and labour insertion (Beltrán Tapia & Martinez-Galarraga, 2018; Espín-Sánchez et al., 2019; Sadjadi & Fernández, 2023; Sargani et al., 2025; Tewathia et al., 2020). Within the interesting evolution of the digital era and the introduction of digital technologies, the agriculture-vulnerable workforce with limited qualifications in terms of education and digital skills lost its competitiveness in the market and is even facing the risk of activity destruction (amplified by the digital era and climate change issues) (Pontones-Rosa et al., 2021; Sadjadi & Fernández, 2023).

Indeed, the MENA region remains among the most vulnerable places facing a serious attack from climate change, threatening therefore its economy, nature, and even decades to come (Tsan et al., 2019; UNEP Finance, 2023). From an economic perspective, the climate change issues in the MENA region can lead to a loss of GDP ranging between 0.4% to 1.3% (UNEP Finance, 2023). The situation can be even worse and multiplied by 14 times if policymakers do not opt for adequate decisions and mitigation actions (Peridy et al., 2013). As the most water-scarce region, with a higher reliance on agriculture, the MENA region could face serious impacts if it does not react properly, such as losing 6 – 14% of GDP in 2050 with a drop of more than 15% of its freshwater availability (UNEP Finance, 2023). Indeed, each increase of 1% in winter temperature leads to a -1.12% in agricultural production in the MENA region (Alboghdady & El-Hendawy, 2016). Inappropriate intervention could lead to higher migration, food crises, economic consequences, and more debt in the MENA region. With the common cultural preference of MENA citizens to move towards developed countries considered as rich in opportunities but also providing better living conditions, the situation will surely be worse due to weaknesses in agriculture as one of the GDP's reliable contributors, except for traditional agriculture facing barriers or cultural uninterest in immigration. As a solution, some researchers have faith in digital technologies. They believe that digitalization will accelerate the movement towards a green future if used properly (Ben Youssef, 2020). However, if digital technologies are not involved correctly, by moving towards a sustainability policy based on the use of renewable energies, the carbon emissions would become bigger, making the environmental situation even worse and climate change more challenging (Ben Youssef, 2023; Singh, 2025). Drone monitoring to check and address plots diseases, big data and data analytics, better devices usage due to agricultural digital platforms, real-time decision-making and many other options provided by digital transformation can help within achieving a green and sustainable future if well used within opting for the correct business-model updates (Contreras-Medina et al., 2022; Dayioğlu & Turker, 2021; Finger, 2023; Herrero et al., 2020; Ruiz-Real et al., 2020; Tubiello et al., 2022; Wepner et al., 2025; Wineman et al., 2024).

Precision agriculture, the Internet of Things (IoT), artificial intelligence (AI), satellite imaging, remote sensing, mobile applications, blockchain, and big data analytics are just a few examples of digital technologies that have the potential to completely transform agricultural practices through increased market accessibility, improved productivity, improved climate adaptation, and resource optimization. With the help of these tools, farmers can make data-driven decisions, predict pest outbreaks, optimise fertiliser and irrigation use, track crop health, and access real-time weather and market data. However, the adoption and integration of digital agriculture in MENA continue to be uneven and face several systemic obstacles, such as poor digital literacy, a lack of infrastructure, restricted access to funding, disjointed policy frameworks, and sociocultural limitations, especially for women. Despite these challenges, several countries in the region are piloting and scaling innovative digital solutions. Egypt, for example, has launched the "Digital Field" project and the Agricultural Knowledge Bank to disseminate information to farmers. Morocco's Green Generation 2020–2030 strategy promotes agritech development and precision irrigation, while Jordan utilizes Earth observation and remote sensing for drought management and agricultural planning. Tunisia is supporting AI-driven startups focusing on soil health and pest detection. Gulf Cooperation Council (GCC) countries such as the UAE and Saudi Arabia are investing heavily in agritech startups and vertical farming, leveraging advanced technologies to address water scarcity and land constraints. At the regional level, organizations like the Food and Agriculture Organization (FAO), the International Center for Agricultural Research in the Dry Areas (ICARDA), and the World Bank play key roles in funding digital agriculture initiatives, building institutional capacity, and promoting knowledge-sharing. To fully retain the benefits of digital transformation for climate-resilient agriculture in MENA, policy coherence and intersectoral collaboration are critical. Governments must integrate digital agriculture into national development, climate adaptation, and food security strategies. Investment in rural digital infrastructure, particularly broadband internet and mobile connectivity, is essential, as is support for innovation ecosystems, agritech startups, and public-private partnerships. Education and capacity building are equally crucial: training programs must equip farmers, extension workers, and policymakers with the digital skills required to use and scale technologies effectively. Moreover, inclusive policies that address gender disparities, youth unemployment, and rural marginalization can ensure that digital agriculture contributes to social equity as well as environmental sustainability. Data governance and digital rights must be safeguarded through clear regulations that protect data ownership and privacy.

Ultimately, the convergence of digital transformation and climate adaptation offers a pathway to revitalizing MENA's agriculture sector, making it more sustainable, productive, and resilient in the face of climate change. It is through a combination of technological innovation, strategic investment, inclusive governance, and international cooperation that the region can navigate its complex challenges and secure a food-secure future. While the road ahead is fraught with obstacles, the transformative potential of digital tools offers a beacon of hope for a more adaptive and robust agricultural paradigm in the MENA region.

Moving forward, the integration of digital agriculture into national climate strategies must be complemented by regionally coordinated action that leverages the unique capabilities and resources of each MENA country. Establishing cross-border platforms for knowledge exchange, data harmonization, and collaborative innovation can significantly amplify the impact of national efforts. Regional centres of excellence in digital agriculture, backed by robust public-private partnerships, can drive research, pilot new technologies, and provide tailored training across linguistic and socio-cultural contexts. Additionally, incorporating local knowledge systems and indigenous practices into digital frameworks can enhance their relevance and adoption. Strengthening monitoring and evaluation mechanisms to track the effectiveness of digital interventions will also help refine policies and investment priorities over time. As climate risks continue to intensify, a unified regional approach that aligns technological innovation with inclusive, climate-smart agricultural development will be essential to securing sustainable food systems for current and future generations in the MENA region.

To deepen this regional transformation, it is vital to expand the scope of digital agriculture beyond production efficiency and climate resilience. Emphasis should be placed on value chain digitization to link farmers with processors, distributors, and consumers through smart contracts and transparent logistics systems. This can reduce post-harvest losses, increase traceability, and improve pricing mechanisms, thereby boosting farmer incomes and consumer trust. Furthermore, fostering interoperability between national and regional digital platforms can improve resource allocation and responsiveness in times of crisis, such as during droughts or pandemics. A concerted effort is also required to develop multilingual and culturally relevant digital content that resonates with diverse rural populations, ensuring inclusive access to information and services. Policymakers must engage not only agricultural ministries but also those overseeing digital development, environment, trade, and education to create a cohesive governance model. Long-term resilience will depend on creating adaptive institutions that can evolve alongside emerging digital trends and climate realities.

Globally, including digital transformation in agriculture to address climate change issues and attend to sustainability is welcomed in different countries (Ancín et al., 2022; Barrett et al., 2020). Nevertheless, relying fully on digital transformation can lead to the entire destruction of the sector, given the employment decrease, the black box, and cybersecurity issues.

KNOWLEDGE GAPS

The digital transformation is seen as an elixir to all the issues we are facing nowadays. It has gathered the interest of many researchers since always, which has been translated into the number of publications and scientific events that consider it as the key theme. However, the current literature deals with developing countries or a single country (with contexts very different from the MENA region). Moreover, very few researchers addressed the African countries, but there are none, to the author's knowledge, scientific publications addressing the MENA region. Only institutional reports (not scientific ones) address a fragmented analysis of this region. Given that the MENA region is among the most affected by climate change issues and its impact on agriculture, these facts show the population and key research gaps addressed in this chapter. Moreover, the existing literature deals with this paper either from a technical perspective or an economic one, oriented towards the macro-level analysis. Additionally, the current publications deal mainly with developed topics related to digital transformation and agriculture and forget to address the basics for a better understanding of the cultural differences first. Thus, our chapter provides an integrative approach (instead of the fragmented analysis in the current literature).

CONCLUSION

Climate change and sustainability are two buzzwords we hear anywhere and anytime in our daily lives. From academic discussions, political speeches, fields of study, and economic reports, policymakers and the community are actively interested but also aware of the importance of climate change in achieving sustainability. Indeed, the basic idea of sustainability relies on the enhancement of the actual population and ensuring the continuity of human existence. However, the actual climate issues are alarming and require, therefore, serious changes. On the other hand, digital technologies can represent both a blessing and a curse depending on their usage. They represent two key issues all Economies are facing. Despite technological improvements, traditional practices are still dominating. Taking them seriously to optimize the link between them is of great importance.

Drought, water scarcity, energy consumption, and temperature upheaval with a tendency towards interesting high temperatures are all a serious reality in the MENA region. These key elements are surely having a huge impact on agriculture, which represents the basis of several economies in the MENA region. Indeed, agriculture is also making an interesting contribution to employment, especially in rural areas. The socioeconomic and environmental landscape of the Middle East and North Africa (MENA) region is being shaped by the interconnected issues of digital transformation, climate change, and agriculture. Due to its arid to semi-arid climate, limited water supplies, and reliance on imported food, the MENA region is particularly vulnerable to climate change, with severe weather, rising temperatures, and water scarcity having a significant impact on agricultural livelihoods and food security. According to climate projections, the region's already fragile agricultural productivity could be made worse by an increase in heatwaves and droughts, a decrease in precipitation, and an increase in average temperatures by 2–4°C by the end of the century. In light of this, digital transformation appears to be a viable and essential solution to the climate and agricultural issues facing the area.

This book chapter relies on an integrative approach to summarise the current literature, but also provides a better understanding of these interesting issues that influence our lives every day. It also provides clear evidence of the link between digital transformation and agriculture and includes some recommendations that can serve as inspiration for public policies. Given the interesting gap on this topic, especially in one of the most impacted regions by climate change issues, researchers can focus on their upcoming work on conducting more studies on the MENA region. A comparative analysis between its countries or sharing their return on experience can be of great benefit for inspiration.

Another interesting point involves energy consumption: we keep investing in digital technologies for a better future, but we are destroying it on the other side, regarding the impact of these technologies on the environment. So why don't we adjust the technologies we are investing in from the beginning instead of being in a rush in the end?

REFERENCES

1. Abbate, S., Centobelli, P., & Cerchione, R. (2023). The digital and sustainable transition of the agri-food sector. *Technological Forecasting and Social Change, 187*, 122222. DOI: 10.1016/j.techfore.2022.122222

Abebaw, S. E. (2025). A Global Review of the Impacts of Climate Change and Variability on Agricultural Productivity and Farmers' Adaptation Strategies. *Food Science & Nutrition, 13*(5), e70260. DOI: 10.1002/fsn3.70260

Aguilera, E., Díaz-Gaona, C., García-Laureano, R., Reyes-Palomo, C., Guzmán, G. I., Ortolani, L., Sánchez-Rodríguez, M., & Rodríguez-Estévez, V. (2020). Agroecology for adaptation to climate change and resource depletion in the Mediterranean region. A review. *Agricultural Systems, 181*, 102809. DOI: 10.1016/j.agsy.2020.102809

Alboghdady, M., & El-Hendawy, S. E. (2016). Economic impacts of climate change and variability on agricultural production in the Middle East and North Africa region. *International Journal of Climate Change Strategies and Management, 8*(3), 463–472. DOI: 10.1108/IJCCSM-07-2015-0100

Alibekova, G., Medeni, T., Panzabekova, A., & Mussayeva, D. (2020). Digital Transformation Enablers and Barriers in the Economy of Kazakhstan. *The Journal of Asian Finance, Economics and Business, 7*(7), 565–575. DOI: 10.13106/jafeb.2020.vol7.no7.565

Alkhatib, S. F., & Momani, R. A. (2023). Supply Chain Resilience and Operational Performance: The Role of Digital Technologies in Jordanian Manufacturing Firms. *Administrative Sciences, 13*(2), 40. DOI: 10.3390/admsci13020040

Ancín, M., Pindado, E., & Sánchez, M. (2022). New trends in the global digital transformation process of the agri-food sector: An exploratory study based on Twitter. *Agricultural Systems, 203*, 103520. DOI: 10.1016/j.agsy.2022.103520

Annosi, M. C., Marchegiani, L., Magnusson, M., & Roidis, M. (2023). Is technology neutral for MSEs? Interdependencies, information transparency and power imbalances in e-commerce ecosystems. *International Journal of Technology Management, 91*(3–4), 190–218. *Scopus*. Advance online publication. DOI: 10.1504/IJTM.2022.10052666

Barrett, C. B., Benton, T. G., Cooper, K. A., Fanzo, J., Gandhi, R., Herrero, M., James, S., Kahn, M., Mason-D'Croz, D., Mathys, A., Nelson, R. J., Shen, J., Thornton, P., Bageant, E., Fan, S., Mude, A. G., Sibanda, L. M., & Wood, S. (2020). Bundling innovations to transform agri-food systems. *Nature Sustainability, 3*(12), 974–976. DOI: 10.1038/s41893-020-00661-8

Bekkhus, R. (2016). Do KPIs used by CIOs Decelerate Digital Business Transformation? The Case of ITIL. *DIGIT 2016 Proceedings*. https://aisel.aisnet.org/digit2016/16

Beltrán Tapia, F. J., & Martinez-Galarraga, J. (2018). Inequality and education in pre-industrial economies: Evidence from Spain. *Explorations in Economic History*, *69*, 81–101. DOI: 10.1016/j.eeh.2017.12.003

Ben Youssef, A. (2020). How Can Industry 4.0 Contribute to Combatting Climate Change? *Revue d'Economie Industrielle*, •••, 161–193. DOI: 10.4000/rei.8911

Ben Youssef, A. (2023). *Digitalization for the Green Transition in the Mediterranean* (IEMed. Mediterranean Yearbook). Economic Research Forum (ERF); IEMed. Mediterranean Yearbook. https://www.iemed.org/publication/digitalization-for-the-green-transition-in-the-mediterranean/?lang=fr

Benazzouz, S., & Sadok, H. (2025). Analysis of the Relationship Between Innovation, Digitalization and Economic Growth in North African Countries. In Bach Tobji, M. A., Jallouli, R., Sadok, H., Lajfari, K., Mafamane, D., & Mahboub, H. (Eds.), *Digital Economy. Emerging Technologies and Business Innovation* (Vol. 531, pp. 183–198). Springer Nature Switzerland., DOI: 10.1007/978-3-031-76368-7_13

Besson, P., & Rowe, F. (2012). Strategizing information systems-enabled organizational transformation: A transdisciplinary review and new directions. *The Journal of Strategic Information Systems*, *21*(2), 103–124. DOI: 10.1016/j.jsis.2012.05.001

Cha, K. J., & Lee, Z. (2013). *What Do We Mean by Information Technology Enabled Organisational Transformation?*

Chanias, S. (2017). MASTERING DIGITAL TRANSFORMATION: THE PATH OF A FINANCIAL SERVICES PROVIDER TOWARDS A DIGITAL TRANSFORMATION STRATEGY. *Research Papers*. https://aisel.aisnet.org/ecis2017_rp/2

Chanias, S., & Hess, T. (2016). *Understanding Digital Transformation Strategy Formation: Insights from Europe's Automotive Industry*.

Contreras-Medina, D. I., Medina-Cuéllar, S. E., & Rodríguez-García, J. M. (2022). Roadmapping 5.0 Technologies in Agriculture: A Technological Proposal for Developing the Coffee Plant Centered on Indigenous Producers' Requirements from Mexico, via Knowledge Management. *Plants*, *11*(11), 1502. DOI: 10.3390/plants11111502

Costa, F., Frecassetti, S., Rossini, M., & Portioli-Staudacher, A. (2023). Industry 4.0 digital technologies enhancing sustainability: Applications and barriers from the agricultural industry in an emerging economy. *Journal of Cleaner Production, 408*, 137208. DOI: 10.1016/j.jclepro.2023.137208

Dahlman, C., Mealy, S., & Wermelinger, M. (2016). *Harnessing the digital economy for developing countries*. OCDE., DOI: 10.1787/4adffb24-

Dayioğlu, M. A., & Turker, U. (2021). Digital Transformation for Sustainable Future - Agriculture 4.0: A review. *Tarim Bilimleri Dergisi*. Advance online publication. DOI: 10.15832/ankutbd.986431

De Bernardi, P. (with Azucar, D.). (2020). *Innovation in Food Ecosystems: Entrepreneurship for a Sustainable Future*. Springer.

Deloitte. (2020). *What is Digital Economy? Deloitte*. https://www2.deloitte.com/mt/en/pages/technology/articles/mt-what-is-digitaleconomy.html

Dörr, J., & Nachtmann, M. (Eds.). (2022). *Handbook Digital Farming: Digital Transformation for Sustainable Agriculture*. Springer Berlin Heidelberg., DOI: 10.1007/978-3-662-64378-5

Dougherty, D., & Dunne, D. D. (2012). Digital Science and Knowledge Boundaries in Complex Innovation. *Organization Science, 23*(5), 1467–1484. DOI: 10.1287/orsc.1110.0700

Ebert, C., & Duarte, C. H. C. (2018). Digital Transformation. *IEEE Software, 35*(4), 16–21. DOI: 10.1109/MS.2018.2801537

Efremova, N. (2023). Gamification resources in the system of specialist training for the agricultural complex. *E3S Web of Conferences, 431*, 01052. DOI: 10.1051/e3sconf/202343101052

Espín-Sánchez, J.-A., Gil-Guirado, S., Giraldo-Paez, W. D., & Vickers, C. (2019). Labor income inequality in pre-industrial Mediterranean Spain: The city of Murcia in the 18th century. *Explorations in Economic History, 73*, 101274. DOI: 10.1016/j.eeh.2019.05.002

FAO. (2021). *The State of the World's Land and Water Resources for Food and Agriculture – Systems at breaking point (SOLAW 2021)*. FAO., DOI: 10.4060/cb7654en

Farah, A. A., Mohamed, M. A., Musse, O. S. H., & Nor, B. A. (2025). The multifaceted impact of climate change on agricultural productivity: A systematic literature review of SCOPUS-indexed studies (2015–2024). *Discover Sustainability, 6*(1), 397. DOI: 10.1007/s43621-025-01229-2

Finger, R. (2023). Digital innovations for sustainable and resilient agricultural systems. *European Review of Agriculture Economics, 50*(4), 1277–1309. DOI: 10.1093/erae/jbad021

Galanakis, C. M. (2020). The Food Systems in the Era of the Coronavirus (COVID-19) Pandemic Crisis. *Foods, 9*(4), 523. DOI: 10.3390/foods9040523

Gemini. (2011). Digital transformation: A roadmap for billion-dollar organizations. *MIT Sloan.*

Goedde, L., & McCullough, R. Ooko-Ombaka, & Pais, G. (2021). *How digital tools can help transform African agri-food systems* (p. 9). McKinsey & Company. https://africa-agricultural-transformation.org/wp-content/uploads/2022/12/How-digital-tools-can-help-transform-African-agri-food-systems.pdf

Goerzig, D., & Bauernhansl, T. (2018). Enterprise Architectures for the Digital Transformation in Small and Medium-sized Enterprises. *Procedia CIRP, 67,* 540–545. DOI: 10.1016/j.procir.2017.12.257

Gong, C., & Ribiere, V. (2021). Developing a unified definition of digital transformation. *Technovation, 102,* 102217. DOI: 10.1016/j.technovation.2020.102217

Gray, J., & Rumpe, B. (2017). Models for the digital transformation. *Software & Systems Modeling, 16*(2), 307–308. DOI: 10.1007/s10270-017-0596-7

Güler, M., & Büyüközkan, G. (2019). Analysis of Digital Transformation Strategies with an Integrated Fuzzy AHP-Axiomatic Design Methodology. *IFAC-PapersOnLine, 52*(13), 1186–1191. DOI: 10.1016/j.ifacol.2019.11.359

Haffke, I., Kalgovas, B., & Benlian, A. (2016). *The Role of the CIO and the CDO in an Organization's Digital Transformation.*

Hassoun, A., Kamiloglu, S., Garcia-Garcia, G., Parra-López, C., Trollman, H., Jagtap, S., Aadil, R. M., & Esatbeyoglu, T. (2023). Implementation of relevant fourth industrial revolution innovations across the supply chain of fruits and vegetables: A short update on Traceability 4.0. *Food Chemistry, 409,* 135303. DOI: 10.1016/j.foodchem.2022.135303

Hemamalini, V., Mishra, A. K., Tyagi, A. K., & Kakulapati, V. (2024). Artificial Intelligence–Blockchain-Enabled–Internet of Things-Based Cloud Applications for Next-Generation Society. In Tyagi, A. K. (Ed.), *Automated Secure Computing for Next-Generation Systems* (1st ed., pp. 65–82). Wiley., DOI: 10.1002/9781394213948.ch4

Herrero, M., Thornton, P. K., Mason-D'Croz, D., Palmer, J., Benton, T. G., Bodirsky, B. L., Bogard, J. R., Hall, A., Lee, B., Nyborg, K., Pradhan, P., Bonnett, G. D., Bryan, B. A., Campbell, B. M., Christensen, S., Clark, M., Cook, M. T., De Boer, I. J. M., Downs, C., & West, P. C. (2020). Innovation can accelerate the transition towards a sustainable food system. *Nature Food*, *1*(5), 266–272. DOI: 10.1038/s43016-020-0074-1

Hong, Z., & Xiao, K. (2024). Digital economy structuring for sustainable development: The role of blockchain and artificial intelligence in improving supply chain and reducing negative environmental impacts. *Scientific Reports*, *14*(1), 3912. DOI: 10.1038/s41598-024-53760-3

Hou, J., Zhang, M., & Li, Y. (2024). Can digital economy truly improve agricultural ecological transformation? New insights from China. *Humanities & Social Sciences Communications*, *11*(1), 153. DOI: 10.1057/s41599-023-02593-y

IPCC. (2022). *Global Warming of 1.5°C: IPCC Special Report on Impacts of Global Warming of 1.5°C above Pre-industrial Levels in Context of Strengthening Response to Climate Change, Sustainable Development, and Efforts to Eradicate Poverty* (1st ed.). Cambridge University Press., DOI: 10.1017/9781009157940

Ismail, M. H., Khater, M., & Zaki, M. (2018). Digital Business Transformation and Strategy: What Do We Know So Far? DOI: 10.13140/RG.2.2.36492.62086

Ivanov, D., Blackhurst, J., & Das, A. (2021). Supply chain resilience and its interplay with digital technologies: Making innovations work in emergency situations. *International Journal of Physical Distribution & Logistics Management*, *51*(2), 97–103. DOI: 10.1108/IJPDLM-03-2021-409

Ji, X., & Li, W. (2022). *Digital Transformation: A Review and Research Framework.*

Jinghua, X. (2020). Cross-system Digital Transformation and Adaptive Changes of Management. *Reform*, *2020*(4), 37–49.

Kabir, M., Habiba, U. E., Khan, W., Shah, A., Rahim, S., Rios-Escalante, P. R. D. L., Farooqi, Z.-U.-R., Ali, L., & Shafiq, M. (2023). Climate change due to increasing concentration of carbon dioxide and its impacts on environment in 21st century; a mini review. *Journal of King Saud University. Science*, *35*(5), 102693. DOI: 10.1016/j.jksus.2023.102693

Kane, G. C., Palmer, D., & Phillips, A. N. (2017). *Achieving Digital Maturity* [Technical Report]. MIT Sloan Management Review. https://sadil.ws/handle/123456789/1453

Lee, C.-H., Liu, C.-L., Trappey, A. J. C., Mo, J. P. T., & Desouza, K. C. (2021). Understanding digital transformation in advanced manufacturing and engineering: A bibliometric analysis, topic modeling and research trend discovery. *Advanced Engineering Informatics*, *50*, 101428. DOI: 10.1016/j.aei.2021.101428

Li, F., Nucciarelli, A., Roden, S., & Graham, G. (2016). How smart cities transform operations models: A new research agenda for operations management in the digital economy. *Production Planning and Control*, *27*(6), 514–528. DOI: 10.1080/09537287.2016.1147096

Liu, L., Cavaye, J., & Ariyawardana, A. (2022). Supply chain responsibility in agriculture and its integration with rural community development: A review of issues and perspectives. *Journal of Rural Studies*, *93*, 134–143. DOI: 10.1016/j.jrurstud.2020.07.003

Loebbecke, C., & Picot, A. (2015). Reflections on societal and business model transformation arising from digitization and big data analytics: A research agenda. *The Journal of Strategic Information Systems*, *24*(3), 149–157. DOI: 10.1016/j.jsis.2015.08.002

Lucas, V., Gasselin, P., & Van Der Ploeg, J. D. (2019). Local inter-farm cooperation: A hidden potential for the agroecological transition in northern agricultures. *Agroecology and Sustainable Food Systems*, *43*(2), 145–179. DOI: 10.1080/21683565.2018.1509168

Mahboub, H., & Sadok, H. (2022). Towards a Better Digital Transformation: Learning from the Experience of a Digital Transformation Project. In Bach Tobji, M. A., Jallouli, R., Strat, V. A., Soares, A. M., & Davidescu, A. A. (Eds.), *Digital Economy. Emerging Technologies and Business Innovation* (pp. 203–214). Springer International Publishing., DOI: 10.1007/978-3-031-17037-9_15

Mahboub, H., & Sadok, H. (2024). Digital transformation of the moroccan economy: A performance comparative approach. *Maghreb-Mashreq International*, *4/2023*, 45–62. DOI: 10.54695/mmi.4.0045

Manucharyan, M. (2025). Climate change impacts on sustainable agriculture: Evidence from Armenia. *Unconventional Resources*, *6*, 100159. DOI: 10.1016/j.uncres.2025.100159

Matt, C., Hess, T., Benlian, A., & Wiesbock, F. (2016). Options for Formulating a Digital Transformation Strategy. *MIS Quarterly Executive*, *15*(2). https://aisel.aisnet.org/misqe/vol15/iss2/6

McFadden, J., Casalini, F., Griffin, T., & Anton, J. (2022). *The digitalisation of agriculture: A literature review and emerging policy issues* (OECD Food, Agriculture and Fisheries Papers 176; OECD Food, Agriculture and Fisheries Papers, Vol. 176). DOI: 10.1787/18156797

Morakanyane, R., Grace, A., & O'Reilly, P. (2017). Conceptualizing Digital Transformation in Business Organizations: A Systematic Review of Literature. *BLED 2017 Proceedings*. https://aisel.aisnet.org/bled2017/21

Mourtzis, D., Angelopoulos, J., & Panopoulos, N. (2022). A Literature Review of the Challenges and Opportunities of the Transition from Industry 4.0 to Society 5.0. *Energies*, *15*(17), 6276. DOI: 10.3390/en15176276

Nishimoto, K., Sampaio, C. M. P., Vale, R. J., & Ruggeri, F. (2021). Numerical Simulation of Hybrid Platform Supply Vessel (PSV) Fuel Consumption for the Pre-Salt Layer in Brazil. In Okada T., Kawamura Y., & Suzuki K. (Eds.), *Lect. Notes Civ. Eng.: Vol. 65 LNCE* (p. 497). Springer Science and Business Media Deutschland GmbH; Scopus. DOI: 10.1007/978-981-15-4680-8_34

Nwankpa, J. K., & Roumani, Y. (2016). IT Capability and Digital Transformation: A Firm Performance Perspective. *International Conference of Information Systems*.

OECD & Food and Agriculture Organization of the United Nations. (2023). *OECD-FAO Business Handbook on Deforestation and Due Diligence in Agricultural Supply Chains*. OECD. DOI: 10.1787/c0d4bca7-en

Parra-López, C., Ben Abdallah, S., Garcia-Garcia, G., Hassoun, A., Sánchez-Zamora, P., Trollman, H., Jagtap, S., & Carmona-Torres, C. (2024). Integrating digital technologies in agriculture for climate change adaptation and mitigation: State of the art and future perspectives. *Computers and Electronics in Agriculture*, *226*, 109412. DOI: 10.1016/j.compag.2024.109412

Parra-López, C., Reina-Usuga, L., Carmona-Torres, C., Sayadi, S., & Klerkx, L. (2021). Digital transformation of the agrifood system: Quantifying the conditioning factors to inform policy planning in the olive sector. *Land Use Policy, 108. Land Use Policy*, *108*, 105537. Advance online publication. DOI: 10.1016/j.landusepol.2021.105537

Peridy, . (2013). *The Economic Costs of Climate Change in MENA countries: A Micro-Spatial Quantitative Assessment and a Survey of Policies*. FEMISE.

Piccinini, E., Hanelt, A., Gregory, R., & Kolbe, L. (2015). *Transforming Industrial Business: The Impact of Digital Transformation on Automotive Organizations*.

Pontones-Rosa, C., Pérez-Morote, R., & Santos-Peñalver, J. F. (2021). ICT-based public policies and depopulation in hollowed-out Spain: A survey analysis on the digital divide and citizen satisfaction. *Technological Forecasting and Social Change*, *169*, 120811. DOI: 10.1016/j.techfore.2021.120811

Remane, G., Hanelt, A., Nickerson, R. C., & Kolbe, L. M. (2017). Discovering digital business models in traditional industries. *The Journal of Business Strategy*, *38*(2), 41–51. DOI: 10.1108/JBS-10-2016-0127

Román-Vázquez, J., Carbonell-Bojollo, R. M., Veroz-González, Ó., Maraschi Da Silva Piletti, L. M., Márquez-García, F., Cabeza-Ramírez, L. J., & González-Sánchez, E. J. (2025). Global Trends in Conservation Agriculture and Climate Change Research: A Bibliometric Analysis. *Agronomy (Basel)*, *15*(1), 249. DOI: 10.3390/agronomy15010249

Rotz, S., Gravely, E., Mosby, I., Duncan, E., Finnis, E., Horgan, M., LeBlanc, J., Martin, R., Neufeld, H. T., Nixon, A., Pant, L., Shalla, V., & Fraser, E. (2019). Automated pastures and the digital divide: How agricultural technologies are shaping labour and rural communities. *Journal of Rural Studies*, *68*, 112–122. DOI: 10.1016/j.jrurstud.2019.01.023

Rudrakar, S., & Rughani, P. (2024). IoT based Agriculture (Ag-IoT): A detailed study on Architecture, Security and Forensics. *Information Processing in Agriculture*, *11*(4), 524–541. DOI: 10.1016/j.inpa.2023.09.002

Ruggerio, C. A. (2021). Sustainability and sustainable development: A review of principles and definitions. *The Science of the Total Environment*, *786*, 147481. DOI: 10.1016/j.scitotenv.2021.147481

Ruiz-Real, J. L., Uribe-Toril, J., Torres Arriaza, J. A., & De Pablo Valenciano, J. (2020). A Look at the Past, Present and Future Research Trends of Artificial Intelligence in Agriculture. *Agronomy (Basel)*, *10*(11), 1839. DOI: 10.3390/agronomy10111839

Sadjadi, E. N., & Fernández, R. (2023). Challenges and Opportunities of Agriculture Digitalization in Spain. *Agronomy (Basel)*, *13*(1), 259. DOI: 10.3390/agronomy13010259

Sakalasooriya, N. (2021). Conceptual Analysis of Sustainability and Sustainable Development. *Open Journal of Social Sciences*, *09*(03), 396–414. DOI: 10.4236/jss.2021.93026

Santos, R. M., & Bakhshoodeh, R. (2021). Climate change/global warming/climate emergency versus general climate research: Comparative bibliometric trends of publications. *Heliyon*, *7*(11), e08219. DOI: 10.1016/j.heliyon.2021.e08219

Sargani, G. R., Wang, B., Leghari, S. J., & Ruan, J. (2025). Is digital transformation the key to agricultural strength? A novel approach to productivity and supply chain resilience. *Smart Agricultural Technology*, *10*, 100838. DOI: 10.1016/j.atech.2025.100838

Shen, Z., Wang, S., Boussemart, J.-P., & Hao, Y. (2022). Digital transition and green growth in Chinese agriculture. *Technological Forecasting and Social Change*, *181*, 121742. DOI: 10.1016/j.techfore.2022.121742

Singh, R. (2025). *Industry 5. 0: Sustainable Business Practices for a Bright Future* (1st ed.). Nova Science Publishers, Incorporated. DOI: 10.52305/UYKO4001

Tewathia, N., Kamath, A., & Ilavarasan, P. V. (2020). Social inequalities, fundamental inequities, and recurring of the digital divide: Insights from India. *Technology in Society*, *61*, 101251. DOI: 10.1016/j.techsoc.2020.101251

Tripathi, S., Srivastava, P., Devi, R. S., & Bhadouria, R. (2020). Influence of synthetic fertilizers and pesticides on soil health and soil microbiology. In *Agrochemicals Detection, Treatment and Remediation* (pp. 25–54). Elsevier. DOI: 10.1016/B978-0-08-103017-2.00002-7

Tsan, M., Totapally, D. S., Hailu, D. M., & K Addom, B. (2019). *The digitalization of African agriculture report* (978-92-9081-657–7). The Technical Centre for Agricultural and Rural Cooperation (CTA).

Tubiello, F. N., Karl, K., Flammini, A., Gütschow, J., Obli-Laryea, G., Conchedda, G., Pan, X., Qi, S. Y., Halldórudóttir Heiðarsdóttir, H., Wanner, N., Quadrelli, R., Rocha Souza, L., Benoit, P., Hayek, M., Sandalow, D., Mencos Contreras, E., Rosenzweig, C., Rosero Moncayo, J., Conforti, P., & Torero, M. (2022). Pre- and post-production processes increasingly dominate greenhouse gas emissions from agri-food systems. *Earth System Science Data*, *14*(4), 1795–1809. DOI: 10.5194/essd-14-1795-2022

UNDP. (2025). *New Homes for Vulnerable Households in Yen Bai Ahead of Storm Season, Thanks to Post-Typhoon Yagi Recovery Support*. https://www.undp.org/vietnam/press-releases/new-homes-vulnerable-households-yen-bai-ahead-storm-season-thanks-post-typhoon-yagi-recovery-support

UNEP finance. (2023). *Adapting to a New Climate in the MENA Region: An assessment of physical risk management and climate adaptation finance in the MENA region*. https://www.uncclearn.org/wp-content/uploads/library/Adapting-to-a-new-climate-MENA.pdf

Usigbe, M. J., Asem-Hiablie, S., Uyeh, D. D., Iyiola, O., Park, T., & Mallipeddi, R. (2023). Enhancing resilience in agricultural production systems with AI-based technologies. *Environment, Development and Sustainability*, *26*(9), 21955–21983. DOI: 10.1007/s10668-023-03588-0

Verhoef, P. C., Broekhuizen, T., Bart, Y., Bhattacharya, A., Qi Dong, J., Fabian, N., & Haenlein, M. (2021). Digital transformation: A multidisciplinary reflection and research agenda. *Journal of Business Research*, *122*, 889–901. DOI: 10.1016/j.jbusres.2019.09.022

Vial, G. (2019). Understanding digital transformation: A review and a research agenda. *The Journal of Strategic Information Systems*, *28*(2), 118–144. DOI: 10.1016/j.jsis.2019.01.003

Walz, A., Lardelli, C., Behrendt, H., Grêt-Regamey, A., Lundström, C., Kytzia, S., & Bebi, P. (2007). Participatory scenario analysis for integrated regional modelling. *Landscape and Urban Planning*, *81*(1–2), 114–131. DOI: 10.1016/j.landurbplan.2006.11.001

Wepner, B., Neuberger, S., Hörlesberger, M., Molin, E. M., Lampert, J., & Koch, H. (2025). How can digitalisation support transformation towards sustainable agri-food systems? Scenario development in Lower Austria. *Agricultural Systems*, *224*, 104251. DOI: 10.1016/j.agsy.2024.104251

Westerman, G., Bonnet, D., & McAfee, A. (2014). *Leading digital: Turning technology into business transformation*. Harvard Business Review Press.

Wineman, A., Zanello, G., Degefaw, D. M., Samuel, F., Yates, J., Shankar, B., Adeyemi, O., Anderson, A. K., Awasthi, A., Bell, W., Chege, C. K., Choo, E., Degefaw, D. M., Deitchler, M., Freeman, M. C., Frongillo, E. A., Gelli, A., Havelaar, A. H., Guitian, J., & Zewdu, A.Nutritious Food Systems Consortium. (2024). A call to action for transformation towards nutritious food systems. *Nature Food*, *5*(1), 2–3. DOI: 10.1038/s43016-023-00896-6

Wu, G., Yang, W., Hou, X., & Tian, Y. (2023). Agri-food supply chain under live streaming and government subsidies: Strategy selection of subsidy recipients and sales agreements. *Computers & Industrial Engineering*, *185*, 109647. DOI: 10.1016/j.cie.2023.109647

Zhong, J., Cheng, H., & Jia, F. (2024). Supply chain resilience capability factors in agri-food supply chains. *Operations Management Research*. DOI: 10.1007/s12063-024-00470-8

Zimmermann, A., Schmidt, R., Sandkuhl, K., El-Sheikh, E., Jugel, D., Schweda, C., Möhring, M., Wißotzki, M., & Lantow, B. (2016). Leveraging Analytics for Digital Transformation of Enterprise Services and Architectures. In El-Sheikh, E., Zimmermann, A., & Jain, L. C. (Eds.), *Emerging Trends in the Evolution of Service-Oriented and Enterprise Architectures* (Vol. 111, pp. 91–112). Springer International Publishing., DOI: 10.1007/978-3-319-40564-3_6

Chapter 13

Impact of Biodiversity Loss on the Consumer Right to Food:

Sustainable Agriculture as a Solution to Climate Change

Mariam Ahmed Alsandal
https://orcid.org/0000-0003-0869-1208
University of Dhaid, UAE

ABSTRACT

One of the core principles of consumer rights is satisfying basic needs, including the right to food: A legal obligation at both international and national levels. As human sustenance depends primarily on agricultural products, access to adequate and nutritious food is closely tied to the quantity and quality of those products—factors directly influenced by biodiversity. Biodiversity loss weakens the capacity to achieve food security, especially given the varying abilities of states to produce agricultural goods locally due to multiple factors. Accordingly, international legislation alone is insufficient; national legislation must support and complement global efforts. This research examines the relevant international and national legal frameworks in the UAE that address biodiversity loss in the context of food security and environmental sustainability. It concludes by emphasizing the need to link international food-related instruments with the Convention on Biological Diversity (CBD), integrate the right to food into agricultural practices, and promote eco-farming.

DOI: 10.4018/979-8-3373-3346-5.ch013

INTRODUCTION

Biodiversity is of utmost importance regarding food obtained from plants, and it refers to the interaction among living organisms within a specific ecosystem, including animals, microorganisms, and plants (DeLong Jr, 1996). Many factors have contributed to the decline of biodiversity related to food, including human activities such as deforestation, the intensive use of chemical pesticides, the use of chemical fertilizers, and urban expansion at the expense of agricultural land. Other non-human factors include desertification, which leads to the drying of arable lands, in addition to climate change. Therefore, "humans play a major role in the environmental and ecological crisis, which is a result of human activities that have led to the existence of different forms of life on planet Earth" (Rocha et al., 2023).

One study has proposed that the trend of importing agricultural foods from tropical regions rich in biodiversity may serve as a solution to the problem of biodiversity loss in temperate and arid areas. Although this approach may contribute to addressing the issue at the national level for importing countries, it exacerbates the problem in the exporting countries. Biodiversity loss thus remains a global concern, even if reduced in the countries importing agricultural foods, resulting in a significantly high cumulative global extinction rate. This necessitates the adoption of sustainable sourcing policies and the protection of habitats in tropical regions (Cabernard et al., 2024).

The loss of habitats and biodiversity is considered one of the most pressing scientific challenges, due to the complexity of determining the extent to which biodiversity is affected by current human and climate-related threats. This affects ecological and economic productivity, as well as the role of biodiversity in promoting well-being and social justice. It is therefore essential to strike a balance between short-term private interests and broader public interests by integrating biodiversity protection into national economies and governance frameworks (Bridle et al., 2025). States must also contribute through their legal systems to the protection of biodiversity, given its critical role in achieving various Sustainable Development Goals (SDGs), including the preservation of terrestrial and aquatic ecosystems, which consequently supports climate regulation and the provision of clean water and food.

At the international level, many countries have sought to contribute to the preservation of biodiversity by joining various agreements and strategies dedicated to biodiversity protection, most notably the Convention on Biological Diversity (CBD), which came into force on December 29, 1993. Two protocols associated with the CBD are the Nagoya Protocol on Access to Genetic Resources and the Fair and Equitable Sharing of Benefits Arising from Their Utilization, and the Cartagena Protocol on Biosafety (The Convention on Biological Diversity, 2025). Complementing these efforts are other international initiatives, including the European Union Biodiversity

Strategy for 2030, aimed at protecting biodiversity within Europe (The European Commission, 2025), and the Food and Agriculture Organization's (FAO) Strategy on Mainstreaming Biodiversity across Agricultural Sectors (Food and Agriculture Organization of the United Nations, 2025), adopted in 2019. On December 18, 2022, the parties to the CBD adopted the Kunming-Montreal Global Biodiversity Framework (GBF), which replaced the Strategic Plan for Biodiversity 2011–2020 and its Aichi Targets (Kunming-Montreal Global Biodiversity Framework, 2025).

In this regard, the UAE has taken steps to protect its biodiversity, having signed the aforementioned CBD on November 24, 1999. At the national level, the UAE adopted the Biodiversity Strategy 2031, which has made the sustainability of biodiversity a central vision. To achieve this goal, the strategy incorporates several projects and measures, including the preparation of the Red List project (2018–2020), through which the UAE collaborated with the International Union for Conservation of Nature (IUCN) to achieve global indicators and sustainable development goals by studying and assessing the status of certain species of animals and plants, evaluating the effectiveness of existing policies in protecting them, and identifying further measures required for their continued protection. The National Invasive Species Strategies and Action Plans (NISSAP) (2022–2026) seeks to combat certain types of invasive non-native animal and plant species that pose environmental, health, economic, and social risks. Additionally, the establishment of nature reserves has been undertaken to address the challenges affecting terrestrial and marine biodiversity.

The aim of this study is to identify the consumer's right to food and clarify its boundaries at the global level and within the UAE from a legal standpoint, through ensuring the achievement of sustainable food security. This involves recognizing the important role of sustainable agriculture, while considering the health and safety of consumers and the ecosystem.

Accordingly, within the framework of studying the issue of biodiversity loss and its impact on the consumer's right to food, as well as the role of sustainable agriculture in addressing this challenge, this study will discuss the legal framework for achieving food security for the consumer. It will then examine the legal aspects of activating the role of sustainable agriculture as one of the solutions to the problem of biodiversity loss, which contributes to achieving food security for the consumer. This will lead to results and recommendations that contribute to adopting measures aimed at preserving biodiversity, which reflect positively on the availability of food for the consumer while simultaneously protecting the environment.

1. LEGAL FRAMEWORK FOR ACHIEVING FOOD SECURITY FOR THE CONSUMER

The definition of food security has evolved over time. Initially, it referred to the availability of global food supplies, later expanding to encompass the accessibility of food at all times by all individuals. The concept further developed to include the dimensions of food availability, accessibility, and the quality of consumption. In its most recent formulation, the concept of social acceptance has been incorporated, resulting in a comprehensive definition that includes four key components: availability, accessibility, adequacy, and acceptance (Sulaiman et al., 2010). Accordingly, food insecurity is defined as the lack of access to food at all times for all individuals.

Food insecurity affects humans in multiple dimensions, including physical, social, and health aspects. Malnutrition, whether due to excessive or insufficient nutrition, can result in physical ailments such as obesity, stunted growth, and chronic diseases like heart disease and diabetes. Socially, malnutrition perpetuates the cycle of hunger and poverty, which contributes to poor health outcomes. Psychologically, malnutrition is associated with anxiety, stress, and depression (Tohit et al., 2025).

Therefore, the right to food is considered one of the most fundamental consumer entitlements, universally recognized by both international and national legal systems. Since food is a central element within the SDGs, which aim to "end hunger, achieve food security and improved nutrition, and promote sustainable agriculture," achieving these objectives requires a fundamental transformation of global food and agricultural systems. This paradigm shift involves a set of solutions, including eradicating hunger, ensuring access to sufficient and safe food for all, and achieving the sustainability of food production systems (United Nations, 2025a). Biodiversity plays a crucial role in realizing these goals. Consequently, the consumer's right to food entails specific conditions, including its sufficiency in quantity and safety in quality (1.1). Naturally, the responsibility for achieving food security for consumers falls upon states (1.2).

1.1. Consumer's Right to Adequate and Proper Food:

For a right to be recognized as a legal privilege that grants entitlement, it must be accompanied by corresponding obligations imposed on a specific person or entity (Skogly, 2007), applicable at both national and international levels.

1.1.1. The international legal framework for the right to food:

One of the most important means of promoting and ensuring sustainable food security is the recognition of the individual's right to adequate food and its enforcement as a legal entitlement (Raponi, 2017). Accordingly, international legal systems emphasize the necessity of providing sufficient food—the minimum necessary to sustain life—for all, in order to prevent starvation, as this constitutes a fundamental pillar for safeguarding human dignity and existence. This right has been recognized at the national level in various forms, with some countries explicitly enshrining it in their national laws, while others recognize it implicitly (Milian Gómez, 2024).

At the international level, the right to adequate food has been enshrined in several instruments, including the United Nations Charter of 1945, the Universal Declaration of Human Rights (UDHR) of 1948, the Declaration of the Rights of the Child of 1959, and the International Covenant on Economic, Social, and Cultural Rights (ICESCR) of 1966. The Committee on Economic, Social and Cultural Rights (CESCR) has emphasized the significance of the right to adequate food, asserting that the ICESCR addresses this right more comprehensively than any other international instrument (United Nations, 2025b). In its clarification of the normative content of Article 11 of the ICESCR, the CESCR stated that "The right to adequate food is realized when every man, woman and child, alone or in community with others, has physical and economic access at all times to adequate food or means for its procurement." The CESCR also highlighted the intrinsic link between the concept of sustainability and that of adequate food or food security, emphasizing that food security entails not only the availability of food but also access to it both in the present and the future (United Nations, 2025b). Accordingly, the adequacy of food is connected to social, economic, cultural, climatic, and environmental conditions, while sustainability relates to ensuring food availability and access over the long term (Becchi, 2011).

It is therefore essential that access to adequate food be recognized as a fundamental human right guaranteed by the international human rights system, constituting a binding obligation in times of both peace and conflict. As a right characterized by continuity, its implementation by states must be assessed under both ordinary and exceptional circumstances to determine the extent of their success or failure in fulfilling this obligation (Bender Smith, 1997).

We emphasize, with regard to the necessity of providing food, that the concept of food adequacy does not merely imply making food available regardless of its impact on the consumer; rather, the food must be safe. The right to safe food means that of obtaining food that does not cause harm to the consumer during its handling or consumption, in accordance with its intended use. The requirement of food safety is closely linked, on the one hand, to the consumer's right to food, and on the other, to their right to health. Furthermore, food safety is intrinsically connected to sustainable development due to its clear role in promoting economic growth, achieving food security, safeguarding human health, and advancing agricultural development.

Thus, food safety at the international level is a necessity rather than a luxury, as affirmed by all international instruments related to food in their various forms.

Therefore, international food safety standards cover the different stages that food undergoes, including production, packaging, wrapping, storage, transportation, and distribution, in addition to the crucial cleanliness requirements at every stage of food processing. The same applies at the national level, with variations according to each country's legislation.

1.1.2. The national legal framework for the right to food

At the national level in the UAE, entitlement to food is a fundamental consumer right. The legislator, in Consumer Protection Law No. 15 of 2020 and its Executive Regulations, affirmed the protection of all consumer rights without explicit mention of the right to adequate food. This approach has been adopted by the UAE legislator with respect to all consumer rights, not just the right to food, and is similarly reflected in most consumer protection laws in Arab countries. The definition of commodity in the aforementioned law includes "Any natural substance or industrial, agricultural, animal, manufactured, intellectual or technological product, including the primary elements of the substances and the components that go into the product."

All national legal systems worldwide have recognized the importance of food safety, with countries incorporating strict food safety regulations into their national legislation, albeit at varying levels depending on their respective economic, scientific, technological, and even religious contexts. Furthermore, international organizations concerned with this matter have established their recommendations regarding food safety. This confirms that food safety is an objective at both the international and national levels, and therefore, it cannot be limited to a specific group or organization; rather, efforts must be combined to achieve this goal (Yun Li & Shi, 2022).

We have previously pointed out that the UAE legislator did not explicitly state in the aforementioned Consumer Protection Law the consumer's right to adequate food. However, it clearly emphasized, within its objectives, the necessity of maintaining the safety and health of the consumer when acquiring a commodity, which naturally includes food. Moreover, it enacted a specific law on food safety, namely Federal Law No. 10 of 2015 and its Executive Regulations, which underscored the need to protect consumer health through the following: First, ensuring the suitability of circulated food by verifying its fitness for human consumption, while monitoring it throughout all stages of the food chain. Second, the necessity to eliminate or at least minimize all risks associated with food. Third, protecting consumers from foods harmful to their health, whether due to spoilage, adulteration, or general unsuitability.

Conversely, Indian legislation is distinguished from others by the existence of Law No. 20 of 2013, which is related to food security and aims to provide good quality food for the Indian people in sufficient quantities and at reasonable prices to live a life with dignity.

The program to eradicate hunger in Brazil is considered a good example of the commitment of countries to achieve food security. The program is not only concerned with eliminating hunger but also with providing food of suitable quality and quantity for all, thereby ensuring their physical and mental health (da Silva et al., 2012).

1.1.3. Is the right to food a necessity or a luxury?

From this perspective, it becomes evident that providing adequate food for all consumers is a necessity emphasized by both international and national legal systems, not a luxury. However, this raises the question: Is the diversity of food and its alignment with consumer choices a legal necessity, or merely a luxury?

We address this question by first clarifying the concept of consumer food choices, which are influenced by numerous factors (Swinburn et al., 2013). The social, cultural, economic, and political environment shapes an individual's food preferences significantly (Chen & Antonelli, 2020). Food consumption depends on personal factors, including health concerns, knowledge, and nutritional status. In addition, environmental factors such as family size, level of urbanization, and seasonality affect these choices. Furthermore, the characteristics of food—such as taste, texture, and price—also play a crucial role in shaping consumer decisions (Randall & Sanjur, 1981).

Secondly, from our perspective, the concept of food diversity and its compatibility with consumer choices depends on several factors. Developing countries suffering from poverty and widespread famine—due to political, economic, or even natural causes such as exposure to natural disasters or other reasons—cannot insist on the necessity of food diversity and its alignment with consumer choices. At such times, the focus is on keeping people alive by providing the minimum basic needs required for survival, including food. The situation differs in politically stable countries where individuals' income levels afford them the luxury of food diversity and multiple options that correspond to choices based on their social and cultural backgrounds, which play a significant role in determining their food preferences. From a legal standpoint, neither international nor national frameworks impose an obligation on states to provide food that aligns with consumer preferences. Rather, the focus has been on ensuring the provision of adequate food, as previously discussed, and on nutritious food, which will be further addressed.

From all the above, we conclude that the consumer's right to adequate and safe food is a necessity, not a luxury. This has been emphasized by international and national legislation. Meanwhile, the issue of food diversity and its compatibility with consumer choices based on social and cultural backgrounds can be seen as a necessity, but also as a luxury; due to multiple considerations, including the political and economic conditions of the country, individual income levels, environmental conditions, and other influencing factors, as previously stated in this regard.

However, another issue raised here is the consumer's lack of sufficient information about the type of food and its effects on their health, such as the impact of genetically modified foods, as well as their effects on the environment. This information is crucial, as it enables consumers to make conscious and informed decisions when choosing products, especially in the absence of legal warnings regarding the use of genetically modified foods and products (Ballona-Castro & Herrera-Mena, 2025). Other studies suggest that the negative impact of genetically modified foods remains an unproven issue and requires further direct research to assess the potential health effects arising from their consumption (Garcia-Alonso et al., 2022). These studies also emphasize the necessity of conducting various biosafety assessments to evaluate the risks associated with genetically modified foods, which should be undertaken by relevant health authorities prior to their approval for human consumption (Akinbo et al., 2021).

One of the most important consumer rights guaranteed under international and national legislation is that of making informed decisions when acquiring products and goods based on sufficient knowledge. This right naturally extends to food, granting the consumer the right to make informed choices about all types of foods, including the right to know whether available foods are genetically modified. The consumer then has full freedom to decide whether to purchase them or not. Since informed decisions rely on consumer awareness, consumer protection laws impose legal obligations on relevant authorities to ensure the consumer's right to education. To address the negative effects of genetically modified foods, Federal Law No. (9) of 2020 on the Biosafety of Genetically Modified Organisms (GMOs) and the Products Thereof and its Executive Regulations were enacted in the UAE, allowing a genetic modification percentage of less than 0.9%.

In this regard, the UAE has incorporated into its national legislation the position of the Food and Agriculture Organization (FAO) concerning protocols for the safety of genetically modified plants, ensuring that national standards for risk assessment and food safety are aligned with international benchmarks (Al-Bagham, 2003).

Regarding the freedom of Arab consumers to choose genetically modified plant-based foods, one study confirms that several factors influence their decisions concerning the consumption of such products. While they acknowledge the significance of genetic engineering technology, they express simultaneous reservations about genetically modified foods due to potential health risks, whether known or unknown (Bakr & Ayinde, 2014).

Since the right to food is closely linked to climate and its changes, the International Court of Justice, in its advisory opinion issued on July 23, 2025, regarding the 'Obligations of States in respect of Climate Change,' acknowledged the duty of states to prevent serious harm to the environment. It deemed this an obligation to exercise due diligence, which ultimately necessitates a tangible assessment of what is reasonable under the circumstances in which a state finds itself. The Court asserted that the standard of due diligence concerning the prevention of harm to the climate system is established and strict; thus, mere diligence and the adoption of appropriate measures are insufficient. It requires a high degree of vigilance in the implementation of those measures, along with the necessity of imposing administrative oversight. Furthermore, the Court emphasized in the same opinion the importance of cooperation among states to preserve the climate system, recognizing it as an endeavor that necessitates a degree of collaboration and cannot be accomplished by states alone. The Court also considered the fundamental human rights treaties, including the International Covenant on Economic, Social and Cultural Rights and the International Covenant on Civil and Political Rights, adopted in 1966 The human rights recognized under customary international law constitute part of the most relevant applicable laws (JUSTICE, 2025).

1.2. Commitment of Countries to Achieve Food Security

Biodiversity is inseparable from agriculture and cannot be considered separate from each other, as biodiversity forms an important foundation for sustainable agriculture (Sunderland, 2011). Food security, which is primarily based on agriculture, cannot be achieved without the maintenance of biodiversity and the implementation of sustainable agricultural practices.

Many factors intersect to influence the commitment of countries to achieving food security, like climate change, economic levels, trade barriers(Zorbas et al., 2024), and more. Still, it's up to countries to take on international and national commitments to make sure they secure their food security.

1.2.1. Commitment to Achieving Food Security Internationally:

Given the importance of food security, the Committee on World Food Security (CFS) was established in 1974 as part of the global governance framework for food security, with the aim of eradicating hunger, eliminating all forms of malnutrition, and making the achievement of food security a global priority.

According to the World Food Summit organized by the FAO, food security is realized when all people—at all times—have physical and economic access to sufficient, safe, and nutritious food that meets their dietary needs and preferences for an active and healthy life (Food and Agriculture Organization of the United Nations, 2025b).

There is no doubt that climate change contributes to biodiversity loss, which affects food security. Transitioning to a sustainable food system is considered an effective method of addressing the climate crisis, even though it remains largely dependent on individual behaviors and cannot be imposed as a global obligation. Changing the global food system faces social challenges, foremost among them being the consumer's freedom to choose their food, as every individual has the absolute freedom to make their own food choices. From this perspective, transitioning to a more sustainable global food system with significant environmental impacts requires a shift in the collective perception of food. Some argue that the individual's freedom to choose their food is a vague concept, as it overlaps with their right to sufficient food in terms of both quantity and quality, and lacks depth. Therefore, the relationship between individuals and food must be linked to the environment, which will support the transition toward a sustainable lifestyle. This shift will contribute to mitigating the effects of climate change on the one hand and adapting to it on the other (Onnis, 2023).

In general, it is the responsibility of states to achieve food security for consumers, given the significant role of food security in supporting economic growth. The FAO has established specific criteria for achieving food security, namely, ensuring a consistent supply of adequate food, whether through imports or local production, and securing sufficient resources to guarantee access to nutritious food (Global Food Security Cluster, 2023). Accordingly, a country is considered to have achieved food security when it meets these criteria successfully.

Countries with large populations, such as India and Egypt, may struggle to achieve food security, unlike countries with smaller populations. In India, despite the increase in agricultural production, as confirmed by various studies and statistics, the challenge lies in providing food sustainably, increasing its production, and the disparity in individual income levels, which continues to impede food security. Moreover, the distribution of food across the Republic of India poses the greatest challenge to achieving food security, necessitating the attention of policymakers, as it cannot be assumed that a rise in individual income and agricultural production alone are sufficient to guarantee food security (Narayanan, 2015). This is what the Indian legislator addressed with Law No. 20 of 2013 regarding nutrition in India. We believe that the population count is not the only reason behind the failure to achieve food security for the population. The nature of the land, the climate conditions, consumer behaviors, and the economic levels of countries and individuals are all factors that significantly contribute to the ability of countries to achieve food security or not.

Climate change, particularly in desert countries such as the United Arab Emirates, may adversely affect the ability of nations to achieve food security. Climate is one of the key factors influencing the life cycles of living organisms, and consequently, biodiversity of species and ecosystems (Hui, 2013). The ongoing rise in global temperatures will lead to the extinction of plants that thrive in cold or temperate environments (El-Keblawy, 2014). The relationship between climate change, rising temperatures, and fluctuations in rainfall, and their impact on agriculture has been scientifically established (Ahmed et al., 2011).

1.2.2. Commitment to achieving national food security:

The UAE legislator emphasized the necessity of achieving food security through various provisions of agricultural legislation. It mandated the UAE Ministry of Climate Change and Environment (MOCCAE) to develop the necessary policies, strategies, and regulations to ensure food safety throughout the food supply chain in order to achieve food security, as stipulated in Article (4/1) of Federal Law No. (10) of 2015 on Food Safety. Also, Article (4/4) of the same law requires raising consumer awareness about food safety and announcing the nature of any associated risks, if present. The legislator also incorporated food security as an objective in other pieces of legislation, including Federal Law No. (9) of 2013 Concerning Plant Genetic Resources for Food and Agriculture, and Cabinet Resolution No. (31) of 2018 on Community Agriculture. Under this resolution, the MOCCAE, local authorities, and institutions are required to promote the cultivation of plants that contribute to strengthening food security. Furthermore, Federal Law No. (3) of 2020 regarding the Regulation of the Strategic Stock of Food Commodities was issued to ensure food security during emergencies, emphasizing the necessity of maintaining a safety stock, meaning a specific percentage of the strategic food stock that must not fall below the level determined by that law.

The UAE has established several entities responsible for monitoring and ensuring food security in the country, including, for example, the MOCCAE, the Emirates Food Security Council (EFSC), and the UAE Food Bank.

By way of elaboration on this matter, the Ministry of Climate Change plays a significant role in achieving food security, owing to the tasks and competencies conferred upon it by the Emirati legislator. The Ministry is responsible, for example, for formulating plans, strategies, and policies across various sectors, including environmental protection, biodiversity conservation, agricultural, animal, and fishery development, as well as combating desertification. It is also entrusted with enhancing the qualitative standards of agricultural, animal, and fish production. Furthermore, the Ministry proposes accession to relevant international treaties and conventions that support the fulfillment of its mandates, including those related to the realization of food security.

The Emirates Food Security Council (EFSC) aims to manage food security across various authorities within the country (Ismail and & Alheji, 2023), in addition to achieving the objectives of the National Food Security Strategy. In pursuit of self-sufficiency in domestic agricultural production, the enhancement of agricultural economic returns, and the utilization of significant global technological advancements in food production, the National Sustainable Agriculture System has been launched.

The UAE Food Bank constitutes a local, regional, and international cornerstone in establishing a comprehensive system grounded in humanitarian, social, economic, and cultural dimensions, all aimed at fulfilling the overarching objective of feeding those in need. It is founded on the concept of preserving surplus food, storing it, and distributing it to individuals in need through the competent national authorities, after ensuring that it is packaged and stored in accordance with applicable health standards and regulations. This initiative contributes to the conservation of surplus food that remains fit for consumption and promotes social solidarity. It is noteworthy that the UAE Food Bank enters into agreements and partnerships with relevant entities in the hospitality and food sectors—such as hotels, food manufacturing facilities, food service providers, and farms—which have played a vital role in advancing the achievement of its objectives (Al-Qodsi et al., 2024).

The UAE has also developed several strategies to ensure its food security, including the National Food Security Strategy 2051, which aims to position the UAE as a global leader in food security based on innovation. This strategy includes 38 key initiatives, comprising both short-term and long-term initiatives (Government of the United Arab Emirates, 2024b). In addition, the UAE has adopted the Water Security Strategy 2036, which seeks to ensure the sustainability and continuity of water availability under both normal and emergency conditions (Government of the United Arab Emirates, 2024c). Furthermore, the UAE Green Growth Strategy aims to build a green economy in the UAE under the slogan "Green Economy for Sustainable Development" (Government of the United Arab Emirates, 2024a).

Emphasizing its commitment to achieving national security, the UAE strives continually to develop a national framework with a fundamental grounding in the promotion of sustainable agriculture as an integral component of national security. This commitment is evidenced by the launch of the National Food Security Strategy 2051—as previously noted—as well as the state's accession to numerous international agreements, treaties, and protocols. At the domestic level, the UAE has enacted various legislative measures to support this objective (Al-Qodsi et al., 2024).

According to the Global Food Security Index (GFSI) classification for 2022, the UAE ranks seventh globally in terms of natural resource availability, thirty-first in food affordability, and is among the top sixteen countries worldwide in food quality, safety, and availability. Overall, the GFSI places the UAE first in the Middle East and North Africa and twenty-third globally (Economist Impact, 2022).

Undoubtedly, biodiversity loss has a considerable effect on the achievement of food security, which impacts the availability of sufficient and safe food for consumers and reduces access opportunities, particularly in countries with limited economic resources. Therefore, all necessary measures must be taken to address the decline in biodiversity and mitigate its effects.

In this regard, the UAE has established the National Biodiversity Strategy 2031. An analytical study examining the factors influencing the implementation of this strategy indicated, from a legal perspective, that the multiplicity and unification of legislation at the national level ensure the consistent application of environmental protection laws. Moreover, the enforcement and monitoring of these laws contribute to the protection of citizens and natural species (Ministry of Climate Change and Environment, 2024).

1.2.3. Is achieving food security a commitment to exert due diligence or to achieve a result?

From this perspective, the question arises: Can the obligation of states to achieve food security for consumers be regarded as an obligation of due diligence, meaning that states would be exempt from responsibility if they demonstrate that they have made reasonable efforts to provide food, even if the intended result is not achieved? Or should this obligation be considered one aimed at achieving a specific result, making states responsible for failing to attain that result?

We can begin to answer this question by clarifying the distinction between an obligation to exercise due diligence and an obligation to achieve a result. In the first case, it is sufficient for the obligor to exercise reasonable care in fulfilling their obligation, and they will not incur legal liability if the intended result is not achieved—as is the case with a physician, who is not required to guarantee the patient's recovery. In contrast, an obligation to achieve a result imposes legal responsibility on the obligor if the specific result is not attained; a common example is a contractor's obligation to complete construction, where they may be held liable if they fail to build according to the agreed terms.

Regarding the obligation of states to provide food to their populations, it can be divided into two categories: First, international obligations arising when a state becomes party to international instruments related to human rights in general—such as the ICESCR—or specifically to the right to food, such as the Rome Declaration on World Food Security. The second category consists of national obligations, which arise when legislation is enacted at the national level affirming the consumer's right to food and specifying the legal nature of that right.

It is well established with respect to the first category of obligations that some countries incorporate their international obligations—arising from accession to a convention, treaty, or otherwise—into their domestic legal systems by enacting national legislation. Others regard their obligations under international law and their domestic legal obligations as forming a unified legal framework (Schachter, 2017). In both cases, it is assumed that states are bound to fulfill their international obligations once they have agreed to sign the relevant international instrument. This entails integrating the provisions arising from those obligations into their national legislation, with national laws enacted in accordance with the requirements of international legislation, including those related to the environment (Alegría, 2024). From this, it may be concluded that providing food as an international obligation constitutes a commitment to achieve results for all member states.

For example, the ICESCR imposes different levels of obligations on the states parties regarding the human right to food. These include the obligation to respect, which requires states to refrain from taking any measures that would hinder individuals' access to food; the obligation to protect, which requires states to ensure that no individual or entity deprives others of access to food; and the obligation to fulfill, which requires states to engage actively in promoting food security and to intervene directly to provide food in any situation where individuals or groups are unable to obtain it due to circumstances beyond their control. Consequently, the obligation of states to provide food constitutes an obligation to achieve a result, provided that they are parties to the ICESCR.

Because food is one of the basic human needs, the commitment of countries to provide it under national legislation constitutes an obligation to achieve a result; they cannot be exempt from responsibility simply by exercising due diligence. The fulfillment of this obligation requires the actual provision of food to consumers.

In confirmation of this, some residents of the Torres Strait Islands filed a complaint with the United Nations Human Rights Committee against Australia, arguing that climate change and heavy rainfall have led to the degradation of agricultural land and trees, negatively impacting their ability to obtain food. Additionally, rising sea levels have harmed coconut trees that are integral to the diet of the islanders. On September 23, 2022, the Human Rights Committee acknowledged Australia's violation of the rights of the residents of those islands due to its failure to adequately protect them from the adverse effects of climate change (Lentner and Cenin, 2024).

In contrast, the Indian legislator imposed on the government the obligation to achieve food security under Article 8 of Chapter Three of Law No. 20 of 2013 regarding food security. It obligated the government, in case of failure to achieve this objective, to provide compensation to those entitled according to the aforementioned law, thereby reinforcing the commitment of countries to achieve food security as an obligation to achieve a result.

Despite clear efforts at both the international and national levels to uphold the human right to sufficient and healthy food, challenges remain in achieving food security for consumers (Milian Gómez, 2024). In light of the foregoing, and given that countries do not rely solely on their domestic production but depend to varying degrees on food imports, we believe that some countries will face difficulties, while others may be unable to fulfill their obligation to provide consumers with sufficient and healthy food as an essential need. This is particularly true in the context of population growth, climate change, and the widespread reliance on agricultural methods involving chemical substances and pesticides—favored for being cheaper, faster, and more productive—amidst the decline in global biodiversity. Therefore, serious steps must be taken to address the challenges facing biodiversity to ensure food security.

To avoid blaming climate change alone for its impact on the environment and biodiversity, and consequently on food availability, we emphasize that humans also play a role in this impact by destroying plant life for various reasons like urban expansion, trade, consumption, and other factors. This calls for international and national intervention to stop or at least reduce that impact. A judicial application of this was seen in the ruling of the Brazilian Supreme Federal Court in April 2024 regarding deforestation, published on June 26, 2024, which acknowledged the structural deficiencies in the legal protection of the Amazon region and ordered the government to commit more robustly to combating illegal deforestation by implementing a plan to conduct inspection measures. It also mandated the necessity for the government to ensure transparency regarding the information it provides about the actions and results taken in compliance with the court's decision (June 26, 2024).

2. LEGAL ASPECTS OF ACTIVATING THE ROLE OF SUSTAINABLE AGRICULTURE IN PROVIDING FOOD FOR CONSUMERS

Despite efforts by countries to achieve food security as a global imperative—regardless of political, economic, or social systems—hunger and food insecurity remain significant problems in some developing countries (Sibhatu & Qaim, 2017). This means that the food security risks remain present in developing countries (Simola et al., 2022). This is due to climate change causing a decline in agricultural production alongside a growing population that increases food demand. Other factors exacerbate the situation, including rising food prices and low individual incomes (Pawlak & Kołodziejczak, 2020). This matter requires conducting and implementing studies that work to mitigate the impacts of this problem (Semchuk et al., 2025).

Therefore, the relationship between the right to food and the preservation of the ecosystem is an extremely close one; this connection also reflects on the achievement of sustainable development at the global level. As previously noted, the right to food falls within the scope of both international and national obligations as a fundamental human right, with domestic consumer protection laws affirming the consumer's right to food.

There is a significant and clear link between climate change, biodiversity, and the consumer's right to access food.

Figure 1. The relationship between climate change, biodiversity, and food availability

The relationship between climate change, biodiversity, and the consumer's right to food is a direct Correlation; the more climate change occurs, such as changes in weather patterns, temperature variations, rising sea levels, droughts, storms, floods, and other changes, the more biodiversity, which is linked to all forms of life on Earth, deteriorates. This, in turn, affects the availability of food for humanity.

Accordingly Biodiversity, soil health, and climate change play a crucial role in achieving food security for consumers; therefore, it is essential to integrate all factors affecting agricultural and food systems—including environmental, health, economic, and social considerations—to address these interconnected challenges effectively. This calls for the establishment of policies that embed human rights within agricultural practices and the adoption of eco-farming as a fundamental approach to resolving food-related and ecosystem issues, particularly those concerning biodiversity, soil health, land governance, natural resources, and justice. Such measures will contribute positively to long-term environmental sustainability and safeguard the well-being of both present and future generations (Owuonda & Olembo, 2024).

Given the above, if consumers have the right to obtain sufficient and safe food—as one of the fundamental rights guaranteed by all international and national legislations—and if this right constitutes an obligation of providing results that is incumbent upon states, as previously explained, the fulfillment of this obligation must not come at the expense of the environment as a whole. Therefore, states must, first, integrate the consumer's right to food into agricultural practices to ensure the availability of sufficient and safe food for consumers without causing adverse environmental impacts (1.2); and second, adopt environmentally friendly agricultural practices to address challenges related to both food security and the preservation of the ecosystem (2.2).

2.1. Integrating the Consumer's Right to Food Within Agricultural Procedures

Food security for consumers means fulfilling their nutritional needs at all times (Awoke et al., 2022). Therefore, integrating the consumer's right to food within agricultural procedures means ensuring the protection of the consumer's right to adequate and proper food at all agricultural stages, extending to the post-consumption stage.

2.1.1. The international legal framework:

Internationally, for example, The United Nations Framework Convention on Climate Change (UNFCCC) aims to preserve ecosystems, allow them to naturally adapt to climate changes, and ensure that food production isn't jeopardized, along with the overall goal of moving towards sustainable economic development. The UN Framework Convention has achieved a national response to tackle climate changes (both adaptation and mitigation), which requires adopting various national mechanisms and procedures to address climate changes, while also needing global cooperation on this matter(Tompkins and Amundsen, 2008).

As there exists there is the CBD, which has been discussed in this research and is concerned with preserving biodiversity, including the diversity of living organisms such as plants and various ecosystems like forests, mountains, and deserts, among others, from anything that affects them negatively, whether arising from human activities or natural causes.

The International Plant Protection Convention (IPPC) is also targeted specifically at safeguarding plants against epidemics and diseases. It is regarded as the only international agreement that establishes global standards for plant health, which serve as its primary mechanism for achieving its objectives (Food and Agriculture Organization of the United Nations, n.d.-b). However, the protection provided by the IPPC is limited to the stage of transboundary movement of plants, focusing on implementing measures to ensure plant health during import and export. Importantly, it seeks to achieve this without imposing unnecessary restrictions on international trade. The core role of the IPPC is to prevent the cross-border spread of agricultural diseases and epidemics (Holden, 2024).

In addition, the International Treaty on Plant Genetic Resources for Food and Agriculture (ITPGRFA), adopted under the auspices of the United Nations and linked to the CBD, focuses on the use of all types of agricultural crops, the development of agricultural systems, the protection of farmers' rights, access to resources, and the fair and equitable sharing of benefits (Food and Agriculture Organization of the United Nations, 2001).

It is observed that, at the international level, the concept of ensuring sufficient and safe food for the consumer—and accordingly achieving food security—is present, yet it is not encompassed within a single legal instrument that guarantees the provision of safe food for the consumer throughout all stages that the plant undergoes. Each instrument, by its nature, is limited to a specific stage, such as the initial stage concerning the provision of all primary factors necessary for plant growth, including climate, soil, and water, which is addressed by the CBD; the import and export stage, as seen in the IPPC; and the focus on plant genetic resources, as in the case of the ITPGRFA. Consequently, international instruments related to food collectively encompass, to a certain extent, the concept of integrating the consumer's right to adequate and safe food within agricultural procedures.

2.1.2. The national legal framework:

At the national level in the UAE, the state has sought to ensure the integration of the consumer's right to safe food within agricultural procedures through various legislative measures. One of the earliest of these is the Federal Law Concerning Agricultural Quarantine issued in 1979, which subjects all plants and agricultural products to quarantine regulations in cases of suspected contamination with pests or diseases specified in the law. This legislation ensures that no party may introduce infected plants or agricultural products into the country unless they have been disinfected at the owner's expense.

One study on agricultural quarantine highlights that it faces complex challenges arising from climate change and the expansion of international trade. It underscores the necessity of adhering to quality standards related to food security by adopting key performance indicators (KPIs) to ensure the effectiveness and efficiency of compliance with those standards. Furthermore, the study advocates leveraging technological advancements, including artificial intelligence, to enhance transparency and improve efficiency. This would strengthen the role of quarantine in safeguarding the agricultural sector from threats and contribute to enhancing the competitiveness of national agriculture (Beniawan & Sinambela, 2025).

In 1992, the UAE legislator issued several federal laws related to agriculture. Of particular relevance here are, first, the Federal Law Concerning the Production, Import and Handling of Seeds and Kernels, which prohibits the multiplication, import, or trading of seeds and kernels without obtaining a permit from the competent authority and in accordance with the conditions stipulated in the law. Second is the Federal Law Concerning Plant Nurseries and Organizing, Producing, Importing and Handling Nursery-Plants, which obliges the competent authorities to establish the foundations for creating nurseries, as well as the conditions, specifications, and methods for producing, importing, trading, and advertising seedlings within the country. It is noteworthy that these laws are primarily connected to the stage of importing plants and agricultural products.

To achieve the objectives of the ITPGRFA, to which the UAE is a party, the UAE legislator enacted Federal Law No. (9) of 2013 Concerning Plant Genetic Resources for Food and Agriculture. Among its key objectives, the law seeks to address the depletion of genetic resources for food, plants, and agriculture; ensure their protection and conservation; and regulate their access and trade in a manner that guarantees the sustainability of their benefits and contributes to achieving food security. The law also aims to promote equity and fairness in the sharing of benefits derived from the use of those resources. Pursuant to its provisions, a gene bank for plant genetic resources of food and agriculture was established in the UAE.

In addition to the Food Safety Law promulgated under Federal Law No. (10) of 2015, which was addressed earlier in this study, the legislator emphasized the extension of its provisions to encompass the primary stages of food production. These include the phases preceding food production, processing, and preparation, specifically covering cultivation and harvesting. Furthermore, the Public Health Law, issued under Federal Law No. (13) of 2020, addresses public health in its comprehensive dimensions, including the nutritional aspect, by establishing provisions that promote healthy alternative food choices and mandating adherence to the standards of the World Health Organization (WHO) adopted within the state, in alignment with national legislation.

While all the aforementioned laws have sought to ensure that consumers have access to safe plant-based food, whether produced domestically or imported, the right to adequate food—although generally not posing a challenge in the UAE under normal circumstances—must also be secured during emergencies. To this end, Federal Law No. (3) of 2020 regarding the Regulation of the Strategic Stock of Food Commodities was enacted. Under this law, all suppliers are obligated to maintain and manage the strategic stock of food commodities in accordance with the specifications and standards approved within the UAE.

To establish a comprehensive system for maintaining public health and mitigating all risks that may affect human health at all stages of life—including those related to nutrition—and to ensure the development of public health policies and strategies that encompass all aspects of human and environmental health, as well as to prepare programs and plans aimed at prevention and achieving an immediate response when necessary, the UAE legislator issued Federal Law No. (13) of 2020 on Public Health.

Moreover, to guarantee the consumer's right to knowledge, support informed purchasing decisions, and strengthen oversight over agricultural products, the UAE legislator introduced the Emirates Sustainable Agriculture Label system pursuant to Cabinet Resolution No. (52) of 2020. This established an accredited label by the

Emirates Authority for Standardization and Metrology (ESMA), which sets out the conditions and requirements for its use. The label is granted to products that comply with the sustainable agriculture standards outlined in the resolution.

From all of the above, we can conclude that the UAE legislator has taken multiple steps to integrate the consumer's right to food within agricultural procedures across various stages. To ensure the effective enforcement of the provisions and regulations set forth in the aforementioned legislation, the legislator has emphasized the necessity of compliance by imposing financial and criminal penalties for violations. Moreover, some laws provide for administrative penalties, while others stipulate deprivation of liberty to address breaches of their provisions.

2.2. Adoption of Eco-Farming to Address Food and Ecosystem Issues: Challenges and Solutions

The significance of eco-farming lies in its pivotal role in developing food and agricultural systems based on essential principles such as biodiversity, connecting rural and urban communities, affirming food sovereignty, and supporting family farming. It also contributes to providing traditional characteristics to organic agriculture instead of relying on modern agriculture, which has a negative impact on the environment and human health (Michelato & Pacini, 2025).

2.2.1. The challenges

The high costs and indirect burdens associated with sustainable agriculture may contribute to the persistence of unsustainable agricultural practices, despite their various adverse consequences and negative impacts, including deforestation, the spread of drought, the proliferation of hard-to-control pest infestations, and disruptions to food production systems in certain agricultural areas. Nonetheless, the role of sustainable agriculture in addressing other agricultural challenges cannot be denied, such as the effective use of organic materials, including crop residues and natural fertilizers, the cultivation of cover crops, and other practices (Harwood, 2020).

Organic farming refers to a system that relies on crop residues, animal manure, and organic waste from outside farms, while shunning industrial inputs such as growth hormones in feed, synthetic fertilizers, and chemical pesticides (Gamage et al., 2023), which contributes to activating eco-farming.

Given the growing importance of biopesticides derived from microbes and plant extracts as sustainable pest control solutions, it has become essential to develop flexible policies and regulations that facilitate the availability of biopesticides for pest management by encouraging responsible innovation (Mawcha et al., 2025).

Climate change refers to ""a change of climate which is attributed directly or indirectly to human activity,", " as defined in Article 1(2) of the United Nations Framework Convention on Climate Change (UNFCCC) (1992).It is indisputable that climate change impacts food systems significantly and in multiple ways, ranging from direct effects on crop production—such as variations in rainfall and rising temperatures—to changes affecting markets, the prices of plant-based food products, and supply chain infrastructure. The effects of climate change on food security naturally vary across regions, influenced by diverse political, economic, social, and environmental factors. These variations result in differing capacities to adapt food systems and reduce their vulnerability to climate change (Gregory et al., 2005). Accordingly, the development of adaptive strategies to address current climatic changes is essential to enhance resilience against future climate-related challenges (Slingo et al., 2005).

2.2.2. The solutions

Eco-farming represents a comprehensive approach to food and agriculture; it is a vital strategy aimed at mitigating the impacts of climate change and adapting to its effects, thus influencing agricultural production positively. This approach relies primarily on crop and pasture rotation alongside organic farming methods to achieve food security and promote a healthy environment (Bezner Kerr et al., 2023). Consequently, eco-farming supports and enhances biodiversity (Christmann, 2020).

Crop rotation is considered one of the simple, environmentally friendly techniques that contribute to meeting the increasing demands for agricultural food while preserving the environment.

Figure 2. Crop rotation in agriculture - grains and legumes

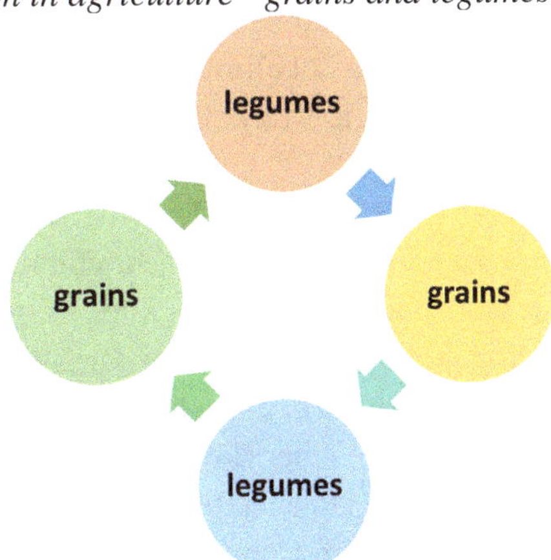

Figure 3. Crop rotation in the cultivation of vegetables - root and leafy

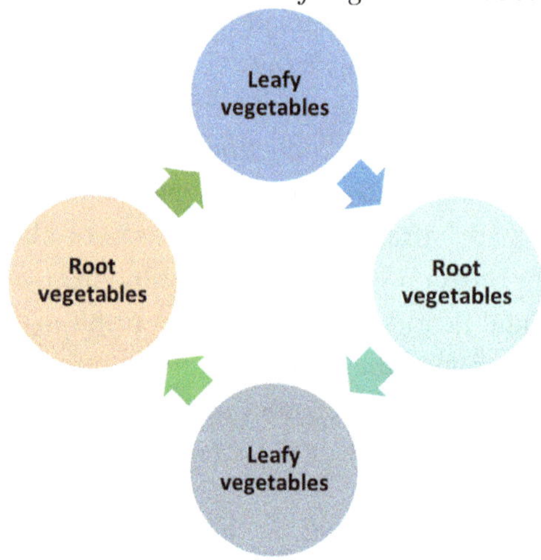

The previous Figure illustrate the contribution of crop rotation to environmental preservation; this mechanism relies on diversifying agricultural crops and not repeating a single type of crop in each season, as the needs of each plant species differ from others. Consequently, this system depends on avoiding the repetition of the same type in farming, which enhances the efficiency of agricultural soil, prevents pests, and maintains nutrient levels in the soil.

Organic farming relies on the living organisms present in the soil to aerate the soil and naturally enhance plant nutrients, as well as on organic fertilizers. This, of course, contributes to the preservation of soil and groundwater from the harmful effects of chemical fertilizers, which positively reflects on biodiversity.

Furthermore, Using RS/GIS to Assess Climate Change Effect on Desert Ecosystems(Al-Blooshi et al., 2020), and Expanding the applications of artificial intelligence will contribute to mitigating the effects of climate change by identifying current and anticipated future climate variations, as well as determining the necessary responses to address them (Shobanke et al., 2025)

To protect consumer health from potential risks associated with agricultural foods produced using fertilizers and pesticides and to ensure food safety, an international food code was established to promote the principles of sustainable agriculture. This code sets food specifications and related standards based on internationally recognized texts, with the dual purpose of ensuring fair food trade on one hand and safeguarding consumer health on the other (Food and Agriculture Organization of the United Nations & World Health Organization, n.d.).

These efforts include the Rotterdam Convention, which addresses the Prior Informed Consent (PIC) procedure for certain hazardous chemicals and pesticides in international trade, and the Stockholm Convention on Persistent Organic Pollutants (POPs), along with the outcomes of joint meetings between the FAO and the WHO concerning pesticide regulation (Food and Agriculture Organization of the United Nations, n.d.-a).

To address the impacts of climate change—which clearly affect the reliance on eco-farming—and to resolve issues related to food and the ecosystem as a whole, the United Nations Framework Convention on Climate Change (UNFCCC) was established at the international level. The UNFCCC defines the adverse effects of climate change as changes that have harmful impacts on natural ecosystems, economic and social systems, or human health and well-being, to which the natural or biological environment is exposed as a result of climate change. The objective of the UNFCCC is to prevent dangerous human interference with the climate system by stabilizing greenhouse gas concentrations in the atmosphere at a certain level within a specified period, in a manner that ensures sustainable economic development, allows ecosystems to adapt naturally to climate change, and guarantees that food production is not exposed to danger.

In line with that, the Supreme Court of Madhya Pradesh in India issued a ruling to cancel the exemption notification dated 24.09.2015 and its subsequent amendments, which allowed the removal of 63 types of trees and plants in forests. The court considered this to have a significant impact on worsening issues related to climate change, given the role these trees and plants play in mitigating the harmful effects of climate change. The court also emphasized the importance of implementing regulatory measures aimed at protecting forests and all natural resources, ensuring that no exemptions contribute to worsening environmental degradation(March 1, 2025).

As a general legal framework for preserving the ecosystem, the UAE legislator issued Federal Law No. (24) of 1999 concerning the Protection and Development of the Environment, with the aim of safeguarding the environment, maintaining its diversity and natural balance, developing the country's natural resources, and preserving its biodiversity. The law also seeks to combat all forms of pollution and protect human health, as well as all other living beings, from anything that may harm the environment or affect its legitimate use. Furthermore, it affirms the implementation of all obligations arising from international or regional agreements related to its objectives. Complementing these efforts, a General Environmental Policy was issued in the UAE in 2020, aimed at enhancing quality of life, supporting sustainability goals, fostering prosperity and economic growth, protecting ecosystems, and ensuring the sustainability of the country's resources—while considering climate change and biodiversity.

In pursuit of establishing a legal framework for agricultural fertilizers to support the goals of sustainable agricultural development, the UAE legislator issued Federal Law No. (39) of 1992. This law defines organic fertilizers as "residuals of plants or animals or a mixture of them, and contain feeding elements for plants and organic materials, which are necessary to improve the characteristics of soil and nature." The law also grants the competent authority the power to determine the types, specifications, conditions, and procedures for the production, manufacture, and use of fertilizers and agricultural ameliorants, as well as their importation, trade, and advertisement within the UAE.

The UAE legislator also addressed the issue of pesticides, given their significant role in achieving sustainable agricultural development. This matter was regulated by Cabinet Resolution No. (37) of 2021, which specified the prohibited pesticides and set out the procedures for importing, manufacturing, and trading in pesticides, restricting them to those authorized by the MOCCAE. The same resolution also specified the conditions, specifications, and information related to pesticides.

Recognizing the impact of climate change on eco-farming and its commitment to mitigating these effects, the UAE issued Federal Decree-Law No. (11) of 2024. This decree reaffirms the goals of the UNFCCC and further aims to leverage international cooperation mechanisms to support the UAE's efforts in climate change mitigation and adaptation. These efforts include capacity building, enhancing private sector contributions, promoting innovation, research and development, and the application of modern technologies. Moreover, the decree supports the UAE's efforts related to sustainable development, strengthens global competitiveness indicators, economic diversification, the green economy, and the circular economy—all through aligning national and local plans and strategies related to climate change.

To promote environmental improvement and the concept of sustainable cities, as well as to strengthen food security by meeting part of families' agricultural needs, Cabinet Resolution No. (31) of 2018 was issued to encourage community agriculture in all its forms and to provide a comprehensive regulatory framework for it. While the resolution encourages individuals and families to utilize residential spaces to grow part of their food needs, and encourages educational institutions to engage in community agriculture for cultivating specific plants identified in the resolution—engaging students in the process and urging coordination between educational institutions and the competent authorities to market the resulting produce within these institutions—the same resolution obligates government buildings to utilize their available spaces for community agriculture, leaving no discretion in this regard.

In reality, the United Arab Emirates has adopted several technological agricultural methods that have achieved great success in facing challenging environmental conditions such as high temperatures and droughts. These include hydroponics, which relies on nutrient-rich water with little or no soil, organic farming that depends on natural and biological materials, and vertical farming which utilizes vertical farming instead of traditional horizontal farming. The UAE is home to the largest vertical farm in the world known as (Bustanica) to supply the aircraft, occupying an area of

330,000 square feet without the use of pesticides and providing 95% of the water needed. This supports the National Food Strategy 2051 of the United Arab Emirates and the United Arab Emirates Water Security Strategy 2036, in addition to smart agriculture that relies on the use of a net house powered by solar energy.

CONCLUSION

This study has demonstrated that the consumer's right to adequate and safe food is guaranteed at both the international and national legislative levels in the UAE. This obligation rests on the shoulders of states, which must remain committed to fulfilling it. The biodiversity loss restricts the ability of countries to meet this obligation, given that humanity relies primarily on agriculture for its food supply. This is not a national problem that can be contained within the borders of a single country or a specific group of countries; rather, it is a global issue whose effects extend to all nations, albeit to varying degrees. The efforts made by the UAE to achieve agricultural and environmental sustainability through national legislation alone are insufficient to provide adequate and safe food within the country, particularly in light of its desert environment and its reliance on agricultural food imports above local production. Addressing this challenge can be supported by considering the following points:

- Since addressing the problem of biodiversity loss cannot be achieved without global cooperation, it is essential to link international instruments related to food, agriculture, and the environment to the Convention on Biological Diversity (CBD). Such integration will help strike a balance between the obligation to ensure food provision on the one hand, and environmental preservation on the other, which guarantees the achievement of sustainability goals concerning both food and the environment.
- National legislation must be in place to help reduce biodiversity loss, as is the case with the UAE's legislation. However, the need for such legislation varies from one country to another depending on political, economic, social, and cultural factors, in addition to other relevant considerations.
- There is a pressing need for international legislation that supports the integration of the right to food into agricultural procedures at all stages of agricultural production. Such legislation should require signatory states to enact national laws to achieve this integration, especially in countries whose agricultural production forms the basis for exporting products to other states.
- It is possible to ensure the provision of adequate and healthy food for consumers, preserve the environment, and maintain biodiversity simultaneously through the adoption of international and national legislation that emphasizes

a reliance on eco-farming, which plays a critical role in mitigating the impacts of climate change and supporting adaptation efforts.

REFERENCES

Ahmed, S. A., Diffenbaugh, N. S., Hertel, T. W., Lobell, D. B., Ramankutty, N., Rios, A. R., & Rowhani, P.AHMED. (2011). Climate volatility and poverty vulnerability in Tanzania. *Global Environmental Change*, *21*(1), 46–55. DOI: 10.1016/j.gloenvcha.2010.10.003

Akinbo, O., Obukosia, S., Ouedraogo, J., Sinebo, W., Savadogo, M., Timpo, S., Mbabazi, R., Maredia, K., Makinde, D., & Ambali, A. (2021). Commercial release of genetically modified crops in Africa: Interface between biosafety regulatory systems and varietal release systems. *Frontiers in Plant Science*, *12*, 605937. DOI: 10.3389/fpls.2021.605937

Al-Bagham, S. H. (2003). Genetically modified organisms in the United Arab Emirates. *National Workshop on Environmental Impact Assessment of the Introduction of Genetically Modified Plant and Animal Species*.

AL-BLOOSHI, L. S., ISSA, S. & KSIKSI, T. (2020). Assessing the environmental impact of climate change on desert ecosystems: A review. *Advances in Ecological and Environmental Research*, *5*, 27–52.

Al-Qodsi, E. M., Jadalhaq, I. M., & El-Maknouzi, M. E. H. (2024). The place of UAE's food security in the national legislation and its role in supporting global food security. *Cogent Social Sciences*, *10*(1), 2319379. Advance online publication. DOI: 10.1080/23311886.2024.2319379

Alegría, P. G. P. (2024). (Mis)adapting domestic law to meet new international environmental and trade rules: How Peru changed its environmental and land use rights laws in response to the European Union deforestation regulation. *Journal of Environmental Law*, *36*(3), 437–444. DOI: 10.1093/jel/eqae019

Awoke, W., Eniyew, K., Agitew, G., & Meseret, B. (2022). Determinants of food security status of household in Central and North Gondar Zone, Ethiopia. *Cogent Social Sciences*, *8*(1), 2040138. Advance online publication. DOI: 10.1080/23311886.2022.2040138

Bakr, S. A., & Ayinde, O. L. (2014). Consumer attitude towards consumption of genetically modified foods in Arab countries. *Middle East Journal of Scientific Research*, *21*(10), 1710–1717.

Ballona-Castro, E. R., & Herrera-Mena, T. K. (2025). Consumer rights in Peruvian territory: An analysis of Law No. 29571 Consumer Protection and Defense Code. Clio. *Revista de Historia. Ciencias Humanas y Pensamiento Crítico*, *5*, 632–658.

Becchi, P. (2011). *Our responsibility towards future generations: Efficiency, sustainability, and justice to future generations*. Springer.

Bender, W., & Smith, M. (1997). Population, food, and nutrition. *Population Bulletin, 51*, 23–32.

Beniawan, A., & Sinambela, L. P. (2025). The relationship between performance management and the quality of animal, fish, and plant quarantine services. *International Journal of Management Science and Information Technology*, *5*(1), 32–38. DOI: 10.35870/ijmsit.v5i1.3621

Bezner Kerr, R., Postigo, J. C., Smith, P., Cowie, A., Singh, P. K., Rivera-Ferre, M., Tirado-von der Pahlen, M. C., Campbell, D., & Neufeldt, H. (2023). Agroecology as a transformative approach to tackle climatic, food, and ecosystemic crises. *Current Opinion in Environmental Sustainability*, *62*, 101294. DOI: 10.1016/j.cosust.2023.101275

Bridle, J., Balmford, A., Durant, S. M., Gregory, R. D., Pearson, R., & Purvis, A. (2025). How should we bend the curve of biodiversity loss to build a just and sustainable future? *Philosophical Transactions of the Royal Society of London. Series B, Biological Sciences*, *380*(1917), 20230205. DOI: 10.1098/rstb.2023.0205

Cabernard, L., Pfister, S., & Hellweg, S. (2024). Biodiversity Impacts of Recent Land-Use Change Driven by Increases in Agri-Food Imports. *Nature Sustainability*, *7*(11), 1512–1524. DOI: 10.1038/s41893-024-01433-4

Chen, P. J., & Antonelli, M. (2020). Conceptual models of food choice: Influential factors related to foods, individual differences, and society. *Foods, 9*(12). *Foods, 1898*(12), s9121898. DOI: 10.3390/foods9121898

Christmann, S. (2020). Climate change enforces to look beyond the plant – The example of pollinators. *Current Opinion in Plant Biology*, *56*, 162–167. DOI: 10.1016/j.pbi.2019.11.001

DA SILVA, J. G., DEL GROSSI, M. E. & DE FRANÇA, C. G. 2012. The Program to Eliminate Hunger: The Brazilian Experience.

Delong, D. C.Jr. (1996). Defining biodiversity. *Wildlife Society Bulletin*, •••, 738–749.

EL-KEBLAWY, A. (2014). Impact of climate change on biodiversity loss and extinction of endemic plants of arid land mountains. *J Biodivers Endanger Species*, *2*, 2.

Food and Agriculture Organization of the United Nations. (2001). *International Treaty on Plant Genetic Resources for Food and Agriculture*. FAO. https://www.fao.org/plant-treaty/overview/texts-of-the-treaty/en/

Food and Agriculture Organization of the United Nations. (2025a). *The Food and Agriculture Organization of the United Nations.* https://www.fao.org/biodiversity/our-work/strategy-and-action-plan/ar

Food and Agriculture Organization of the United Nations. (2025b). *The World Food Summit and its follow-up.* https://www.fao.org/4/X2051a/x2051a00.htm

Food and Agriculture Organization of the United Nations. (n.d.-a). *Guidelines & standards – Pest and pesticide management.* FAO. https://www.fao.org/pest-and-pesticide-management/guidelines-standards/en/

Food and Agriculture Organization of the United Nations. (n.d.-b). *International Plant Protection Convention (IPPC).* FAO. https://www.fao.org/plant-production-protection/in-action/governance-and-normative-work/ippc/en

Food and Agriculture Organization of the United Nations, & the World Health Organization. (n.d.). *About Codex Alimentarius.* FAO-WHO Codex Alimentarius. https://www.fao.org/fao-who-codexalimentarius/about-codex/en/

Gamage, A., Gangahagedara, R., Gamage, J., Jayasinghe, N., Kodikara, N., Suraweera, P., & Merah, O. (2023). Role of organic farming for achieving sustainability in agriculture. *Farming System, 1*(1), 100005. DOI: 10.1016/j.farsys.2023.100005

Garcia-Alonso, M., Novillo, C., Kostolaniova, P., Martinez Parrilla, M., Alcalde, E., & Podevin, N. (2022). The EU's GM crop conundrum: Did the EU policy strategy to convert EFSA GMO guidance into legislation deliver on its promises? *EMBO Reports, 23*(5), e54529. DOI: 10.15252/embr.202154529

Global Food Security Cluster. (2023). *The four pillars of food security.* https://handbook.fscluster.org/docs/231-the-four-pillars-of-food-security

Government of the United Arab Emirates. (2024a). *Green Economy for Sustainable Development.* The Official Portal of the UAE Government. https://u.ae/en/about-the-uae/economy/green-economy-for-sustainable-development

Government of the United Arab Emirates. (2024b). *National Food Security Strategy 2051.* The Official Portal of the UAE Government. https://u.ae/ar-ae/about-the-uae/strategies-initiatives-and-awards/strategies-plans-and-visions/environment-and-energy/national-food-security-strategy-2051

Government of the United Arab Emirates. (2024c). *The UAE Water Security Strategy 2036.* The Official Portal of the UAE Government. https://u.ae/en/about-the-uae/strategies-initiatives-and-awards/strategies-plans-and-visions/environment-and-energy/the-uae-water-security-strategy-2036

Gregory, P. J., Ingram, J. S., & Brklacich, M. (2005). Climate change and food security. *Philosophical Transactions of the Royal Society of London. Series B, Biological Sciences, 360*(1463), 2139–2148. DOI: 10.1098/rstb.2005.1745

Harwood, R. R. (2020). A history of sustainable agriculture. In Edwards, C. A. (Ed.), *Sustainable agricultural systems*. CRC Press. DOI: 10.1201/9781003070474-2

Holden, L. (2024). From farm to fork? Brexit and the International Plant Protection Convention. *Environmental Law Review, 26*(2), 185–202. DOI: 10.1177/14614529241259879

HUI, D. (2013). Global climate change and biodiversity: Issues and future research. *Journal of Biodiversity and Endangered Species, 1*, 1–2.

Impact, E. (2022). *Global Food Security Index 2022: Global report.* Economist Impact. https://impact.economist.com/sustainability/project/food-security-index/resources/Economist_Impact_GFSI_2022_Global_Report_Sep_2022.pdf

Ismail, S. M., & Alheji, A. I. (2023). The Justice Against Sponsors of Terrorism Act (JASTA) from a Civil Procedure Perspective. *Laws, 12*(1), 15. DOI: 10.3390/laws12010015

JUSTICE, I. C. O. 2025. Obligations of States in respect of Climate Change Summary of the Advisory Opinion of 23 July 2025 INTERNATIONAL COURT OF JUSTICE

Kunming-Montreal Global Biodiversity Framework. (2025). *Kunming-Montreal Global Biodiversity Framework.* https://unctad.org/topic/trade-and-environment/biotrade/kunming-montreal-global-biodiversity-framework

LENTNER, G. M. & CENIN, W. 2024. Daniel Billy et al v Australia (Torres Strait Islanders Petition): Climate change inaction as a human rights violation. Review of European, Comparative & International Environmental Law, 33, 136–143.

Li, Y. O., & Shi, X.-M. (2022). Chapter 5 – Food safety regulations within countries of increasing global supplier impact. In A. Martinović, S. Oh, & H. Lelieveld (Eds.), Ensuring global food safety (2nd ed., pp. 151-158). Academic Press. June 26, 2024. The unconstitutional Federal Government omission to protect the Amazon from deforestation. Brazil Federal Supreme Court. March 1, 2025. Vivek Kumar Sharma vs The State Of Madhya Pradesh. India High Court of Madhya Pradesh.

Mawcha, K. T., Kyampaire, D., Marciale, C., Simiyu-Wafukho, S., Chinyama, C., Babalola, O. O., Kinyanjui, G., & Ndolo, D. (2025). An overview of biopesticide regulatory frameworks in selected countries in Southern Africa. *Frontiers in Sustainable Food Systems, 9*, 1522526. Advance online publication. DOI: 10.3389/fsufs.2025.1522526

Michelato, A., & Pacini, G. C. (2025). Agroecology in Brazil and Italy: Comparative analysis of the historical formation process of agroecology. *Studia Ecologiae et Bioethicae, 23*(1), 89–99. DOI: 10.21697/seb.5836

Milian Gómez, J. F. (2024). Rethinking the human right to food from a single perspective to a four-fold legal interpretation. *Journal of Human Rights Practice, 16*(2), 589–602. DOI: 10.1093/jhuman/huad047

Ministry of Climate Change and Environment. (2024). *National Air Quality Agenda 2031.* Government of the United Arab Emirates. https://u.ae/en/about-the-uae/strategies-initiatives-and-awards/strategies-plans-and-visions/environment-and-energy/the-national-air-quality-agenda-2031

Narayanan, S.NARAYANAN. (2015). Food security in India: The imperative and its challenges. *Asia & the Pacific Policy Studies, 2*(1), 197–209. DOI: 10.1002/app5.62

Onnis, E. (2023). Food and freedom of choice: Towards new food narratives compatible with climate mitigation. *Rivista di Estetica*, (74), 123–144. DOI: 10.4000/estetica.15447

Owuonda, S., & Olembo, S. (2024). Balancing the right to food and environmental sustainability: A call for holistic transformation. *African Journal of Food, Agriculture, Nutrition and Development, 24*(1), 24790–24806. DOI: 10.18697/ajfand.135.25405

Pawlak, K., & Kołodziejczak, M. (2020). The role of agriculture in ensuring food security in developing countries: Considerations in the context of the problem of sustainable food production. *Sustainability (Basel), 12*(13), 5488. DOI: 10.3390/su12135488

Randall, E., & Sanjur, D. (1981). Food preferences—Their conceptualization and relationship to consumption. *Ecology of Food and Nutrition, 11*(3), 151–161. DOI: 10.1080/03670244.1981.9990671

Raponi, S. (2017). A defense of the human right to adequate food. *Res Publica (Liverpool, England), 23*(1), 99–115. DOI: 10.1007/s11158-015-9315-9

Rocha, J. R., Martín, M. P., & Martín, M. V. (2023). The Paradox of Biodiversity Loss and the Emergence of New Forms of Life, Linked to Anthropogenic Effects. *Astragalo, 1*, 203–222.

Schachter, O. (2017). Towards a theory of international obligation. In Evans, M. D. (Ed.), *Sources of international law*. Routledge. (Original work published 2000) DOI: 10.4324/9781315087795-1

Semchuk, I., Mishchenko, V., Petrukha, N., Strochenko, N., & Kapelista, I. (2025). Agricultural practices' impact on achieving the sustainable development goal: Eliminating world hunger. *Grassroots Journal of Natural Resources*, *8*(3), 737–758. DOI: 10.33002/nr2581.6853.080131

SHOBANKE, M., BHATT, M. & SHITTU, E. (2025). Advancements and future outlook of Artificial Intelligence in energy and climate change modeling. *Advances in Applied Energy*, •••, 100211.

Sibhatu, K. T., & Qaim, M. (2017). Rural food security, subsistence agriculture, and seasonality. *PLoS One*, *12*(10), e0186406. DOI: 10.1371/journal.pone.0186406

Simola, A., Boysen, O., Ferrari, E., Nechifor, V., & Boulanger, P. (2022). Economic integration and food security – The case of the AfCFTA. *Global Food Security*, *35*, 100651. DOI: 10.1016/j.gfs.2022.100651

Skogly, S. I. (2007). Right to adequate food: National implementation and extraterritorial obligations. *Max Planck Yearbook of United Nations Law*, *11*(1), 339–358. DOI: 10.1163/18757413-90000013

Slingo, J. M., Challinor, A. J., Hoskins, B. J., & Wheeler, T. R. (2005). Introduction: Food crops in a changing climate. *Philosophical Transactions of the Royal Society of London. Series B, Biological Sciences*, *360*(1463), 1983–1989. DOI: 10.1098/rstb.2005.1755

Sulaiman, N., Zalilah, M. S., & Asnarulkhadi, A. S. (2010). Food security: Concepts and definitions. *Journal of Community Health*, *16*(2), 2–9. http://journalarticle.ukm.my/651/1/Vol16%282%29-Norhasmah.pdf

Sunderland, T. C. H.SUNDERLAND. (2011). Food security: Why is biodiversity important? *International Forestry Review*, *13*(3), 265–274. DOI: 10.1505/146554811798293908

Swinburn, B., Sacks, G., Vandevijvere, S., Kumanyika, S., Lobstein, T., Neal, B., Barquera, S., Friel, S., Hawkes, C., Kelly, B., L'Abbé, M., Lee, A., Ma, J., Macmullan, J., Mohan, S., Monteiro, C., Rayner, M., Sanders, D., Snowdon, W., & Walker, C. (2013). INFORMAS (International Network for Food and Obesity/non-communicable diseases Research, Monitoring and Action Support): Overview and key principles. *Obesity Reviews*, *14*(S1), 1–12. DOI: 10.1111/obr.12087

The Convention on Biological Diversity. (2025). *The Convention on Biological Diversity*. https://www.cbd.int

The European Commission. (2025). *The European Commission*. https://environment.ec.europa.eu/index_en

Tohit, N. F. M., Ya, R. M., & Haque, M. (2025). Unveiling the impacts of food security on community health: A scoping review. *Bangladesh Journal of Medical Science*, *24*, 307–330.

Tompkins, E. L., & Amundsen, H.TOMPKINS. (2008). Perceptions of the effectiveness of the United Nations Framework Convention on Climate Change in advancing national action on climate change. *Environmental Science & Policy*, *11*(1), 1–13. DOI: 10.1016/j.envsci.2007.06.004

United Nations. (2025a). *Food.* https://www.un.org/ar/global-issues/food

United Nations. (2025b). *Human Rights Treaty Body Documents: International Covenant on Economic, Social and Cultural Rights.* https://tbinternet.ohchr.org/_layouts/15/treatybodyexternal/TBSearch.aspx?Lang=en&TreatyID=9&DocTypeID=11

ZORBAS, C., RESNICK, D., JONES, E., SURI, S., IRUHIRIYE, E., HEADEY, D. D., MARTIN, W., VOS, R., ARNDT, C. & MENON, P. 2024. From promises to action: Analyzing global commitments on food security and diets since 2015, Intl Food Policy Res Inst.

KEY TERMS AND DEFINITIONS

Biodiversity: The interaction among living organisms within a specific ecosystem, encompassing animals, microorganisms, and plants (DeLong Jr, 1996).

Climate Change: According to the definition set forth in Article 1(2) of the United Nations Framework Convention on Climate Change (UNFCCC) (1992), climate change refers to changes in climate caused by human activities, whether directly or indirectly.

Consumer's Right to Food: The right to adequate food as a fundamental human right guaranteed by the international human rights system, constituting a binding obligation in times of both peace and conflict. As a right characterized by continuity, its implementation by states must be assessed under both ordinary and exceptional circumstances to determine the extent of their success or failure in fulfilling this obligation (Bender & Smith, 1997).

Eco-Farming: A comprehensive approach to food and agriculture; it is a vital strategy aimed at mitigating the impacts of climate change and adapting to its effects, thus influencing agricultural production positively. This approach relies primarily on crop and pasture rotation alongside organic farming methods to achieve food security and promote a healthy environment (Bezner Kerr et al., 2023).

Food Security: The definition of food security has evolved over time. Initially, it referred to the availability of global food supplies, later expanding to encompass the accessibility of food at all times by all individuals. The concept further developed to include the dimensions of food availability, accessibility, and the quality of consumption. In its most recent formulation, the concept of social acceptance has been incorporated, resulting in a comprehensive definition that includes four key components: availability, accessibility, adequacy, and acceptance (Sulaiman et al., 2010).

Sustainable Agriculture: It is based on the principle of fulfilling current agricultural food needs without compromising the ability of future generations to satisfy their own agricultural food requirements (Brodt et al., 2011).

Chapter 14

Expert's Opinion on the UAE's Progress Toward Sustainable Development Goal 13, Climate Action

Duha Alazaizeh
https://orcid.org/0009-0007-9337-1722
Abu Dhabi University, UAE

Ismail Badraoui
Zayed University, UAE

Rahaf Ajaj
Abu Dhabi University, UAE

ABSTRACT

Climate change remains a critical barrier to sustainable development globally, with energy production responsible for approximately 60% of greenhouse gas (GHG) emissions. The United Arab Emirates (UAE), a country with substantial coastal infrastructure in an arid region, is particularly vulnerable to climate-related risks, including rising sea levels and extreme heat events. Despite substantial efforts—including investments in renewable energy, a net-zero carbon target by 2050, and the implementation of comprehensive policies outlined in its Nationally Determined Contribution (NDC)—progress toward Sustainable Development Goal 13 (Climate Action) remains below targeted levels. This study examines the UAE's progress on SDG 13 objectives and employs the Delphi methodology to identify key barriers hindering advancement. These include inadequate public regulatory frameworks,

DOI: 10.4018/979-8-3373-3346-5.ch014

insufficient private sector incentives, and limited climate finance mechanisms. The paper also explores actionable insights by benchmarking international best practices and offering strategic recommendations.

INTRODUCTION

Sustainable development can be defined as 'development that meets the needs of the present without compromising the ability of future generations to meet their own needs (Lafortune et al., 2020). Following the expiration of the Millennium Development Goals, the UN introduced 17 Sustainable Development Goals (SDGs) in 2015 as part of the UN's 2030 Agenda (Lee et al., 2016; Kochskämper et al., 2024). Among these goals, climate change (SDG13) is a pressing concern. The widespread use of non-renewable energy resources for power generation, such as oil and natural gas, is a primary reason for the increase in greenhouse gases (GHG), especially carbon dioxide (CO_2), which is considered a major contributor to the dramatic climate change on Earth (Arman, 2015). In response to this critical issue, many countries have undertaken initiatives and actions to mitigate the impact of climate change (Krantz & Gustafsson, 2021). However, despite the general efforts and actions to tackle climate change, countries worldwide are still lagging in achieving the set objectives (Sachs et al., 2022). Considering the pressing issue of climate change and the positive impact of GHG emissions reduction on other SDGs (e.g., 3, 6, 12, 14, 15) (Campbell et al., 2018), this research aims to understand the challenges countries face in implementing actions required to address climate change, with the United Arab Emirates (UAE) serving as the study context.

SDG 13 can be simplified as 'The urgent action to combat climate change and its impact' (Doni et al., 2020). The Earth's climate has changed naturally; however, the rapid changes are mainly due to human activities (Trenberth, 2018). As one of the world's major oil and gas producers, the UAE plays a significant role in achieving the SDG 13 objectives. The country, which holds around 10% of the world's proven oil reserves, has experienced rapid economic development accompanied by population increase since the discovery of fossil fuels (Arman, 2015). Since the early 2000s, the UAE government has become increasingly aware of the climate crisis and has actively sought solutions. For instance, the UAE has launched numerous initiatives to mitigate its impact on climate change, including establishing the Green Building Council and a large-scale afforestation project (Salam, 2015). The UAE has also developed numerous sustainable cities that utilize renewable energy to reduce their reliance on fossil fuels, including Masdar City, Dubai Sustainable City, and Sharjah Sustainable City. Moreover, Abu Dhabi was selected to host the International Renewable Energy Agency IRENA in 2009 (Krzymowski, 2020). In

2023, the UAE hosted COP28, thus further showing its commitment to SDG13. However, the UAE's progress towards achieving SDG 13 has been evaluated as moderate (Sachs et al., 2022), with significant challenges remaining. Globally, several obstacles impede climate action, such as public skepticism about climate change, concerns about the economic costs of climate policies, political divisions, and the need to address social justice issues for vulnerable communities affected by climate change (Bergquist et al., 2020)

Considering the discussion above, this research aims to (i) identify the significant challenges that hinder the UAE's achievement of SDG13 and (ii) suggest solutions to overcome these challenges. To achieve these objectives, a sequential two-step research design is employed. First, a Delphi Study is conducted with a pool of climate change experts to identify the primary barriers hindering the implementation of SDG 13 in the UAE. Second, a benchmark analysis is conducted using secondary data (government reports, organizational publications, articles, etc.) to suggest potential solutions to the identified barriers. By following this approach, this study provides answers to the following two research questions:

RQ1: What obstacles impede the UAE's progress toward its objectives?
RQ2: What measures should the UAE take to achieve SDG13?

The remaining sections of the paper are divided into the following sections: Section 2 presents an overview of SDG 13 and its implementation in the UAE. Section 3 outlines the research method. The results are presented in Section 4, followed by a discussion in Section 5. Finally, Section 6 concludes with conclusions and future research directions.

CONTEXT

SDG 13, climate action

In 2015, the global community made a significant leap forward with two major international agreements: the Sustainable Development Goals (SDGs) under the 2030 Agenda for Sustainable Development and the Paris Agreement within the United Nations Framework Convention on Climate Change (UNFCCC) (Kobayakawa, 2021). The UN declared the Paris Climate Agreement at COP21 in 2015, where member states committed to achieving five targets and eight indicators related to SDG 13. These targets set the objectives, while the indicators track progress and help countries assess whether they are on the right path (Doni et al., 2020). The 2030 Agenda emphasizes the economy, environment, and society as the three main pillars of sustainability (Fuller et al., 2023), while the Paris Agreement was the first international treaty with legal power to unite developed and developing nations in limiting global warming to below two degrees Celsius (preferably 1.5 degrees) compared to pre-industrial levels (Gao et al., 2017). The Paris Agreement and the SDGs are interlinked, as climate change threatens economic growth, and achieving a low-carbon transition depends on progress in economic, social, and environmental areas (Markkanen & Anger-Kraavi, 2019).

Countries that are parties to the UNFCCC Paris Climate Agreement must submit Nationally Determined Contributions (NDCs), which are voluntary commitments to reduce emissions and undertake other measures to address climate change (Mills-Novoa and Liverman, 2019). The content, scope, and length of NDCs vary widely, ranging from 3 to 57 pages, and countries are required to update them every five years (Pauw et al., 2018). However, current Nationally Determined Contributions (NDCs) are insufficient to meet the temperature goals of the Paris Agreement. Instead, these NDCs will cause a minimum of 3°C of warming by 2100, endangering the accomplishment of the SDGs (Honegger et al., 2021). The main challenges in integrating new climate policies include (i) the limited availability of climate-related data and the science-based knowledge needed to translate this data into actionable policy information, (ii) a lack of awareness and climate knowledge among governments and stakeholders, which affects the prioritization and behavior related to climate action, and (iii) potential trade-offs between economic development and climate action goals, where economic benefits are often prioritized SDG 13 implementation in the UAE (Al-Sarihi & Mason, 2020).

The UAE, a young and progressive nation since its establishment in 1971, has undergone significant societal and economic transformations. Over the past few decades, the UAE's economy has grown impressively; nevertheless, due to this rapid growth, development requirements have often overshadowed environmental considerations. This rapid expansion was initially fueled by its reliance on oil exports; however, laws have since been implemented to address sustainability and energy use in many sectors (Locke et al., 2023). With approximately 85% of its population and over 90% of its infrastructure located in coastal regions, the UAE is highly vulnerable to the adverse effects of climate change. It is classified as a non-Annex I country, meaning it hasn't been a major contributor to greenhouse gas emissions and isn't obligated to specific emissions reduction targets in international climate agreements. The country, facing challenges such as low-lying coastal areas and risks of desertification and drought, has warmed by 1.8 °C in recent years, exceeding the global average of 1.1 degrees (MOCCAE, 2023).

As a key player in the global energy market, the UAE is uniquely positioned to balance economic growth with its climate commitments. Its strategic initiatives, such as hosting IRENA and setting ambitious Nationally Determined Contributions

(NDCs), demonstrate its dedication to achieving Sustainable Development Goal 13. Recognizing the importance of climate change to its sustainability and economic progress, the UAE established the Ministry of Climate Change and Environment (Locke et al., 2023). The nation is committed to reaching net-zero emissions by 2050 and has launched a comprehensive strategy to achieve this goal (MOCCAE, 2023). After submitting its initial Nationally Determined Contribution (NDC) in 2015, the UAE increased its ambitions with a second NDC in 2020, aligning with international standards and the Paris Agreement. Ahead of COP27, the UAE further updated its Nationally Determined Contribution (NDC), setting an enhanced emission reduction target of 31% below the projected 2030 business-as-usual (BAU) scenario—one of the few nations to take this step (Climate Action Tracker, 2023). The updated NDC aims to limit emissions to 208 million metric tons of CO_2 equivalent by 2030, representing a reduction of 93.2 million metric tons compared to the business-as-usual (BAU) projection of 301.2 million metric tons of CO2 equivalent (MOCCAE, 2022).

The UAE's economy is projected to grow by 24% by 2030, alongside a 14% increase in population over the same period. Despite this economic and population expansion, the UAE has intensified its efforts to address the climate crisis and reduce emissions, particularly after setting a new reduction target in its third National Determined Contribution (NDC) update. In July 2023, the UAE submitted its third update and second NDC, which included enhanced strategies, targets, and policies. The updated target aims to reduce GHG emissions to 182 $MtCO_2e$ by 2030, down from the previously set target of 208 $MtCO_2e$ —a 40% reduction compared to the projected 2030 business-as-usual (BAU) scenario outlined in the previous NDC update in 2022 (MOCCAE, 2023). Additionally, the UAE Energy Strategy 2050 aims to generate 50% of the electricity using renewable energy resources by 2050, reducing carbon emissions by 70%. These plans aim to reduce greenhouse gas (GHG) emissions, ensure continued economic growth, mitigate climate risks, and strengthen the ability to adapt to climate change (Locke et al., 2023). In 2023, the UAE presided over the COP28 climate negotiations.

According to the MIT Technology Review (2023), the Green Future Index assesses countries' progress across five key pillars: energy transition, clean innovation, green society, carbon emissions, and climate policy. The UAE has improved its ranking, moving from 41st place in 2022 to 36th place worldwide, and is the top-ranked Arab country. The new ranking indicates that the country is making progress, thanks to remarkable advancements resulting from its dedicated efforts. The UAE's climate policy focuses on combating climate change through emissions management, enhancing resilience, and promoting a green economy. The country has established specific targets and indicators to advance its climate change initiatives, including increasing the contribution of clean energy to 50% of its energy mix by 2050 and reducing energy consumption by 40% at both individual and institutional levels by

2050, which indicates the enormous efforts that the country must put to achieve these remarkable targets (MOCCAE, 2021).

METHODOLOGY

This section outlines the methodological approach employed in the research, including the tools, data collection methods, and data analysis techniques. It was essential to address the primary challenges hindering the UAE's progress towards SDG13 from an expert's perspective. The Delphi method was selected due to its structured process involving diverse experts, which is particularly valuable when addressing complex issues like climate change. The Delphi method is an iterative process that gathers expert opinions through multiple rounds of questionnaires and feedback. The process refines these views by providing summarized feedback to experts (Skulmoski et al., 2007). The Delphi methodology comprises three sequential rounds or phases (Figure 1).

Figure 1. Research methodology.

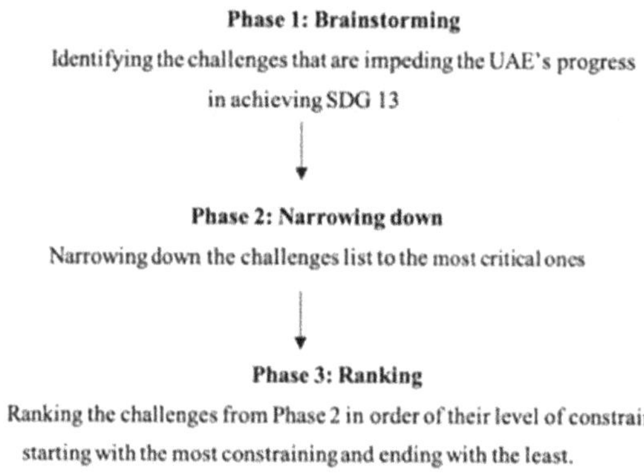

In Phase 1, selected panelists receive surveys or questionnaires asking them to list points or issues related to a specific topic. In Phase 2, panelists are tasked with choosing the top issues from the compiled list generated in Phase 1. Finally, in Phase 3, panelists are invited to rank the top issues identified in Round 2 (Schmidt, 1997). We contacted potential experts through email and LinkedIn invitations to construct the panel for this research. The objective was to assemble diverse climate change professionals from various sectors, including private and public industries, and academics based in the United Arab Emirates (UAE). In total, invitations were extended to 37 professionals, among whom 10 experts expressed their interest in participating as panelists for the Delphi study. Table 1 provides details about the panelist industry and work experience. The questionnaire used in the study is presented in Appendix 1.

Table 1. Climate change professionals' details.

Participant reference	Occupation	Years of experience
1	Academic	18
2	ESG & Climate change advisor	9
3	ESG & Sustainability Consultant	3
4	ESG & Climate change advisor	14
5	Academic	15
6	Sustainability Market Leader	8
7	Climate Analyst	5
8	ESG & Sustainability Consultant	13
9	Academic	16
10	Climate change & policy advisor	12

In the first phase, considered a brainstorming phase, the panelists were asked to identify the challenges they believe are impeding the UAE's progress in achieving SDG 13. The identified challenges were then compiled, and the terminology was unified, with duplicates removed. In the second phase, the panelists were requested to select half of the compiled list of challenges resulting from Phase 1. After completing this phase, only the challenges chosen by over 50% of the panelists were retained for phase 3. To assess the level of agreement between the panelists at this stage, Fleiss Kappa (κ) was used (Randolph, 2005). Values from 0 to 0.2 indicate slight agreement, while values between 0.2 and 0.4 indicate fair agreement. Values between 0.4 and 0.6 indicate moderate agreement, whereas values between 0.6 and 0.8 indicate substantial agreement, and values above 0.8 indicate perfect agreement (Zuehlke et al., 2009). To evaluate the level of consensus among the panelists.

$$\kappa = \frac{P_o - P_e}{1 - P_e}$$

$$P_o = \frac{1}{Nn(n-1)}\left(\sum_{i=1}^{N}\sum_{j=1}^{K} n_{ij^2} - Nn\right)$$

$$P_e = \sum_{j=1}^{K}\left(\frac{1}{Nn}\sum_{i=1}^{N} n_{ij}\right)^2$$

In the third phase, the panelists were asked to rank the challenges from phase 2, from most important to least important, according to their impact on climate action in the UAE. Kendall's *W* test was used to measure the level of agreement in their rankings. The range for *W* is limited between 0 (no agreement) and 1 (perfect agreement) (Field, 2005), with values above 0.7 indicating a high level of consensus (Badraoui et al., 2024). Kendall's coefficient of concordance, W, was measured using the following equations (Field, 2005):

$$SS = \sum_{i=1}^{n} (x_i - \bar{x})^2$$

$$W = \frac{12 \times SS_{Rank\ Totals}}{k^2(n^3 - n)}$$

RESULTS

The study was conducted over 7 weeks, with the initial two phases requiring approximately 2 weeks for completion. A reminder was sent one week after the questionnaire was sent. The third phase was extended to 3 weeks as a second ranking round was required due to the low agreement resulting from the first ranking round. The first phase of the Delphi study involved ten panelists. They were asked to provide at least six main challenges that they believe are impeding the UAE's progress toward climate action. A total of sixty challenges were received from the panelists, and after unifying the terminology and removing duplicates, a final list of 17 challenges was compiled. All ten panelists participated again in the second phase and were asked to select half of the challenges from the compiled list from Phase 1. The level of agreement among the panelists on selected elements was evaluated using Fleiss Kappa (κ). The findings indicate slight agreement, with a κ value of 0.16. This result reflects diverse opinions among the panelists, suggesting that achieving a consensus during the ranking phase might require multiple attempts. This diversity can be attributed to the panelists' varying sectors, which can have different priorities. Nine out of seventeen challenges were selected by at least half of the panelists and were retained for the next phase.

In the third phase of the Delphi study, the panelists ranked the challenges obtained from phase 2 in descending order based on their perceived importance. The agreement between the panelists' rankings was then evaluated using Kendall's *W*. As shown in Figure 2, in the first round of ranking, the obtained *W* coefficient of concordance was 0.23 with a p-value of 0.037, indicating a statistically significant but low level of agreement. As such, the panelists were asked to rank the factors a second time while having access to the mean rank for each challenge from the first attempt.

Figure 2. Round 1 challenges ranking results from most to least important (Kendall's W = 0.23).

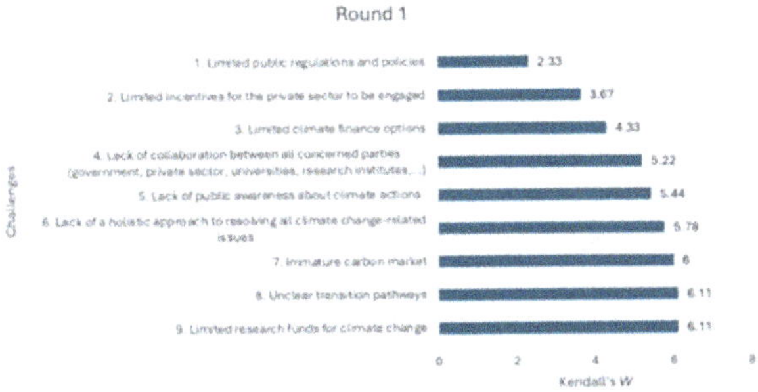

The revised ranking, as shown in Figure 3, demonstrated a statistically significant consensus among the panelists with a p-value of 0.003; however, it remains low with a W value 0.32. Considering the level of agreement at this stage, the panelists' willingness to re-rank the challenges for the third time, and the fact that not all panelists participated in the second ranking (with concerns about receiving responses beyond this point), the research team has decided to conclude the Delphi study at this stage.

Figure 3. Round 2 challenges ranking results from most to least important (Kendall's W = 0.32).

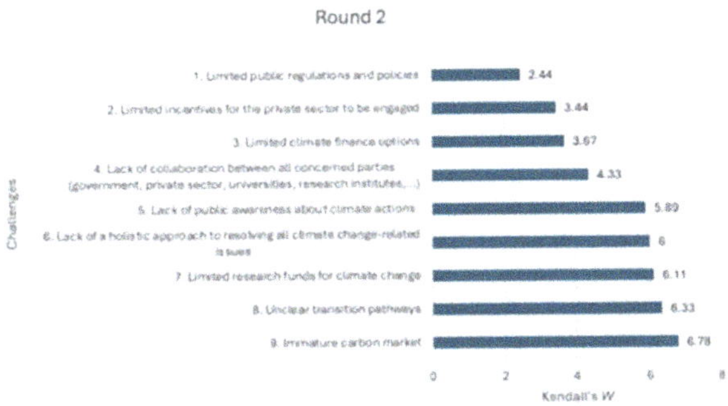

DISCUSSION

In this section, we discuss the first three challenges and examine the actions taken by various countries in these areas.

Public regulation and policies

Climate policies aim to capitalize on the beneficial outcomes of climate effects, such as favorable growth seasons, while minimizing future adverse consequences, like lowered vulnerability to floods. These strategies consider the financial and non-financial pros and cons of different approaches. Climate policy primarily centers on two response categories: mitigation and adaptation. At the same time, climate resilience combines adaptation and mitigation strategies to achieve sustainable development (Pielke Jr., 1998; Field & Barros, 2014). The role of law in fostering climate resilience is primarily demonstrated through facilitating climate adaptation, enhancing adaptive capacity, promoting market-based incentive approaches, and establishing liability frameworks (Du et al., 2023).

The UAE has adopted numerous targets and initiatives, implementing new policies to adapt to and mitigate climate change. The UAE's efforts in recent years to address the climate crisis and mitigate its impact have been significantly accelerated, and years are needed to see the full impact of the new policies and strategies. In the electricity generation sector, the USA and China have introduced tax incentives for renewable energy technologies. Canada has released targets to phase out coal power plants that are not supported by Carbon Capture and Storage (CCS) by 2030 (Fekete et al., 2021). The EU has adopted the Renewable Energy Roadmap (2009) (Rayner and Jordan, 2016) and the Energy and Vehicle Tax by Japan (Gokhale, 2021). In terms of the oil production sector, Norway set a CO_2 tax on gas and oil production in 1991 and promoted CCS options, the USA set a methane target which aims to reduce methane emissions (40-45%) from 2012 levels by 2025, and Russia aims to reduce flaring by %5. Regarding transportation, notable initiatives include California's Zero-Emission Vehicle program, the USA's Clean Cars 2040 Act, and efforts within EU member states, such as France and the Netherlands, to phase out the use of combustion engines. Additionally, concerning policies related to F-gas emissions, there are significant regulations at the EU level, as well as the Indian HCFC Phase-out Management Plan (Fekete et al., 2021).

Incentives for the private sector

Involving the private sector as a partner in acknowledging and adapting to climate change is essential, which can be achieved by incentivizing this sector and improving Public-Private Partnerships (PPPs). This will boost climate resilience and open new business prospects (Biagini & Miller, 2013). Companies' climate change adaptation incentives are influenced by uncertainties in climate impacts, operational flexibility, regulatory environments, and business planning horizons. Some companies may hesitate to invest in adaptation due to unclear benefits and upfront costs associated with it. Regulatory factors can either support or impede adaptation efforts. Additionally, the length of a company's business planning horizon can affect its engagement in climate adaptation, with shorter planning horizons potentially limiting consideration of long-term climate impacts (Agrawala et al., 2011). The private sector must participate in climate adaptation efforts, as climate change has steadily increased the financial impacts associated with high temperatures, droughts, floods, hurricanes, and other weather-related disasters.

These incentives include compensation or economic rewards, market-based mechanisms such as payments for ecosystem services, regulation, and adjustments to property rights to enforce adaptation requirements. Transfers of public resources are also suggested to encourage private individuals and businesses to participate in climate action efforts. By utilizing these diverse incentives, policymakers aim to harness the self-interest of private actors to generate public adaptation benefits and address the challenges posed by climate change (Tompkins & Eakin, 2012). Furthermore, voluntary agreements between the government and the private sector to foster cooperation, encourage early involvement in policy development, and promote sustainable practices from the beginning will drive meaningful climate action and achieve more robust climate change mitigation and adaptation outcomes (de Oliveira 2009). Additionally, incentives such as providing funding to reward emission reductions and increasing costs for emission-intensive goods and services (Michaelowa & Michaelowa, 2019) can be employed.

Climate finance options

UNFCCC defines Climate Finance as "local, national, or transnational financing drawn from public, private, and alternative sources of financing that seeks to support mitigation and adaptation actions that will address climate change." It involves the financial commitments that governments, households, and corporations must make to shift the global economy towards a low-carbon trajectory, decrease greenhouse gas levels, and enhance countries' ability to withstand the impacts of climate change (Hong et al., 2020). In 2021/2022, global climate finance reached USD 1.3 trillion, showing growth but still falling short of the needed scale. By 2030, funding must reach USD 9 trillion; however, current progress is slow, necessitating a five-fold increase annually to avoid severe climate impacts. Delays will lead to higher costs for both mitigation and adaptation efforts (Buchner et al., 2023).

Climate finance options such as:

- Green Bonds: Green bonds are simply bonds that finance environmentally friendly or climate change-related activities that have a positive impact on the environment. Green bonds provide an opportunity to finance a range of smaller projects; they can be issued by various entities, including multilateral development banks, local governments, and companies seeking to enhance their environmental image. Green bond markets are gaining momentum in emerging economies such as China, India, Brazil, Mexico, Korea, and South Africa. The Chinese government is focusing on green investments, aiming to establish a green bond market. The Export-Import Bank of India has released green bonds in India. Similarly, in 2014, Johannesburg, South Africa, introduced the first municipal green bond within an emerging market context. Within the past three years, green bonds have also been issued by Mexico's state-owned development bank Nafin, the Brazilian food producer BRF, and the Export-Import Bank of Korea (KEXIM) (Clapp & Pillay, 2017).

- Climate Insurance: G7 leaders and the Paris Agreement have suggested that 2015 is the year of climate insurance, as they believe it can serve as a climate adaptation instrument (Surminski et al., 2016). Extreme weather events have adverse economic consequences, affecting millions of individuals and exerting significant financial pressure on government budgets. To address these events, many countries, including EU countries, have adopted climate insurance to mitigate current and future disasters (Porrini & Schwarze, 2014).
- PPPs in climate finance: Climate finance involves many actors, and it's not only governmental responsibility, but also private sector involvement is essential; therefore, PPPs are important to fill the gap in climate finance. A report by Green Growth Best Practice has suggested that public-private collaboration in climate finance can be achieved in three areas: firstly, natural resource management by sharing the ownership of natural resources. Secondly, green infrastructure enhances the efficiency of large infrastructure projects and supports them. Lastly, innovation investment by fostering environmentally conscious entrepreneurship and building networks that promote innovative achievement (Gardiner et al., 2015).

Table 3 outlines the three primary challenges, their corresponding SDG13 targets, actions taken by other countries, and the measures implemented by the UAE to address these challenges. (-) refers to no action or no available information, while (+) is in progress and (++) is achieved.

Table 2. Summary table

Challenge	Actions taken by other countries	Score	Measures taken by UAE
Public regulation and policies	Tax incentive for renewable energy technologies	-	
	CO_2 tax on gas and oil production	-	
	F-gases emissions regulation	+	The UAE's new emissions target covers all national GHG emissions except for F-gases, but the UAE promised in the third update of its second NDC to include F-gases in their future NDC.
	Passing laws on the Prevention and Control of Atmospheric Pollution	+	
	Setting methane emission reduction targets	+	
	Tax incentive for electric vehicle	-	
	Energy and Vehicle Tax	-	
	Reducing flaring	++	The Abu Dhabi National Oil Company (ADNOC) is recognized for its lower-carbon energy production, driven by practices such as zero-carbon grid power, a strict no-flaring policy, and pioneering large-scale carbon capture in the region.
	Phasing out non-CCS-supported power plants	-	
	Renewable Energy Roadmap	++	Abu Dhabi's 2035 Clean Energy Target and the UAE Hydrogen Leadership Roadmap are pivotal. Masdar, a global leader in renewable energy, is actively involved, with projects set to cut a significant 95 million tonnes of global CO2e emissions yearly. In addition to many large-scale clean energy project
Incentives for the private sector	Incentives include compensation or economic rewards	+	In the third updated second NDC, the UAE mentioned that it's working on implementing policy measures to facilitate private market participation in providing the necessary financing, alongside the government, thereby reducing the need for additional international assistance to achieve these objectives.
	Promoting sustainable practices from the outset	-	
	Market-based mechanisms like payments for ecosystem services, regulation	-	
	Government-private sector agreements to foster cooperation	+	The UAE launched the National Dialogue for Climate Ambition (NDCA), a platform that brings together NGOs, the private sector, and the public sector to promote climate collaboration in diverse fields, including manufacturing, waste, cement, transportation, and energy.
	Encouraging early private sector involvement in policy development	-	

continued on following page

Table 2. Continued

Challenge	Actions taken by other countries	Score	Measures taken by UAE
Climate finance options	Green bonds finance environmentally beneficial projects and are issued by governments, companies, and banks Emerging economies are increasingly involved in this market to support green investments and environmental sustainability.	+	• Masdar initiates its first $750 million Green Bond with a 10 year term • The UAE commits $4.5 billion to fund climate projects in Africa
	Climate insurance is viewed as a vital tool for adapting to climate change by addressing the increasing frequency of extreme weather events. These events have significant economic impacts, prompting countries like EU nations to adopt climate insurance to manage these financial consequences.	-	
	Public-Private Partnerships (PPPs) are crucial for climate finance, involving both government and the private sector. They focus on resource sharing, green infrastructure, and innovation	+	The third update of the second NDC outlines the government's goal to foster investments in industrial and building sectors. Major infrastructure projects will receive funding through public-private partnerships to enhance cost efficiency and risk control. Public funds will additionally back investments in sectors like remote charging infrastructure, even if they don't offer immediate profits for private investors.

CONCLUSION

The UAE is at the forefront of regional efforts to combat the climate crisis, with concrete proactive initiatives to achieve SDG 13 and comply with the Paris Agreement. Despite the country's numerous accomplishments in this area, several challenges still need to be addressed in the ongoing fight against climate change. By conducting a Delphi study with a panel of climate change experts, this research highlights the main challenges hindering the UAE's progress towards SDG 13: limited public regulations and policies, incentives for the private sector, and climate finance options.

The obtained results contribute uniquely to the existing literature by exploring an area that has been underrepresented in previous research, particularly in oil-producing countries. They also pave the way for future research. First, this study has highlighted policy frameworks from other countries that have successfully advanced their climate action goals. Future research can explore how these can be adapted to the UAE's context. Second, empirically studying successful models (e.g., public-private partnerships, tax incentives, etc.) that encourage businesses to adopt green technologies and sustainable practices can lead to tailored government strategies and policies. Finally, future research could explore the compatibility of alternative green financing mechanisms with the country's financial regulations. This could include examining the potential benefits and challenges, and the role of policy in integrating green finance options and their impact on both economic stability and environmental sustainability, and additionally, exploring the long-term impact and effectiveness of new policies and initiatives to understand and track the UAE's ongoing commitment to achieving SDG 13 and enhancing climate resilience. Importantly, future research should take an interdisciplinary approach, given the complexity of climate change, integrating environmental, economic, policy, and social science perspectives to develop holistic solutions.

Despite the interesting results and contributions to the literature, the study also suffers from a low consensus between the panel experts on the relative importance of the identified barriers. The limited level of agreement, materialized by low Fleiss' Kappa ($\kappa = 0.16$) and Kendall's W ($W = 0.32$), stems from the diverse professional backgrounds of the panelists and their sectoral priorities. As such, future studies can focus on each sector specifically to identify its needs and derive concrete strategies.

DISCLOSURE STATEMENT

No potential conflict of interest was reported by the author(s).

REFERENCES

Agrawala, S., Carraro, M., Kingsmill, N., Lanzi, E., Mullan, M., & Prudent-Richard, G. (2011). *Private sector engagement in adaptation to climate change: Approaches to managing climate risks*. https://doi.org/DOI: 10.1787/19970900

Al-Sarihi, A., & Mason, M. (2020). Challenges and opportunities for climate policy integration in oil-producing countries: The case of the UAE and Oman. *Climate Policy*, *20*(10), 1226–1241. DOI: 10.1080/14693062.2020.1781036

Arman, H. (2015). Uncertainties, risks and challenges relating to CO_2 emissions and its possible impact on climate change in the United Arab Emirates. *International Journal of Global Warming*, *8*(1), 1–17. DOI: 10.1504/IJGW.2015.071575

Badraoui, I., van der Lans, I. A. M. C., Boulaksil, Y., & van der Vorst, J. G. A. J. (2024). Horizontal logistics collaboration success factors: Expectations versus reality. *Benchmarking*, *31*(1), 29–52. DOI: 10.1108/BIJ-04-2022-0274

Bergquist, P., Mildenberger, M., & Stokes, L. C. (2020). Combining climate, economic, and social policy builds public support for climate action in the US. *Environmental Research Letters*, *15*(5), 054019. DOI: 10.1088/1748-9326/ab81c1

Biagini, B., & Miller, A. (2013). Engaging the private sector in adaptation to climate change in developing countries: Importance, status, and challenges. *Climate and Development*, *5*(3), 242–252. DOI: 10.1080/17565529.2013.821053

Buchner, B., Naran, B., Padmanabhi, R., Stout, S., Strinati, C., Wignarajah, D., Miao, G., Connolly, J., & Marini, N. (2023). *Global landscape of climate finance 2023*. Climate Policy Initiative. https://www.climatepolicyinitiative.org/wp-content/uploads/2023/11/Global-Landscape-of-Climate-Finance-2023.pdf

Campbell, B. M., Hansen, J., Rioux, J., Stirling, C. M., & Twomlow, S. (2018). Urgent action to combat climate change and its impacts (SDG 13): Transforming agriculture and food systems. *Current Opinion in Environmental Sustainability*, *34*, 13–20. DOI: 10.1016/j.cosust.2018.06.005

Clapp, C., & Pillay, K. (2017). Green bonds and climate finance. In Klein, R. J. T., O'Brien, K. L., & Wolf, J. (Eds.), *Climate finance: Theory and practice* (pp. 79–105). Routledge., DOI: 10.1142/9789814641814_0005

Climate Action Tracker. (2023, April 5). *Climate Action Tracker, UAE*. https://climateactiontracker.org/countries/uae/2023-04-05/

de Oliveira, J. A. P. (2009). The implementation of climate change related policies at the subnational level: An analysis of three countries. *Habitat International*, *33*(3), 253–259. DOI: 10.1016/j.habitatint.2008.10.006

Doni, F., Gasperini, A., & Soares, J. T. (2020). *SDG13–Climate action: Combating climate change and its impacts*. Emerald Publishing Limited., DOI: 10.1108/9781787569157

Du, H., Dang, K. K., Nguyen, H. Q., & van Rijswick, H. F. M. W. (2023). A framework for reviewing laws and policies for climate resilience: The case of the Vietnamese Mekong Delta. *Journal of Environmental Planning and Management*, *66*(6), 1280–1304. DOI: 10.1080/09640568.2022.2026308

Fekete, H., Kuramochi, T., Roelfsema, M., den Elzen, M., Forsell, N., Höhne, N., & Luna, L.. (2021). A review of successful climate change mitigation policies in major emitting economies and the potential of global replication. *Renewable & Sustainable Energy Reviews*, *137*, 110602. DOI: 10.1016/j.rser.2020.110602

Field, A. P. (2005). Kendall's coefficient of concordance. In Everitt, B. S., & Howell, D. C. (Eds.), Vol. 2, pp. 1010–1011). Encyclopedia of statistics in behavioral science, DOI: 10.1002/0470013192.bsa327

Field, C. B., & Barros, V. R. (Eds.). (2014). *Climate change 2014–Impacts, adaptation and vulnerability: Regional aspects*. Cambridge University Press. DOI: 10.1017/CBO9781107415379

Fuller, J. L., van Putten, I., Kraan, M., Bjørkan, M., & Dankel, D. J. (2023). 'Sustainability is not a vegan coffee shop': Eliciting citizen attitudes and perspectives to localize the UN Sustainable Development Goals. *Journal of Environmental Planning and Management*, •••, 1–22. DOI: 10.1080/09640568.2023.2223761

Gao, Y., Gao, X., & Zhang, X. (2017). The 2 °C global temperature target and the evolution of the long-term goal of addressing climate change—From the United Nations Framework Convention on Climate Change to the Paris Agreement. *Engineering (Beijing)*, *3*(2), 272–278. DOI: 10.1016/J.ENG.2017.01.022

Gardiner, A., Bardout, M., Grossi, F., & Dixson-Declève, S. (2015). *Public-private partnerships for climate finance*. Nordic Council of Ministers.

Gokhale, H. (2021). Japan's carbon tax policy: Limitations and policy suggestions. *Current Research in Environmental Sustainability*, *3*, 100082. DOI: 10.1016/j.crsust.2021.100082

Honegger, M., Michaelowa, A., & Roy, J. (2021). Potential implications of carbon dioxide removal for the Sustainable Development Goals. *Climate Policy, 21*(5), 678–698. DOI: 10.1080/14693062.2020.1843388

Hong, H., Karolyi, G. A., & Scheinkman, J. A. (2020). Climate finance. *Review of Financial Studies, 33*(3), 1011–1023. DOI: 10.1093/rfs/hhz146

Kobayakawa, T. (2021). Country diagnostics for low carbon development: Can developing countries pursue simultaneous implementation of the Sustainable Development Goals and the Paris Agreement? *Business Strategy & Development, 4*(3), 294–312. DOI: 10.1002/bsd2.159

Kochskämper, E., Glass, L.-M., Haupt, W., Malekpour, S., & Grainger-Brown, J. (2024). Resilience and the Sustainable Development Goals: A scrutiny of urban strategies in the 100 Resilient Cities Initiative. *Journal of Environmental Planning and Management*, •••, 1–27. DOI: 10.1080/09640568.2023.2297648

Krantz, V., & Gustafsson, S. (2021). Localizing the Sustainable Development Goals through an integrated approach in municipalities: Early experiences from a Swedish forerunner. *Journal of Environmental Planning and Management, 64*(14), 2641–2660. DOI: 10.1080/09640568.2021.1877642

Krzymowski, A. (2020). Sustainable development goals in Arab region–United Arab Emirates' case study. *Problemy Ekorozwoju, 15*(1), 211–220. DOI: 10.35784/pe.2020.1.22

Lafortune, G., Fuller, G., Schmidt-Traub, G., & Kroll, C. (2020). How is progress towards the Sustainable Development Goals measured? Comparing four approaches for the EU. *Sustainability (Basel), 12*(18), 7675. DOI: 10.3390/su12187675

Lee, B. X., Kjaerulf, F., Turner, S., Cohen, L., Donnelly, P. D., Muggah, R., & Gilligan, J. (2016). Transforming our world: Implementing the 2030 Agenda through Sustainable Development Goal indicators. *Journal of Public Health Policy, 37*(S1), 13–31. DOI: 10.1057/s41271-016-0002-7

Locke, J., Dsilva, J., & Zarmukhambetova, S. (2023). Decarbonization strategies in the UAE built environment: An evidence-based analysis using COP26 and COP27 recommendations. *Sustainability (Basel), 15*(15), 11603. DOI: 10.3390/su151511603

Markkanen, S., & Anger-Kraavi, A. (2019). Social impacts of climate change mitigation policies and their implications for inequality. *Climate Policy, 19*(7), 827–844. DOI: 10.1080/14693062.2019.1596873

Michaelowa, K., & Michaelowa, A. (2019). Transnational climate governance initiatives: Designed for effective climate change mitigation? In Bulkeley, H., & Andonova, L. B. (Eds.), *The comparative politics of transnational climate governance* (pp. 146–178). Routledge., DOI: 10.4324/9781351135436-6

Mills-Novoa, M., & Liverman, D. M. (2019). Nationally determined contributions: Material climate commitments and discursive positioning in the NDCs. *Wiley Interdisciplinary Reviews: Climate Change, 10*(5), e589. DOI: 10.1002/wcc.589

Ministry of Climate Change & Environment (MOCCAE). (2021). *The United Arab Emirates general environmental policy*. Ministry of Climate Change & Environment.

Ministry of Climate Change & Environment (MOCCAE). (2022). *Updated second nationally determined contribution of the United Arab Emirates*. Ministry of Climate Change & Environment.

Ministry of Climate Change & Environment (MOCCAE). (2023). *Third update of second NDC for the UAE*. Ministry of Climate Change & Environment.

MIT Technology Review. (2023). *The Green Future Index 2023*. https://mittrinsights.s3.amazonaws.com/GFI23report.pdf

Tompkins, E. L., & Eakin, H. (2012). Managing private and public adaptation to climate change. *Global Environmental Change, 22*(1), 3–11. https://doi.org/DOI: 10.1016/j.gloenvcha.2011.09.010

Pauw, W. P., Klein, R. J. T., Mbeva, K., Dzebo, A., Cassanmagnago, D., & Rudloff, A. (2018). Beyond headline mitigation numbers: We need more transparent and comparable NDCs to achieve the Paris Agreement on climate change. *Climatic Change, 147*(1-2), 23–29. DOI: 10.1007/s10584-017-2122-x

Pielke, R. A.Jr. (1998). Rethinking the role of adaptation in climate policy. *Global Environmental Change, 8*(2), 159–170. DOI: 10.1016/S0959-3780(98)00011-9

Porrini, D., & Schwarze, R. (2014). Insurance models and European climate change policies: An assessment. *European Journal of Law and Economics, 38*(1), 7–28. DOI: 10.1007/s10657-012-9376-6

Randolph, J. J. (2005). Free-marginal multirater kappa (multirater K [free]): An alternative to Fleiss' fixed-marginal multirater kappa. *Online Submission*. https://eric.ed.gov/?id=ED490661

Rayner, T., & Jordan, A. (2016). Climate change policy in the European Union. In *Oxford research encyclopedia of climate science*. https://doi.org/DOI: 10.1093/acrefore/9780190228620.013.47

Sachs, J., Kroll, C., Lafortune, G., Fuller, G., & Woelm, F. (2022). *Sustainable Development Report 2022*. Cambridge University Press. DOI: 10.1017/9781009210058

Salam, A. (2015). Climate change: The challenges for public health and environmental effects in UAE. *WIT Transactions on Ecology and the Environment, 193*, 457–466. DOI: 10.2495/SDP150391

Schmidt, R. C. (1997). Managing Delphi surveys using nonparametric statistical techniques. *Decision Sciences, 28*(3), 763–774. DOI: 10.1111/j.1540-5915.1997.tb01330.x

Skulmoski, G. J., Hartman, F. T., & Krahn, J. (2007). The Delphi method for graduate research. *Journal of Information Technology Education, 6*(1), 1–21. DOI: 10.28945/199

Surminski, S., Bouwer, L. M., & Linnerooth-Bayer, J. (2016). How insurance can support climate resilience. *Nature Climate Change, 6*(4), 333–334. DOI: 10.1038/nclimate2979

Surminski, S., Bouwer, L. M., & Linnerooth-Bayer, J. (2016). How insurance can support climate resilience. *Nature Climate Change, 6*(4), 333–334. DOI: 10.1038/nclimate2979

Trenberth, K. E. (2018). Climate change caused by human activities is happening and it already has major consequences. *Journal of Energy & Natural Resources Law, 36*(4), 463–481. DOI: 10.1080/02646811.2018.1450895

Zuehlke, D., Geweniger, T., Heimann, U., & Villmann, T. (2009). Fuzzy Fleiss-kappa for comparison of fuzzy classifiers. In *Proceedings of the European Symposium on Artificial Neural Networks (ESANN)*.

APPENDIX 1

Delphi Study Tools

Phase 1: Brainstorming

Dear participant,

We would like to thank you for accepting our invitation to participate in our research study. Your contribution is invaluable to the success of this study, and we appreciate your time and effort.

The UAE has made significant progress towards achieving its commitments under the Paris Agreement and SDG 13, Climate Action. The country has set ambitious targets to reduce its greenhouse gas emissions and transition to a low-carbon economy, with a particular focus on renewable energy and energy efficiency measures. Over the years, the UAE has launched many initiatives to decrease its impact on climate change, however, the UAE's progress towards achieving SDG 13 has been evaluated as moderate with major challenges remaining according to Sustainable Development Report 2022.

As part of the study, we kindly request your assistance in identifying the challenges that the UAE achieving SDG 13, Climate Action. In the first phase of the study, we would like you to engage in a brainstorming exercise and list at least six challenges that you believe are impeding the UAE's progress in achieving these climate change goals.

Question: Please provide a list of at least six challenges that you believe are impeding the UAE's progress in achieving SDG 13 using the attached link.

Phase 2: Narrowing down

Dear participant,

Thank you for taking part in the initial phase of the research. Based on the responses received from you and other panelists, a list of challenges impeding the UAE's progress in achieving SDG 13 (Climate Action) has been compiled. For this phase, we kindly request that you select half of the compiled challenges resulting from Phase 1.

Question: Please select 9 challenges that you believe are the key factors impeding the UAE's progress towards SDG 13, Climate Action. You can find the list of challenges in the provided link.

Phase 3: Ranking

Dear participant,

Thank you for taking part in the second phase of the research. After considering your responses and those of the other panelists, we have narrowed down the list of challenges to those deemed most significant by a majority of the panel members.

In this phase, we kindly ask you to rank these challenges according to their impact on climate action in the UAE. Please prioritize them from the most important to the least important.

Question: Please rank these 9 challenges in decreasing order of priority according to their impact on climate action in the UAE using the provided link.

Chapter 15
Legal Safeguards for Consumers in the UAE's Agricultural Sector Amid Climate Change

Enas Mohammed Alqodsi
https://orcid.org/0000-0003-4643-9660
United Arab Emirates University, UAE

Imad El Din Abdul Hay
https://orcid.org/0000-0001-6854-2605
University of Sharjah, UAE

Iyad Mohammad Jadalhaq
University of Sharjah, UAE

Saher Alwalid
Sultan Qaboos University, Oman

Mohammed El Hadi El Maknouzi
https://orcid.org/0000-0003-3396-8985
University of Al Dhaid, UAE

Ibrahim Al Shawabkeh
https://orcid.org/0000-0001-6548-258X
United Arab Emirates University, UAE

ABSTRACT

This chapter explores the legal safeguards available to consumers in the UAE's agricultural sector in the face of climate change and increasing reliance on biotechnology. It analyzes the regulatory framework that governs food safety and biosafety, focusing on Federal Law No. (9) of 2020 on Biosafety of Genetically Modified Organisms and its Executive Regulations issued under Cabinet Resolution No. (84) of 2022. These legal instruments are assessed alongside Federal Law No. (10) of 2015 on Food Safety to evaluate how the UAE ensures consumer protection amid environmental pressures and evolving agricultural practices. The chapter highlights the UAE's proactive approach to regulating genetically modified products through labeling, import controls, and environmental risk assessments. By examining the intersection of climate adaptation, technological innovation, and legal regulation,

DOI: 10.4018/979-8-3373-3346-5.ch015

this chapter contributes to the understanding of how national legal systems in the MENA region can uphold consumer rights and food security in a changing climate (Alqodsi, 2021b).

INTRODUCTION

The accelerating impacts of climate change have brought unprecedented stress to agricultural systems, especially in regions already burdened by extreme climatic conditions, such as the United Arab Emirates (UAE). With its arid climate, limited freshwater resources, and dependence on food imports, the UAE has recognized the urgent need to strengthen domestic agricultural production and ensure long-term food security. In this context, climate change is not merely an environmental concern but a critical driver of legal and policy reform. As the country explores climate-resilient agricultural strategies, including the use of biotechnology and genetically modified organisms (GMOs), legal safeguards for consumers have become central to national sustainability efforts(Al-Shawabkeh, 2020).

Modern agricultural technologies offer innovative solutions to climate-related challenges such as soil degradation, pest outbreaks, and crop yield instability. However, they also introduce potential risks to human health, environmental biodiversity, and consumer autonomy. To manage these risks, the UAE has adopted a robust legislative framework that aligns with international biosafety standards and emphasizes the precautionary principle. At the core of this framework is Federal Law No. (9) of 2020 concerning Biosafety of Genetically Modified Organisms and Their Products(Gbashi et al., 2021), which provides a legal foundation for the safe handling, use, and commercialization of GMOs. This law is further elaborated in Cabinet Resolution No. (84) of 2022, which serves as its executive regulation and outlines the specific procedures for risk assessment, traceability, labeling, approval, and monitoring of GMO-related activities.

These legislative instruments establish a comprehensive biosafety regime designed to balance technological advancement with environmental protection and consumer rights(Alqodsi, 2021a). For instance, they require prior approval from competent national authorities before the release or marketing of GMOs, and they mandate the labeling of genetically modified products to ensure consumers are adequately informed. Moreover, they include provisions for environmental risk assessments and post-market monitoring to detect any unforeseen effects of genetically engineered organisms on ecosystems and public health.

In addition to biosafety-specific regulations, the UAE's broader food safety framework, particularly Federal Law No. (10) of 2015 on Food Safety(Nasser & Ghloom, 2023), plays a vital role in safeguarding consumers. This law governs all stages of food production and distribution, including hygiene standards, food traceability, recall procedures, and penalties for violations. It reinforces the obligations of food producers and importers to maintain high safety standards and protect consumers from unsafe or deceptive practices.

By integrating biosafety and food safety legislation, the UAE presents a legal model that responds effectively to the twin pressures of climate change and technological innovation. This model reflects a proactive governance approach that prioritizes transparency, consumer trust, and environmental responsibility. It also resonates with the country's strategic vision for sustainable development(Alqodsi & Abdallah, 2024) and its commitment to the United Nations Sustainable Development Goals, particularly Goal 2 (Zero Hunger) and Goal 13 (Climate Action).

This chapter investigates how the UAE's legal system addresses the implications of climate change on agriculture through regulatory controls over biotechnology and food safety. It aims to provide a comprehensive analysis of the legal mechanisms that protect consumers in an evolving agro-environmental landscape. By situating the UAE experience within regional and global contexts, the chapter offers insights into the potential for legal reform and harmonization in MENA countries facing similar environmental and technological challenges.

REGULATORY GOVERNANCE OF GENETICALLY MODIFIED ORGANISMS AND BIOSAFETY IN THE UAE

This section examines the UAE's legislative and regulatory approach to the use and control of genetically modified organisms (GMOs) within its agricultural sector, particularly in response to climate adaptation. It analyzes Federal Law No. (9) of 2020 on Biosafety and its Executive Regulations under Cabinet Resolution No. (84) of 2022, highlighting their alignment with international biosafety standards such as the Cartagena Protocol. The section explores key legal requirements related to risk assessment, prior consent, product labeling, environmental release, and post-market surveillance. Special attention is given to the role of regulatory authorities in approving and monitoring GMO imports and local cultivation, ensuring that such practices do not compromise environmental integrity or public health. The section also discusses the legal implications of traceability systems and liability provisions for damage caused by GMOs, emphasizing the state's commitment to precautionary and preventive legal principles.

Biosafety Governance of Genetically Modified Organisms in the UAE: A Legal Analysis of Federal Law No. (9) of 2020

This subsection offers an in-depth analysis of the UAE's legislative foundation for biosafety governance in the agricultural sector, with a particular focus on Federal Law No. (9) of 2020 on Biosafety of Genetically Modified Organisms (GMOs) and Their Products(Al-Mssallem et al., 2024). This law constitutes a cornerstone of the UAE's efforts to align environmental, agricultural, and public health policies within a coherent legal framework, especially in light of climate change, which necessitates adaptive strategies in food production, including the cautious integration of genetic modification technologies.

At the heart of the law lies the principle of precaution, aimed at managing scientific uncertainty surrounding the environmental and health effects of genetically modified organisms. The legislation adopts a preventive regulatory model, imposing rigorous controls on all stages of GMO handling—from development and production to importation and distribution—while also emphasizing the protection of biodiversity, the integrity of local ecosystems, and the health of consumers.

The law opens with a comprehensive set of definitions (Article 1) that reflect international best practices and ensure legal clarity. These include critical terms such as "biosafety," "genetically modified organism," "genetic modification," "environmental release," and "risk assessment," among others(Entine et al., 2021). By clearly defining concepts like *genetically modified component* and *genetically modified product*, the law draws sharp lines between natural and engineered entities, facilitating regulatory oversight and enforcement(McHughen, 2016).

The objectives of the law (Article 2) are both ambitious and multidimensional, seeking to:

- Protect human health from potential hazards posed by GMOs and their byproducts;
- Preserve environmental quality and prevent ecological degradation resulting from GMO use;
- Safeguard local *genetic resources* and *biological diversity* from contamination or displacement;
- Introduce effective control and remediation mechanisms for actual or potential GMO-related damage;
- Regulate the scope and conduct of scientific research on genetically modified organisms.

The scope of application (Article 3) is notable for its comprehensiveness. It extends to all GMO-related activities, including import, export, re-export, transit, development, production, and trade of organisms or products containing more than 0.9% genetically modified components, whether directly or indirectly. This threshold mirrors global standards and ensures that even small-scale commercial uses are not exempt from oversight. At the same time, the law provides for specific exemptions, such as GMO content below the threshold and pharmaceutical applications intended for human use, reflecting a balanced approach that avoids unnecessary regulatory burdens while still maintaining essential safety protocols(GEBREYES, 2023).

Importantly, Article 4 prohibits the importation of GMOs or their products into the UAE unless a prior permit is obtained from the competent authority in each emirate, and where it is a first-time import, an additional approval from the Emirates Drug Establishment (EDE) is required(Albedwawi, 2025). This dual-approval mechanism enhances regulatory scrutiny and allows for centralized scientific and administrative evaluation of GMO entry into the national market. The same article bans the import of genetically modified animals or their derivatives altogether, indicating a heightened level of concern regarding animal genetic engineering and its potential long-term effects on ecosystems and food chains(Corona, 2015).

Article 5 extends similar restrictions to export and re-export activities, requiring prior authorization from both the EDE and local regulatory authorities. This demonstrates the UAE's commitment not only to controlling the inward flow of GMOs but also to its responsibilities as a transit and exporting country under international biosafety obligations.

Meanwhile, Article 6 regulates the *release* of GMOs into the environment and *isolated use* (such as within research facilities), requiring both prior approval and strict compliance with safety protocols(Madi, Barafi, & Abu Saleh, 2020). This provision underscores the recognition that even controlled environments carry potential risks if not adequately monitored and secured. Accordingly, Article 7 mandates the creation of a national biosafety register to record all facilities, products, and transactions involving GMOs, enabling traceability, legal accountability, and transparency(Khater, Imad, Abouhaiba, & Tannar).

A key provision lies in Article 8, which outlines the obligations of the person legally responsible for the facility—referred to as the "establishment manager." These include mandatory registration of the facility(Gutiérrez González, 2019), obtaining licenses for any stage of GMO involvement, and registering the GMOs being handled. This assigns legal responsibility directly to the actors involved in biotechnological activities, thereby reinforcing the principle of legal accountability.

Central to the risk governance system is Article 9, which details the requirement for scientific risk assessment prior to any GMO activity. The article stipulates that the risk assessment must cover hazard identification, exposure analysis, and risk characterization. This science-based approach enables regulators to evaluate potential threats before they materialize, in line with international best practices in environmental risk management(Premanandh, Maruthamuthu, Sabbagh, & Al Muhairi, 2012).

The law also addresses data confidentiality (Article 10), allowing entities to request that sensitive commercial or technical information they submit remain classified. This provision attempts to balance the need for transparency in public health and environmental matters with the protection of legitimate commercial interests, particularly in research and development(Jadalhaq & Russi, 2020).

Moreover, Article 11 introduces a mandatory labeling system for all GMO-containing shipments and products, ensuring that consumers are clearly informed of the presence of genetically modified materials. This aligns with international consumer rights principles and promotes informed decision-making(Newton, 2014).

The law provides for emergency response obligations (Article 12) and waste disposal procedures (Article 13), compelling parties to notify authorities immediately in the event of an incident and to assume financial responsibility for the containment, disposal(Holst-Jensen et al., 2012), or repatriation of any unauthorized or non-compliant GMOs(Bergkamp, 2000).

Enforcement mechanisms are outlined in Article 14, which empowers the competent authority and the Emirates Drug Establishment to conduct surveillance, monitoring, and inspections. This is complemented by Article 25, which grants certain officers judicial powers to enforce compliance and investigate violations, strengthening the institutional capacity of regulatory agencies(Al-Worafi, 2020).

The legal regime is further reinforced by civil liability provisions (Article 16) and a graduated system of administrative and criminal penalties (Articles 17–24). These range from temporary or permanent closure of non-compliant facilities to heavy fines and imprisonment for violations such as importing GM animals, unauthorized environmental release, or failure to obtain approvals. Notably, Article 24 allows for the doubling of penalties in cases of repeat offenses, thereby adding a deterrent dimension to the law.

In sum, Federal Law No. (9) of 2020 represents more than a conventional legislative text; it embodies a strategic legal response to the intricate and multifaceted challenges arising from the integration of genetic technologies into an agricultural sector increasingly exposed to climate stressors. In a region characterized by aridity, water scarcity, and fragile biodiversity, the introduction of genetically modified organisms (GMOs) presents both opportunity and risk. On one hand, biotechnology offers solutions for enhancing crop resilience, increasing yields, and reducing dependency on chemical inputs; on the other hand, it raises legitimate concerns about ecological disruption, transboundary contamination, long-term health implications, and loss of genetic heritage. This law seeks to navigate these tensions through a well-structured, precautionary, and science-based legal framework.

What distinguishes this legislation is its comprehensive life-cycle approach, whereby GMOs are regulated at every stage—from laboratory research and genetic modification, to commercial production, cross-border movement, marketing, labeling, environmental release, surveillance, and eventual disposal. Each phase is subject to defined standards, institutional oversight, and liability mechanisms, ensuring that no aspect of GMO governance falls through regulatory gaps. This cradle-to-grave legal architecture reflects an advanced understanding of the risks associated with biotechnology, particularly in the context of environmental fragility and consumer vulnerability.

Moreover, the law demonstrates an integrated model of biosafety governance that balances technological advancement with ethical imperatives and public interest. It harmonizes environmental concerns with food security goals(Al-Qodsi, Jadalhaq, & El-Maknouzi, 2024), positions scientific risk assessment as a cornerstone of regulatory decisions, and embeds liability and enforcement provisions to ensure compliance and deterrence. The explicit inclusion of civil liability for damages(Dahiyat, 2011), criminal penalties for violations, and institutional monitoring powers provides teeth to the regulatory system, distinguishing it from softer, guideline-based approaches adopted elsewhere.

Equally important is the law's alignment with international norms, particularly the Cartagena Protocol on Biosafety, which emphasizes informed consent, cross-border notification, and the sovereign right of states to restrict or regulate the entry of GMOs based on risk assessment. The UAE's legislation embodies these principles while tailoring them to national priorities such as environmental conservation, human health protection, and the safeguarding of traditional agricultural systems.

Furthermore, Federal Law No. (9) of 2020 plays a critical role in shaping public trust and consumer confidence in the use of modern biotechnologies. By mandating labeling, ensuring traceability, and requiring transparency in all transactions involving GMOs, the law empowers consumers with knowledge and choice—two pillars of rights-based regulation. This is particularly relevant in societies undergoing rapid technological transformation, where public skepticism can hinder innovation if not accompanied by robust legal assurances(Nobanee, El Hadi El Maknouzi, Sadok, & Alodat, 2024).

In essence, the law serves as a cornerstone for sustainable agricultural transformation in the UAE, offering a model for other MENA countries seeking to modernize their food systems without compromising environmental and public health standards. It reflects a governance vision that is adaptive, legally rigorous, and environmentally responsible, signaling the UAE's commitment to leveraging science and law in concert to address one of the 21st century's most complex intersections: the convergence of climate change, food security, and biotechnology.

From Policy to Practice: Cabinet Resolution (84/2022) on GMOs Regulation in the UAE

The issuance of Cabinet Resolution No. (84) of 2022 represents a significant regulatory development in the practical enforcement of the UAE's Federal Law No. (9) of 2020 concerning biosafety with respect to Genetically Modified Organisms (GMOs) and their products. As the executive regulation for this law, the Resolution converts its broad legal provisions into detailed regulatory measures that govern the importation, transit, domestic circulation, export, release, and contained use of GMOs. It introduces robust institutional responsibilities, reinforces risk-based management systems, and establishes a clear framework for licensing, oversight, labeling, emergency response, and legal accountability. Through this legislative instrument, the UAE demonstrates its commitment to both international biosafety norms and local environmental and public health priorities, setting the stage for responsible biotechnological advancement within a precautionary and controlled regulatory environment.

Legal Framework and Core Provisions of the Executive Regulation

The Executive Regulation issued through Cabinet Resolution No. (84) of 2022 presents a detailed and integrated regulatory framework that operationalizes Federal Law No. (9) of 2020 concerning biosafety from Genetically Modified Organisms (GMOs) and their products in the UAE(Laith, Alnemri, Ali, Alkhawaldeh, & Mihyar, 2022). Beginning with Article (1), the Resolution affirms the adoption of the same definitions established in the Federal Law, thereby ensuring legal continuity and terminological consistency throughout the implementation process(Albedwawi, 2025).

A cornerstone of the regulation is set forth in Article (2), which governs the import, transit, and circulation of GMOs(Madi et al., 2020). The article introduces a purpose-based approach, where the regulatory obligations vary according to the intended use—be it for environmental release, commercial distribution, food or feed production, industrial processing, or contained laboratory use. Each category is supported by specialized technical annexes that outline the documentation and biosafety requirements(Al-Mssallem et al., 2024). For initial importation, entities must submit comprehensive risk assessment reports from laboratories approved by the Ministry of Climate Change and Environment (MOCCAE), in addition to emergency response plans, monitoring strategies, and records of prior export authorizations where applicable(Nawaz, Chung, Golokhvast, & Tsatsakis, 2022). Transit operations are tightly controlled, mandating the use of sealed containers, biosafety-compliant transport vehicles, and full customs traceability(Sasani, 2024). Domestically, entities engaging in the circulation of GMOs must acquire official permits, store products in their original unopened packaging, and maintain accurate transaction records for at least ten years. Any act of re-packaging, partial opening, or fragmentation of GMO shipments is expressly prohibited, reinforcing the traceability and integrity of GMO handling(Laith & Ali, 2025).

Article (3) addresses the export and re-export of GMOs or their derivatives. The article stipulates that all exporting entities and products must be registered with the ministry, comply with the regulations of recipient countries, and obtain prior written approvals. These approvals are limited in both time (valid for six months) and usage (single-use only)(Paremoer, 2018), thereby minimizing the risk of unmonitored or unintended GMO dissemination across borders.

Further advancing biosafety governance, Article (4) introduces distinct regulatory tracks for environmental release and contained use of GMOs. Environmental release is permitted only upon ministry authorization and is subject to strict spatial confinement, risk evaluation, and ongoing reporting. Release locations must be appropriately isolated and pose no foreseeable harm to public health or ecosystems(Aspinall, 2001). Conversely, contained use—relevant for scientific research, testing, or industrial development—requires verified risk assessments, approved containment facilities, waste management systems, and full adherence to the national biosafety strategy(Khater et al.). Exemptions are narrowly defined and restricted to pre-approved organisms or those explicitly listed for research and educational purposes under ministerial resolutions.

In a move to centralize oversight, Article (5) mandates the creation of a National Biosafety Register, administered by MOCCAE. This register serves as the official repository for all institutional data, product information, and GMO-related activities conducted within the UAE(Thomsen et al., 2023). It enables inter-agency coordination, enhances transparency, and supports long-term tracking of all genetically modified inputs and outputs across the supply chain.

Article (6) assigns explicit responsibilities to establishment managers, requiring them to register their operations with the ministry, report any incidents of leakage or unintended release, train staff in biosafety protocols, and ensure full compliance with usage restrictions. Research institutions, in particular, are subject to supplementary regulatory scrutiny and must follow ministerial guidance specific to their field.

To ensure public awareness and product accountability, Article (7) mandates the use of bilingual labeling (Arabic and English) on all GMO products. Labels must clearly state the presence and proportion of genetically modified content, specify the country of origin, and include the product's scientific name and international identifiers. The labels must be affixed in a permanent, tamper-resistant manner to support informed consumer choices and streamline regulatory checks.

Article (8) strengthens the national biosafety response system by detailing emergency protocols in the event of actual or potential harm caused by GMOs. Entities are obligated to immediately notify MOCCAE and other competent authorities, submit updated risk assessments, and activate predefined containment and remediation procedures. Furthermore, the article empowers the authorities to recover the cost of corrective actions from the entity responsible for the breach, thus reinforcing accountability and the polluter-pays principle.

To uphold the enforceability of the regulation, Article (9) establishes a system for monitoring and inspection, authorizing MOCCAE and relevant agencies to conduct scheduled and surprise audits of GMO storage, transport, and processing sites. The inspection mechanisms are informed by technical annexes and are designed to ensure continued adherence to biosafety rules.

Disputes arising from administrative decisions can be addressed under Article (10), which lays out the grievance and appeal procedure. Affected parties have 15 days to submit a formal appeal, and the competent authority must issue a final, non-contestable decision within 30 days.

Finally, Article (11) empowers the ministry and concerned authorities to suspend or revoke approvals if scientific evidence suggests that GMOs or their products pose actual or potential harm to human health or the environment. The same applies in cases of non-compliance with regulatory obligations. In such instances, all related products must be withdrawn from the market, and production or trade activities immediately halted.

In the face of escalating climate change impacts across the MENA region—marked by extreme weather, water scarcity, and declining agricultural productivity—there is an urgent need to harness advanced technologies, including AI-driven genetic modification, to develop climate-resilient plant varieties. Within this context, Cabinet Resolution No. (84) of 2022 plays a pivotal role in enabling the safe and responsible deployment of genetically modified organisms (GMOs), including those bred using artificial intelligence, as part of a broader national strategy to secure food systems and environmental sustainability.

This Executive Regulation not only operationalizes the UAE's Federal Law No. (9) of 2020 on Biosafety, but also creates a robust legal infrastructure that can accommodate future biotechnological innovations, such as AI-assisted crop breeding(Mohanan, 2025). Through its stringent procedures for risk assessment, environmental monitoring, containment, labeling, and institutional accountability, the Resolution ensures that biotechnological tools are deployed in a way that aligns with ecological and public health priorities—a necessity in climate-vulnerable regions.

Furthermore, by delineating clear pathways for the research, circulation, and environmental release of GMOs, including climate-adaptive plant varieties, the Resolution positions the UAE as a regional model for regulating emerging agritech. It balances regulatory caution with scientific ambition, enabling researchers, developers, and policy-makers to collaboratively address the dual challenge of climate resilience and agricultural innovation(Nguyen, Nguyen, Vrontis, & Vo, 2025).

As the MENA region increasingly looks toward AI and biotechnology to overcome the agricultural burdens of climate change, the UAE's biosafety framework—as articulated in Cabinet Resolution No. (84) of 2022—offers a replicable model for other states. It demonstrates that legal certainty and environmental safeguards are not barriers to innovation, but prerequisites for sustainable progress.

Ultimately, the Resolution represents more than a set of administrative rules—it embodies a strategic governance tool that supports the ethical, scientific, and climate-aware integration of biotechnology into agricultural systems. In doing so, it reinforces the role of legal systems in shaping a food-secure, climate-resilient future for the UAE and the broader MENA region.

LEGAL MECHANISMS FOR CONSUMER PROTECTION AND FOOD SAFETY AMID CLIMATE CHALLENGES

As the MENA region—and the world at large—grapples with the compounding effects of climate change, from rising temperatures to prolonged droughts and increasing food insecurity, the need for resilient and sustainable food systems has never been more urgent. In this context, the United Arab Emirates (UAE) has emerged as a regional and global model in integrating climate-aware strategies into its food governance policies, placing food security and consumer protection at the heart of its sustainable development agenda. Anchored in the National Food Security Strategy 2051, the UAE's approach reflects a comprehensive vision to transform traditional food production models into sustainable, technology-driven ecosystems that can withstand climate-induced pressures. This long-term strategy outlines clear goals: enhancing local food production capacity, minimizing food waste, ensuring the availability and affordability of healthy diets, and becoming a leading global exporter of climate-smart agricultural solutions. Notably, the UAE is one of only four countries in the world where 100% of the population has access to affordable healthy diets, as recognized by the UN Food and Agriculture Organization (FAO).

To achieve these ambitions, the UAE has adopted a multi-tiered legal and policy framework that not only regulates food production and biotechnology (such as GMOs and AI-bred crops), but also protects consumers through rigorous food safety laws, labelling obligations, traceability systems, and sustainability standards. These mechanisms are vital in promoting transparency and restoring consumer trust in a food system increasingly impacted by global supply chain disruptions and ecological volatility.

The country's legal infrastructure is further supported by Cabinet Resolution No. (84) of 2022—a landmark regulation that operationalizes Federal Law No. (9) of 2020 on Biosafety. It addresses the importation, circulation, environmental release, and export of genetically modified organisms (GMOs) through strict licensing, monitoring, and containment requirements. This regulation ensures that climate-resilient biotechnologies are introduced responsibly, with full regard for human health and environmental protection. Moreover, the UAE has launched an array of national initiatives and platforms aimed at embedding sustainability in food systems, including:

The National Dialogue for Food Security, initiated in 2023 to promote innovation and sustainable consumption;The Sustainable Agriculture System, introduced in 2020 to improve crop yield and reduce water use; Financial tools such as the AED 100 million fund by the Emirates Development Bank to support agri-tech and food security projects; The launch of digital platforms like the UAE Open Food Research Portal, which democratizes access to cutting-edge agri-tech data; Strategic partnerships with innovation hubs like Hub71 and the adoption of international agreements such as the Milan Urban Food Policy Pact.

At the institutional level, a Food Security Council was formed to coordinate inter-agency collaboration and implement the goals of the National Strategy. Other community-driven programs, such as the "Ne'ma" Initiative, aim to cut food waste by 50% by 2030, aligning national practices with the UN Sustainable Development Goals.

These legal, institutional, and technological frameworks collectively reflect a preventive and responsive model to climate-induced risks, in which consumer rights, public health, environmental protection, and food innovation converge. By prioritizing consumer protection through sustainability, the UAE's food governance architecture exemplifies how legal systems can serve as engines of resilience and adaptation in the face of global climate uncertainty.

As the country hosted COP28, these regulatory and policy achievements positioned the UAE as a global leader in climate-conscious food security, setting a precedent for other nations in the MENA region to emulate.

National Legal Instruments Ensuring Consumer Rights and Food Safety in the UAE: A Climate-Responsive Perspective

In tandem with these statutory instruments, the National Food Security Strategy 2051 serves as a policy anchor for aligning legal frameworks with climate-responsive agricultural practices(Rodrigues, 2025). The strategy emphasizes increasing domestic food production using smart and sustainable technologies, diversifying sources of imported food, and building resilient supply chains capable of withstanding global shocks. This long-term vision interweaves food security with consumer protection, making it clear that the legal right to safe and sufficient food must be protected even in the context of environmental instability(Luu, Yadav, Srivastava, & Gopal, 2025).

The practical implementation of these legal safeguards relies on a coordinated network of regulatory institutions. The Ministry of Economy acts as the primary authority on consumer rights enforcement, while MOCCAE plays a central role in shaping and enforcing food safety policies through environmental risk monitoring and agricultural oversight. The Emirates Authority for Standardization and Metrology (ESMA)(Nihayati, 2025), meanwhile, is responsible for ensuring compliance with product safety and quality standards, including food labeling, packaging, and storage protocols. These institutions work in concert to deploy proactive tools such as digital consumer complaint platforms, early warning systems for contaminated food products, awareness campaigns on sustainable consumption, and capacity-building programs for food sector stakeholders.

By establishing a legislative ecosystem that incorporates climate science, environmental ethics, and consumer welfare, the UAE sets a notable precedent for legal innovation in the Arab region. Rather than reacting to environmental risks after they materialize, the country's legal framework anticipates such risks through adaptive governance mechanisms that are preventive, inclusive, and forward-looking. The cumulative effect of these instruments is a resilient legal structure that upholds consumer rights to health, information, and food safety—even amid the unpredictability of global climate systems.

Thus, through the fusion of modern consumer protection laws, specialized food safety regulations, and national strategies grounded in sustainability, the UAE exemplifies how legal systems can evolve to meet the dual demands of environmental change and public interest. This comprehensive approach ensures that consumer protection is not merely a market concern, but a cornerstone of climate-responsive governance—where legal safeguards are not only preserved but continuously strengthened in the pursuit of equitable and sustainable food systems.

CONCLUSION

As climate change continues to pose complex and evolving threats to global food systems, the United Arab Emirates has emerged as a legislative pioneer in the MENA region—crafting a legal framework that not only anticipates environmental disruptions but also embeds sustainability, consumer rights, and biosafety at its core. This chapter has examined how the UAE's legal and institutional architecture has responded to the dual challenges of climate vulnerability and technological innovation, particularly in the agricultural sector.

Through the integration of Federal Law No. (9) of 2020 on Biosafety of Genetically Modified Organisms and its Executive Regulation Cabinet Resolution No. (84) of 2022, alongside Federal Law No. (10) of 2015 on Food Safety and Federal Law No. (15) of 2020 on Consumer Protection, the UAE has established a comprehensive regulatory regime. This regime governs the introduction and monitoring of genetically modified products, ensures food quality and traceability, and upholds consumer autonomy and health in an increasingly complex agro-environmental landscape. The legislation is complemented by institutional mechanisms, such as the Ministry of Climate Change and Environment (MOCCAE), the Ministry of Economy, and the Emirates Authority for Standardization and Metrology (ESMA), which collaborate to enforce compliance, conduct risk assessments, and promote public awareness.

A particularly notable feature of the UAE's approach is its integration of climate resilience into legal frameworks—anchored in national strategies like the National Food Security Strategy 2051. By aligning food safety and biotechnology laws with climate adaptation goals, the UAE ensures that consumer protection is not treated as a standalone function, but as an integral part of a broader sustainability agenda. This fusion of law, science, and policy reflects a preventive and participatory governance model—one that supports innovation while maintaining safeguards against environmental and health risks.

Furthermore, the UAE's insistence on labeling, traceability, and ethical oversight strengthens transparency and reinforces public trust in biotechnology and food production systems. These measures enable consumers to make informed choices, empower regulators to act swiftly in the face of emerging risks, and foster an environment where technological progress and societal well-being are not in conflict but in balance.

Ultimately, the UAE's legislative experience illustrates how law can serve as both a shield and a compass: a shield that protects consumers from the uncertainties of a changing climate and a rapidly evolving food system, and a compass that guides national development toward sustainability, equity, and resilience. For other countries in the MENA region and beyond, the UAE offers a replicable model—one that proves legal systems, when thoughtfully designed and strategically implemented, can play a transformative role in securing the future of food and public health in an age of environmental uncertainty.

REFERENCES

Al-Mssallem, M. Q., Al-Khayri, J. M., Almagthali, B. H., Abdelkader, H. S., Alnabbat, K. I., & Sattar, M. N. (2024). Genetically modified food: potentiality for food and nutritional security in Saudi Arabia. In *Food and Nutrition Security in the Kingdom of Saudi Arabia, Vol. 2: Macroeconomic Policy and Its Implication on Food and Nutrition Security* (pp. 359-392): Springer. DOI: 10.1007/978-3-031-46704-2_16

Al-Qodsi, E. M., Jadalhaq, I. M., & El-Maknouzi, M. E. H. (2024). The place of UAE's food security in the national legislation and its role in supporting global food security. *Cogent Social Sciences, 10*(1), 2319379. Advance online publication. DOI: 10.1080/23311886.2024.2319379

Al-Shawabkeh, I. K. (2020). Legal guarantees for objective performance evaluation of the federal public employee in the uae legislation. *Lawyer Quarterly, 10*(3), 217-229. Retrieved from https://www.scopus.com/inward/record.uri?eid=2-s2.0-85090291804&partnerID=40&md5=e35ade47f008985468403b808b9d3170

Al-Worafi, Y. M. (2020). Drug safety in United Arab Emirates. In *Drug safety in developing countries* (pp. 419–428). Elsevier. DOI: 10.1016/B978-0-12-819837-7.00031-5

Albedwawi, A. (2025). Food and Pharmaceutical Transformation in Islamic Jurisprudence: A Comparison with UAE Law. *Journal of International Crisis and Risk Communication Research*, *8*(S1), 61.

Alqodsi, E. M. (2021a). Analyzing the Implementation of Usufruct Rights and Obligations in the UAE Civil Transactions Law. *International Journal of Criminal Justice Sciences*, *16*(2), 202–219. https://www.scopus.com/inward/record.uri?eid=2-s2.0-85134491872&partnerID=40&md5=9f4e39444da0981dc669dd27dbf128f7

Alqodsi, E. M. (2021b). THE RIGHT TO PRE-CONTRACTUAL INFORMATION IN E-COMMERCE CONSUMER CONTRACTS: UAE LAW AND COMPARATIVE PERSPECTIVES. *Journal of Legal, Ethical and Regulatory Issues, 24*(2), 1-16. Retrieved from https://www.scopus.com/inward/record.uri?eid=2-s2.0-85103711308&partnerID=40&md5=b795381746e49b8e94a5e798ac846ec9

. Alqodsi, E. M., & Abdallah, A. K. (2024). *Legal frameworks and educational strategies for sustainable development.*

. Aspinall, S. (2001). Environmental Development and Protection in the UAE. *United Arab Emirates: a new perspective, 278.*

Bergkamp, L. (2000). Allocating unknown risk: liability for environmental damages caused by deliberately released genetically modified organisms. *Available at SSRN* 223068. DOI: 10.2139/ssrn.223068

. Corona, D. (2015). Allowing or banning genetically modified food and feed? Morocco in search of a synthesis between conflicting regulatory models. *Building Sustainable Agriculture for Food Security in the Euro-Mediterranean Area: Challenges and Policy Options*, 85.

Dahiyat, E. A. R. (2011). Towards new recognition of liability in the digital world: Should we be more creative? *International Journal of Law and Information Technology*, *19*(3), 224–242. DOI: 10.1093/ijlit/ear006

Entine, J., Felipe, M. S. S., Groenewald, J.-H., Kershen, D. L., Lema, M., McHughen, A., & Parrott, W. A. (2021). Regulatory approaches for genome edited agricultural plants in select countries and jurisdictions around the world. *Transgenic Research*, *30*(4), 551–584. DOI: 10.1007/s11248-021-00257-8

Gbashi, S., Adebo, O., Adebiyi, J. A., Targuma, S., Tebele, S., Areo, O. M., Olopade, B., Odukoya, J. O., & Njobeh, P. (2021). Food safety, food security and genetically modified organisms in Africa: A current perspective. *Biotechnology & Genetic Engineering Reviews*, *37*(1), 30–63. DOI: 10.1080/02648725.2021.1940735

GEBREYES. A. (2023). *SCHOOL OF LAW.* Wachemo University

Gutiérrez González, A. (2019). The right to a clean and healthy environment: GMOs in Mexico and the European Union. *Mexican Law Review*, *11*(2), 91–113. DOI: 10.22201/iij.24485306e.2019.1.13129

Holst-Jensen, A., Bertheau, Y., De Loose, M., Grohmann, L., Hamels, S., Hougs, L., & Van den Bulcke, M. (2012). Detecting un-authorized genetically modified organisms (GMOs) and derived materials. *Biotechnology Advances*, *30*(6), 1318–1335. DOI: 10.1016/j.biotechadv.2012.01.024

Jadalhaq, I. M., & Russi, L. (2020). Finding Direction at the Edge of Law and Life: Islamic Fiqh, Correspondence, and UAE Takāful Insurance Regulation. *Canadian Journal of Law and Society*, *35*(3), 477–497. DOI: 10.1017/cls.2020.19

Khater, M. H., Imad, D., Abouhaiba, N., & Tannar, H. Legal Environmental Protection and Sustainable Development in the United Arab Emirates: A Comparative Study with International Frameworks. *Available at SSRN* 5119294. DOI: 10.2139/ssrn.5119294

Laith, A.-E., & Ali, H. (2025). Balancing Innovation, Biosecurity, and Biosafety in Dual-Use Research in the Middle East: Dual-Use Research in the Middle East. *Global Biosecurity*.

Laith, A.-E., Alnemri, M., Ali, H., Alkhawaldeh, M., & Mihyar, A. (2022)... *Journal of Biosafety and Biosecurity*.

Luu, T. M. N., Yadav, M., Srivastava, A. P., & Gopal, K. (2025). Innovation and Technology to Address the Challenges of Climate Change. In *Effects of Climate Change on Social and Economic Factors* (pp. 423–464). IGI Global.

Madi, R., Barafi, J., & Abu Saleh, M. (2020). Releasing GMOs into the environment or launching its products into markets: Regional and international legislations. *Industrial Biotechnology (New Rochelle, N.Y.)*, *16*(6), 359–367. DOI: 10.1089/ind.2020.29229.rma

McHughen, A. (2016). A critical assessment of regulatory triggers for products of biotechnology: Product vs. process. *GM Crops and Food: Biotechnology in Agriculture and the Food Chain*, *7*(3-4), 125–158. DOI: 10.1080/21645698.2016.1228516

. Mohanan, P. (2025). *Artificial Intelligence and Biological Sciences*: CRC Press, Taylor & Francis Group.

Nasser, A. E., & Ghloom, K. A. (2023). Ensuring Environmental Safety through Sustainable Food Safety. *Ajrsp*, *5*(54), 141–155. DOI: 10.52132/Ajrsp.e.2023.54.6

Nawaz, M. A., Chung, G., Golokhvast, K. S., & Tsatsakis, A. M. (2022). *GMOs and Political Stance: Global GMO Regulation, Certification, Labeling, and Consumer Preferences*. Academic Press.

. Newton, D. E. (2014). GMO food.

Nguyen, N. T. T., Nguyen, P. V., Vrontis, D., & Vo, N. T. T. (2025). Enhancing organizational sustainable performance through green innovation: The roles of knowledge application, government policy, and green market orientation. *Journal of Knowledge Management*, *29*(3), 870–890. DOI: 10.1108/JKM-06-2024-0688

Nihayati, A. (2025). *Integrating Halal Certification in Free Trade Agreements: Indonesia's Modest Fashion and Cosmetics* (Vol. 18). Jurnal Hubungan Internasional.

Nobanee, H., El Hadi El Maknouzi, M., Sadok, H., & Alodat, A. Y. (2024). Analysis of insurance entrepreneurship as a hedge in times of crisis: A literature review. *Sustainable Technology and Entrepreneurship*, *3*(2), 100065. Advance online publication. DOI: 10.1016/j.stae.2023.100065

. Paremoer, T. (2018). Regional value chains: exploring linkages and opportunities in the agro-processing sector across five SADC countries.

Premanandh, J., Maruthamuthu, M., Sabbagh, A., & Al Muhairi, S. (2012). Prevalence of genetically modified foods (GM foods) in the United Arab Emirates. *Food Control*, *25*(1), 10–12. DOI: 10.1016/j.foodcont.2011.10.001

. Rodrigues, E. (2025). About Transitions Research.

Sasani, M. S. (2024). The importance of biosecurity in emerging biotechnologies and synthetic biology. *Avicenna Journal of Medical Biotechnology*, *16*(4), 223.

Thomsen, J., Abdulrazzaq, N. M., AlRand, H., Consortium, U. A. S., Elhag Ahmed, A., Yousef, A. F., & Enshasy, D. A. M. (2023). Surveillance of antimicrobial resistance in the United Arab Emirates: The early implementation phase. *Frontiers in Public Health*, *11*, 1247627. DOI: 10.3389/fpubh.2023.1247627

Chapter 16
Towards Responsible Marine Aquaculture:
Integrating the Blue Economy and Sustainable Development

Omar Er-Rousse
https://orcid.org/0009-0007-5303-6356
Ibn Tofail University, Morocco

ABSTRACT

Marine aquaculture, as a key sector for blue growth, plays a crucial role in the economic development of marine resources. Integrating sustainable practices into this activity is becoming an imperative to preserve marine ecosystems while meeting the growing demand and consumption of aquatic products. This approach aims to align aquaculture with the PCI of sustainability by optimising the use of marine resources, minimising environmental impacts, and supporting blue communities. The blue economy, based on the sustainable use of oceans, seas, offers a framework that integrates both the preservation of marine biodiversity and economic innovation. The challenges are numerous: reducing carbon emissions, avoiding overexploitation of resources, and guaranteeing economic development that benefits coastal populations while preserving the health of the oceans. However, innovative solutions, such as closed-circuit aquaculture techniques, integrated marine culture, and research into sustainable genetics, are opening up promising prospects for reconciling productivity and sustainability.

DOI: 10.4018/979-8-3373-3346-5.ch016

INTRODUCTION

In the face of continuous population growth and the overexploitation of fishery resources, marine aquaculture is emerging as a strategic solution to ensure food security, stimulate coastal economies, and protect marine ecosystems. In this context, marine aquaculture in the southern Mediterranean, particularly in the MENA region, is set to play a pivotal role, especially in coastal countries such as Egypt, Morocco and Lebanon that are grappling with environmental, socio-economic and institutional challenges.

Meanwhile, the concept of the blue economy* (Gunter, P., 2010) — the sustainable and inclusive management of marine resources — is gaining ground on the international political agenda. This approach is based on innovation, resilience, and social inclusion as the pillars of sustainable growth in the marine environment. Integrating this paradigm into aquaculture policies and practises represents a significant opportunity to rethink the sector's development with sustainability in mind.

However, despite the stated ambitions, research into the blue economy and marine aquaculture in the MENA region has several significant shortcomings. First, the lack of reliable data and effective monitoring systems limits understanding of maritime resources. In addition, investment in research and development is often insufficient, leading to delays in innovation in sustainable farming technologies. Regulatory frameworks are disparate, and lack harmonisation, complicating resource management. Furthermore, specialised training is limited, creating a skills gap, and awareness among local communities of the benefits of the blue economy is low. Aquaculture practises are not always sustainable and their environmental impact is often overlooked. Finally, the lack of regional cooperation hinders the sharing of knowledge and best practises, isolating research efforts and limiting their effectiveness. To overcome these challenges, it is essential to strengthen data collection, increase investment, develop coherent policies and promote training and awareness. In this context, Mediterranean marine aquaculture faces a difficult choice between intensive production and the need for a sustainable, circular and equitable model.

The development of marine aquaculture depends on a thorough understanding of the biology of the species being farmed. This approach maximises yields while minimising environmental impacts and health risks. The biology of farmed marine species encompasses knowledge of their life cycle, behaviour, nutrition, reproduction, health, and adaptation to farming conditions.

Thus, in the context of marine aquaculture, sustainable development refers to a system of principles, criteria, and indicators (PCI) that can be used to evaluate and promote production and consumption patterns that respect the environment, the economy, and society.

This chapter focuses on four main areas: the history and classifications of aquaculture; the biology of farmed marine species; the principles, criteria and indicators (PCI) for marine aquaculture; and aquaculture development policies in Egypt, Lebanon, and Morocco. The work explores the conditions for implementing responsible marine aquaculture based on historical and biological approaches and the convergence of sustainable development goals and blue economy principles. The main question this study seeks to answer is how the principles of the blue economy can contribute to the development of responsible, sustainable, and inclusive marine aquaculture in Mediterranean countries in the MENA region, such as Egypt, Morocco, and Lebanon. Through a comparative analysis of these countries, the aim is to identify the levers, obstacles, and good practises that can guide the blue transformation of the sector towards a more sustainable model.

The scientific data comes mainly from national institutions, such as the National Agency for the Development of Aquaculture in Morocco and the GAFRD in Egypt, as well as international institutions such as the FAO, the World Bank, and the Barcelona Convention (UNEP/MAP). Recent studies have also been used to supplement the data. This approach will reveal the aquaculture strategies employed by each country.

1. MARINE AQUACULTURE: DEFINITIONS AND HISTORICAL DEVELOPMENT.

Marine aquaculture is the practice of farming marine organisms in a controlled environment. It plays a crucial role in ensuring global food security and preserving ocean resources. This section explores the various definitions of this activity, which is expanding rapidly, as well as its historical evolution. By tracing the origins and transformations of marine aquaculture, we shed light on the technical innovations, economic factors, social considerations, ecological impacts and management policies that have shaped this industry. Our aim is to provide an in-depth understanding of marine aquaculture in terms of both its foundations and its future prospects.

1.1. Definitions

Aquaculture, also known as fish farming, was first used around the start of the 20th century and recommended by the French Academy. It is defined as the "art of multiplying and raising aquatic animals. A more controlled environment is required for the latter, which includes fish, crustaceans, molluscs, and algae. This technique is used to produce food for human consumption. It is also used to preserve endangered species. Farmers can streamline aquatic animal growth and health through practises such as pond stocking, aeration, water filtration, and constant monitoring of the environment.

The term ‘aquaculture’ refers to all human activities related to the farming of aquatic animals and, less commonly, the cultivation of aquatic plants. Furthermore, it indicates that this term encompasses all activities related to the production, processing (packaging) and marketing of aquatic species, whether plants or animals from freshwater, brackish water or saltwater, under conditions controlled or semi-controlled by humans.

Translated with DeepL.com (free version)According to France's National Institute of Statistics and Economic Studies (INSEE), aquaculture is “the cultivation of plants and the breeding of animals in continental or marine waters, with the aim of increasing production, and involving the individual or legal possession of the breeding stock.”. This includes fish farming, shellfish farming (including oysters, mussels, clams, and cockles), seaweed farming, and carcinoculture (the farming of crustaceans, primarily shrimp and crayfish).

Broadly speaking, aquaculture can be defined as the farming of aquatic organisms, including fish, molluscs, crustaceans and aquatic plants. However, in order to improve production, certain forms of support are necessary in the implementation of these processes, such as regular stocking, feeding and protection from predators. As with livestock farming, ownership of the products is also required.

G. Burton Ayles (2013) defines aquaculture as the cultivation and harvesting of freshwater or marine organisms by humans. Synonyms of aquaculture include fish farming, mariculture, and sea-based grazing. Aquaculture farms play a vital role in fisheries and the management of aquatic resources worldwide. The species farmed or cultivated include fish such as trout, carp, and tuna, as well as crustaceans (e.g. shrimp), molluscs (e.g. oysters), and algae. These species are reared or cultivated in ponds, reservoirs, or nets using saltwater, brackish water, or freshwater.

In accordance with the strategic choices made for the implementation of its Blue Plan, Morocco is committed to the responsible and sustainable development of marine aquaculture as an economic activity that contributes to food and energy security, environmental protection, and the conservation of fishery resources. To this end, the country has adopted the following definition: marine aquaculture encompasses all breeding, cultivation, and conservation activities involving living marine organisms, with the exception of the conservation of adult tuna in marine waters for fattening purposes.

Furthermore, Law No. 84-21 on Marine Aquaculture[†](B.O. No. 7174, 9 Chaabane 1444 [2 March 2023]) clarifies key terms to prevent misunderstandings and ambiguity surrounding the activities of the marine aquaculture sector. In this context, the following definitions are provided:

1. **Marine aquaculture**: all activities relating to the breeding, cultivation and/or conservation of live marine aquatic organisms, except for the conservation of adult tuna caught at sea for fattening in marine waters.
2. **An aquaculture farm** is defined as any fenced area, structure, or fixed or mobile installation used for the practise of marine aquaculture, whether on land or at sea. This excludes establishments used for the packaging, shipping, and purification of live shellfish, as well as nurseries and aquariums.
3. **A tank** is defined as any container used to temporarily house marine aquatic organisms in a living state without feeding them, including during transport.
4. **A marine aquaculture auxiliary vessel** is any support vessel registered as such and used for operating a fish farm. This type of vessel is used to transport personnel and equipment, as well as feed for farmed fish species and harvested marine aquaculture products. It is also used to carry out any other work related to the operation of the farm.
5. **An aquarium** is any receptacle equipped with an open or closed circuit facility that allows fish species to be conserved in a living state for the purpose of being presented to the public in a cultural setting, for play or demonstration, and can be sold in a living state directly to a final buyer.
6. **Aquaculture structure**: any arrangement of an aquaculture space determined by its geographical location, dimensions and the nature and technical specifics of the aquaculture activities carried out there.
7. **Aquaculture scientific research**: marine aquaculture activity aimed at studying any fish species in their breeding, cultivation or conservation environment while they are alive, or studying any aquaculture production process.
8. **Experimental aquaculture**: a marine aquaculture activity that tests and controls the technical and economic feasibility and viability of an aquaculture project on a small scale, involving the breeding, cultivation or conservation of new fish species in a living state, or the adoption of new aquaculture processes.
9. **Aquaculture for restocking**: Marine aquaculture activity involving the production of aquatic organisms for restocking, environmental protection, or decontamination purposes.
10. **Aquaculture training**: Marine aquaculture activity aimed at training people for aquaculture professions.

We distinguish between three types of aquaculture: commercial, experimental and repopulating. Experimental aquaculture is a scientific research activity that aims to test and control, on a small scale, the technical and economic feasibility and viability of an aquaculture project involving the breeding or cultivation of new fish species or the adoption of new processes. Conversely, restocking aquaculture refers to any aquaculture activity aimed at producing aquatic organisms for restocking, environmental protection, or pollution abatement purposes.

1.2. History of aquaculture

The breeding of aquatic animals is a very old economic activity like humanity (de Marignan, Claire, 2017). It is difficult to precisely date its origin. The discovery, a few years ago, of a Theban tomb whose bas-relief depicted a fishing scene with Nile tilapia (Geletu, T. T., & Zhao, J. (2023) (Oreochromis niloticus (Linnaeus 1758)) in an equipped basin, al-lowed affirming that it was already practiced more than 4,000 years ago in Pharaonic Egypt (Soliman, N. F., & Yacout, D. M. (2016)) . In addition, far from being confined to the kingdom of Pharaoh, it spreads to the other shore of the Mediterranean: the Etruscans and the Romans thus develop fish farming in the lagoon, while the Greeks are already embarking on oyster farming. In addition, Asia is no exception to the practice of aquaculture in managed ponds, notably in China, Vietnam, Thailand...A treaty on fish farming, also entitled 'The Breeding of Fish by Tao Zhu Gong' was written by Fan Li, a Chinese politician from the Spring and Autumn period, in 475 BC: it is the oldest written document related to fish farming known to date (hao, Y., Gozlan, R. E., & Zhang, C. (2015)).

China is the oldest country in the world with a history of fish farming. It has a long history, and extensive experience in aquaculture production, as well as a high level of scientific expertise. According to the Fan Li Treaty on Fish Farming, there are many references to the successful cultivation of various aquatic species, particularly carp. Although the text is not long, it contains a great deal of information. Important production methods and techniques, such as farming purposes, fish pond structures, breeding standards, and seasons, female-to-male fish proportions, intensive breeding and alternative fishing, have reached a high level.

One of the most notable aspects of the Fan Li Treaty is its presentation of carp as a model for pond-based aquaculture. This demonstrates its significant advantages from the perspectives of production and economy. If we choose to raise carp, it is because they do not devour their peers, they grow easily, and they fetch a high price. The rule of natural regeneration in ponds can be exploited to solve the problem of supplying alevins. The method recommended by Fan Li for perpetuating fish intended for reproduction is simple, effective, and easy to implement. Indeed, carp farming has been highly profitable and has laid the technical foundations for the development of fish farming. The Romans introduced the Danube carp to Europe from the 1st to the 4th century AD (Hoffman, R. C. (1995). Monks developed aquaculture techniques between the 14th and 16th centuries. Trout were first raised on a fish farm in Germany in 1741 (Gage, S. D. (1978)), and the first attempts at commercial farming began in the United States as early as 1853. The rainbow trout, which is native to North America, is now farmed worldwide. Other species of fish and shrimp that were once impossible to farm are now cultivated in several tropical and subtropical regions. Several other species that may one day be suitable for aquaculture are currently being studied.

In total, we summarise the history of aquaculture according to two approaches: the diachronic approach and the species-specific approach, as follows:

Table 1. The history of aquaculture.

Period	Summary of Development	Examples/Details
- 2000 BC	rudimentary fish farming	Carp in China Tilapia in Egypt
- 600	Oyster farming	Greece
15th Century	Valliculture (keeping fish caught during their ascent to brackish waters in an enclosed space)	Italy
18th Century	Discovery of artificial fertilization applied to Salmonids	
19th Century	Transplantation applied to salmonids; development of oyster farming	Europe
1950	Eel farming	Japan
1960	Explosion of Rainbow Trout production (TAC)	Europe, North America
1970	Ranching or marine cage farming with aquaculture of Amberjack, Catfish, and Pectinidae (scallops)	
1980	"New aquaculture" with the production of Salmon, Shrimp, Bass, Sea Bream	
1990	Emergence of Turbot, Sturgeon, and tropical species like Grouper, Wolf fish, or Umbrine (meagre)	
2000	Mouth of the Thon	

Source: developed by the author

Marine aquaculture has undergone significant changes since its beginnings several centuries ago, with this evolution accelerating particularly during the 20th century. Initially practised for subsistence purposes, it has evolved into a global industry in response to the growing demand for seafood products. This transformation has been characterised by technological innovations, scientific research, and heightened environmental awareness. Today, marine aquaculture plays a key role in ensuring food security and supporting the economies of many countries, while presenting challenges related to sustainability and resource management.

2. DIFFERENT CLASSIFICATIONS OF AQUACULTURE

2.1. Sustainable Aquaculture and Commercial Aquaculture

Sustainable marine aquaculture is a form of aquaculture that aims to produce fish, shellfish, and molluscs in a way that minimizes the impact on the marine environment while ensuring the long-term sustainability of the industry. Indicators of sustainable aquaculture are at the forefront of specialized literature. They encompass economic, social, and environmental factors. This can include practices such as using quality feed and high-quality seeds, managing the marine environment to minimize negative impacts on ecosystems, and utilizing clean and eco-friendly technologies. Sustainable marine aquaculture may also involve social sustainability management practices, such as considering the interests of local communities and involving them in the decision-making process. In general, sustainable marine aquaculture aims to produce seafood while meeting industry needs, preserving the health of the marine environment, and respecting the interests of local communities.

Commercial marine aquaculture involves raising fish, shellfish, molluscs and algae in facilities located in marine environments. This can be done in cages, ponds or enclosures at sea. Although its primary aim is to produce species intended for human consumption in order to maximize profits, commercial aquaculture may not be sustainable. However, it can also be used to restock marine ecosystems or to produce ornamental species for aquariums. It has become an important source of animal protein in many countries, contributing to the growing demand for seafood. However, it can pose environmental challenges such as pollution and impacts on wild species. Therefore, it is important to implement sustainable management measures to minimize these impacts.

Commercial aquaculture contributes directly to economic growth by increasing added value in the form of labor income, activity rents and tax revenues, as well as creating jobs. Indirectly, it contributes to economic growth through infrastructure development, human capital, increased productivity and foreign currency earnings in developing countries.

2.2. Classification of Aquaculture According to Intensity

Marine aquaculture involves raising marine organisms in controlled environments, and can be classified according to production intensity. These methods are divided into three categories: extensive, intensive, and semi-intensive aquaculture. Semi-intensive aquaculture falls between extensive and intensive practises. These methods play a crucial role in meeting the growing global demand for seafood products while presenting distinct advantages and challenges.

Extensive aquaculture includes systems of cultivation and breeding where human intervention is limited, in addition to capture, to the reproduction of the raised species.

Experts note that external inputs in extensive aquaculture are limited, costs remain low, capital investment is restricted, and the quantity produced of raised species per unit area is low (yield up to a maximum of 500 kg/ha/year). Production factors are controlled and maintained at a low level. Labor yield is high. This type of cultivation depends on the natural ecosystem with low stocking density and little (or no) feed input. Extensive farming allows for the production of animals in large environments, potentially on the scale of an ocean (for example, the North Pacific for salmon ranching from hatcheries in western North America or eastern Asia, Japan, and Russia). Notably, seaweed cultivation and traditional shellfish farming are considered extensive cultures.

Traditionally, this type of aquaculture (intensive fish farming) is called commercial (FAO. 2016), and requires a greater human intervention. The production of organisms is carried out on small areas, ponds, enclosures, tanks, cages, etc., but with high yields. For example, for shrimp, yields range from 150 to 800 kilograms per hectare per year; the feed input is from 3 to 25 tons per hectare per year. Feed input is essential, and it is private companies, sometimes very powerful multinationals, that engage in these activities.

Currently, the use of robots in intensive aquaculture is a developing area aimed at improving the efficiency and sustainability of this practice. Robots can be used in several aspects of aquaculture, such as monitoring water conditions, distributing feed, cleaning ponds or cages, and even sorting fish.

By utilizing advanced sensors, robots can collect real-time data on environmental parameters such as temperature, salinity, and water quality. This allows operators to monitor and adjust conditions to optimize fish growth and health while reducing the risks of disease or pollution. Robots can also be programmed to perform routine tasks such as regular feed distribution, freeing up time and resources for farmers.

In summary, the use of robots in intensive aquaculture presents several potential advantages, including improved productivity, reduced costs, and minimized environmental impacts. However, it should be noted that this technology is still developing, and challenges in terms of cost, user-friendliness, and reliability may need to be addressed before it is widely adopted.

The semi-intensive marine aquaculture system is a method of raising fish, shellfish, or other marine organisms in controlled environments such as ponds, floating cages, or tanks. Unlike intensive systems that use extremely controlled artificial conditions, semi-intensive systems allow for greater interaction with the natural marine environment.

Figure 1. Systems related to fisheries and aquaculture

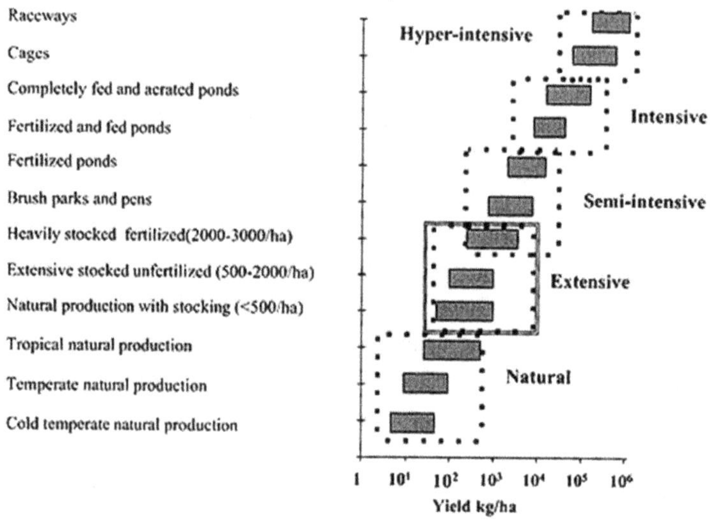

Source: *Welcomme, R. L., and D. M. Bartley. "An evaluation of present techniques for the enhancement of fisheries." FAO Fisheries Technical Paper (1998): 1-36.*

2.3. Integrated Multi-Trophic Aquaculture (IMTA): A New Approach

Integrated Multi-Trophic Aquaculture (IMTA) is defined as the combination of different marine cultures using species from various taxonomic groups within the same physical system or installation, aiming to improve the environmental quality of the habitat and maximize benefits.

IMTA is a highly flexible concept, as it can be applied in both open sea and freshwater systems. It represents a relevant solution that promotes better environmental management while increasing economic returns for producers and communities.

In IMTA systems, the by-products (wastes) from the farming of certain species serve as raw materials (fertilizers, feed, etc.) for other species. IMTA borrows this concept from nature (biomimicry) (Thierry Chopin. 2023), primarily the idea that within the food chain, one species always finds a food source in the waste of another. IMTA mimics a natural ecosystem. It does this by combining the farming of species that require feed. These are fish and crustaceans. It also includes the aquaculture of inorganic extractors. These are marine algae. And organic extractors are mollusks. This achieves a balanced ecosystem. This offers mutual benefits to co-cultivated species. This promotes environmental sustainability (bio-mitigation), economic stability (product diversification and risk reduction) and social acceptance.

By considering the synergistic interactions between the cultivated species and the ecosystem services they provide, the goal is to design and implement aquaculture systems and practices ecologically to ensure increased economic expansion, environmental sustainability, and social acceptability, supported by facilitative regulations and within a holistic and circular economic approach. Integrated Multi-Trophic Aquaculture positively contributes to achieving the United Nations Sustainable Development Goals (SDGs), particularly SDG 14 (life below water) and other sustainable development objectives.

The concept of Integrated Multi-Trophic Aquaculture (IMTA) is very broad, flexible, and evolving. IMTA is practiced worldwide, with systems in freshwater or marine environments (such as in aquaponics) and in temperate or tropical zones. IMTA is not limited to a short bureaucratic definition; it is valued for its versatility.

A framework for classifying various projects is provided by a branching structure of Integrated Multi-Trophic Aquaculture (IMTA), on which research works are based.

In contrast to integrated aquaculture, aquaponics operates in closed environments with recirculated water. Numerous methods exist, often referred to as "aquaponics." For example, Chinampas or fish farming in rice paddies.

3. THE BIOLOGY OF CULTIVATED MARINE SPECIES: A PREREQUISITE FOR SUSTAINABLE AQUACULTURE

Marine aquaculture, or the breeding of aquatic organisms in the sea, is an economically viable and growing practice. This industry meets the increasing demand for seafood while ensuring sustainable use of marine resources. The costs of marine aquaculture can vary depending on several key factors, such as the size of the aquaculture operation, the species raised, environmental conditions, and current regulations. However, many marine aquaculture operators have succeeded in achieving profitability by optimizing their practices and innovating in the field.

Indeed, while marine aquaculture can be economically viable, it requires careful planning and effective management to ensure financial success. To be financially successful in "marine aquaculture," which encompasses all types of breeding of animal or plant species living in the sea, it is crucial that this farming is based on a complete or partial mastery of the biological cycle of the cultivated species. This involves an in-depth understanding of egg or larval recovery, juvenile bivalves or fish imports, growth, feeding, and environmental needs of the species in question, capture of juveniles from the marine environment, and controlled growth to increase the production of high-value species while minimizing losses and preserving the marine environment.

We aim to present the dimensions of aquaculture projects that must consider various parameters related to environmental conditions, the requirements and zootechnical standards of the species to be raised, and the economic indicators of the aquaculture sector.

3.1. Potential Species in the MENA Region

Sea Bream (Sparus aurata) (Linnaeus, 1758) – Sparidae: This is a benthic predatory species with a wide geographical distribution. It is well represented in the Mediterranean and the Moroccan Atlantic, and can be raised in sea cages or in land-based ponds. Breeders of different age groups, ranging from one year for males (2 to 4 years) to 5 years for females (2 to 6 years), are kept under long-term storage conditions. Breeders may also come from breeding farms or the natural environment. It can withstand high densities, between 25 to 40 kg / m^3 (at sea) and up to 45 kg/m^3 in land-based ponds. Optimal temperature (Topt) = 22 to 28 °C and optimal salinity (Sopt) = 30-37 ‰, conversion index (IC) = 1.7 to 2. The production cycle lasts 16 months, and the commercial size is between 350 and 400 g.

Sea Bass: The term "bass" essentially refers to two species, the common bass (Dicentrarchus labrax) and the spotted bass (Dicentrarchus punctatus) (Rodríguez-García, C., Toro-Podadera, A., Sarmiento-Carbajal, J., & Cabrera-Castro, R. (2024)).

It is an adaptive species with strong growth. It can be raised in sea cages or in land-based ponds. To ensure a reliable and sufficient supply of high-quality eggs, most hatcheries have established their own broodstock units, where breeders of different age groups are maintained long-term (FAO, 2009). The optimal age for females is between 5 and 8 years, while for males it is between 2 and 4 years. Management of broodstock in captivity includes natural maturation, induction of ovulation by manipulating photoperiod or hormonal treatment, fertilization in breeding tanks, and incubation in an open-flow water system. It thrives under conditions where Topt varies between 22 to 28 °C, and Sopt between 30-37 ‰, with an IC of 2. Density ranges from 25 to 35 kg/m^3 (at sea) and can reach up to 45 kg/m^3 in land-based ponds. The production cycle lasts 18 months, and the commercial size varies between 350 – 1500 g.

3.2. Zootechnical Management of a Sea Bass Farm

The zootechnical parameters are (FAO. 2014):

- Biomass refers to the total mass of living fish, expressed in kg, and is equal to average weight * number of fish (representing the total weight of the aquaculture stock).

- The volume of breeding, expressed in m^3, equals the capacity of the cage to accommodate fish.
- The breeding density, expressed in kg/m^3, is equal to biomass (total weight of living fish) / available volume.
- The final density at the end of the production cycle, expressed in kg/m^3, equals biomass at the end of the production cycle / available volume.
- The mortality rate, expressed in %, is the fraction of fish that will be lost (dead, escaped) during the production cycle.
- The production potential, expressed in kg, is equal to final density * breeding volume.

3.3 Oyster Farming on Intertidal Zones

The oyster is a species introduced to Morocco and Europe in the 1970s. According to scientific standards, in a farm of Pacific oysters (Crassostrea gigas) on intertidal zones, seeding is done once a year, covering an area of 2 hectares, and the techniques used are bags on tables, long-lines in more exposed areas, with an initial density of 2 to 3 kg of juveniles per bag (mesh < 9 mm). The growth cycle lasts between 12 to 18 months with a mortality rate of 30 to 40% at the end of the cycle. Optimal temperature ranges between 14 to 20 °C, and optimal salinity between 20 to 25 ‰. Harvest begins in October of the first year, with a density of 10 kg/bag, and a commercial size of 30 to 150 g (depending on calibers), with total production varying between 40 to 60 tons/ha, and average yield from 20 to 30 tons/ha (NADA.2017).

The duration of the harvest of Pacific oysters along the Atlantic coast of Morocco varies depending on local zones and conditions. Oyster farming in Oualidia generally takes between 2 to 3 years before the oysters are ready for harvest, while in the Dakhla region, this period is 7.5 to 9 months.

Specific marine aquaculture principles, criteria, and indicators must be adhered to when farming fish such as sea bass and sea bream, in order to ensure sustainable and responsible production.

4. THE SYSTEM OF PRINCIPLES, CRITERIA, AND INDICATORS OF SUSTAINABLE MARINE AQUACULTURE

Marine aquaculture has multiple impacts on blue growth. This is the sustainable development of activities related to oceans, seas, and marine resources. These impacts can occur during the installation phase or the production phase and can be economic, social, or environmental in nature.

4.1. Impact of Marine Aquaculture on Blue Growth

Marine aquaculture significantly contributes to blue growth, supporting the sustainable development of activities associated with the ocean, seas, and marine resources. These impacts can be economic, social, and environmental.

Economic Impact: Marine aquaculture is a vital part of the blue economy, contributing considerably to food security, income generation, job creation, diversification of aquatic products and markets, technological innovation, and competitiveness within the aquaculture sector. Currently, marine aquaculture accounts for over 50% of global fish production intended for human consumption and offers growth opportunities in many regions of the world. It can also promote the development of other sectors related to the blue economy, such as tourism, renewable energy, and biotechnology.

Social impact: Marine aquaculture can have a positive social impact, particularly within blue communities. It can reduce poverty, improve living conditions, promote gender equality, value local knowledge and enhance the participation of local communities. It can also contribute to preserving cultural heritage and fostering social cohesion in coastal and rural areas. Sustainable and competitive marine aquaculture can enhance the training and skills of stakeholders in the aquaculture sector, especially in the areas of farming system management, understanding biological cycles, disease prevention and treatment, adherence to health and environmental standards, and fostering technological innovation. Professional training and skill development in aquaculture are essential to attract younger generations to the sector and meet the growing global demand for aquatic products.

Environmental Impact: Environmentally, aquaculture can have both negative and positive effects depending on the species cultivated, production systems, and management practices. Negative effects may include habitat modification, nutrient dis-charge, transmission of infectious diseases, escape of farmed fish, chemical pollution, or competition with wild species. Positive effects may include ecosystem restoration, carbon sequestration, climate regulation, and the pro-vision of ecosystem services.

To maximize the positive effects and minimize the negative impacts of marine aquaculture on blue growth, it is essential to adopt an inclusive and sustainable approach that respects principles of good governance, adherence to environmental and social standards, risk prevention, and innovation promotion. Strengthening cooperation among various public and private stakeholders, international partners, and all relevant parties is also critical.

4.2. Economic Indicators of the Aquaculture Sector

The global food security and economic development of the world are significantly impacted by the aquaculture sector. As demand for seafood continues to increase, it is essential for stakeholders, including policymakers, investors, and producers, to understand the economic indicators of this sector.

These indicators help to assess the health and sustainability of aquaculture operations, guiding decisions relating to investment, production practises, and market strategies. Key indicators such as production volume, market value, employment levels, and sustainability metrics provide insights into the sector's performance and its impact on local and global economies.

This overview will examine the primary economic indicators relevant to the aquaculture sector, emphasising their importance for future growth and sustainability.

Table 2. Socio-Economic Indicators

Sea bass (Dicentrarchus labrax), Sea bream (Sparus aurata)	Investment	25 to 35 million DHS
	Turnover	80 to 120 million DHS
	Permanent jobs	20 to 20
Crustose oyster (Crassostrea gigas)	Investment	0.5 to 1.5 million DHS
	Turnover	1 to 1.5 million DHS
	Permanent jobs	4 to 6
Mussels (Mytilus galloprovincialis and Perna perna)	Investment	6 to 8 million DHS
	Turnover	4 to 6 million DHS
	Permanent jobs	17

Source: Omar ER-ROUSSE. «Challenges of blue growth and sustainable development: empirical study of marine aquaculture in Morocco», doctoral thesis defended at ITU, Kenitra, Morocco. 2024. p. 128.

4.3. Indicators of Sustainability in Marine Aquaculture

Indicators are quantitative or qualitative measures used to evaluate criteria. For instance, the indicator "the volume and value of aquaculture production by region." The sustainable development indicators for marine aquaculture were developed based on a study conducted between 2005 and 2008 in six Mediterranean countries (France, Spain, Italy, Greece, Tunisia, and Morocco) to create a battery of sustainability indicators based on a participatory approach involving various stakeholders in marine aquaculture, scientists, producers, public authorities, NGOs, etc.

Additionally, the participatory initiative to develop sustainable marine aquaculture indicators launched by the FAO in 2013 under the EAF-Nansen project aims to support the implementation of an ecosystem-based approach to fisheries and marine aquaculture in partner countries in Africa and Asia. This project involves integrating local actors in identifying priorities, collecting data, and monitoring indicators related to marine biodiversity, productivity, and food security.

These indicators are grouped into five categories: environmental, economic, social, governance, and product quality.

These tools must be developed through participation from various stakeholders throughout the marine aquaculture value chain, including producers, consumers, public authorities, scientists and NGOs. This approach fosters collective learning, dialogue and consultation among stakeholders, as well as their ownership of sustainable development. It also considers the diversity of the economic, social and environmental contexts in which marine aquaculture operates.

To summarise, if marine aquaculture is to be sustainable in the long term, it is vital that it is integrated into a policy framework that promotes sustainability, innovation, and collaboration.

5. POLICIES FOR THE DEVELOPMENT OF MARINE AQUACULTURE IN EGYPT, LEBANON, AND MOROCCO

The marine aquaculture policies of the MENA region countries are influenced by Mediterranean policies. These policies are structured at multiple governance levels: international, regional (notably via the European Union and the General Fisheries Commission for the Mediterranean), and national.

5.1. International and regional Framework

The GFCM, as a FAO body coordinating regional fisheries and aquaculture policies, adopts non-binding resolutions to promote sustainable aquaculture and implement the 2021-2025 Strategy for the Sustainable Development of Aquaculture. This strategy focuses on environmental sustainability, food security, social and economic integration, and improved governance.

At the regional level, the Barcelona Convention for the Protection of the Mediterranean Sea Against Pollution has led the Union for the Mediterranean (UfM) to adopt a strategy aimed at changing the region's development model towards a more sustain-able and inclusive approach. While the blue economy plays a crucial role in improving the livelihoods of communities in the region, it must be regulated and controlled through an effective participatory governance approach, considering the limits to ocean exploitation.

The ecosystem-based approach, marine spatial planning, and Integrated Coastal Zone Management (ICZM) are essential tools for promoting an inclusive and sustainable blue economy. To this end, Mediterranean states adopted the ICZM Proto-col in 2008. The regional dimension of the blue economy is crucial for the Mediterranean basin, as it is within this framework that the Sustainable Development Goals (SDGs) of the 2030 Agenda are translated into actions at national and local levels. For this reason, collaboration among different stakeholders is a priority. The Protocol on Specially Protected Areas may affect the location of aquaculture farms.

5. 2. The national marine aquaculture policies in Morocco, Egypt and Lebanon

Most countries, such as Morocco, Egypt, and Lebanon, adopt national marine aquaculture development plans, which are a rapidly growing practice. They encourage public-private partnerships (PPPs) to foster investment, which is a crucial topic for the economic development of blue economy sectors, primarily marine aquaculture. To consolidate governance and technical skills, they also receive support from renowned international organizations such as the FAO, the European Union, and the World Bank.

In this context, we present the various aquaculture policies of these three MENA countries, highlighting the principles, criteria, indicators, and institutional frameworks specific to each country.

Morocco: A pioneer in North Africa in terms of sectoral organization. The National Agency for the Development of Aquaculture (NADA) plays a leading role in planning, site allocation, and supporting project holders, making it a key player in the region's economic development.

Egypt: The country's aquaculture policy is well-established and clear in its objectives: to increase national fish production and reduce dependence on imports. Egypt's focus is on marine aquaculture, particularly for species such as mullet, grouper, sea bass, sole, shrimp and sea urchins. Production methods include recirculating aquaculture systems (RAS) and marine farms. The General Authority for Fish Resources Development (GAFRD) plays a vital role in managing and promoting the nation's fishery resources. GAFRD's research focuses on enhancing fishery practices and improving aquaculture techniques.

Lebanon: The aquaculture sector in Lebanon is still in its development phase, with a focus on freshwater aquaculture of rainbow trout. The country's development of marine aquaculture is driven by a desire to enhance food security, generate employment opportunities, and decrease reliance on imported fish. Key challenges hindering agricultural production include a dearth of suitable infrastructure and the necessity to enhance the technical capabilities of producers..

We have attempted to create a summary table of aquaculture policies for Morocco, Egypt, and Lebanon, based on the principles, criteria, and indicators defined by the FAO/GFCM as well as national strategies.

Table 3. A summary table of marine aquaculture policies in Morocco, Egypt and Lebanon

Category	Morocco	Egypt	Lebanon
Guiding principles	- Blue growth -Environmental sustainability - Local job creation	- Food safety -Economic development -Technological innovation	- Rural development -Food self- sufficiency - Support for small producers
Main institution (s)	- ANDA (National Agency for the Development of Aquaculture) -Ministry of agriculture	- GAFRD (General Authority for fish Resources Development) -Ministry of Agriculture	-Ministry of Agriculture -Department of Animal Resources
Legal framework	- Halieutis strategy (2009-2020), extended via the New Aquaculture Roadmap 2030 - Law n° 84-21 on marine aquaculture	- Law n°124/1983 on fishing and aquaculture - Environment law n°4/1994	- No specific law on marine aquaculture (general laws on fisheries)
Strategic objectives	- Produce 200000 tonnes/ year by 2030 - Developing 6 regional aquaculture clusters	- Reaching 1.5 M tonnes/year of production - Developing marine aquaculture (mullet, sea bass, prawns, etc.)	- Diversification into marine aquaculture - Creating jobs and improving nutrition
Raised species	Sea bass, sea bream, seaweed, shellfish	Mullet, sea bass, sea bream, tilapia, prawns, grouper, sole	Trout (dominant), sea bass trials
Systems used	Marine cages, onshore basins, multi-trophic pilot projects	Onshore tanks, floating cages at sea, RAS, IAA	Freshwater farms (trout), small coastal marine trials
Indicators monitored	- Annual volume - Number of jobs created - Aquaculture area under production	- Annual volume -Self-sufficiency rate - Water quality - Employment rate in the sector	- Production by site - Fish mortality rate - Number of active farms
Issues/Challenges	- Social acceptability - Access to land and authorisations - Need for private investment	- Pressure on water resources - Limited EIA regulations - Biosafety challenges	- Lack of a clear regulatory framework -Weak infrastructure -Lack of systematic data
International partners	FAO, EU, ADB, bilateral cooperation (e.g. Norway, Spain)	FAO, World Bank, USAID	FAO, EU, UNDP

Source: elaborated by the author from the literature review

Table 4. Global comparative SWOT.

Element	Morocco	Egypt	Lebanon
Current Position	Structured development	Maturity in freshwater, emerging in marine	Emerging phase, poorly structured
Growth Capacity	High (Six plans, coastline, supportive policy)	High if diversification succeeds	Moderate, dependent on stability
Main-Risk	Administrative and environmental	Pollution and water quality	Instability and pollution

Source: elaborated by the author from the literature review

This table 4. Gives a brief summary of the current situation with regard to marine aquaculture in Morocco, Egypt and Lebanon. It focuses on three main points: the current position, growth capacity and the main risks.

Current position:

Morocco is in a structured development phase, indicating organised efforts and institutional support for advancing marine aquaculture.

Egypt's freshwater aquaculture sector is mature, while its marine aquaculture sector is still in its infancy. This suggests potential for growth, but further investment and diversification are required.

Lebanon is characterised as being in an emerging phase with a poorly structured approach, indicating significant room for improvement and the need for better organisation, and planning.

Growth capacity:

Morocco shows high growth potential due to its strategic plans, extensive coastline, and supportive policies that facilitate aquaculture development.

Egypt also has high growth capacity, particularly if it can successfully diversify into marine species, which would boost its aquaculture output, and resilience.

Lebanon has moderate growth capacity, which is heavily reliant on political stability, and investment, both of which are currently uncertain.

Main risks:

Morocco faces administrative and environmental risks, suggesting that bureaucratic hurdles and ecological challenges could hinder progress.

Egypt is primarily threatened by pollution and water quality issues, which could affect production and the health of aquatic ecosystems.

Lebanon is vulnerable to instability and pollution, so it is crucial to address these issues in order to foster a sustainable aquaculture sector.

The table shows the strengths and challenges of each country in the area of marine aquaculture. It also shows the need for strategies that are made especially for each country to improve growth and sustainability in the sector.

Morocco stands out as a pioneer in North Africa regarding the sectoral organization of aquaculture activities. NADA plays a crucial role in planning, site allocation, and project holder support.

Conversely, Egypt is the continental leader in aquaculture, particularly in freshwater, and is making significant investments in marine systems (as evidenced by the Berket Ghalioun project, the largest integrated aquaculture farm in Africa) (M Dickson - 2022). However, in Lebanon, the development of marine aquaculture is in its early stages, mainly due to political, legal and economic constraints. There is still significant untapped potential.

In the context of interactions between the pillars of sustainable development and the blue economy in marine aquaculture, we have established a list of strategic conclusions for each country. Morocco adopts a more balanced model between sustainable development (SDD) and blue economy (EB), demonstrating a strong commitment to sustainable integration. Egypt is economically advanced but remains less inclusive and has ecological gaps. Finally, Lebanon has high potential but has much to build in terms of governance, infrastructure, training, and legal frameworks.

6. CONCLUSION

Marine aquaculture represents a strategic lever to address the growing challenges of food security, coastal job creation, and sustainable economic development in the Mediterranean. In the face of multiple environmental and social pressures, integrating the principles of the blue economy—based on innovation, resilience, and inclusion—constitutes a promising pathway for ensuring the sustainable transformation of the sector. The comparative analysis conducted between Egypt, Morocco, and Lebanon reveals that national trajectories differ, but a common foundation of priorities is emerging:

- The necessity for a clear and incentive-based regulatory framework to govern marine aquaculture practices.
- The urgent development of sustainable and adapted technologies (e.g., offshore systems, IMTA, circular aquaculture).
- The need to train, include, and support small producers and young entrepreneurs.

- Most importantly, the imperative to anchor projects in an ecosystemic and territorial vision, where economic, social, and environmental dimensions are addressed integratively.

Morocco stands out for its structured strategy, articulated around the Halieutis initiative and ANDA, focused on sustainability and regional inclusion. Egypt, which is advanced in overall production, must now transfer its successes to marine aquaculture, which is still marginal and centralized. Lebanon, meanwhile, faces significant structural constraints but can rely on pilot initiatives and international partnerships to lay the groundwork for a more equitable and sustainable model.

Thus, marine aquaculture can become a pillar of the Mediterranean blue economy and the MENA region, provided it is conceived as a living and interdependent system, managed long-term in respect of ecosystems and coastal populations. Through integrating advances in genetics, nutrition, reproduction, and disease management, marine aquaculture has become a major player in global food security. It also respects the principles of sustainable development and contributes to ocean preservation.

The principles, criteria and indicators (PCI) of marine aquaculture in relation to the blue economy and blue growth are essential to ensure aquaculture practices are sustainable and beneficial to coastal communities and ecosystems. Integrating PCI enables marine aquaculture to contribute to a sustainable blue economy by promoting economic development while preserving the environment. Indeed, sustainable marine development is vital for the health of the oceans and the well-being of coastal communities. An integrated, collaborative approach is necessary, as are appropriate policies and a long-term commitment from all aquaculture stakeholders, both institutional and non-institutional.

REFERENCES

Bendag, M. (1995). Production systems for sea bass and sea bream. Intensive breeding in tanks in Tunisia. Cahiers Options Mediterraneennes (CIHEAM), 14.

Claire De Marignan. (2017). "Aquaculture: Or when the man domesticates the sea." 19-30

Dickson, M. (2022). *Regional review on status and trends in aquaculture development in the Near East and North Africa–2020*. Food & Agriculture Org.

Dickson, M. (2022). *Regional review on status and trends in aquaculture development in the Near East and North Africa–2020*. Food & Agriculture Org.

FAO. (2009). Dicentrarchus labrax. In Cultured aquatic species fact sheets. Text by Bagni, M. Edited and compiled by Valerio Crespi and Michael New. CD-ROM (multilingual).

FAO. (2009). Sparus aurata. In Cultured aquatic species fact sheets. Text by Colloca, F. & Cerasi, S. Edited and compiled by Valerio Crespi and Michael New. CD-ROM (multilin-gual). https://www.fao.org/fishery/docs/CDrom/aquaculture/I1129m/file/fr/fr_giltheadseabr.htm

Gage, S. D. (1978). Aquaculture: A new force in social ecology (Doctoral dissertation, Union Institute and University).

Geletu, T. T., & Zhao, J. (2023). Genetic resources of Nile tilapia (Oreochromis niloticus Linnaeus, 1758) in its native range and aquaculture. *Hydrobiologia, 850*(10), 2425–2445. DOI: 10.1007/s10750-022-04989-4

Hoffman, R. C. (1995). Environmental change and the culture of common carp in medieval Europe. *Guelph Ichthyology Reviews*, •••, 3.

La loi n°84-21 relative à l'aquaculture marine, B.O N°7174, du 9 chaabane 1444 (2 mars 2023).

Rodríguez-García, C., Toro-Podadera, A., Sarmiento-Carbajal, J., & Cabrera-Castro, R. (2024). Coexisting in the Surf Zone: Age and Feeding Habits of the Spotted Seabass (Dicentrarchus punctatus) and European Seabass (Dicentrarchus labrax) on the Gulf of Cádiz Beaches (Southwest Iberian Peninsula). *Fishes*, *9*(5), 173. DOI: 10.3390/fishes9050173

Soliman, N. F., & Yacout, D. M. (2016). Aquaculture in Egypt: Status, constraints and potentials. *Aquaculture International*, *24*(5), 1201–1227. DOI: 10.1007/s10499-016-9989-9

ENDNOTES

* Gunter Pauli defines the "Blue Economy" as an economic model that emphasizes the sustainable use of ocean resources for economic growth, improved livelihoods, and jobs while preserving the health of ocean ecosystems. It promotes innovation and the efficient use of resources, aiming to create a balance between economic development and environmental sustainability

† 1 Law No. 84-21 on marine aquaculture was promulgated by the Kingdom of Morocco. It was published in Official Gazette No. 7174 of 9 Chaabane 1444 (2 March 2023). It can be found on pages 749 to 762. This text was promulgated by Dahir No. 1-22-81 of 18 Joumada I 1444 (13 December 2022).

Compilation of References

Abd-Elgawad, M. M. (2025). Integrated Nematode Management Strategies: Optimization of Combined Nematicidal and Multi-Functional Inputs. *Plants*, *14*(7), 1004.

Abdelli, R., Yasmine, S., Kaouche, S., & Benhacine, R. (2021). Etat des lieux de la filière laitière en Algérie et perspectives de développement. *Journal Algérien des Milieux Arides*, *11*(1), 4–14.

Abebaw, S. E. (2025). A Global Review of the Impacts of Climate Change and Variability on Agricultural Productivity and Farmers' Adaptation Strategies. *Food Science & Nutrition*, *13*(5), e70260. DOI: 10.1002/fsn3.70260

ABHSM. (2021). *Water stress in Morocco's basins*. Agence du Bassin Hydraulique du Souss-Massa.

Abidoye, B. O., Kurukulasuriya, P., Reed, B., & Mendelsohn, R. (2017). Structural Ricardian analysis of south-east Asian agriculture. *Climate Change Economics (Singapore)*, *8*(03), 1740005. DOI: 10.1142/S201000781740005X

Abidoye, B. O., & Odusola, A. F. (2015). Climate change and economic growth in Africa: An econometric analysis. *Journal of African Economies*, *24*(2), 277–301. DOI: 10.1093/jae/eju033

Abouelmagd, A., & Ahmed, M. (2024). Groundwater in North Africa: Effects of climatic and anthropogenic pressures on groundwater availability. In *The Geology of North Africa* (pp. 305–321). Springer International Publishing. DOI: 10.1007/978-3-031-48299-1_11

Abunnour, M. A., Hashim, N. M., & Bin Jaafar, M. (2016). Agricultural water demand, water quality and crop suitability in Souk-Alkhamis Al-Khums, Libya. *IOP Conference Series. Earth and Environmental Science*, *37*, 012045. Advance online publication. DOI: 10.1088/1755-1315/37/1/012045

AFDB. (2025). *Defining a new approach to water management in Libya: Issues & options*. AFDB.

Affoh, R., Zheng, H., Dangui, K., & Dissani, B. M. (2022). The Impact of Climate Variability and Change on Food Security in Sub-Saharan Africa: Perspective from Panel Data Analysis. *Sustainability (Basel)*, *14*(2), 759. DOI: 10.3390/su14020759

African Development Bank (AfDB). (2018). *Morocco: National Irrigation Water Saving Programme Support Project (PAPNEEI) – Project Completion Report. Regional Department North (RDGN)*. African Development Bank.

Agrawala, S., Carraro, M., Kingsmill, N., Lanzi, E., Mullan, M., & Prudent-Richard, G. (2011). *Private sector engagement in adaptation to climate change: Approaches to managing climate risks*. https://doi.org/DOI: 10.1787/19970900

Aguilera, E., Díaz-Gaona, C., García-Laureano, R., Reyes-Palomo, C., Guzmán, G. I., Ortolani, L., Sánchez-Rodríguez, M., & Rodríguez-Estévez, V. (2020). Agroecology for adaptation to climate change and resource depletion in the Mediterranean region. A review. *Agricultural Systems*, *181*, 102809. DOI: 10.1016/j.agsy.2020.102809

Ahmed, S. A., Diffenbaugh, N. S., Hertel, T. W., Lobell, D. B., Ramankutty, N., Rios, A. R., & Rowhani, P.AHMED. (2011). Climate volatility and poverty vulnerability in Tanzania. *Global Environmental Change*, *21*(1), 46–55. DOI: 10.1016/j.gloenvcha.2010.10.003

Aieb, A., Lefsih, K., Scarpa, M., Bonaccorso, B., Cicero, N., Mimeche, O., & Madani, K. (2020). Statistical modeling of monthly rainfall variability in Soummam watershed of Algeria, between 1967 and 2018. *Natural Resource Modeling*, *33*(4), e12288. Advance online publication. DOI: 10.1111/nrm.12288

Akesbi, N. (2012). Une nouvelle stratégie pour l'agriculture marocaine: Plan Maroc Vert. *New Medit*, *11*(2), 12.

Akinbo, O., Obukosia, S., Ouedraogo, J., Sinebo, W., Savadogo, M., Timpo, S., Mbabazi, R., Maredia, K., Makinde, D., & Ambali, A. (2021). Commercial release of genetically modified crops in Africa: Interface between biosafety regulatory systems and varietal release systems. *Frontiers in Plant Science*, *12*, 605937. DOI: 10.3389/fpls.2021.605937

Akinyi, D. P., Ng'ang'a, S. K., Ngigi, M., Mathenge, M., & Girvetz, E. (2022). Cost-benefit analysis of prioritized climate-smart agricultural practices among smallholder farmers: Evidence from selected value chains across sub-Saharan Africa. *Heliyon*, *8*(4), e09228. DOI: 10.1016/j.heliyon.2022.e09228

Al-Addous, M., Bdour, M., Rabaiah, S., Boubakri, A., Schweimanns, N., Barbana, N., & Wellmann, J. (2024). Innovations in solar-powered desalination: A comprehensive review of sustainable solutions for water scarcity in the Middle East and North Africa (MENA) region. *Water (Basel), 16*(13), 1877. DOI: 10.3390/w16131877

Al-Bagham, S. H. (2003). Genetically modified organisms in the United Arab Emirates. *National Workshop on Environmental Impact Assessment of the Introduction of Genetically Modified Plant and Animal Species.*

Albedwawi, A. (2025). Food and Pharmaceutical Transformation in Islamic Jurisprudence: A Comparison with UAE Law. *Journal of International Crisis and Risk Communication Research, 8*(S1), 61.

AL-BLOOSHI, L. S., ISSA, S. & KSIKSI, T. (2020). Assessing the environmental impact of climate change on desert ecosystems: A review. *Advances in Ecological and Environmental Research, 5*, 27–52.

Alboghdady, M., & El-Hendawy, S. E. (2016). Economic impacts of climate change and variability on agricultural production in the Middle East and North Africa region. *International Journal of Climate Change Strategies and Management, 8*(3), 463–472. DOI: 10.1108/IJCCSM-07-2015-0100

Alegría, P. G. P. (2024). (Mis)adapting domestic law to meet new international environmental and trade rules: How Peru changed its environmental and land use rights laws in response to the European Union deforestation regulation. *Journal of Environmental Law, 36*(3), 437–444. DOI: 10.1093/jel/eqae019

Aleksandrova, M., Lamers, J. P. A., Martius, C., & Tischbein, B. (2014). Rural vulnerability to environmental change in the irrigated lowlands of Central Asia and options for policy-makers: A review. *Environmental Science & Policy, 41*, 77–88. DOI: 10.1016/j.envsci.2014.03.001

Alfieri, L., Burek, P., Feyen, L., & Forzieri, G. (2015). Global warming increases the frequency of river floods in Europe. *Hydrology and Earth System Sciences, 19*(5), 2247–2260. DOI: 10.5194/hess-19-2247-2015

Alibekova, G., Medeni, T., Panzabekova, A., & Mussayeva, D. (2020). Digital Transformation Enablers and Barriers in the Economy of Kazakhstan. *The Journal of Asian Finance, Economics and Business, 7*(7), 565–575. DOI: 10.13106/jafeb.2020.vol7.no7.565

Al-Jawaldeh, A., Nabhani, M., Taktouk, M., & Nasreddine, L. (2022). Climate change and nutrition: Implications for the Eastern Mediterranean Region. *International Journal of Environmental Research and Public Health, 19*(24), 17086. DOI: 10.3390/ijerph192417086

Aljeddani, G. S., Hamouda, R. A., Abdelsattar, A. M., & Heikal, Y. M. (2024). Stress-Responsive Gene Expression, Metabolic, Physiological, and Agronomic Responses by Consortium Nano-Silica with Trichoderma against Drought Stress in Bread Wheat. *International Journal of Molecular Sciences, 25*(20), 10954.

Alkhatib, S. F., & Momani, R. A. (2023). Supply Chain Resilience and Operational Performance: The Role of Digital Technologies in Jordanian Manufacturing Firms. *Administrative Sciences, 13*(2), 40. DOI: 10.3390/admsci13020040

Allani, M., Frija, A., Nemer, R., Ribbe, L., & Sahli, A. (2022). Farmers' perceptions on an irrigation advisory service: Evidence from Tunisia. *Water (Basel), 14*(22), 3638. DOI: 10.3390/w14223638

Almazroui, M., Saeed, F., Saeed, S., Islam, M. N., Ismail, M., Klutse, N. B., & Siddiqui, M. H. (2020). Projected Change in Temperature and Precipitation Over Africa from CMIP6. *Earth Systems and Environment, 4*(3), 455–475. DOI: 10.1007/s41748-020-00161-x

Al-Mssallem, M. Q., Al-Khayri, J. M., Almagthali, B. H., Abdelkader, H. S., Alnabbat, K. I., & Sattar, M. N. (2024). Genetically modified food: potentiality for food and nutritional security in Saudi Arabia. In *Food and Nutrition Security in the Kingdom of Saudi Arabia, Vol. 2: Macroeconomic Policy and Its Implication on Food and Nutrition Security* (pp. 359-392): Springer. DOI: 10.1007/978-3-031-46704-2_16

Alomari-Mheidat, M., Corell, M., Martín-Palomo, M. J., Castro-Valdecantos, P., Medina-Zurita, N., de Sosa, L. L., & Moriana, A. (2024). Moderate water stress impact on yield components of greenhouse tomatoes in relation to plant water status. *Plants, 13*(1), 128. DOI: 10.3390/plants13010128

Alqodsi, E. M. (2021b). THE RIGHT TO PRE-CONTRACTUAL INFORMATION IN E-COMMERCE CONSUMER CONTRACTS: UAE LAW AND COMPARATIVE PERSPECTIVES. *Journal of Legal, Ethical and Regulatory Issues, 24*(2), 1-16. Retrieved from https://www.scopus.com/inward/record.uri?eid=2-s2.0-85103711308&partnerID=40&md5=b795381746e49b8e94a5e798ac846ec9

Alqodsi, E. M. (2021a). Analyzing the Implementation of Usufruct Rights and Obligations in the UAE Civil Transactions Law. *International Journal of Criminal Justice Sciences, 16*(2), 202–219. https://www.scopus.com/inward/record.uri?eid=2-s2.0-85134491872&partnerID=40&md5=9f4e39444da0981dc669dd27dbf128f7

Al-Qodsi, E. M., Jadalhaq, I. M., & El-Maknouzi, M. E. H. (2024). The place of UAE's food security in the national legislation and its role in supporting global food security. *Cogent Social Sciences*, *10*(1), 2319379. Advance online publication. DOI: 10.1080/23311886.2024.2319379

Alrteimei, H. A., Ash'aari, Z. H., & Muharram, F. M. (2022). Last decade assessment of the impacts of regional climate change on crop yield variations in the Mediterranean region. *Agriculture*, *12*(11), 1787. DOI: 10.3390/agriculture12111787

Al-Sarihi, A., & Mason, M. (2020). Challenges and opportunities for climate policy integration in oil-producing countries: The case of the UAE and Oman. *Climate Policy*, *20*(10), 1226–1241. DOI: 10.1080/14693062.2020.1781036

Al-Shawabkeh, I. K. (2020). Legal guarantees for objective performance evaluation of the federal public employee in the uae legislation. *Lawyer Quarterly*, *10*(3), 217-229. Retrieved from https://www.scopus.com/inward/record.uri?eid=2-s2.0-85090291804&partnerID=40&md5=e35ade47f008985468403b808b9d3170

Altieri, M. A. (2004). *Agroecology: The science of sustainable agriculture*. Westview Press.

Altieri, M. A., & Nicholls, C. I. (2017). Agroecology: A brief account of its origins and currents of thought in Latin America. *Agroecology and Sustainable Food Systems*, *41*(3-4), 231–237. DOI: 10.1080/21683565.2017.1287147

Altieri, M. A., & Nicholls, C. I. (2017). The adaptation and mitigation potential of traditional agriculture in a changing climate. *Climatic Change*, *140*(1), 33–45. DOI: 10.1007/s10584-013-0909-y

Altieri, M. A., Nicholls, C. I., Montalba, R., Vieli, L., & Vazquez, L. L. (2025). Agroecology and the limits to resilience: Extending the adaptation capacity of agroecosystems to drought. *Frontiers in Agronomy*, *7*, 1534370. DOI: 10.3389/fagro.2025.1534370

Al-Worafi, Y. M. (2020). Drug safety in United Arab Emirates. In *Drug safety in developing countries* (pp. 419–428). Elsevier. DOI: 10.1016/B978-0-12-819837-7.00031-5

Anas, M., Khalid, A., Saleem, M. H., Ali Khan, K., Ahmed Khattak, W., & Fahad, S. (2025). Symbiotic synergy: Unveiling Plant-Microbe interactions in stress adaptation. *Journal of Crop Health*, *77*(1), 18.

Ancín, M., Pindado, E., & Sánchez, M. (2022). New trends in the global digital transformation process of the agri-food sector: An exploratory study based on Twitter. *Agricultural Systems*, *203*, 103520. DOI: 10.1016/j.agsy.2022.103520

Andrade, C., Contente, J., & Santos, J. A. (2021). Climate change projections of aridity conditions in the Iberian peninsula. *Water (Basel)*, *13*(15), 2035. DOI: 10.3390/w13152035

Angelsen, A., & Dokken, T. (2018). Climate exposure, vulnerability and environmental reliance: A cross-section analysis of structural and stochastic poverty. *Environment and Development Economics*, *23*(3), 257–278. DOI: 10.1017/S1355770X18000013

Annosi, M. C., Marchegiani, L., Magnusson, M., & Roidis, M. (2023). Is technology neutral for MSEs? Interdependencies, information transparency and power imbalances in e-commerce ecosystems. *International Journal of Technology Management, 91*(3–4), 190–218. *Scopus.* Advance online publication. DOI: 10.1504/IJTM.2022.10052666

AOAD (2024). Agricultural Statistics Yearbook. Arab Organization for Agricultural Development. Arab League.

APIA. (2023). *Rapports statistiques.* APIA, Ministère de l'Agriculture.

APS. (2023a). Les stations de dessalement contribueront à hauteur de 60% à l'alimentation en eau potable d'ici 2030. Publié le 27 février. https://www.aps.dz/economie/152381

APS. (2023b). Agriculture: création de plus de 160 nouvelles coopératives depuis 2020. Publié le 2 juillet. https://www.aps.dz/economie/142312

Ardakani, Z., Bartolini, F., & Brunori, G. (2020). New evaluation of small farms: Implication for an analysis of food security. *Agriculture*, *10*(3), 74. DOI: 10.3390/agriculture10030074

Arfasa, G. F., Sekyere, E. O., & Doke, D. A. (2023). Temperature and precipitation trend analysis using the CMIP6 model in the Upper East region of Ghana. *All Earth*, *36*(1), 1–14. DOI: 10.1080/27669645.2023.2290966

Arman, H. (2015). Uncertainties, risks and challenges relating to CO_2 emissions and its possible impact on climate change in the United Arab Emirates. *International Journal of Global Warming*, *8*(1), 1–17. DOI: 10.1504/IJGW.2015.071575

Aronson, J., Goodwin, N., Orlando, L., Eisenberg, C., & Cross, A. T. (2020). A world of possibilities: Six restoration strategies to support the United Nation's Decade on Ecosystem Restoration. *Restoration Ecology*, *28*(4), 730–736. DOI: 10.1111/rec.13170

Atkinson, K. (2023). Libya's Soil Resources and Their Potential for Sustained Agricultural Production. In *The Economic Development of Libya* (pp. 134–156). Routledge. DOI: 10.4324/9781003420552-8

Attar, O., Brouziyne, Y., Bouchaou, L., El Bilali, A., Ait Brahim, Y., & Chehbouni, A. (2024). Understanding the trade-offs between climate change-induced aridity and agricultural water demand in the Souss basin, Morocco. *Frontiers in Water, 6*, 1270078. DOI: 10.3389/frwa.2024.1270078

Attiaoui, I., & Boufateh, T. (2019). Impacts of climate change on cereal farming in Tunisia: A panel ARDL-PMG approach. *Environmental Science and Pollution Research International, 26*(13), 13334–13345. DOI: 10.1007/s11356-019-04867-y

Auffhammer, M., & Schlenker, W. (2014). Empirical studies on agricultural impacts and adaptation. *Energy Economics, 46*, 555–561. DOI: 10.1016/j.eneco.2014.09.010

Awoke, W., Eniyew, K., Agitew, G., & Meseret, B. (2022). Determinants of food security status of household in Central and North Gondar Zone, Ethiopia. *Cogent Social Sciences, 8*(1), 2040138. Advance online publication. DOI: 10.1080/23311886.2022.2040138

Azzarri, C., & Signorelli, S. (2020). Climate and poverty in Africa South of the Sahara. *World Development, 125*, 104691. DOI: 10.1016/j.worlddev.2019.104691 PMID: 31902973

Baarsch, F., Granadillos, J. R., Hare, W., Knaus, M., Krapp, M., Schaeffer, M., & Lotze-Campen, H. (2020). The impact of climate change on incomes and convergence in Africa. *World Development, 126*, 104699. DOI: 10.1016/j.worlddev.2019.104699

Bachta, M. S., & Ben Nasr, J. (2020). *Water policy in Tunisia. Water policies in MENA countries*, 161-184.

Badraoui, I., van der Lans, I. A. M. C., Boulaksil, Y., & van der Vorst, J. G. A. J. (2024). Horizontal logistics collaboration success factors: Expectations versus reality. *Benchmarking, 31*(1), 29–52. DOI: 10.1108/BIJ-04-2022-0274

Bakr, S. A., & Ayinde, O. L. (2014). Consumer attitude towards consumption of genetically modified foods in Arab countries. *Middle East Journal of Scientific Research, 21*(10), 1710–1717.

Ballona-Castro, E. R., & Herrera-Mena, T. K. (2025). Consumer rights in Peruvian territory: An analysis of Law No. 29571 Consumer Protection and Defense Code. Clio. *Revista de Historia. Ciencias Humanas y Pensamiento Crítico, 5*, 632–658.

Bandarra, N.-J. (2009). Un développement durable pour la Méditerranée. *Revue du Marché commun et de l'Union européenne,* (529).

Banderas Tarabay, A. G., & González-Villela, R. (2018). The environmental added value of water: A proposal. *International Journal of Sustainable Development and Planning, 13*(3), 406–417. DOI: 10.2495/SDP-V13-N3-406-417

Barrett, C. B., Benton, T. G., Cooper, K. A., Fanzo, J., Gandhi, R., Herrero, M., James, S., Kahn, M., Mason-D'Croz, D., Mathys, A., Nelson, R. J., Shen, J., Thornton, P., Bageant, E., Fan, S., Mude, A. G., Sibanda, L. M., & Wood, S. (2020). Bundling innovations to transform agri-food systems. *Nature Sustainability, 3*(12), 974–976. DOI: 10.1038/s41893-020-00661-8

Bastin, J. F., Berrahmouni, N., Grainger, A., Maniatis, D., Mollicone, D., Moore, R., & Castro, R. (2017). The extent of forest in dryland biomes. *Science, 356*(6338), 635–638. DOI: 10.1126/science.aam6527

Becchi, P. (2011). *Our responsibility towards future generations: Efficiency, sustainability, and justice to future generations.* Springer.

Bekkhus, R. (2016). Do KPIs used by CIOs Decelerate Digital Business Transformation? The Case of ITIL. *DIGIT 2016 Proceedings.* https://aisel.aisnet.org/digit2016/16

Bekkis, S., Benmehaia, M. A., & Kaci, A. (2022). Les enjeux de la dépendance de la filière de blé en Algérie: Analyse par asymétries de réponses de l'offre dans la chaîne de valeur. *New Medit, 21*(1), 133–147. DOI: 10.30682/nm2201h

Bekkis, S., Benmehaia, M. A., & Kaci, A. (2023). Price transmission in the wheat market in Algeria: Threshold cointegration approach. *International Journal of Food and Agricultural Economics, 11*(1), 17–32.

Belaidi, S., & Benmehaia, A. M. (2024). The selection of modern irrigation technologies within a rural community in Algeria: Stated preferences approach. *Journal of Rural and Community Development, 19*(2).

Belaidi, S., Chehat, F., & Benmehaia, M. A. (2022). The adoption of water-saving irrigation technologies in the Mitidja plain, Algeria: An econometric analysis. *New Medit, 21*(1), 53–73. DOI: 10.30682/nm2201d

Beltrán Tapia, F. J., & Martinez-Galarraga, J. (2018). Inequality and education in pre-industrial economies: Evidence from Spain. *Explorations in Economic History, 69*, 81–101. DOI: 10.1016/j.eeh.2017.12.003

Ben Youssef, A. (2023). *Digitalization for the Green Transition in the Mediterranean* (IEMed. Mediterranean Yearbook). Economic Research Forum (ERF); IEMed. Mediterranean Yearbook. https://www.iemed.org/publication/digitalization-for-the-green-transition-in-the-mediterranean/?lang=fr

Ben Youssef, A. (2020). How Can Industry 4.0 Contribute to Combatting Climate Change? *Revue d'Economie Industrielle*, •••, 161–193. DOI: 10.4000/rei.8911

Benabdelkader, M., Saifi, R., & Saifi, H. (2021). Sustainable Agriculture in Some Arab Maghreb Countries (Morocco, Algeria, Tunisia). In *Agro-environmental sustainability in MENA regions* (pp. 233–261). Springer International Publishing. DOI: 10.1007/978-3-030-78574-1_10

Benazzouz, S., & Sadok, H. (2025). Analysis of the Relationship Between Innovation, Digitalization and Economic Growth in North African Countries. In Bach Tobji, M. A., Jallouli, R., Sadok, H., Lajfari, K., Mafamane, D., & Mahboub, H. (Eds.), *Digital Economy. Emerging Technologies and Business Innovation* (Vol. 531, pp. 183–198). Springer Nature Switzerland., DOI: 10.1007/978-3-031-76368-7_13

Bendag, M. (1995). Production systems for sea bass and sea bream. Intensive breeding in tanks in Tunisia. Cahiers Options Mediterraneennes (CIHEAM), 14.

Benderradji, M. E. H., Alatou, D., Arfa, A. M. T., & Benachour, K. (2006). Problèmes de dégradation de l'environnement par la désertification et la déforestation. Impact du phénomène en Algérie. *New Medit.* N 4.

Bender, W., & Smith, M. (1997). Population, food, and nutrition. *Population Bulletin*, *51*, 23–32.

Beniawan, A., & Sinambela, L. P. (2025). The relationship between performance management and the quality of animal, fish, and plant quarantine services. *International Journal of Management Science and Information Technology*, *5*(1), 32–38. DOI: 10.35870/ijmsit.v5i1.3621

Benmehaia, A. M., & Bekkis, S. (2024). Agri-food policy trends in Algeria: Selected explorations. *New Medit*, *23*(3), 49–62. DOI: 10.30682/nm2403d

Benmehaia, A. M., Merniz, N., & Oulmane, A. (2020). Spatiotemporal analysis of rainfed cereal yields across the eastern high plateaus of Algeria: An exploratory investigation of the effects of weather factors. *Euro-Mediterranean Journal for Environmental Integration*, *5*(3), 54. DOI: 10.1007/s41207-020-00191-x

Benmehaia, M. A. (2023). Assessing asymmetrical effects of climate change on cereal yields in Algeria: The NARDL-AEC approach. *Environment, Development and Sustainability*, *27*(2), 1–22. DOI: 10.1007/s10668-023-04079-y

Benmehaia, M. A., & Benmehaia, R. (2018, March). Socioeconomic analysis of date palm sector: The case of Biskra region (Algeria). *Sixth International Date Palm Conference*. Abu Dhabi, United Arab Emirates (pp. 19-21).

Bergkamp, L. (2000). Allocating unknown risk: liability for environmental damages caused by deliberately released genetically modified organisms. *Available at SSRN* 223068. DOI: 10.2139/ssrn.223068

Bergquist, P., Mildenberger, M., & Stokes, L. C. (2020). Combining climate, economic, and social policy builds public support for climate action in the US. *Environmental Research Letters, 15*(5), 054019. DOI: 10.1088/1748-9326/ab81c1

Berhanu, M., & Wolde, A. O. (2019). Review on Climate Change Impacts and its Adaptation strategies on Food Security in Sub-Saharan Africa. *Agricultural Social Economic Journal, 19*(3), 145–154. DOI: 10.21776/ub.agrise.2019.019.3.3

Berninger, T., González López, Ó., Bejarano, A., Preininger, C., & Sessitsch, A. (2018). Maintenance and assessment of cell viability in formulation of non-sporulating bacterial inoculants. *Microbial Biotechnology, 11*(2), 277–301.

Besson, P., & Rowe, F. (2012). Strategizing information systems-enabled organizational transformation: A transdisciplinary review and new directions. *The Journal of Strategic Information Systems, 21*(2), 103–124. DOI: 10.1016/j.jsis.2012.05.001

Bezner Kerr, R. (2010). Reconceptualizing food sovereignty: Peasant women, agroecology and feminist politics. *Development, 53*(1), 112–117. DOI: 10.1057/dev.2009.47

Bezner Kerr, R., Postigo, J. C., Smith, P., Cowie, A., Singh, P. K., Rivera-Ferre, M., Tirado-von der Pahlen, M. C., Campbell, D., & Neufeldt, H. (2023). Agroecology as a transformative approach to tackle climatic, food, and ecosystemic crises. *Current Opinion in Environmental Sustainability, 62*, 101294. DOI: 10.1016/j.cosust.2023.101275

Bhattarai, K., & Yousef, M. (2025). Water Scarcity and Climate Change in MENA: Challenges, Innovations, and Geopolitical Impacts. In *The Middle East: Past, Present, and Future* (pp. 105–136). Springer Nature Switzerland.

Biagini, B., & Miller, A. (2013). Engaging the private sector in adaptation to climate change in developing countries: Importance, status, and challenges. *Climate and Development, 5*(3), 242–252. DOI: 10.1080/17565529.2013.821053

Bloch, P. (2005). Présentation synthétique du partenariat Public-Privé. In *Partenariat pubic-privé* (Dossier). *Revue de Droit Bancaire et Financier*.

Bogale, G. A., & Erena, Z. B. (2022). Drought vulnerability and impacts of climate change on livestock production and productivity in different agro-Ecological zones of Ethiopia. *Journal of Applied Animal Research, 50*(1), 471–489.

Boorboori, M. R., & Lackóová, L. (2025). Arbuscular mycorrhizal fungi and salinity stress mitigation in plants. *Frontiers in Plant Science, 15*, 1504970.

Bouchakour, R., & Saad, M. (2020). Farm and farmer characteristics and off-farm work: Evidence from Algeria. *The Australian Journal of Agricultural and Resource Economics, 64*(2), 455–476. DOI: 10.1111/1467-8489.12349

Bougnoux, N., Wodon, Q., Liverani, A., & Joseph, G. (2014). *Climate change and migration: Evidence from the Middle East and North Africa* (World Bank Publications No. 18929). The World Bank.

Boussaada, D., Yerou, H., Benabdelli, K., & Djelailia, S. (2022). Evaluation des potentialités pastorales des parcours steppiques algériennes: Cas de M'sila (Algérie). *Livestock Research for Rural Development, 34*(4).

Bouzid, A., Lazereg, M., Bédrani, S., Behnassi, M., & Baig, M. B. (2022). Facteurs naturels et réglementaires sous-jacents à la dépendance alimentaire en Algérie. In Behnassi, M., Baig, M. B., Sraïri, M. T., Alsheikh, A. A., & Abu Risheh, A. W. A. (Eds.), *Sécurité alimentaire et systèmes alimentaires climato-intelligents*. Springer., DOI: 10.1007/978-3-030-92738-7_17

Bouznit, M., & Aïssaoui, R. (2023). The impacts of climate change factors and innovative capabilities on food production in Algeria: Evidence from ARDL model. *Environment, Development and Sustainability, 26*(9), 23889–23908. DOI: 10.1007/s10668-023-03627-w

Brabez, F., & Bédrani, S. (2015). Les coopératives agricoles de services: Étude de cas. *Les cahiers du CREAD*, Volume 31. *Numéro, 113*, 193–217.

Bridle, J., Balmford, A., Durant, S. M., Gregory, R. D., Pearson, R., & Purvis, A. (2025). How should we bend the curve of biodiversity loss to build a just and sustainable future? *Philosophical Transactions of the Royal Society of London. Series B, Biological Sciences, 380*(1917), 20230205. DOI: 10.1098/rstb.2023.0205

Buchner, B., Naran, B., Padmanabhi, R., Stout, S., Strinati, C., Wignarajah, D., Miao, G., Connolly, J., & Marini, N. (2023). *Global landscape of climate finance 2023*. Climate Policy Initiative. https://www.climatepolicyinitiative.org/wp-content/uploads/2023/11/Global-Landscape-of-Climate-Finance-2023.pdf

Burke, M., Hsiang, S. M., & Miguel, E. (2015). Global non-linear effect of temperature on economic production. *Nature*, *527*(7577), 235–239. DOI: 10.1038/nature15725 PMID: 26503051

Byers, E., Gidden, M., Leclère, D., Balkovic, J., Burek, P., Ebi, K., Greve, P., Grey, D., Havlik, P., Hillers, A., Johnson, N., Kahil, T., Krey, V., Langan, S., Nakicenovic, N., Novak, R., Obersteiner, M., Pachauri, S., Palazzo, A., & Riahi, K. (2018). Global exposure and vulnerability to multi-sector development and climate change hotspots. *Environmental Research Letters*, *13*(5), 055012. DOI: 10.1088/1748-9326/aabf45

Cabernard, L., Pfister, S., & Hellweg, S. (2024). Biodiversity Impacts of Recent Land-Use Change Driven by Increases in Agri-Food Imports. *Nature Sustainability*, *7*(11), 1512–1524. DOI: 10.1038/s41893-024-01433-4

Campbell, B. M., Hansen, J., Rioux, J., Stirling, C. M., & Twomlow, S. (2018). Urgent action to combat climate change and its impacts (SDG 13): Transforming agriculture and food systems. *Current Opinion in Environmental Sustainability*, *34*, 13–20. DOI: 10.1016/j.cosust.2018.06.005

Carbonell, I. M. (2016). The ethics of big data in agriculture. *Internet Policy Review*, *5*(1), 1–13. DOI: 10.14763/2016.1.405

Carleton, T. A., & Hsiang, S. M. (2016). Social and economic impacts of climate. *Science*, *353*(6304), aad9837. DOI: 10.1126/science.aad9837 PMID: 27609899

Castle, J. L., & Hendry, D. F. (2020). Climate Econometrics: An Overview. *Foundations and Trends® in Econometrics, 10*(3–4), 145–322. https://doi.org/DOI: 10.1561/0800000037

CAT. (2023). Policies & action: Morocco. Policy overview. Climate Action Tracker. https://climateactiontracker.org/countries/morocco/policies-action/

Cerdà, A., Novara, A., & Moradi, E. (2021). Long-term non-sustainable soil erosion rates and soil compaction in drip-irrigated citrus plantation in Eastern Iberian Peninsula. *The Science of the Total Environment*, *787*, 147549. DOI: 10.1016/j.scitotenv.2021.147549

Cevik, S., & Jalles, J. T. (2022). *For whom the bell tolls: Climate change and inequality* (IMF Working Paper No. 103). International Monetary Fund. https://doi.org/DOI: 10.5089/9798400208126.001

Cha, K. J., & Lee, Z. (2013). *What Do We Mean by Information Technology Enabled Organisational Transformation?*

Chaami, M., Bédrani, S., Bessaad, A., & Lambarraa-Lehnhardt, F. (2025). Why is there a shortage of agricultural labor in Algeria? Evidence from M'zirâa, southern Algeria's leading vegetable market. *New Medit, 25*(1). Advance online publication. DOI: 10.30682/nm2501h

Chambers, R., Pacey, A., & Thrupp, L. A. (1989). *Farmer First: Farmer Innovation and Agricultural Research.* Intermediate Technology Publications. DOI: 10.3362/9781780440149

Chancel, L., Bothe, P., & Voituriez, T. (2023). *Climate inequality report 2023: Fair shares in a warming world.* World Inequality Lab.

Chanias, S. (2017). MASTERING DIGITAL TRANSFORMATION: THE PATH OF A FINANCIAL SERVICES PROVIDER TOWARDS A DIGITAL TRANSFORMATION STRATEGY. *Research Papers.* https://aisel.aisnet.org/ecis2017_rp/2

Chanias, S., & Hess, T. (2016). *Understanding Digital Transformation Strategy Formation: Insights from Europe's Automotive Industry.*

Chaudhry, S., & Sidhu, G. P. S. (2022). Climate change regulated abiotic stress mechanisms in plants: A comprehensive review. *Plant Cell Reports, 41*(1), 1–31. DOI: 10.1007/s00299-021-02759-5 PMID: 34351488

Chaves, M. M., Flexas, J., & Pinheiro, C. (2009). Photosynthesis under drought and salt stress: Regulation mechanisms from whole plant to cell. *Annals of Botany, 103*(4), 551–560. DOI: 10.1093/aob/mcn125 PMID: 18662937

Chebil, A., Souissi, A., Frija, A., & Stambouli, T. (2019). Estimation of the economic loss due to irrigation water use inefficiency in Tunisia. *Environmental Science and Pollution Research International, 26*(11), 11261–11268. DOI: 10.1007/s11356-019-04566-8

Chenchouni, H., & Neffar, S. (2022). Soil organic carbon stock in arid and semi-arid steppe rangelands of North Africa. *Catena, 211*, 106004. DOI: 10.1016/j.catena.2021.106004

Chen, P. J., & Antonelli, M. (2020). Conceptual models of food choice: Influential factors related to foods, individual differences, and society. *Foods, 9*(12). *Foods, 1898*(12), s9121898. DOI: 10.3390/foods9121898

Chevallier, R. (2023). Strengthening Africa's climate-smart agriculture and food systems through enhanced policy coherence and coordinated action. *The South African Journal of International Affairs, 30*(4), 595–618. DOI: 10.1080/10220461.2024.2318712

Chieb, M., & Gachomo, E. W. (2023). The role of plant growth promoting rhizobacteria in plant drought stress responses. *BMC Plant Biology, 23*(1), 407.

Choudhary, T. F., & Gupta, M. (2024). Impact of Climate Change on Agriculture: Evidence from Major Crop Production in India. The Indian Economic Journal, 72(3), 442-459. (Original work published 2024)DOI: 10.1177/00194662241230655

Choukr-Allah, R., Mouridi, Z. E., Benbessis, Y., & Shahid, S. A. (2023). Salt-affected soils and their management in the Middle East and North Africa (MENA) region: a holistic approach. In *Biosaline agriculture as a climate change adaptation for food security* (pp. 13–45). Springer International Publishing.

Christmann, S. (2020). Climate change enforces to look beyond the plant – The example of pollinators. *Current Opinion in Plant Biology, 56*, 162–167. DOI: 10.1016/j.pbi.2019.11.001

Claire De Marignan. (2017). "Aquaculture: Or when the man domesticates the sea." 19-30

Clapp, J. (2021). Food: Governance challenges for a hot and hungry planet. In: *Global Governance Futures* (pp. 207-221). Routledge.

Clapp, C., & Pillay, K. (2017). Green bonds and climate finance. In Klein, R. J. T., O'Brien, K. L., & Wolf, J. (Eds.), *Climate finance: Theory and practice* (pp. 79–105). Routledge., DOI: 10.1142/9789814641814_0005

Cleveland, D. A., & Soleri, D. (2002). *Farmers, scientists and plant breeding: Integrating knowledge and practice*. CABI Publishing. DOI: 10.1079/9780851995854.0000

Climate Action Tracker. (2023, April 5). *Climate Action Tracker, UAE*. https://climateactiontracker.org/countries/uae/2023-04-05/

Closas, A., & Villholth, K.G. (2016). Aquifer contracts: a means to solving groundwater over-exploitatin in Morocco?. Internatinal Water Management Instiute (IWMI). Groundwater Solutins Inititie for Policy and Practie – GRIPP. Case Profie Series 01).

Cohen, J. I., & Paarlberg, R. (2004). Unlocking crop biotechnology in developing countries—A report from the field. *World Development, 32*(9), 1563–1577. DOI: 10.1016/j.worlddev.2004.05.003

Conseil d'État. (2009, 9 décembre). *Motivation de dévolution en marché global et insertion d'une clause de révision de prix.* Recueil Lebon – Recueil des décisions du Conseil d'État, (6).

Conseil d'État. (2010, 9 avril). *Le juge de cassation contrôle l'interprétation d'un cahier des charges.* Recueil Lebon – Recueil des décisions du Conseil d'État, (4).

Contreras-Medina, D. I., Medina-Cuéllar, S. E., & Rodríguez-García, J. M. (2022). Roadmapping 5.0 Technologies in Agriculture: A Technological Proposal for Developing the Coffee Plant Centered on Indigenous Producers' Requirements from Mexico, via Knowledge Management. *Plants*, *11*(11), 1502. DOI: 10.3390/plants11111502

Costa, F., Frecassetti, S., Rossini, M., & Portioli-Staudacher, A. (2023). Industry 4.0 digital technologies enhancing sustainability: Applications and barriers from the agricultural industry in an emerging economy. *Journal of Cleaner Production*, *408*, 137208. DOI: 10.1016/j.jclepro.2023.137208

Curriero, F. C., Heiner, K. S., Samet, J. M., Zeger, S. L., Strug, L., & Patz, J. A. (2002). Temperature and mortality in 11 cities of the eastern United States. *American Journal of Epidemiology*, *155*(1), 80–87. DOI: 10.1093/aje/155.1.80 PMID: 11772788

Czipf, C. (2025). The impact of changing energy prices, interest rates, and investment costs on the net present value and internal rate of return for alternative energy projects. *Discover Sustainability*, *6*(1), 161. Advance online publication. DOI: 10.1007/s43621-025-00921-7

DA SILVA, J. G., DEL GROSSI, M. E. & DE FRANÇA, C. G. 2012. The Program to Eliminate Hunger: The Brazilian Experience.

Dahiyat, E. A. R. (2011). Towards new recognition of liability in the digital world: Should we be more creative? *International Journal of Law and Information Technology*, *19*(3), 224–242. DOI: 10.1093/ijlit/ear006

Dahlman, C., Mealy, S., & Wermelinger, M. (2016). *Harnessing the digital economy for developing countries*. OCDE., DOI: 10.1787/4adffb24-

Darwish, T., Shaban, A., Faour, G., Jomaa, I., Moubarak, P., & Khadra, R. (2025). Sustainable and Traditional Irrigation and Fertigation Practices for Potato and Zucchini in Dry Mediterranean Regions. *Sustainability (Basel)*, *17*(5), 1860. Advance online publication. DOI: 10.3390/su17051860

Dasgupta, S., Huq, M., Khan, Z. H., Ahmed, M. M. Z., Mukherjee, N., Khan, M., & Pandey, K. D. (2010). Vulnerability of Bangladesh to cyclones in a changing climate: Potential damages and adaptation cost. *World Bank Policy Research Working Paper*, (5280).

Dasthagir, K. G. (2024). Hydraulic bureaucracy and irrigation management transfer: Analysing the efficacy of two models in South-India. *International Journal of Water Resources Development*, *40*(5), 885–905. DOI: 10.1080/07900627.2024.2341992

Data and statistics. (2023). International Labour Organization. https://www.ilo.org/data-and-statistics

Dayioğlu, M. A., & Turker, U. (2021). Digital Transformation for Sustainable Future - Agriculture 4.0: A review. *Tarim Bilimleri Dergisi*. Advance online publication. DOI: 10.15832/ankutbd.986431

De Bernardi, P. (with Azucar, D.). (2020). *Innovation in Food Ecosystems: Entrepreneurship for a Sustainable Future*. Springer.

de Oliveira, J. A. P. (2009). The implementation of climate change related policies at the subnational level: An analysis of three countries. *Habitat International*, *33*(3), 253–259. DOI: 10.1016/j.habitatint.2008.10.006

De Schutter, O. (2009). Seed policies and the right to food: Enhancing agrobiodiversity and encouraging innovation. Report of the UN Special Rapporteur on the Right to Food, A/64/170.

Debbabi, O. S., Amar, F. B., Rahmani, S. M., Taranto, F., Montemurro, C., & Miazzi, M. M. (2022). The status of genetic resources and olive breeding in Tunisia. *Plants*, *11*(13), 1759. DOI: 10.3390/plants11131759

Deeb, M., Smagin, A. V., Pauleit, S., Fouché-Grobla, O., Podwojewski, P., & Groffman, P. M. (2024). The urgency of building soils for Middle Eastern and North African countries: Economic, environmental, and health solutions. *The Science of the Total Environment*, *917*, 170529. DOI: 10.1016/j.scitotenv.2024.170529

Del Pozo, A., Brunel-Saldias, N., Engler, A., Ortega-Farias, S., Acevedo-Opazo, C., Lobos, G. A., Jara-Rojas, R., & Molina-Montenegro, M. A. (2019). Climate change impacts and adaptation strategies of agriculture in Mediterranean-climate regions (MCRs). *Sustainability (Basel)*, *11*(10), 2769. Advance online publication. DOI: 10.3390/su11102769

Dell, M., Jones, B. F., & Olken, B. A. (2014). What do we learn from the weather? The new climate–economy literature. *Journal of Economic Literature*, *52*(3), 740–798. DOI: 10.1257/jel.52.3.740

Deloitte. (2020). *What is Digital Economy? Deloitte*. https://www2.deloitte.com/mt/en/pages/technology/articles/mt-what-is-digitaleconomy.html

Delong, D. C.Jr. (1996). Defining biodiversity. *Wildlife Society Bulletin*, •••, 738–749.

Dercon, S., & Christiaensen, L. (2011). Consumption risk, technology adoption and poverty traps: Evidence from Ethiopia. *Journal of Development Economics*, *96*(2), 159–173. DOI: 10.1016/j.jdeveco.2010.08.003

Deryugina, T., & Hsiang, S. M. (2014). Does the environment still matter? *Daily temperature and income in the United States* (No. w20750). National Bureau of Economic Research.

Deschenes, O. (2014). Temperature, human health, and adaptation: A review of the empirical literature. *Energy Economics*, *46*, 606–619. DOI: 10.1016/j.eneco.2013.10.013

Deschênes, O., & Greenstone, M. (2011). Climate change, mortality, and adaptation: Evidence from annual fluctuations in weather in the United States. *American Economic Journal. Applied Economics*, *3*(4), 152–185. DOI: 10.1257/app.3.4.152

Deschênes, O., & Moretti, E. (2009). Extreme weather events, mortality, and migration. *The Review of Economics and Statistics*, *91*(4), 659–681. DOI: 10.1162/rest.91.4.659

Deutsch, C. A., Tewksbury, J. J., Tigchelaar, M., Battisti, D. S., Merrill, S. C., Huey, R. B., & Naylor, R. L. (2018). Increase in crop losses to insect pests in a warming climate. *Science*, *361*(6405), 916–919. DOI: 10.1126/science.aat3466 PMID: 30166490

Devarajan, S., Go, D. S., Maliszewska, M., Osorio-Rodarte, I., & Timmer, H. (2015). Stress-testing Africa's recent growth and poverty performance. *Journal of Policy Modeling*, *37*(4), 521–547. DOI: 10.1016/j.jpolmod.2015.04.006

DGF. (2024). Conservation des eaux et des sols et lutte contre la désertification. http://www.dgf.org.dz/fr/content/conservation-des-eaux-et-des-sols-et-lutte-contre-la-d%C3%A9sertification

Dhaouadi, L., Ben Maachia, S., Namsi, A., Abouthief, F., Nagaz, K., & Masmoudi, M. M. (2022). Literature review on water productivity of date palm trees in Tunisian agro systems. *Journal of Oasis Agriculture and Sustainable Development*, *2*(4), 30–44. DOI: 10.56027/JOASD.222022

Dickson, M. (2022). *Regional review on status and trends in aquaculture development in the Near East and North Africa–2020*. Food & Agriculture Org.

Dirnat, C., & Souffir, J. (2005). Approche contractuelle de la dimension financière du contrat de partenariat: un déconcentrant ménage à trois. In *Partenariat pubic-privé* (Dossier). *Revue de Droit Bancaire et Financier.*

Doni, F., Gasperini, A., & Soares, J. T. (2020). *SDG13–Climate action: Combating climate change and its impacts.* Emerald Publishing Limited., DOI: 10.1108/9781787569157

Dörr, J., & Nachtmann, M. (Eds.). (2022). *Handbook Digital Farming: Digital Transformation for Sustainable Agriculture*. Springer Berlin Heidelberg., DOI: 10.1007/978-3-662-64378-5

Dougherty, D., & Dunne, D. D. (2012). Digital Science and Knowledge Boundaries in Complex Innovation. *Organization Science*, *23*(5), 1467–1484. DOI: 10.1287/orsc.1110.0700

Drine, I. (2011). *Climate change compounding risks in North Africa* (WIDER Working Paper No. 2011/32). United Nations University World Institute for Development Economics Research.

Du, H., Dang, K. K., Nguyen, H. Q., & van Rijswick, H. F. M. W. (2023). A framework for reviewing laws and policies for climate resilience: The case of the Vietnamese Mekong Delta. *Journal of Environmental Planning and Management*, *66*(6), 1280–1304. DOI: 10.1080/09640568.2022.2026308

Dumitrescu, E.-I., & Hurlin, C. (2012). Testing for Granger Non-Causality in Heterogeneous Panels. *Economic Modelling*, *29*(4), 1450–1460. DOI: 10.1016/j.econmod.2012.02.014

Dzanku, F. M., Jirström, M., & Marstorp, H. (2015). Yield gap-based poverty gaps in rural Sub-Saharan Africa. *World Development*, *67*, 336–362. DOI: 10.1016/j.worlddev.2014.10.030

Dzvene, A. R., Zhou, L., Slayi, M., & Dirwai, T. L. (2025). A scoping review on challenges and measures for climate change in arid and semi-arid agri-food systems. *Discover Sustainability*, *6*(1), 151. Advance online publication. DOI: 10.1007/s43621-025-00945-z

Ebert, C., & Duarte, C. H. C. (2018). Digital Transformation. *IEEE Software*, *35*(4), 16–21. DOI: 10.1109/MS.2018.2801537

EFFIS. (2023). Forest fires in Europe, Middle East and North Africa 2023. https://forest-fire.emergency.copernicus.eu

Efremova, N. (2023). Gamification resources in the system of specialist training for the agricultural complex. *E3S Web of Conferences, 431*, 01052. DOI: 10.1051/e3sconf/202343101052

Eid, H. M., El-Marsafawy, S. M., & Ouda, S. A. (2007). Assessing the economic impacts of climate change on agriculture in Egypt: a Ricardian approach. World Bank Policy Research Working Paper, (4293).

El-Agha, D. E., Molle, F., Metwally, M. I., Emara, S. R., Shalby, A., Armanuos, A. M., & Gado, T. A. (2024). Toward sustainable management of groundwater in the deserts of Egypt. *Hydrogeology Journal, 32*(3), 663–678. DOI: 10.1007/s10040-023-02738-y

ElFakih, S., Ben Abdallah, S., & Elkadri, A. (2023). Water accounting for food security: The case of rainfed and irrigated olive growing in arid regions in Tunisia. *Journal of Arid Arboriculture and Olive Growing, 2*(1), 12–26. https://iosfax.agrinet.tn/annonce/revue/Volume2-N1/Article%20saida.pdf

EL-KEBLAWY, A. (2014). Impact of climate change on biodiversity loss and extinction of endemic plants of arid land mountains. *J Biodivers Endanger Species, 2*, 2.

Elkhalfi, O., Chaabita, R., Ghoujdam, M., Zehraoui, K., El Alaoui, H., & Belhaj, I. (2025). The impact of climate change on food security in the Middle East and North Africa: Challenges and adaptation strategies. *Journal of Agriculture and Food Research, 21*, 101963. DOI: 10.1016/j.jafr.2025.101963

Elloumi, M. (2016). La gouvernance des eaux souterraines en Tunisie (Report No. 7). International Water Management Institute (IWMI). https://gw-mena.iwmi.org/wp-content/uploads/sites/3/2017/04/Rep.7-Groundwater-governance-in-Tunisia_final_cover.pdf

El-Malky, M. G., Abdelwahab, M. M., Bryant, C., & Tantawi, S. (2016). VULNERABILITY AND ADAPTATION ASSESSMENT OF AGRICULTURE TO CLIMATE CHANGE IN EGYPT USING MATHEMATICAL MODELS. *Journal of Environmental Sciences (China), 35*(1), 1–19. DOI: 10.21608/jes.2016.27758

Elsaie, Y., Ismail, S., Soussa, H., Gado, M., & Balah, A. (2023). Water desalination in Egypt; literature review and assessment. *Ain Shams Engineering Journal, 14*(7), 101998. DOI: 10.1016/j.asej.2022.101998

Emadodin, I., Corral, D. E. F., Reinsch, T., Kluß, C., & Taube, F. (2021). Climate Change Effects on Temperate Grassland and Its Implication for Forage Production: A Case Study from Northern Germany. *Agriculture, 11*(3), 232. DOI: 10.3390/agriculture11030232

Emery, C. (2006). L'exécution des conventions de délégation de service public (Chap. 3, Folio 3223). In Dalloz (Ed.), *Encyclopédie des collectivités locales*. Dalloz.

Entine, J., Felipe, M. S. S., Groenewald, J.-H., Kershen, D. L., Lema, M., McHughen, A., & Parrott, W. A. (2021). Regulatory approaches for genome edited agricultural plants in select countries and jurisdictions around the world. *Transgenic Research, 30*(4), 551–584. DOI: 10.1007/s11248-021-00257-8

Epule, T. E., Ford, J. D., Lwasa, S., & Lepage, L. (2017). Climate change adaptation in the Sahel. *Environmental Science & Policy*, *75*, 121–137. DOI: 10.1016/j.envsci.2017.05.018

Espín-Sánchez, J.-A., Gil-Guirado, S., Giraldo-Paez, W. D., & Vickers, C. (2019). Labor income inequality in pre-industrial Mediterranean Spain: The city of Murcia in the 18th century. *Explorations in Economic History*, *73*, 101274. DOI: 10.1016/j.eeh.2019.05.002

European Union. (2000). Directive 2000/60/EC of the European Parliament and of the Council of 23 October 2000 establishing a framework for Community action in the field of water policy. https://data.europa.eu/eli/dir/2000/60/oj

Evenson, R. E., & Gollin, D. (2003). Assessing the impact of the Green Revolution, 1960 to 2000. *Science*, *300*(5620), 758–762. DOI: 10.1126/science.1078710 PMID: 12730592

Evers, H. D., & Benedikter, S. (2009). Hydraulic bureaucracy in a modern hydraulic society-strategic group formation in the Mekong delta, Vietnam. *Water Alternatives*, *2*(3), 416.

Ezrari, S., Radouane, N., Tahiri, A., El Housni, Z., Mokrini, F., Özer, G., Lazraq, A., Belabess, Z., Amiri, S., & Lahlali, R. (2022). Dry root rot disease, an emerging threat to citrus industry worldwide under climate change: A review. *Physiological and Molecular Plant Pathology*, *117*, 101753. DOI: 10.1016/j.pmpp.2021.101753

Fafchamps, M., Udry, C., & Czukas, K. (1998). Drought and saving in West Africa: Are livestock a buffer stock? *Journal of Development Economics*, *55*(2), 273–305. DOI: 10.1016/S0304-3878(98)00037-6

Falkenmark, M., Lundqvist, J., & Widstrand, C. (1989). Macro-scale water scarcity requires micro-scale approaches: Aspects of vulnerability in semi-arid development. [). Oxford, UK: Blackwell Publishing Ltd.]. *Natural Resources Forum*, *13*(4), 258–267. DOI: 10.1111/j.1477-8947.1989.tb00348.x

FAO, IFAD, WHO, WFP and UNICEF. (2024). The State of Food Security and Nutrition in the World 2024. Financing to eliminate hunger, food insecurity and all forms of malnutrition. Rome. https://doi.org/DOI: 10.4060/cd1254fr

FAO. (2005). Utilisation des engrais par culture en Algérie. Service de la gestion des terres et de la nutrition des plantes. Division de la mise en valeur des terres et des eaux. Organisation des nations unies pour l'alimentation et l'agriculture, Rome.

FAO. (2009). Dicentrarchus labrax. In Cultured aquatic species fact sheets. Text by Bagni, M. Edited and compiled by Valerio Crespi and Michael New. CD-ROM (multilingual).

FAO. (2009). Sparus aurata. In Cultured aquatic species fact sheets. Text by Colloca, F. & Cerasi, S. Edited and compiled by Valerio Crespi and Michael New. CD-ROM (multilin-gual). https://www.fao.org/fishery/docs/CDrom/aquaculture/I1129m/file/fr/fr_giltheadseabr.htm

FAO. (2011). *La pratique de la gestion durable des terres. Directives et bonnes pratiques en Afrique subsaharienne. TerrAfrica, Panorama mondial des approches et technologies de conservation (WOCAT) et Organisation des Nations Unies pour l'alimentation et l'agriculture*. FAO.

FAO. (2018). *Scaling up agroecology to achieve the sustainable development goals: Summary for policy makers*. Food and Agriculture Organization of the United Nations. https://www.fao.org/documents/card/en/c/I9049EN

FAO. (2018). *The right to food and the rights of peasants and other people working in rural areas*. Food and Agriculture Organization of the United Nations.

FAO. (2018). Water efficiency, productivity and sustainability in the NENA regions (WEPS-NENA). https://www.fao.org/in-action/water-efficiency-nena/countries/tunisia/background-and-sites/ar/?

FAO. (2019). Proceedings from the Regional Workshop on Climate-Smart Agriculture in the Near East and North Africa. Cairo. 68 pp. Licence: CC BY-NC-SA 3.0 IGO. https://openknowledge.fao.org/items/a7cad295-ed2d-47b7-8c27-52450f4a7ac0

FAO. (2019). *Proceedings from the Regional Workshop on Climate-Smart Agriculture in the Near East and North Africa*. FAO, Cairo.

FAO. (2021). *The State of the World's Land and Water Resources for Food and Agriculture – Systems at breaking point (SOLAW 2021)*. FAO., DOI: 10.4060/cb7654en

FAO. (2022). *Water efficiency, productivity and sustainability in the NENA regions*. FAO-UN.

FAO. (2024). Statistics on the agricultural sector. FAO-UN. https://fao.org/faostat/en/#data

FAO. (2025). Statistics on productions and trade. FAO: https://fao.org/faostat/en/#data

Farah, A. A., Mohamed, M. A., Musse, O. S. H., & Nor, B. A. (2025). The multifaceted impact of climate change on agricultural productivity: A systematic literature review of SCOPUS-indexed studies (2015–2024). *Discover Sustainability, 6*(1), 397. Advance online publication. DOI: 10.1007/s43621-025-01229-2

Farhad, M., Kumar, U., Tomar, V., Bhati, P. K., & Krishnan, J, N., Barek, V., ... & Hossain, A. (2023). Heat stress in wheat: A global challenge to feed billions in the current era of the changing climate. *Frontiers in Sustainable Food Systems, 7*, 1203721.

Farooq, M., Wahid, A., Kobayashi, N. S. M. A., Fujita, D. B. S. M. A., & Basra, S. M. (2009). Plant drought stress: effects, mechanisms and management. In *Sustainable agriculture* (pp. 153–188). Springer Netherlands. DOI: 10.1007/978-90-481-2666-8_12

Feher, A., Raicov, M., Brad, I., Toader, C. S., Sirbulescu, C. E., & Ben Saleh, R. (2022). Food security in a changing climate world. *Review on Agriculture and Rural Development, 11*(1-2), 32–38. DOI: 10.14232/rard.2022.1-2.32-38

Fekad, M., & Bekalu, Y. (2020). Climate change impacts on household food security and its adaptation options in rural Ethiopia: A systematic review. [IJRD]. *EPRA International Journal of Research & Development, 5*, 2455–7838. DOI: 10.36713/epra4142

Fekete, H., Kuramochi, T., Roelfsema, M., den Elzen, M., Forsell, N., Höhne, N., & Luna, L.. (2021). A review of successful climate change mitigation policies in major emitting economies and the potential of global replication. *Renewable & Sustainable Energy Reviews, 137*, 110602. DOI: 10.1016/j.rser.2020.110602

Fellag, M., Achite, M., & Walega, A. (2021). Spatial-temporal characterization of meteorological drought using the Standardized precipitation index. Case study in Algeria. *Acta Scientiarum Polonorum. Formatio Circumiectus, 20*(1), 19–31. DOI: 10.15576/ASP.FC/2021.20.1.19

Feng, S., Oppenheimer, M., & Schlenker, W. (2012). *Climate change, crop yields, and internal migration in the United States* (NBER Working Paper No. 17734). National Bureau of Economic Research. https://www.nber.org/papers/w17734

Ferchichi, I., Zairi, A., & Marlet, S. (2022). Water scarcity conflicts in a community-managed irrigation system in Northern Tunisia: Supporting dialogue and negotiation. IAHS-AISH Scientific Assembly, Montpellier, France, 385, 253–258. DOI: 10.5194/piahs-385-253-2024

Fernández-Lobato, L., López-Sánchez, Y., Baccar, R., Fendri, M., & Vera, D. (2022). Life cycle assessment of the most representative virgin olive oil production systems in Tunisia. *Sustainable Production and Consumption*, *32*, 908–923. DOI: 10.1016/j.spc.2022.06.002

Field, A. P. (2005). Kendall's coefficient of concordance. In Everitt, B. S., & Howell, D. C. (Eds.), Vol. 2, pp. 1010–1011). Encyclopedia of statistics in behavioral science, DOI: 10.1002/0470013192.bsa327

Field, C. B., & Barros, V. R. (Eds.). (2014). *Climate change 2014–Impacts, adaptation and vulnerability: Regional aspects*. Cambridge University Press. DOI: 10.1017/CBO9781107415379

Filahi, S., Tanarhte, M., Mouhir, L., El Morhit, M., & Tramblay, Y. (2016). Trends in indices of daily temperature and precipitations extremes in Morocco. *Theoretical and Applied Climatology*, *124*(3-4), 959–972. DOI: 10.1007/s00704-015-1472-4

Filchev, L., & Chanev, M. (2025). Remote Sensing in Climate Change Research. In Dobrinkova, N., & Fidanova, S. (Eds.), *Environmental Protection and Disaster Risks (EnviroRisks 2024). EnviroRISKs 2024. Lecture Notes in Networks and Systems* (Vol. 883). Springer., DOI: 10.1007/978-3-031-74707-6_17

Finger, R. (2023). Digital innovations for sustainable and resilient agricultural systems. *European Review of Agriculture Economics*, *50*(4), 1277–1309. DOI: 10.1093/erae/jbad021

Folke, C., Biggs, R., Norström, A. V., Reyers, B., & Rockström, J. (2016). Social-ecological resilience and biosphere-based sustainability science. *Ecology and Society*, *21*(3), art41. DOI: 10.5751/ES-08748-210341

Food and Agriculture Organization of the United Nations, & the World Health Organization. (n.d.). *About Codex Alimentarius*. FAO-WHO Codex Alimentarius. https://www.fao.org/fao-who-codexalimentarius/about-codex/en/

Food and Agriculture Organization of the United Nations. (2001). *International Treaty on Plant Genetic Resources for Food and Agriculture*. FAO. https://www.fao.org/plant-treaty/overview/texts-of-the-treaty/en/

Food and Agriculture Organization of the United Nations. (2020). AQUASTAT database. https://www.fao.org/aquastat

Food and Agriculture Organization of the United Nations. (2025a). *The Food and Agriculture Organization of the United Nations*. https://www.fao.org/biodiversity/our-work/strategy-and-action-plan/ar

Food and Agriculture Organization of the United Nations. (2025b). *The World Food Summit and its follow-up.* https://www.fao.org/4/X2051a/x2051a00.htm

Food and Agriculture Organization of the United Nations. (n.d.-a). *Guidelines & standards – Pest and pesticide management.* FAO. https://www.fao.org/pest-and-pesticide-management/guidelines-standards/en/

Food and Agriculture Organization of the United Nations. (n.d.-b). *International Plant Protection Convention (IPPC).* FAO. https://www.fao.org/plant-production-protection/in-action/governance-and-normative-work/ippc/en

Formetta, G., & Feyen, L. (2019). Empirical evidence of declining global vulnerability to climate-related hazards. *Global Environmental Change, 57*, 101920. DOI: 10.1016/j.gloenvcha.2019.05.004 PMID: 31417231

Fourgoux, J.-L. (2020). La clause de révision de prix dans les contrats d'affaires. *Actualité Juridique Contrats d'affaires,* (12).

Francis, D., & Fonseca, R. (2024). Recent and projected changes in climate patterns in the Middle East and North Africa (MENA) region. *Scientific Reports, 14*(1), 10279.

Frija, A., Oulmane, A., Chebil, A., & Makhlouf, M. (2021). Socio-Economic implications and potential structural adaptations of the Tunisian agricultural sector to climate change. *Agronomy (Basel), 11*(11), 2112. DOI: 10.3390/agronomy11112112

Frimpong, T. D., Atchadé, M. N., & Tona Landu, T. (2024). Assessing the impact of CO2 emissions, food security and agriculture expansion on economic growth: A panel ARDL analysis. *Discover Sustainability, 5*(1), 424. DOI: 10.1007/s43621-024-00630-7

Fuentes-Peñailillo, F., Gutter, K., Vega, R., & Silva, G. C. (2024). Transformative technologies in digital agriculture: Leveraging Internet of Things, remote sensing, and artificial intelligence for smart crop management. *Journal of Sensor and Actuator Networks., 13*(4), 39. DOI: 10.3390/jsan13040039

Fuglie, K. O., Heisey, P. W., King, J. L., & Schimmelpfennig, D. (2011). *Research investments and market structure in the food processing, agricultural input, and biofuel industries worldwide.* USDA Economic Research Service. DOI: 10.2139/ssrn.2027051

Fuglie, K., Gautam, M., Goyal, A., & Maloney, W. F. (2019). *Harvesting Prosperity: Technology and Productivity Growth in Agriculture*. World Bank Group.

Fuller, J. L., van Putten, I., Kraan, M., Bjørkan, M., & Dankel, D. J. (2023). 'Sustainability is not a vegan coffee shop': Eliciting citizen attitudes and perspectives to localize the UN Sustainable Development Goals. *Journal of Environmental Planning and Management*, •••, 1–22. DOI: 10.1080/09640568.2023.2223761

Fusco, G. (2022). Climate Change and Food Security in the Northern and Eastern African Regions: A Panel Data Analysis. *Sustainability (Basel)*, *14*(19), 12664. DOI: 10.3390/su141912664

Fusco, G., Melgiovanni, M., Porrini, D., & Ricciardo, T. M. (2020). How to Improve the Diffusion of Climate-Smart Agriculture: What the Literature Tells us. *Sustainability (Basel)*, *12*(12), 5168. DOI: 10.3390/su12125168

Gabr, M. E. (2023). Impact of climatic changes on future irrigation water requirement in the Middle East and North Africa's region: A case study of upper Egypt. *Applied Water Science*, *13*(7), 158. Advance online publication. DOI: 10.1007/s13201-023-01961-y

Gage, S. D. (1978). Aquaculture: A new force in social ecology (Doctoral dissertation, Union Institute and University).

Gaies, B., Nakhli, M. S., & Sahut, J. (2021). Is financial globalization polluting? – Symmetric and asymmetric effects of external liabilities on CO2 emissions in the MENA region. *Research Square (Research Square)*. https://doi.org/DOI: 10.21203/rs.3.rs-467034/v1

Galanakis, C. M. (2020). The Food Systems in the Era of the Coronavirus (COVID-19) Pandemic Crisis. *Foods*, *9*(4), 523. DOI: 10.3390/foods9040523

Gamage, A., Gangahagedara, R., Gamage, J., Jayasinghe, N., Kodikara, N., Suraweera, P., & Merah, O. (2023). Role of organic farming for achieving sustainability in agriculture. *Farming System*, *1*(1), 100005. DOI: 10.1016/j.farsys.2023.100005

Gannouni, S., Riahi, R., Kefi, M., Fehri, R. M., & Rebai, N. (2023). Identification and monitoring of groundwater potential zones using geospatial techniques (Siliana, Tunisia). *Euro-Mediterranean Journal for Environmental Integration*, *8*(4), 1015–1034. DOI: 10.1007/s41207-023-00395-x

Gao, Y., Gao, X., & Zhang, X. (2017). The 2 °C global temperature target and the evolution of the long-term goal of addressing climate change—From the United Nations Framework Convention on Climate Change to the Paris Agreement. *Engineering (Beijing)*, *3*(2), 272–278. DOI: 10.1016/J.ENG.2017.01.022

Garcia-Alonso, M., Novillo, C., Kostolaniova, P., Martinez Parrilla, M., Alcalde, E., & Podevin, N. (2022). The EU's GM crop conundrum: Did the EU policy strategy to convert EFSA GMO guidance into legislation deliver on its promises? *EMBO Reports*, *23*(5), e54529. DOI: 10.15252/embr.202154529

Gardiner, A., Bardout, M., Grossi, F., & Dixson-Declève, S. (2015). *Public-private partnerships for climate finance*. Nordic Council of Ministers.

Gbashi, S., Adebo, O., Adebiyi, J. A., Targuma, S., Tebele, S., Areo, O. M., Olopade, B., Odukoya, J. O., & Njobeh, P. (2021). Food safety, food security and genetically modified organisms in Africa: A current perspective. *Biotechnology & Genetic Engineering Reviews*, *37*(1), 30–63. DOI: 10.1080/02648725.2021.1940735

Gebeyhu, B., Dagalo, S., & Muluneh, M. (2024). Soil moisture-based irrigation interval and irrigation performance evaluation: In the case of lower kulfo catchment, Ethiopia. *Heliyon*, *10*(16), e36089. DOI: 10.1016/j.heliyon.2024.e36089

GEBREYES. A. (2023). *SCHOOL OF LAW.* Wachemo University

Geletu, T. T., & Zhao, J. (2023). Genetic resources of Nile tilapia (Oreochromis niloticus Linnaeus, 1758) in its native range and aquaculture. *Hydrobiologia*, *850*(10), 2425–2445. DOI: 10.1007/s10750-022-04989-4

Gemini. (2011). Digital transformation: A roadmap for billion-dollar organizations. *MIT Sloan*.

Gengenbach, C., Palm, F. C., & Urbain, J. (2006). Cointegration Testing in Panels with Common Factors*. *Oxford Bulletin of Economics and Statistics*, *68*(s1), 683–719. DOI: 10.1111/j.1468-0084.2006.00452.x

Gentle, P., & Maraseni, T. N. (2012). Climate change, poverty and livelihoods: Adaptation practices by rural mountain communities in Nepal. *Environmental Science & Policy*, *21*, 24–34. DOI: 10.1016/j.envsci.2012.03.007

Gerald, A. B., Vibhute, A. D., Aitouhanni, I., Mouniane, Y., El-Khadir, I., Chriqui, A., Mehdioui, S., & Hmouni, D. (2025). Precision agriculture and artificial intelligence: Towards sustainable resource management in the face of climate challenges. In Vibhute, A. D. (Eds.), *Generative adversarial networks for remote sensing* (pp. 227–256). IGI Global., DOI: 10.4018/979-8-3693-6900-5.ch009

Ghadge, A., Wurtmann, H., & Seuring, S. (2020). Managing climate change risks in global supply chains: A review and research agenda. *International Journal of Production Research*, *58*(1), 44–64. DOI: 10.1080/00207543.2019.1629670

Gibson, J. (2009). *Intellectual property, traditional knowledge and the UN*. Cambridge University Press.

Giwa, A., Ademola, H., & Yusuf, A. O. (2025). Machine learning and application for modeling and prediction of desalination cost globally. *Desalination*, *608*, 118829. Advance online publication. DOI: 10.1016/j.desal.2025.118829

GIZ. (2020). Regional cooperation in the water sector in the Maghreb (CREM II). Deutsche Gesellschaft für Internationale Zusammenarbeit (GIZ). https://www.giz.de/en/worldwide/24690.html

Global Food Security Cluster. (2023). *The four pillars of food security*. https://handbook.fscluster.org/docs/231-the-four-pillars-of-food-security

Goedde, L., & McCullough, R. Ooko-Ombaka, & Pais, G. (2021). *How digital tools can help transform African agri-food systems* (p. 9). McKinsey & Company. https://africa-agricultural-transformation.org/wp-content/uploads/2022/12/How-digital-tools-can-help-transform-African-agri-food-systems.pdf

Goerzig, D., & Bauernhansl, T. (2018). Enterprise Architectures for the Digital Transformation in Small and Medium-sized Enterprises. *Procedia CIRP*, *67*, 540–545. DOI: 10.1016/j.procir.2017.12.257

Gokhale, H. (2021). Japan's carbon tax policy: Limitations and policy suggestions. *Current Research in Environmental Sustainability*, *3*, 100082. DOI: 10.1016/j.crsust.2021.100082

Gomez-Zavaglia, A., Mejuto, J. C., & Simal-Gandara, J. (2020). Mitigation of emerging implications of climate change on food production systems. *Food Research International*, *134*, 109256. DOI: 10.1016/j.foodres.2020.109256

Gong, C., & Ribiere, V. (2021). Developing a unified definition of digital transformation. *Technovation*, *102*, 102217. DOI: 10.1016/j.technovation.2020.102217

Government of the United Arab Emirates. (2024a). *Green Economy for Sustainable Development*. The Official Portal of the UAE Government. https://u.ae/en/about-the-uae/economy/green-economy-for-sustainable-development

Government of the United Arab Emirates. (2024b). *National Food Security Strategy 2051*. The Official Portal of the UAE Government. https://u.ae/ar-ae/about-the-uae/strategies-initiatives-and-awards/strategies-plans-and-visions/environment-and-energy/national-food-security-strategy-2051

Government of the United Arab Emirates. (2024c). *The UAE Water Security Strategy 2036*. The Official Portal of the UAE Government. https://u.ae/en/about-the-uae/strategies-initiatives-and-awards/strategies-plans-and-visions/environment-and-energy/the-uae-water-security-strategy-2036

Grar K, 2023. L'Algérie atteint son autosuffisance en huile de table. *L'économiste maghrébin*, du 23-3-2023.

Gray, J., & Rumpe, B. (2017). Models for the digital transformation. *Software & Systems Modeling*, *16*(2), 307–308. DOI: 10.1007/s10270-017-0596-7

Gregory, N. G. (2010). How climatic changes could affect meat quality. *Food Research International*, *43*(7), 1866–1873. DOI: 10.1016/j.foodres.2009.05.018

Gregory, P. J., Ingram, J. S., & Brklacich, M. (2005). Climate change and food security. *Philosophical Transactions of the Royal Society of London. Series B, Biological Sciences*, *360*(1463), 2139–2148. DOI: 10.1098/rstb.2005.1745

Gruère, G. P., Mehta-Bhatt, P., & Sengupta, D. (2008). Bt cotton and farmer suicides in India: Reviewing the evidence. *IFPRI Discussion Paper No. 00808*.

Güler, M., & Büyüközkan, G. (2019). Analysis of Digital Transformation Strategies with an Integrated Fuzzy AHP-Axiomatic Design Methodology. *IFAC-PapersOnLine*, *52*(13), 1186–1191. DOI: 10.1016/j.ifacol.2019.11.359

Guo, H., Chen, X., Dong, Z., & Zhang, H. (2019). Construction method of water right trading based on water-saving management contract. *Water Resources Protection*, *35*(3), 33-38and62. DOI: 10.3880/j.issn.1004-6933.2019.03.005

Gutiérrez González, A. (2019). The right to a clean and healthy environment: GMOs in Mexico and the European Union. *Mexican Law Review*, *11*(2), 91–113. DOI: 10.22201/iij.24485306e.2019.1.13129

Haffke, I., Kalgovas, B., & Benlian, A. (2016). *The Role of the CIO and the CDO in an Organization's Digital Transformation*.

Hallegatte, S. (2015). The indirect cost of natural disasters and an economic definition of macroeconomic resilience. *World Bank Policy Research Working Paper*, •••, 7357. DOI: 10.1596/1813-9450-7357

Hallegatte, S., Rogelj, J., Allen, M., Clarke, L., Edenhofer, O., Field, C. B., Friedlingstein, P., van Kesteren, L., Knutti, R., Mach, K. J., Mastrandrea, M., Michel, A., Minx, J., Oppenheimer, M., Plattner, G.-K., Riahi, K., Schaeffer, M., Stocker, T. F., & Van Vuuren, D. P. (2016). Mapping the climate change challenge. *Nature Climate Change*, *6*(7), 663–668. DOI: 10.1038/nclimate3057

Hallegatte, S., & Rozenberg, J. (2017). Climate change through a poverty lens. *Nature Climate Change*, *7*(4), 250–256. DOI: 10.1038/nclimate3253

Hamed, M. M., Al-Hasani, A. A. J., Nashwan, M. S., Sa'adi, Z., & Shahid, S. (2024). Assessing the growing threat of heat stress in the North Africa and Arabian Peninsula region connected to climate change. *Journal of Cleaner Production*, *447*, 141639. DOI: 10.1016/j.jclepro.2024.141639

Hamed, M. M., Ullah, S., Jasim Al-Hasani, A. A., Nashwan, M. S., & Shahid, S. (2025). Rising Threats of Compound Droughts and Heatwaves in MENA: Implications for Population and Agriculture under Paris Agreement Scenarios. *Earth Systems and Environment*, •••, 1–20.

Hameed, M., Moradkhani, H., Ahmadalipour, A., Moftakhari, H., Abbaszadeh, P., & Alipour, A. (2019). A review of the 21st century challenges in the food-energy-water security in the Middle East. *Water (Basel)*, *11*(4), 682. DOI: 10.3390/w11040682

Hamitouche, Y., Zeroual, A., Meddi, M., Assani, A. A., & Alkama, R. (2024). Changes in extreme precipitation indices across Algeria climate zones. *International Journal of Climatology*, *44*(8), 2537–2560. DOI: 10.1002/joc.8467

Hanjra, M. A., & Williams, T. O. (2020). Global change and investments in smallholder irrigation for food and nutrition security in Sub-Saharan Africa. In *The role of smallholder farms in food and nutrition security* (pp. 99–131). Springer International Publishing. DOI: 10.1007/978-3-030-42148-9_6

Haouari, M. (2024). Évolution récente des températures moyennes en Algérie. [On onm-blog.meteo.dz.]. *Journal of the American Medical Association*, *8*, 1–09.

Harsonowati, W., Manohara, D., Dwiastuti, M. E., Widawati, S., Sodiq, A. H., Khastini, R. O., & Purwani, J. (2024). Microbial-based stimulants on plant adaptation to climate change. In *Biostimulants in Plant Protection and Performance* (pp. 369–383). Elsevier.

Harwood, R. R. (2020). A history of sustainable agriculture. In Edwards, C. A. (Ed.), *Sustainable agricultural systems*. CRC Press. DOI: 10.1201/9781003070474-2

Hassoun, A., Kamiloglu, S., Garcia-Garcia, G., Parra-López, C., Trollman, H., Jagtap, S., Aadil, R. M., & Esatbeyoglu, T. (2023). Implementation of relevant fourth industrial revolution innovations across the supply chain of fruits and vegetables: A short update on Traceability 4.0. *Food Chemistry*, *409*, 135303. DOI: 10.1016/j.foodchem.2022.135303

HCP (Haut Commissariat au Plan). (2022)available at:https://www.hcp.ma/Evolution-des-inegalites-sociales-dans-un-contexte-marque-par-les-effets-de-la-COVID-19-et-de-la-hausse-des-prix_a3588.html

Heal, G., & Park, J. (2016). Reflections—temperature stress and the direct impact of climate change: A review of an emerging literature. *Review of Environmental Economics and Policy*, *10*(2), 347–362. DOI: 10.1093/reep/rew007

Hemamalini, V., Mishra, A. K., Tyagi, A. K., & Kakulapati, V. (2024). Artificial Intelligence–Blockchain-Enabled–Internet of Things-Based Cloud Applications for Next-Generation Society. In Tyagi, A. K. (Ed.), *Automated Secure Computing for Next-Generation Systems* (1st ed., pp. 65–82). Wiley., DOI: 10.1002/9781394213948.ch4

Herrero, M., Thornton, P. K., Mason-D'Croz, D., Palmer, J., Benton, T. G., Bodirsky, B. L., Bogard, J. R., Hall, A., Lee, B., Nyborg, K., Pradhan, P., Bonnett, G. D., Bryan, B. A., Campbell, B. M., Christensen, S., Clark, M., Cook, M. T., De Boer, I. J. M., Downs, C., & West, P. C. (2020). Innovation can accelerate the transition towards a sustainable food system. *Nature Food*, *1*(5), 266–272. DOI: 10.1038/s43016-020-0074-1

Hertel, T. W. (2015). The challenges of sustainably feeding a growing planet. *Food Security*, *7*(2), 185–198. DOI: 10.1007/s12571-015-0440-2

Hoffman, R. C. (1995). Environmental change and the culture of common carp in medieval Europe. *Guelph Ichthyology Reviews*, ●●●, 3.

Hofisi, C., Chigavazira, B., Mago, S., & Hofisi, M. (2013). "Climate Finance Issues": Implications for climate change adaptation for food security in Southern Africa. *Mediterranean Journal of Social Sciences*. Advance online publication. DOI: 10.5901/mjss.2013.v4n6p47

Holden, L. (2024). From farm to fork? Brexit and the International Plant Protection Convention. *Environmental Law Review*, *26*(2), 185–202. DOI: 10.1177/14614529241259879

Holst-Jensen, A., Bertheau, Y., De Loose, M., Grohmann, L., Hamels, S., Hougs, L., & Van den Bulcke, M. (2012). Detecting un-authorized genetically modified organisms (GMOs) and derived materials. *Biotechnology Advances*, *30*(6), 1318–1335. DOI: 10.1016/j.biotechadv.2012.01.024

Honegger, M., Michaelowa, A., & Roy, J. (2021). Potential implications of carbon dioxide removal for the Sustainable Development Goals. *Climate Policy*, *21*(5), 678–698. DOI: 10.1080/14693062.2020.1843388

Hong, H., Karolyi, G. A., & Scheinkman, J. A. (2020). Climate finance. *Review of Financial Studies*, *33*(3), 1011–1023. DOI: 10.1093/rfs/hhz146

Hong, Z., & Xiao, K. (2024). Digital economy structuring for sustainable development: The role of blockchain and artificial intelligence in improving supply chain and reducing negative environmental impacts. *Scientific Reports*, *14*(1), 3912. DOI: 10.1038/s41598-024-53760-3

Hossain, M. M., Ahmed, S., Alam, M. S., & Hossain, A. (2024). Adverse effects of heat shock in rice (Oryza sativa L.) and approaches to mitigate it for sustainable rice production under the changing climate: A comprehensive review. Heliyon.

Hossain, M. S., Arshad, M., Qian, L., Zhao, M., Mehmood, Y., & Kächele, H. (2019). Economic impact of climate change on crop farming in Bangladesh: An application of Ricardian method. *Ecological Economics*, *164*, 106354. DOI: 10.1016/j.ecolecon.2019.106354

Hou, J., Zhang, M., & Li, Y. (2024). Can digital economy truly improve agricultural ecological transformation? New insights from China. *Humanities & Social Sciences Communications*, *11*(1), 153. DOI: 10.1057/s41599-023-02593-y

Howard, P. H. (2009). Visualizing consolidation in the global seed industry: 1996–2008. *Sustainability (Basel)*, *1*(4), 1266–1287. DOI: 10.3390/su1041266

Howard, P. H. (2015). Intellectual property and consolidation in the seed industry. *Crop Science*, *55*(6), 2489–2495. DOI: 10.2135/cropsci2014.09.0669

Huang, W., Liu, Q., & Abu Hatab, A. (2023). Is the technical efficiency green? The environmental efficiency of agricultural production in the MENA region. *Journal of Environmental Management*, *327*, 116820. DOI: 10.1016/j.jenvman.2022.116820

HUI, D. (2013). Global climate change and biodiversity: Issues and future research. *Journal of Biodiversity and Endangered Species*, *1*, 1–2.

Huynh, C. M., & Hoang, H. H. (2024). Climate change and income inequality in Asia: How does institutional quality matter? *Journal of the Asia Pacific Economy*. Advance online publication. DOI: 10.1080/13547860.2024.2315700

ICARDA. (2018). *Contributing to Egypt's 1.5 Million Feddan project*. ICARDA Communication Team.

ICARDA. (2020). Ultra-low-energy drip irrigation helps small farmers save water, energy, and money! https://icarda.org/impact/impact-stories/ultra-low-energy-drip-irrigation-helps-small-farmers-save-water-energy-and

ICRC. (2022). Libya: Conflict weakens the abilities of farmers to mitigate climate risks. International Committee of the Red Cross.

IEMed. (2021). *Climate change in the MENA region: Environmental risks, socio-economic effects and policy challenges for the future.* IEMed. https://www.iemed.org/wp-content/uploads/2021/01/Climate-Change-in-the-MENA-Region.pdf

IFAD (2022). The State of Libya: Country strategy note 2023 – 2024, Main report and appendices. No. 5683-LY.

ILO (International Report Organization). (2019) "Annual Report 2029" https://www.ilo.org/wcmsp5/groups/public/---ed_emp/documents/publication/wcms_739991.pdf (Accessed30 October 2023)

Imache, A., Bouarfa, S., Kuper, M., Hartani, T., & Dionnet, M. (2009). Integrating "invisible" farmers in a regional debate on water productivity: The case of informal water and land markets in the Algerian Mitidja plain. *Irrigation and Drainage*, *58*(3), 264–272. DOI: 10.1002/ird.523

Impact, E. (2022). *Global Food Security Index 2022: Global report.* Economist Impact. https://impact.economist.com/sustainability/project/food-security-index/resources/Economist_Impact_GFSI_2022_Global_Report_Sep_2022.pdf

IPCC. (2021). *Climate change 2021: The physical science basis. Contribution of Working Group I to the Sixth Assessment Report of the Intergovernmental Panel on Climate Change* (Masson-Delmotte, V., Eds.). Cambridge University Press., https://www.ipcc.ch/report/ar6/wg1/

IPCC. (2022). AR6 Synthesis Report. *Climatic Change*, *2023*, •••. https://www.ipcc.ch/assessment-report/ar6/

IPCC. (2022). *Climate change 2022: Impacts, adaptation, and vulnerability. Contribution of Working Group II to the Sixth Assessment Report of the Intergovernmental Panel on Climate Change*. Cambridge University Press.

IPCC. (2022). *Global Warming of 1.5°C: IPCC Special Report on Impacts of Global Warming of 1.5°C above Pre-industrial Levels in Context of Strengthening Response to Climate Change, Sustainable Development, and Efforts to Eradicate Poverty* (1st ed.). Cambridge University Press., DOI: 10.1017/9781009157940

IPCC. (2023). [Impacts, Adaptation and Vulnerability.]. *Climatic Change*, •••, 2023.

IPCC. (2023). *Climate change 2023: Synthesis report. Contribution of Working Groups I, II and III to the Sixth Assessment Report of the Intergovernmental Panel on Climate Change* (Core Writing Team, H. Lee, & J. Romero, Eds.). IPCC. https://www.ipcc.ch/report/ar6/syr/

IPCC. 2022: *Climate Change 2022: Impacts, Adaptation, and Vulnerability.* Contribution of Working Group II to the Sixth Assessment Report of the Intergovernmental Panel on Climate Change [H.-O. Pörtner, D.C. Roberts, M. Tignor, E.S. Poloczanska, K. Mintenbeck, A. Alegría, M. Craig, S. Langsdorf, S. Löschke, V. Möller, A. Okem, B. Rama (eds.)]. Cambridge University Press. Cambridge University Press, Cambridge, UK and New York, NY, USA, 3056 pp., DOI: 10.1017/9781009325844

Islam, S. N., & Winkel, J. (2017). *Climate change and social inequality* (UN DESA Working Paper No. 152). United Nations Department of Economic and Social Affairs. https://www.un.org/esa/desa/papers/2017/wp152_2017.pdf

Ismail, M. H., Khater, M., & Zaki, M. (2018). Digital Business Transformation and Strategy: What Do We Know So Far? DOI: 10.13140/RG.2.2.36492.62086

Ismail, S. M., & Alheji, A. I. (2023). The Justice Against Sponsors of Terrorism Act (JASTA) from a Civil Procedure Perspective. *Laws, 12*(1), 15. DOI: 10.3390/laws12010015

Ivanov, D., Blackhurst, J., & Das, A. (2021). Supply chain resilience and its interplay with digital technologies: Making innovations work in emergency situations. *International Journal of Physical Distribution & Logistics Management, 51*(2), 97–103. DOI: 10.1108/IJPDLM-03-2021-409

Jacob, B. A., Lefgren, L., & Moretti, E. (2007). The dynamics of criminal behavior: Evidence from weather shocks. *The Journal of Human Resources, 42*(3), 489–527. DOI: 10.3368/jhr.XLII.3.489

Jadalhaq, I. M., & Russi, L. (2020). Finding Direction at the Edge of Law and Life: Islamic Fiqh, Correspondence, and UAE Takāful Insurance Regulation. *Canadian Journal of Law and Society, 35*(3), 477–497. DOI: 10.1017/cls.2020.19

Jagannathan, N. V., Mohamed, A. S., & Kremer, A. (Eds.). (2009). *Water in the Arab world: Management perspectives and innovations.* World Bank., https://documents.worldbank.org/curated/en/590691468052797602

Jarecki, W. (2022). Physiological response of soybean plants to seed coating and inoculation under pot experiment conditions. *Agronomy (Basel), 12*(5), 1095.

Jatuporn, C., & Takeuchi, K. (2023). Assessing the impact of climate change on the agricultural economy in Thailand: An empirical study using panel data analysis. *Environmental Science and Pollution Research International, 30*(3), 8123–8132. DOI: 10.1007/s11356-022-22743-0

Jayachandran, S. (2006). Selling labor low: Wage responses to productivity shocks in developing countries. *Journal of Political Economy, 114*(3), 538–575. DOI: 10.1086/503579

Ji, X., & Li, W. (2022). *Digital Transformation: A Review and Research Framework.*

Jiang, X., & Chen, Y. (2020). The potential of absorbing foreign agricultural investment to improve food security in developing countries. *Sustainability (Basel), 12*(6), 2481. DOI: 10.3390/su12062481

Ji, H., Lee, H. S., hyun Moon, J., Thu, K., Kim, Y.-D., & Jeon, W.-J. (2025). Experimental and theoretical evaluation of a novel hybrid multi-effect adsorption desalination process. *Separation and Purification Technology, 361*, 131522. Advance online publication. DOI: 10.1016/j.seppur.2025.131522

Jinghua, X. (2020). Cross-system Digital Transformation and Adaptive Changes of Management. *Reform, 2020*(4), 37–49.

Joseph, J., & Pandey, A. (2025a). Green Industrialization for the Sustainable Future of Developing Countries: A Critical Study. In *Adaptive Strategies for Green Economy and Sustainability Policies* (pp. 227-244). IGI Global Scientific Publishing.

Joseph, J., & Pandey, A. (2025b). Gender Equality: A Catalyst for Global Sustainable Future. In *Social System Reforms to Achieve Global Sustainability* (pp. 101-126). IGI Global Scientific Publishing. Brush, S. B. (2005). *Farmers' Bounty: Locating Crop Diversity in the Contemporary World.* Yale University Press. DOI: 10.4018/979-8-3373-1280-4.ch005

JUSTICE, I. C. O. 2025. Obligations of States in respect of Climate Change Summary of the Advisory Opinion of 23 July 2025 INTERNATIONAL COURT OF JUSTICE

Kabir, M., Habiba, U. E., Khan, W., Shah, A., Rahim, S., Rios-Escalante, P. R. D. L., Farooqi, Z.-U.-R., Ali, L., & Shafiq, M. (2023). Climate change due to increasing concentration of carbon dioxide and its impacts on environment in 21st century; a mini review. *Journal of King Saud University. Science, 35*(5), 102693. DOI: 10.1016/j.jksus.2023.102693

Kabubo-Mariara, J., & Karanja, F. K. (2007). The economic impact of climate change on Kenyan crop agriculture: A Ricardian approach. *Global and Planetary Change, 57*(3-4), 319–330. DOI: 10.1016/j.gloplacha.2007.01.002

Kamilaris, A., Kartakoullis, A., & Prenafeta-Boldú, F. X. (2018). A review on the practice of big data analysis in agriculture. *Computers and Electronics in Agriculture*, *143*, 23–37. DOI: 10.1016/j.compag.2017.09.037

Kane, G. C., Palmer, D., & Phillips, A. N. (2017). *Achieving Digital Maturity* [Technical Report]. MIT Sloan Management Review. https://sadil.ws/handle/123456789/1453

Kang, Y., Chen, Z., Li, L., & Zhang, Q. (2023). Construction of multidimensional features to identify tea plantations using multisource remote sensing data: A case study of Hangzhou city, China. *Ecological Informatics*, *77*, 102185. DOI: 10.1016/j.ecoinf.2023.102185

Kang, Y., Khan, S., & Ma, X. (2009). Climate change impacts on crop yield, crop water productivity and food security – A review. *Progress in Natural Science*, *19*(12), 1665–1674. DOI: 10.1016/j.pnsc.2009.08.001

Karrou, M., Oweis, T., & Bahri, A. (2011). *Improving water and land productivity in rainfed systems*. ICARDA Policy Briefs.

Kassie, M., Shiferaw, B., & Muricho, G. (2011). Agricultural technology, crop income, and poverty alleviation in Uganda. *World Development*, *39*(10), 1784–1795. DOI: 10.1016/j.worlddev.2011.04.023

Khalifa, J. (2025). The Impacts of Climate Change, Agricultural Productivity, and Food Security on Economic Growth in Tunisia: Evidence from an Econometrics Analysis. *Research on World Agricultural Economy*, 577–599. DOI: 10.36956/rwae.v6i1.1457

Khan, M. A., Tahir, A., Khurshid, N., Husnain, M. I., Ahmed, M., & Boughanmi, H. (2020). Economic Effects of Climate Change-Induced Loss of Agricultural Production by 2050: A Case Study of Pakistan. *Sustainability (Basel)*, *12*(3), 1216. DOI: 10.3390/su12031216

Khan, M. Y., Nadeem, S. M., Sohaib, M., Waqas, M. R., Alotaibi, F., Ali, L., & Al-Barakah, F. N. (2022). Potential of plant growth promoting bacterial consortium for improving the growth and yield of wheat under saline conditions. *Frontiers in Microbiology*, *13*, 958522.

Khater, M. H., Imad, D., Abouhaiba, N., & Tannar, H. Legal Environmental Protection and Sustainable Development in the United Arab Emirates: A Comparative Study with International Frameworks. *Available at SSRN* 5119294. DOI: 10.2139/ssrn.5119294

Kherbache, N. (2020). Water policy in Algeria: Limits of supply model and perspectives of water demand management (WDM). *Desalination and Water Treatment, 180*, 141–155. DOI: 10.5004/dwt.2020.25009

Kherbache, N., & Molle, F. (2021). *La surexploitation des ressources en eau: le cas du bassin de la Macta en Algérie*. IRD.

Kherbache, N., & Molle, F. (2022). Causes and consequences of the Macta basin closure, Algeria. *International Journal of Water Resources Development, 0*, 1–22.

Kloppenburg, J. (2010). Impeding dispossession, enabling repossession: Biological open source and the recovery of seed sovereignty. *Journal of Agrarian Change, 10*(3), 367–388. DOI: 10.1111/j.1471-0366.2010.00275.x

Knippertz, P., Christoph, M., & Speth, P. (2003). Long-term precipitation variability in Morocco and the link to the large-scale circulation in recent and future climates. *Meteorology and Atmospheric Physics, 83*(1), 67–88. DOI: 10.1007/s00703-002-0561-y

Kobayakawa, T. (2021). Country diagnostics for low carbon development: Can developing countries pursue simultaneous implementation of the Sustainable Development Goals and the Paris Agreement? *Business Strategy & Development, 4*(3), 294–312. DOI: 10.1002/bsd2.159

Kochskämper, E., Glass, L.-M., Haupt, W., Malekpour, S., & Grainger-Brown, J. (2024). Resilience and the Sustainable Development Goals: A scrutiny of urban strategies in the 100 Resilient Cities Initiative. *Journal of Environmental Planning and Management*, •••, 1–27. DOI: 10.1080/09640568.2023.2297648

Kopańska, A., Osinski, R., & Korbus, B. (2024). Private entities motivations to participate in public-private partnerships. *Socio-Economic Planning Sciences, 92*, 101841. Advance online publication. DOI: 10.1016/j.seps.2024.101841

Krantz, V., & Gustafsson, S. (2021). Localizing the Sustainable Development Goals through an integrated approach in municipalities: Early experiences from a Swedish forerunner. *Journal of Environmental Planning and Management, 64*(14), 2641–2660. DOI: 10.1080/09640568.2021.1877642

Krzymowski, A. (2020). Sustainable development goals in Arab region–United Arab Emirates' case study. *Problemy Ekorozwoju, 15*(1), 211–220. DOI: 10.35784/pe.2020.1.22

Kunming-Montreal Global Biodiversity Framework. (2025). *Kunming-Montreal Global Biodiversity Framework.* https://unctad.org/topic/trade-and-environment/biotrade/kunming-montreal-global-biodiversity-framework

Kürklü, A., Pearson, S., & Felek, T. (2025). Climate change impacts on tomato production in high-tech soilless greenhouses in Türkiye. *BMC Plant Biology*, *25*(1), 339. Advance online publication. DOI: 10.1186/s12870-025-06307-1 PMID: 40089667

Kurukulasuriya, P., & Mendelsohn, R. O. (2007). A Ricardian analysis of the impact of climate change on African cropland. *World Bank Policy Research Working Paper*, •••, 4305. DOI: 10.1596/1813-9450-4305

Kurukulasuriya, P., Mendelsohn, R., Hassan, R., Benhin, J., Deressa, T., Diop, M., Eid, H. M., Fosu, K. Y., Gbetibouo, G., Jain, S., Mahamadou, A., Mano, R., Kabubo-Mariara, J., El-Marsafawy, S., Molua, E., Ouda, S., Ouedraogo, M., Séne, I., Maddison, D., & Dinar, A. (2006). Will African agriculture survive climate change? *The World Bank Economic Review*, *20*(3), 367–388. DOI: 10.1093/wber/lhl004

La loi n°84-21 relative à l'aquaculture marine, B.O N°7174, du 9 chaabane 1444 (2 mars 2023).

Lachgar, R., Badri, W., & Chlaida, M. (2022). Assessment of future changes in downscaled temperature and precipitation over the Casablanca-Settat region (Morocco). *Modeling Earth Systems and Environment*, *8*(2), 2123–2133. DOI: 10.1007/s40808-021-01213-5

Lacy, W. B. (2006). Agricultural biotechnology, societal implications, and the need for public dialogue. *Agriculture and Human Values*, *23*(3), 273–280.

Lafortune, G., Fuller, G., Schmidt-Traub, G., & Kroll, C. (2020). How is progress towards the Sustainable Development Goals measured? Comparing four approaches for the EU. *Sustainability (Basel)*, *12*(18), 7675. DOI: 10.3390/su12187675

Laith, A.-E., & Ali, H. (2025). Balancing Innovation, Biosecurity, and Biosafety in Dual-Use Research in the Middle East: Dual-Use Research in the Middle East. *Global Biosecurity*.

Laith, A.-E., Alnemri, M., Ali, H., Alkhawaldeh, M., & Mihyar, A. (2022)... *Journal of Biosafety and Biosecurity*.

Lange, M. A. (2019). Impacts of climate change on the Eastern Mediterranean and the Middle East and North Africa region and the water–energy nexus. *Atmosphere (Basel)*, *10*(8), 455. DOI: 10.3390/atmos10080455

Laoubi, K., & Yamao, M. J. W. T. (2009). Management of irrigation schemes in Algeria: An assessment of water policy impact and perspectives on development. *WIT Transactions on Ecology and the Environment*, *125*, 503–514. DOI: 10.2495/WRM090451

Larrieu, D. (2007). L'interprétation des pactes extra-statutaires. *Revue des Sociétés,* (4).

Laurans, Y. (2006). Role of economic analysis in the Water Framework Directive. *Houille Blanche*, (4), 75–80. DOI: 10.1051/lhb:200604012

Le Mouël, C., Forslund, A., Marty, P., Manceron, S., Marajo-Petitzon, E., Caillaud, M. A., Dumas, P., & Schmitt, B. (2023). Can the Middle East-North Africa region mitigate the rise of its food import dependency under climate change? *Regional Environmental Change*, *23*(2), 52. DOI: 10.1007/s10113-023-02045-y

Lee, B. X., Kjaerulf, F., Turner, S., Cohen, L., Donnelly, P. D., Muggah, R., & Gilligan, J. (2016). Transforming our world: Implementing the 2030 Agenda through Sustainable Development Goal indicators. *Journal of Public Health Policy*, *37*(S1), 13–31. DOI: 10.1057/s41271-016-0002-7

Lee, C.-H., Liu, C.-L., Trappey, A. J. C., Mo, J. P. T., & Desouza, K. C. (2021). Understanding digital transformation in advanced manufacturing and engineering: A bibliometric analysis, topic modeling and research trend discovery. *Advanced Engineering Informatics*, *50*, 101428. DOI: 10.1016/j.aei.2021.101428

Lee, S. T., & Jung, S. M. (2025). Social return on investment as a tool for environmental impact and strategic assessments: Evidence from South Korea. *Environmental Impact Assessment Review*, *112*, 107765. Advance online publication. DOI: 10.1016/j.eiar.2024.107765

LENTNER, G. M. & CENIN, W. 2024. Daniel Billy et al v Australia (Torres Strait Islanders Petition): Climate change inaction as a human rights violation. Review of European, Comparative & International Environmental Law, 33, 136–143.

Leray, G. (2023). L' ineffectivité du droit à l'eau. *Recueil Dalloz,* (44).

Levidow, L., Zaccaria, D., Maia, R., Vivas, E., Todorovic, M., & Scardigno, A. (2014). Improving water-efficient irrigation: Prospects and difficulties of innovative practices. *Agricultural Water Management*, *146*, 84–94. DOI: 10.1016/j.agwat.2014.07.012

Li, Y. O., & Shi, X.-M. (2022). Chapter 5 – Food safety regulations within countries of increasing global supplier impact. In A. Martinović, S. Oh, & H. Lelieveld (Eds.), Ensuring global food safety (2nd ed., pp. 151-158). Academic Press. June 26, 2024. The unconstitutional Federal Government omission to protect the Amazon from deforestation. Brazil Federal Supreme Court. March 1, 2025. Vivek Kumar Sharma vs The State Of Madhya Pradesh. India High Court of Madhya Pradesh.

Li, F., Nucciarelli, A., Roden, S., & Graham, G. (2016). How smart cities transform operations models: A new research agenda for operations management in the digital economy. *Production Planning and Control*, *27*(6), 514–528. DOI: 10.1080/09537287.2016.1147096

Li, M., Zhou, S., Shen, S., Wang, J., Yang, Y., Wu, Y., Chen, F., & Lei, Y. (2024). Climate-smart irrigation strategy can mitigate agricultural water consumption while ensuring food security under a changing climate. *Agricultural Water Management*, *292*, 108663. DOI: 10.1016/j.agwat.2023.108663

Linditch, F. (2024). Responsabilité contractuelle – Fait générateur In Dalloz (Ed.), *Répertoire de la responsabilité de la puissance publique* (Vol. R), Dalloz: Paris.

Liu, L., Cavaye, J., & Ariyawardana, A. (2022). Supply chain responsibility in agriculture and its integration with rural community development: A review of issues and perspectives. *Journal of Rural Studies*, *93*, 134–143. DOI: 10.1016/j.jrurstud.2020.07.003

Liu, M., & Yan, A. (2022). Feasibility analysis of water resources market transaction of water diversion project based on evolutionary game. *Water Science and Technology: Water Supply*, *22*(3), 3061–3071. DOI: 10.2166/ws.2021.423

Liu, X., Yin, B., Bao, X., Hou, X., Wang, T., Shang, C., Yang, M., & Zhen, W. (2024). Optimization of irrigation period improves wheat yield by regulating source-sink relationship under water deficit. *European Journal of Agronomy*, *156*, 127164. DOI: 10.1016/j.eja.2024.127164

Locke, J., Dsilva, J., & Zarmukhambetova, S. (2023). Decarbonization strategies in the UAE built environment: An evidence-based analysis using COP26 and COP27 recommendations. *Sustainability (Basel)*, *15*(15), 11603. DOI: 10.3390/su151511603

Loebbecke, C., & Picot, A. (2015). Reflections on societal and business model transformation arising from digitization and big data analytics: A research agenda. *The Journal of Strategic Information Systems*, *24*(3), 149–157. DOI: 10.1016/j.jsis.2015.08.002

Lucas, V., Gasselin, P., & Van Der Ploeg, J. D. (2019). Local inter-farm cooperation: A hidden potential for the agroecological transition in northern agricultures. *Agroecology and Sustainable Food Systems*, *43*(2), 145–179. DOI: 10.1080/21683565.2018.1509168

Lu, M., Lall, U., Robertson, A. W., & Cook, E. (2017). Optimizing multiple reliable forward contracts for reservoir allocation using multitime scale streamflow forecasts. *Water Resources Research*, *53*(3), 2035–2050. DOI: 10.1002/2016WR019552

Luu, T. M. N., Yadav, M., Srivastava, A. P., & Gopal, K. (2025). Innovation and Technology to Address the Challenges of Climate Change. In *Effects of Climate Change on Social and Economic Factors* (pp. 423–464). IGI Global.

Maamri, K., Djerroudi-Zidane, O., Chaabena, A., & Bazile, D. (2021). Preliminary evaluation of Quinoa genotypes introduced in the Saharan zone of Algeria, case of Ouargla. In *Euro-Mediterranean Conference for Environmental Integration* (pp. 283–286). Springer Nature Switzerland.

Macheroum, A., & Chenchouni, H. (2022). Short-term land degradation driven by livestock grazing does not affect soil properties in semiarid steppe rangelands. *Frontiers in Environmental Science*, *10*, 846045. DOI: 10.3389/fenvs.2022.846045

Madi, R., Barafi, J., & Abu Saleh, M. (2020). Releasing GMOs into the environment or launching its products into markets: Regional and international legislations. *Industrial Biotechnology (New Rochelle, N.Y.)*, *16*(6), 359–367. DOI: 10.1089/ind.2020.29229.rma

MADR. (2022). *Statistiques agricoles de l'Algérie: Série B*. Ministère de l'Agriculture et du Développement Rural.

Mahboub, H., & Sadok, H. (2024). Digital transformation of the moroccan economy: A performance comparative approach. *Maghreb-Mashreq International*, *4/2023*, 45–62. DOI: 10.54695/mmi.4.0045

Mahdhi, N., Sghaier, M., & Smida, Z. (2014). Efficiency of the irrigation water user association in the Zeuss-Koutine region, south-eastern Tunisia. *New Medit*, *13*(2), 47–55.

Malkawi, H. I., & Kapiel, T. Y. S. (2024). Microbial biotechnology: A key tool for addressing climate change and food insecurity. *European Journal of Biology and Biotechnology*, *5*(2), 1–15.

Mansourian, S., Dudley, N., & Vallauri, D. (2017). Forest landscape restoration: Progress in the last decade and remaining challenges. *Ecological Restoration*, *35*(4), 281–288. DOI: 10.3368/er.35.4.281

Manucharyan, M. (2025). Climate change impacts on sustainable agriculture: Evidence from Armenia. *Unconventional Resources*, *6*, 100159. DOI: 10.1016/j.uncres.2025.100159

Markkanen, S., & Anger-Kraavi, A. (2019). Social impacts of climate change mitigation policies and their implications for inequality. *Climate Policy*, *19*(7), 827–844. DOI: 10.1080/14693062.2019.1596873

Martínez-Valderrama, J., Ibáñez, J., Del Barrio, G., Alcalá, F. J., Sanjuán, M. E., Ruiz, A., & Puigdefábregas, J. (2018). Doomed to collapse: Why Algerian steppe rangelands are overgrazed and some lessons to help land-use transitions. *The Science of the Total Environment*, *613*, 1489–1497. DOI: 10.1016/j.scitotenv.2017.07.058

Matt, C., Hess, T., Benlian, A., & Wiesbock, F. (2016). Options for Formulating a Digital Transformation Strategy. *MIS Quarterly Executive*, *15*(2). https://aisel.aisnet.org/misqe/vol15/iss2/6

Mauvais, F. (2008). The delegated management: An inventive and efficient way for water and waste water public utilities management. *Revue des Sciences de l'Eau*, *21*(2), 143–153. DOI: 10.7202/018463ar

Mawcha, K. T., Kyampaire, D., Marciale, C., Simiyu-Wafukho, S., Chinyama, C., Babalola, O. O., Kinyanjui, G., & Ndolo, D. (2025). An overview of biopesticide regulatory frameworks in selected countries in Southern Africa. *Frontiers in Sustainable Food Systems*, *9*, 1522526. Advance online publication. DOI: 10.3389/fsufs.2025.1522526

Mazoyer, M., & Roudart, L. (2006). *A History of World Agriculture: From the Neolithic Age to the Current Crisis*. Monthly Review Press.

McDowell, C., & de Hess, M. (2012). Social vulnerability and disaster: Understanding pathways to resilience. *Disasters*, *36*(1), 1–18. DOI: 10.1111/j.1467-7717.2011.01255.x PMID: 21702892

McFadden, J., Casalini, F., Griffin, T., & Anton, J. (2022). *The digitalisation of agriculture: A literature review and emerging policy issues* (OECD Food, Agriculture and Fisheries Papers 176; OECD Food, Agriculture and Fisheries Papers, Vol. 176). DOI: 10.1787/18156797

McHughen, A. (2016). A critical assessment of regulatory triggers for products of biotechnology: Product vs. process. *GM Crops and Food: Biotechnology in Agriculture and the Food Chain*, *7*(3-4), 125–158. DOI: 10.1080/21645698.2016.1228516

McKee, T. B., Doesken, N. J., & Kleist, J. (1993, January). The relationship of drought frequency and duration to time scales. In *Proceedings of the 8th Conference on Applied Climatology* (Vol. 17, No. 22, pp. 179-183).

MEER. (2023). *Premier rapport biennal actualise de l'Algérie: à la convention cadre des nations unies sur le changement climatique*. Ministère de l'Environnement et des Énergies Renouvelables.

Mejri, R., Dhraief, M. Z., Souissi, A., Dhehibi, B., Oueslati, M., Charry, A. C., Frija, A., Ouerghemmi, H., Oumer, A. M., Fendri, M., & Larbi, A. (2025). Empowering smallholder olive growers in northwest Tunisia through an agroecological business model. *Frontiers in Sustainable Food Systems*, *9*, 1587318. Advance online publication. DOI: 10.3389/fsufs.2025.1587318

Mendelsohn, R., & Dinar, A. (2009). *Climate change and agriculture: An economic analysis of global impacts, adaptation, and distributional effects*. Edward Elgar Publishing. DOI: 10.4337/9781849802239

Mendelsohn, R., Nordhaus, W. D., & Shaw, D. (1994). The impact of global warming on agriculture: A Ricardian analysis. *The American Economic Review*, *84*(4), 753–771. https://www.jstor.org/stable/2118029

Merabtene, M. D., Faraoun, F., Mlih, R., Djellouli, R., Latreche, A., & Bol, R. (2021). Forest soil organic carbon stocks of Tessala Mount in North-West Algeria-Preliminary estimates. *Frontiers in Environmental Science*, *8*, 520284. DOI: 10.3389/fenvs.2020.520284

Merniz, N., Tahar, A., & Benmehaia, A. M. (2019). Statistical assessment of rainfall variability and trends in northeastern Algeria. *Journal of Water and Land Development*, *40*(1), 87–96. DOI: 10.2478/jwld-2019-0009

Meyer, E. S., Characklis, G. W., & Brown, C. (2017). Evaluating financial risk management strategies under climate change for hydropower producers on the Great Lakes. *Water Resources Research*, *53*(3), 2114–2132. DOI: 10.1002/2016WR019889

Meyghani, S., Khodaparast Mashadi, M., & Salehnia, N. (2021). The Impacts of Climate Change on Value-Added Agriculture in the MENA Region. *Journal of Economics and Regional Development*, *27*(20), 129–158.

Michaelowa, K., & Michaelowa, A. (2019). Transnational climate governance initiatives: Designed for effective climate change mitigation? In Bulkeley, H., & Andonova, L. B. (Eds.), *The comparative politics of transnational climate governance* (pp. 146–178). Routledge., DOI: 10.4324/9781351135436-6

Michelato, A., & Pacini, G. C. (2025). Agroecology in Brazil and Italy: Comparative analysis of the historical formation process of agroecology. *Studia Ecologiae et Bioethicae*, *23*(1), 89–99. DOI: 10.21697/seb.5836

Migliore, G., Zinnanti, C., Schimmenti, E., Borsellino, V., Schifani, G., Di Franco, C. P., & Asciuto, A. (2019). A Ricardian analysis of the impact of climate change on permanent crops in a Mediterranean region. *New Medit*, *18*(1), 41–52. DOI: 10.30682/nm1901d

Miguel, E., Satyanath, S., & Sergenti, E. (2004). Economic shocks and civil conflict: An instrumental variables approach. *Journal of Political Economy*, *112*(4), 725–753. DOI: 10.1086/421174

Milian Gómez, J. F. (2024). Rethinking the human right to food from a single perspective to a four-fold legal interpretation. *Journal of Human Rights Practice*, *16*(2), 589–602. DOI: 10.1093/jhuman/huad047

Mills-Novoa, M., & Liverman, D. M. (2019). Nationally determined contributions: Material climate commitments and discursive positioning in the NDCs. *Wiley Interdisciplinary Reviews: Climate Change*, *10*(5), e589. DOI: 10.1002/wcc.589

Miltiadou, M., Grieve, S., Ruiz-Benito, P., Astigarraga, J., Cruz-Alonso, V., Tijerín Triviño, J., & Lines, E. R. (2025). PlotToSat: A tool for generating time-series signatures from Sentinel-1 and Sentinel-2 at field-based plots for machine learning applications. *Environmental Modelling & Software*, *188*, 106395. DOI: 10.1016/j.envsoft.2025.106395

Ministère des Ressources en Eau. (2017). *Plan National de l'Eau: PNE*. MRE.

Ministry of Climate Change & Environment (MOCCAE). (2021). *The United Arab Emirates general environmental policy*. Ministry of Climate Change & Environment.

Ministry of Climate Change & Environment (MOCCAE). (2022). *Updated second nationally determined contribution of the United Arab Emirates*. Ministry of Climate Change & Environment.

Ministry of Climate Change & Environment (MOCCAE). (2023). *Third update of second NDC for the UAE*. Ministry of Climate Change & Environment.

Ministry of Climate Change and Environment. (2024). *National Air Quality Agenda 2031*. Government of the United Arab Emirates. https://u.ae/en/about-the-uae/strategies-initiatives-and-awards/strategies-plans-and-visions/environment-and-energy/the-national-air-quality-agenda-2031

Mishra, A. K., & Singh, V. P. (2010). A review of drought concepts. *Journal of Hydrology (Amsterdam)*, *391*(1-2), 202–216.

MIT Technology Review. (2023). *The Green Future Index 2023*. https://mittrinsights.s3.amazonaws.com/GFI23report.pdf

Mittal, S., & Mehar, M. (2016). Socio-economic factors affecting adoption of modern information and communication technology by farmers in India: Analysis using multivariate probit model. *Journal of Agricultural Education and Extension*, *22*(2), 199–212. DOI: 10.1080/1389224X.2014.997255

Mittler, R. (2006). Abiotic stress, the field environment and stress combination. *Trends in Plant Science*, *11*(1), 15–19. DOI: 10.1016/j.tplants.2005.11.002 PMID: 16359910

Mohanavelu, A., Naganna, S. R., & Al-Ansari, N. (2021). Irrigation induced salinity and sodicity hazards on soil and groundwater: An overview of its causes, impacts and mitigation strategies. *Agriculture*, *11*(10), 983. DOI: 10.3390/agriculture11100983

Mokhtari, S., Benmehaia, A. M., Nourani, A., & Hadjeb, A. (2025). Future water requirements of the wheat crop in Northern Algeria under climate change scenarios. *Regional Environmental Change*, *25*(2), 1–14. DOI: 10.1007/s10113-025-02378-w

Monsanto Canada Inc. v. Schmeiser, [2004] 1 S.C.R. 902

Moon, J., Son, S., Kim, J., & Park, K. (2025). Critical challenges in high-salinity seawater reverse osmosis systems: Technical, energy, and environmental reviews. *Desalination*, *607*, 118811. Advance online publication. DOI: 10.1016/j.desal.2025.118811

Morakanyane, R., Grace, A., & O'Reilly, P. (2017). Conceptualizing Digital Transformation in Business Organizations: A Systematic Review of Literature. *BLED 2017 Proceedings*. https://aisel.aisnet.org/bled2017/21

Morsli, B., Habi, M., & Meddi, M. (2013). Dynamique de l'érosion en zone méditerranéenne algérienne: Facteurs explicatifs de variation du ruissellement et de l'érosion sous différentes occupations du sol. *Revue des Sciences de l'Eau*, *26*(2), 89–105. DOI: 10.7202/1016061ar

Mouniane, Y., Chriqui, A., El-Khadir, I., Hbyaj, K., El-Moustaqim, K., Mezouara, M., Zirari, M., Mabrouki, J., & Hmouni, D. (2024). Impact of remote sensing on irrigation management in precision agriculture. In Mabrouki, J., & Mourade, A. (Eds.), *Technical and technological solutions towards a sustainable society and circular economy* (pp. 237–245). Springer., DOI: 10.1007/978-3-031-56292-1_19

Mourtzis, D., Angelopoulos, J., & Panopoulos, N. (2022). A Literature Review of the Challenges and Opportunities of the Transition from Industry 4.0 to Society 5.0. *Energies*, *15*(17), 6276. DOI: 10.3390/en15176276

Moussaoui, A., Abbas, K., Bottalico, F., & Ottomano, G. (2021). Pastoralism in Algeria: transformation, challenges and prospects. *Pastoralism and Sustainable Development*, 189.

Mozas, M., & Ghosn, A. (2013). *État des lieux du secteur de l'eau en Algérie*. Études & Analyses, Ipemed.

Msimbira, L. A., & Smith, D. L. (2020). The roles of plant growth promoting microbes in enhancing plant tolerance to acidity and alkalinity stresses. *Frontiers in Sustainable Food Systems*, *4*, 106. DOI: 10.3389/fsufs.2020.00106

Müller, C., Cramer, W., Hare, W. L., & Lotze-Campen, H. (2011). Climate change risks for African agriculture. *Proceedings of the National Academy of Sciences of the United States of America*, *108*(11), 4313–4315. DOI: 10.1073/pnas.1015078108 PMID: 21368199

Myers-Smith, I. H., Grabowski, M. M., Thomas, H. J., Angers-Blondin, S., Daskalova, G. N., Bjorkman, A. D., Cunliffe, A. M., Assmann, J. J., Boyle, J. S., McLeod, E., McLeod, S., Joe, R., Lennie, P., Arey, D., Gordon, R. R., & Eckert, C. D. (2019). Eighteen years of ecological monitoring reveals multiple lines of evidence for tundra vegetation change. *Ecological Monographs*, *89*(2), e01351. DOI: 10.1002/ecm.1351

Mysiak, J., & Gómez, C. M. (2015). Water pricing and taxes: An introduction. In Lago, M., Mysiak, J., Gómez, C. M., Delacámara, G., & Maziotis, A. (Eds.), *Use of economic instruments in water policy* (Vol. 14, pp. 15–32). Springer., DOI: 10.1007/978-3-319-18287-2_2

Naik, B. M., Singh, A. K., Venkatesan, P., Maji, S., Sunil, J., & Naik, M. R. (2025). Assessing the contribution of climate resilient agricultural technologies in advancing sustainable development goals in Telangana, India. *Discover Sustainability*, *6*(1), 117. Advance online publication. DOI: 10.1007/s43621-025-00870-1

Namdar, R., Karami, E., & Keshavarz, M. (2021). Climate change and vulnerability: The case of MENA countries. *ISPRS International Journal of Geo-Information*, *10*(11), 794. DOI: 10.3390/ijgi10110794

Narayanan, S.NARAYANAN. (2015). Food security in India: The imperative and its challenges. *Asia & the Pacific Policy Studies*, *2*(1), 197–209. DOI: 10.1002/app5.62

Nasser, A. E., & Ghloom, K. A. (2023). Ensuring Environmental Safety through Sustainable Food Safety. *Ajrsp*, *5*(54), 141–155. DOI: 10.52132/Ajrsp.e.2023.54.6

Nawar, B., Scardigno, A., & Sagardoy, J. A. (2024). Treatment and reuse of water: Economic feasibility and assessment of water pricing policies in Ouardanine irrigation district (Tunisia). *New Medit*, *23*(1), 39–74. DOI: 10.30682/nm2401e

Nawaz, M. A., Chung, G., Golokhvast, K. S., & Tsatsakis, A. M. (2022). *GMOs and Political Stance: Global GMO Regulation, Certification, Labeling, and Consumer Preferences*. Academic Press.

Nedjraoui, D., & Bédrani, S. (2008, April). La désertification dans les steppes algériennes: Causes, impacts et actions de lutte. *Vertigo*, *8*(1), •••. http://vertigo.revues.org/index448.html

Nelson, G. C., Rosegrant, M. W., Palazzo, A., Gray, I., Ingersoll, C., Robertson, R. D., Tokgoz, S., Zhu, T., Sulser, T. B., Ringler, C., Msangi, S., & You, L. (2010). Food Security, Farming, and Climate Change to 2050: Scenarios, Results, Policy Options. Washington, DC: International Food Policy Research Institute (IFPRI). https://doi.org/10.2499/9780896291867

Ngure, M. W., Wandiga, S. O., Olago, D. O., & Oriaso, S. O. (2021). Climate change stressors affecting household food security among Kimandi-Wanyaga smallholder farmers in Murang'a County, Kenya. *Open Agriculture*, *6*(1), 587–608. DOI: 10.1515/opag-2021-0042

Nguyen, N. T. T., Nguyen, P. V., Vrontis, D., & Vo, N. T. T. (2025). Enhancing organizational sustainable performance through green innovation: The roles of knowledge application, government policy, and green market orientation. *Journal of Knowledge Management*, *29*(3), 870–890. DOI: 10.1108/JKM-06-2024-0688

Nihayati, A. (2025). *Integrating Halal Certification in Free Trade Agreements: Indonesia's Modest Fashion and Cosmetics* (Vol. 18). Jurnal Hubungan Internasional.

Nikolaou, G., Neocleous, D., Christou, A., Kitta, E., & Katsoulas, N. (2020). Implementing sustainable irrigation in water-scarce regions under the impact of climate change. *Agronomy (Basel)*, *10*(8), 1120. DOI: 10.3390/agronomy10081120

Nishimoto, K., Sampaio, C. M. P., Vale, R. J., & Ruggeri, F. (2021). Numerical Simulation of Hybrid Platform Supply Vessel (PSV) Fuel Consumption for the Pre-Salt Layer in Brazil. In Okada T., Kawamura Y., & Suzuki K. (Eds.), *Lect. Notes Civ. Eng.: Vol. 65 LNCE* (p. 497). Springer Science and Business Media Deutschland GmbH; Scopus. DOI: 10.1007/978-981-15-4680-8_34

Nkedianye, D., de Leeuw, J., Ogutu, J. O., Said, M. Y., Saidimu, T. L., Kifugo, S. C., Kaelo, D. S., & Reid, R. S. (2011). Mobility and livestock mortality in communally used pastoral areas: the impact of the 2005-2006 drought on livestock mortality in Maasailand. *Pastoralism: Research. Pastoralism*, *1*(1), 1–17. DOI: 10.1186/2041-7136-1-17

Noack, F., Wunder, S., Angelsen, A., & Börner, J. (2015). Responses to weather and climate: a cross-section analysis of rural incomes. *World Bank Policy Research Working Paper*, (7478).

Nobanee, H., El Hadi El Maknouzi, M., Sadok, H., & Alodat, A. Y. (2024). Analysis of insurance entrepreneurship as a hedge in times of crisis: A literature review. *Sustainable Technology and Entrepreneurship, 3*(2), 100065. Advance online publication. DOI: 10.1016/j.stae.2023.100065

Nordhaus, W. (2019). Climate Change: The Ultimate Challenge for Economics. *The American Economic Review, 109*(6), 1991–2014. DOI: 10.1257/aer.109.6.1991

North, D. C. (1990). *Institutions, Institutional Change and Economic Performance*. Cambridge University Press. DOI: 10.1017/CBO9780511808678

Nwankpa, J. K., & Roumani, Y. (2016). IT Capability and Digital Transformation: A Firm Performance Perspective. *International Conference of Information Systems*.

O'Neill, B. C., Kriegler, E., Ebi, K. L., Kemp-Benedict, E., Riahi, K., Rothman, D. S., van Ruijven, B. J., van Vuuren, D. P., Birkmann, J., Kok, K., Levy, M., & Solecki, W. (2017). The roads ahead: Narratives for shared socioeconomic pathways describing world futures in the 21st century. *Global Environmental Change, 42*, 169–180. DOI: 10.1016/j.gloenvcha.2015.01.004

OECD & Food and Agriculture Organization of the United Nations. (2023). *OECD-FAO Business Handbook on Deforestation and Due Diligence in Agricultural Supply Chains*. OECD. DOI: 10.1787/c0d4bca7-en

OECD (2020). Climate Impacts on African Hydropower. IEA report.

OECD/FAO. (2018). *OECD-FAO agricultural outlook 2018–2027*. OECD Publishing., DOI: 10.1787/agr_outlook-2018-

Olive Tree Institute (OTI). (2012). Olive tree irrigation manual: Techniques. Ministry of Agriculture, Institution of Agricultural Research and Higher Education. http://www.iosfax.agrinet.tn/annonce/Manuel%20d'irrigation%20agencee.pdf

ONAGRI (National Observatory of Agriculture). (2021). National report of the water sector. Ministry of Agriculture, Water Resources and Fisheries, Office of Planning and Hydraulic Balancing, Republic of Tunisia. http://www.onagri.tn/uploads/secteur-eau/RNE_2021_V-Anglaise.pdf

Onnis, E. (2023). Food and freedom of choice: Towards new food narratives compatible with climate mitigation. *Rivista di Estetica*, (74), 123–144. DOI: 10.4000/estetica.15447

OSS. (2003). *Système aquifère du Sahara septentrional: Une conscience de bassin. 2e édition. Hydrogéologie* (Vol. II). Observatoire du Sahara et du Sahel.

OSS. (2021). *Northwestern Sahara Aquifer System: Diagnostic Report*. Sahara and Sahel Observatory.

Ostrom, E. (1990). *Governing the Commons*. Cambridge University Press. DOI: 10.1017/CBO9780511807763

Ouaadi, N., Jarlan, L., Khabba, S., Le Page, M., Chakir, A., Er-Raki, S., & Frison, P.-L. (2023). Are the C-band backscattering coefficient and interferometric coherence suitable substitutes of NDVI for the monitoring of the FAO-56 crop coefficient? *Agricultural Water Management*, *282*, 108276. DOI: 10.1016/j.agwat.2023.108276

Oubadi, M., & Faci, M. (2025). Analysis of the evolution of heatwaves recordings in Algeria during the period 1979–2023. *Theoretical and Applied Climatology*, *156*(4), 1–9. DOI: 10.1007/s00704-025-05456-5

Ouerghemmi, H., Frija, A., Souissi, A., Carpentier, I., Shiri, Z., Dhehibi, B., & Rejeb, H. (2024). Are farmer organizations effective intermediaries and facilitators of agricultural innovations processes? Evidence from Tunisia. *Frontiers in Sustainable Food Systems*, *8*, 1440086. DOI: 10.3389/fsufs.2024.1440086

Ouessar, M., Sghaier, A., Frija, A., Sghaier, M., & Baig, M. B. (2021). Impacts of climate change on agriculture and food security in Tunisia: challenges, existing policies, and way forward. *Emerging Challenges to Food Production and Security in Asia, Middle East, and Africa: Climate Risks and Resource Scarcity*, 65-99.

Ouessar, M., Sghaier, A., Frija, A., Sghaier, M., & Baig, M. B. (2021). Impacts of climate change on agriculture and food security in Tunisia: challenges, existing policies, and way forward. In *Emerging Challenges to Food Production and Security in Asia, Middle East, and Africa: Climate Risks and Resource Scarcity* (pp. 65–99). Springer International Publishing. DOI: 10.1007/978-3-030-72987-5_3

Oulmane, A., Chebil, A., Frija, A., & Benmehaia, M. A. (2020). Water-saving technologies and total factor productivity growth in small horticultural farms in Algeria. *Agricultural Research*, *9*(4), 585–591. DOI: 10.1007/s40003-019-00446-2

Oulmane, A., Kechar, A., Benmihoub, A., & Amine Benmehaia, M. (2022). Assessment of surface water management institutions: A case of public irrigation schemes in northern Algeria. *Water Policy*, *24*(2), 229–241. DOI: 10.2166/wp.2022.025

Ouraich, I., & Tyner, W. E. (2014) Climate Change Impacts on Moroccan Agriculture and the Whole Economy: An Analysis of the Impacts of the Plan Maroc Vert in Morocco. WIDER Working Paper 2014/083. Helsinki: UNU-WIDER. https://doi.org/DOI: 10.35188/UNU-WIDER/2014/804-9

Oustani, M., Mehda, S., Halilat, M. T., & Chenchouni, H. (2023). Yield, growth development and grain characteristics of seven Quinoa (Chenopodium quinoa Willd.) genotypes grown in open-field production systems under hot-arid climatic conditions. *Scientific Reports, 13*(1), 1991. DOI: 10.1038/s41598-023-29039-4

Owuonda, S., & Olembo, S. (2024). Balancing the right to food and environmental sustainability: A call for holistic transformation. *African Journal of Food, Agriculture, Nutrition and Development, 24*(1), 24790–24806. DOI: 10.18697/ajfand.135.25405

Oxford Analytica (2024). Algeria's water challenges will grow. Emerald Expert Briefings, (oxan-db).

Ozdemir, D. (2022). The impact of climate change on agricultural productivity in Asian countries: A heterogeneous panel data approach. *Environmental Science and Pollution Research International, 29*(6), 8205–8217. DOI: 10.1007/s11356-021-16291-2

Paglialunga, E., Coveri, A., & Zanfei, A. (2022). Climate change and within-country inequality: New evidence from a global perspective. *World Development, 159*, 106030. Advance online publication. DOI: 10.1016/j.worlddev.2022.106030

Pandey, A. (2025a). The Transformative Potential of Artificial Intelligence in Building Sustainable Agricultural Systems: Innovations, Challenges, and Future Directions. In *AI and Ecological Change for Sustainable Development* (pp. 161-182). IGI Global Scientific Publishing.

Pandey, A. (2025b). Agroecological Transformation Through Farmers' Empowerment and IP Reform: A Human Rights-Based Approach. In *Social System Reforms to Achieve Global Sustainability* (pp. 1-22). IGI Global Scientific Publishing.

Pandey, A. (2026). Technological Innovation and Intellectual Property in Agriculture: A Critical Study With Reference to Farmers' Rights. In *Precision and Intelligence in Agriculture: Advanced Technologies for Sustainable Farming* (pp. 29-50). IGI Global Scientific Publishing.

Pandey, A., & Joseph, J. (2025). Leveraging Artificial Intelligence for Sustainable Development in Agriculture: An Exploratory Study. In *Artificial Intelligence and Data Science for Sustainability: Applications and Methods* (pp. 187-212). IGI Global Scientific Publishing.

Pandey, A., & Mishra, A. (2025). Farmers' Rights in the Age of Agricultural Intellectual Property: From Soil to Statute. In *Geostatistical Insights on Mapping Flood Hazards and Wetland Dynamics* (pp. 123-150). IGI Global Scientific Publishing.

Pandey, A., & Mishra, A. (2025a). Intellectual Property and Human Rights of Farmers: Striking a Balance in the Era of Globalization. In *Globalization, Global Security, and New International Realities for Modern Democracies* (pp. 221-246). IGI Global Scientific Publishing.

Pandey, A., & Mishra, A. (2025b). Conflict and Coexistence of Human Rights: An Exploratory Study with Reference to Intellectual Property Rights. *Journal of Human Rights and Social Work, 10*(1), 1–12. DOI: 10.1007/s41134-024-00361-9

Parra-López, C., Ben Abdallah, S., Garcia-Garcia, G., Hassoun, A., Sánchez-Zamora, P., Trollman, H., Jagtap, S., & Carmona-Torres, C. (2024). Integrating digital technologies in agriculture for climate change adaptation and mitigation: State of the art and future perspectives. *Computers and Electronics in Agriculture, 226*, 109412. DOI: 10.1016/j.compag.2024.109412

Parra-López, C., Reina-Usuga, L., Carmona-Torres, C., Sayadi, S., & Klerkx, L. (2021). Digital transformation of the agrifood system: Quantifying the conditioning factors to inform policy planning in the olive sector. *Land Use Policy, 108. Land Use Policy, 108*, 105537. Advance online publication. DOI: 10.1016/j.landusepol.2021.105537

Patel, R. (2009). Food sovereignty. *The Journal of Peasant Studies, 36*(3), 663–706. DOI: 10.1080/03066150903143079

Patel, R. (2013). *Stuffed and Starved: The Hidden Battle for the World Food System.* Melville House.

Pauw, W. P., Klein, R. J. T., Mbeva, K., Dzebo, A., Cassanmagnago, D., & Rudloff, A. (2018). Beyond headline mitigation numbers: We need more transparent and comparable NDCs to achieve the Paris Agreement on climate change. *Climatic Change, 147*(1-2), 23–29. DOI: 10.1007/s10584-017-2122-x

Pawlak, K., & Kołodziejczak, M. (2020). The role of agriculture in ensuring food security in developing countries: Considerations in the context of the problem of sustainable food production. *Sustainability (Basel), 12*(13), 5488. DOI: 10.3390/su12135488

Paxson, C. H. (1992). Using weather variability to estimate the response of savings to transitory income in Thailand. *The American Economic Review*, •••, 15–33.

Pellicone, G., Caloiero, T., & Guagliardi, I. (2019). The De Martonne aridity index in Calabria (Southern Italy). *Journal of Maps, 15*(2), 788–796. DOI: 10.1080/17445647.2019.1673840

Perez, N. D., Kassim, Y., Ringler, C., Thomas, T. S., ElDidi, H., & Breisinger, C. (2021). *Climate-resilience policies and investments for Egypt's agriculture sector: Sustaining productivity and food security*. Intl Food Policy Res Inst.

Peridy, . (2013). *The Economic Costs of Climate Change in MENA countries: A Micro-Spatial Quantitative Assessment and a Survey of Policies*. FEMISE.

Pesaran, M. H. (2006). Estimation and Inference in Large Heterogeneous Panels with a Multifactor Error Structure. *Econometrica, 74*(4), 967–1012. DOI: 10.1111/j.1468-0262.2006.00692.x

Pesaran, M. H. (2007). A simple panel unit root test in the presence of cross-section dependence. *Journal of Applied Econometrics, 22*(2), 265–312. DOI: 10.1002/jae.951

Pesaran, M. H. (2014). Testing Weak Cross-Sectional Dependence in Large Panels. *Econometric Reviews, 34*(6-10), 1089–1117. DOI: 10.1080/07474938.2014.956623

Pesaran, M. H., Shin, Y., & Smith, R. P. (1999). Pooled Mean Group Estimation of Dynamic Heterogeneous Panels. *Journal of the American Statistical Association, 94*(446), 621–634. DOI: 10.1080/01621459.1999.10474156

Pettitt, A. N. (1979). A non-parametric approach to the change-point problem. *Journal of the Royal Statistical Society. Series C, Applied Statistics, 28*(2), 126–135.

Peyrical, J.-M. (2017). Marché de partenariat: exécution et extinction (Chap. 2, folio n°3230). In *Encyclopédie des collectivités locales* (Vol. M). Paris: Dalloz.

Piccinini, E., Hanelt, A., Gregory, R., & Kolbe, L. (2015). *Transforming Industrial Business: The Impact of Digital Transformation on Automotive Organizations*.

Pickson, R. B., & Boateng, E. (2022). Climate change: A friend or foe to food security in Africa? *Environment, Development and Sustainability, 24*(3), 1–26. DOI: 10.1007/s10668-021-01621-8

Pickson, R. B., Boateng, E., Gui, P., & Chen, A. (2024). The impacts of climatic conditions on cereal production: Implications for food security in Africa. *Environment, Development and Sustainability, 26*(7), 18333–18360. DOI: 10.1007/s10668-023-03391-x

Pielke, R. A.Jr. (1998). Rethinking the role of adaptation in climate policy. *Global Environmental Change, 8*(2), 159–170. DOI: 10.1016/S0959-3780(98)00011-9

Pierson, P. (2000). Increasing returns, path dependence, and the study of politics. *The American Political Science Review, 94*(2), 251–267. DOI: 10.2307/2586011

Piketty, T., & Goldhammer, A. (2014). *Capital in the Twenty-First Century*. Harvard University Press. https://www.jstor.org/stable/j.ctt6wpqbc

Pingali, P. (2012). Green Revolution: Impacts, limits, and the path ahead. *Proceedings of the National Academy of Sciences of the United States of America, 109*(31), 12302–12308. DOI: 10.1073/pnas.0912953109 PMID: 22826253

Pontones-Rosa, C., Pérez-Morote, R., & Santos-Peñalver, J. F. (2021). ICT-based public policies and depopulation in hollowed-out Spain: A survey analysis on the digital divide and citizen satisfaction. *Technological Forecasting and Social Change, 169*, 120811. DOI: 10.1016/j.techfore.2021.120811

Porrini, D., & Schwarze, R. (2014). Insurance models and European climate change policies: An assessment. *European Journal of Law and Economics, 38*(1), 7–28. DOI: 10.1007/s10657-012-9376-6

Prakash Khedun, C., Lewis, A. C., & Kaiser, R. A. (2025). Building Resilience into the Public Water Supply System through Targeted Data-Driven Conservation Strategies. *Journal of Irrigation and Drainage Engineering, 151*(3), 04025011. Advance online publication. DOI: 10.1061/JIDEDH.IRENG-10453

Prasad, R. S., & Reddy, D. N. (2020). Farmers' rights in India: A policy perspective. *Journal of Intellectual Property Rights, 25*(4), 152–158.

Premanandh, J., Maruthamuthu, M., Sabbagh, A., & Al Muhairi, S. (2012). Prevalence of genetically modified foods (GM foods) in the United Arab Emirates. *Food Control, 25*(1), 10–12. DOI: 10.1016/j.foodcont.2011.10.001

Priyadarshini, S., Mulage, P., Conchita, S., Pavani, P., & Wasule, D., KB, C.K., Balaji, M.P. and Ashwini, K.S. (2024). Microbial Biotechnology: Innovations for Crop Protection and Improvement. *Journal of Advances in Biology & Biotechnology, 27*(7), 238–256.

Qadir, M., Bahri, A., Sato, T., & Al-Karadsheh, E. (2010). Wastewater production, treatment, and irrigation in Middle East and North Africa. *Irrigation and Drainage Systems, 24*(1), 37–51. DOI: 10.1007/s10795-009-9081-y

Qaim, M. (2020). Role of new plant breeding technologies for food security and sustainable agricultural development. *Applied Economic Perspectives and Policy, 42*(2), 129–150. DOI: 10.1002/aepp.13044

Rahim, H. U., Ali, W., Uddin, M., Ahmad, S., Khan, K., Bibi, H., & Alatalo, J. M. (2024). Abiotic stresses in soils, their effects on plants, and mitigation strategies: A literature review. *Chemistry and Ecology*, •••, 1–34.

Raihan, A., Ridwan, M., & Rahman, M. S. (2024). An exploration of the latest developments, obstacles, and potential future pathways for climate-smart agriculture. *Climate Smart Agriculture*, *1*(2), 100020. DOI: 10.1016/j.csag.2024.100020

Rallo, G., González-Altozano, P., Manzano-Juárez, J., & Provenzano, G. (2017). Using field measurements and FAO-56 model to assess the eco-physiological response of citrus orchards under regulated deficit irrigation. Agricultural Water Management, 180(Part A), 136–147. https://doi.org/DOI: 10.1016/j.agwat.2016.11.011

Randall, E., & Sanjur, D. (1981). Food preferences—Their conceptualization and relationship to consumption. *Ecology of Food and Nutrition*, *11*(3), 151–161. DOI: 10.1080/03670244.1981.9990671

Randolph, J. J. (2005). Free-marginal multirater kappa (multirater K [free]): An alternative to Fleiss' fixed-marginal multirater kappa. *Online Submission*. https://eric.ed.gov/?id=ED490661

Rangel, T., & Galende, J. (2010). Innovation in public-private partnerships (PPPs): The Spanish case of highway concessions. *Public Money & Management*, *30*(1), 49–54. DOI: 10.1080/09540960903492380

Rangnekar, D. (2004). The socio-economics of geographical indications: A review of empirical evidence from Europe. *UNCTAD-ICTSD Project on IPRs and Sustainable Development*.

Rao, N. D., van Ruijven, B. J., Riahi, K., & Bosetti, V. (2017). Improving poverty and inequality modelling in climate research. *Nature Climate Change*, *7*(12), 857–862. DOI: 10.1038/s41558-017-0004-x

Raponi, S. (2017). A defense of the human right to adequate food. *Res Publica (Liverpool, England)*, *23*(1), 99–115. DOI: 10.1007/s11158-015-9315-9

Rashid, U., Yasmin, H., Hassan, M. N., Naz, R., Nosheen, A., Sajjad, M., & Ahmad, P. (2022). Drought-tolerant Bacillus megaterium isolated from semi-arid conditions induces systemic tolerance of wheat under drought conditions. *Plant Cell Reports*, ●●●, 1–21.

Ravallion, M., & Jalan, J. (2001). *Household income dynamics in rural China* (Policy Research Working Paper No. 2612). World Bank. https://openknowledge.worldbank.org/handle/10986/18714

Rayner, T., & Jordan, A. (2016). Climate change policy in the European Union. In *Oxford research encyclopedia of climate science*. https://doi.org/DOI: 10.1093/acrefore/9780190228620.013.47

Razek, A., & Othman, S. (2023). Some Economic Aspects of the Olive Crop in Libya. *Alexandria Science Exchange Journal, 44*(4), 1109–1122. DOI: 10.21608/asejaiqjsae.2023.326999

Remane, G., Hanelt, A., Nickerson, R. C., & Kolbe, L. M. (2017). Discovering digital business models in traditional industries. *The Journal of Business Strategy, 38*(2), 41–51. DOI: 10.1108/JBS-10-2016-0127

Reynolds, J. F., Smith, D. M. S., Lambin, E. F., Turner, B. L., Mortimore, M., Batterbury, S. P., & Walker, B. (2007). Global desertification: Building a science for dryland development. *Science, 316*(5826), 847–851. DOI: 10.1126/science.1131634

Ribot, J. (2014). Cause and response: Vulnerability and climate in the Anthropocene. *The Journal of Peasant Studies, 41*(5), 667–705. DOI: 10.1080/03066150.2014.894911

Ringler, C., Agbonlahor, M., Barron, J., Baye, K., Meenakshi, J. V., Mekonnen, D. K., & Uhlenbrook, S. (2022). The role of water in transforming food systems. *Global Food Security, 33*, 100639. DOI: 10.1016/j.gfs.2022.100639

Ripert, G., & Roblot, R. (2003). *Traité de droit commercial (T. 1, V. 2), Les sociétés commerciales.* L.G.D.J.

Rocha, J. R., Martín, M. P., & Martín, M. V. (2023). The Paradox of Biodiversity Loss and the Emergence of New Forms of Life, Linked to Anthropogenic Effects. *Astragalo, 1*, 203–222.

Rodríguez-García, C., Toro-Podadera, A., Sarmiento-Carbajal, J., & Cabrera-Castro, R. (2024). Coexisting in the Surf Zone: Age and Feeding Habits of the Spotted Seabass (Dicentrarchus punctatus) and European Seabass (Dicentrarchus labrax) on the Gulf of Cádiz Beaches (Southwest Iberian Peninsula). *Fishes, 9*(5), 173. DOI: 10.3390/fishes9050173

Román-Vázquez, J., Carbonell-Bojollo, R. M., Veroz-González, Ó., Maraschi Da Silva Piletti, L. M., Márquez-García, F., Cabeza-Ramírez, L. J., & González-Sánchez, E. J. (2025). Global Trends in Conservation Agriculture and Climate Change Research: A Bibliometric Analysis. *Agronomy (Basel), 15*(1), 249. DOI: 10.3390/agronomy15010249

Rosenzweig, C., & Parry, M. L. (1994). Potential impact of climate change on world food supply. Nature, 367(6459), 133–138. https://doi.org/10.1038/367133a0

Rosenzweig, C., Elliott, J., Deryng, D., Ruane, A. C., Müller, C., Arneth, A., Boote, K. J., Folberth, C., Glotter, M., Khabarov, N., Neumann, K., Piontek, F., Pugh, T. A. M., Schmid, E., Stehfest, E., Yang, H., & Jones, J. W. (2014). Assessing agricultural risks of climate change in the 21st century in a global gridded crop model intercomparison. *Proceedings of the National Academy of Sciences of the United States of America*, *111*(9), 3268–3273. DOI: 10.1073/pnas.1222463110 PMID: 24344314

Rotz, S., & Duncan, E.. (2019). The politics of digital agricultural platforms: Markets, institutions and technology. *Canadian Food Studies*, *6*(3), 48–86.

Rotz, S., Gravely, E., Mosby, I., Duncan, E., Finnis, E., Horgan, M., LeBlanc, J., Martin, R., Neufeld, H. T., Nixon, A., Pant, L., Shalla, V., & Fraser, E. (2019). Automated pastures and the digital divide: How agricultural technologies are shaping labour and rural communities. *Journal of Rural Studies*, *68*, 112–122. DOI: 10.1016/j.jrurstud.2019.01.023

Rudrakar, S., & Rughani, P. (2024). IoT based Agriculture (Ag-IoT): A detailed study on Architecture, Security and Forensics. *Information Processing in Agriculture*, *11*(4), 524–541. DOI: 10.1016/j.inpa.2023.09.002

Ruggerio, C. A. (2021). Sustainability and sustainable development: A review of principles and definitions. *The Science of the Total Environment*, *786*, 147481. DOI: 10.1016/j.scitotenv.2021.147481

Ruiz-Real, J. L., Uribe-Toril, J., Torres Arriaza, J. A., & De Pablo Valenciano, J. (2020). A Look at the Past, Present and Future Research Trends of Artificial Intelligence in Agriculture. *Agronomy (Basel)*, *10*(11), 1839. DOI: 10.3390/agronomy10111839

Saad, M., Abokoura, H., Abdellatif, K., El-Latif, M., & Aly, M. (2020). Application of Alginate Bead Encapsulated N 2-Fixing Bacteriais Improvingwheat Yield under Drought Stress. *Plant Archives*, *20*, 1735–1747.

Sachs, J., Kroll, C., Lafortune, G., Fuller, G., & Woelm, F. (2022). *Sustainable Development Report 2022*. Cambridge University Press. DOI: 10.1017/9781009210058

Sadjadi, E. N., & Fernández, R. (2023). Challenges and Opportunities of Agriculture Digitalization in Spain. *Agronomy (Basel)*, *13*(1), 259. DOI: 10.3390/agronomy13010259

Sahabi-Abed, S., Ayugi, B. O., & Selmane, A. N. E. I. (2023). Spatiotemporal projections of extreme precipitation over Algeria based on CMIP6 global climate models. *Modeling Earth Systems and Environment*, *9*(3), 3011–3028. DOI: 10.1007/s40808-023-01716-3

Sainath, P. (2013). *Everybody loves a good drought: Stories from India's poorest districts*. Penguin Books.

Sakalasooriya, N. (2021). Conceptual Analysis of Sustainability and Sustainable Development. *Open Journal of Social Sciences*, *09*(03), 396–414. DOI: 10.4236/jss.2021.93026

Salam, A. (2015). Climate change: The challenges for public health and environmental effects in UAE. *WIT Transactions on Ecology and the Environment*, *193*, 457–466. DOI: 10.2495/SDP150391

Salami, A. A., & Babatunde, O. R. (2024). Environmental Challenges, The Impacts of Climate Change in North Africa Region: A review. *Earth and Environmental Sciences Library*, 281–294. https://doi.org/DOI: 10.1007/978-3-031-58315-5_14

Salami, A. A., & Babatunde, O. R. (2024). Environmental challenges, the impacts of climate change in North Africa region: A review. *Natural resources deterioration in MENA region: land degradation, soil erosion, and desertification*, 281-294.

Salimi, M., & Al-Ghamdi, S. G. (2020). Climate change impacts on critical urban infrastructure and urban resiliency strategies for the Middle East. *Sustainable Cities and Society*, *54*, 101948. DOI: 10.1016/j.scs.2019.101948

Sambou, M. J. G., Pohl, B., Janicot, S., Landry Famien, A. M., Roucou, P., Badiane, D., & Gaye, A. T. (2021). Heat waves in spring from Senegal to Sahel: Evolution under climate change. *International Journal of Climatology*, *41*(14), 6238–6253. DOI: 10.1002/joc.7176

Sanabria, S., & Torres, J. (2020). Water price: Environment sustainability and resource cost. *Water (Basel)*, *12*(11), 1–18. DOI: 10.3390/w12113176

Sanghi, A., & Mendelsohn, R. (2008). The impacts of global warming on farmers in Brazil and India. *Global Environmental Change*, *18*(4), 655–665. DOI: 10.1016/j.gloenvcha.2008.06.008

Sangsefidi, Y., Rios, A., Bagheri, K., Carroll, M. W., & Davani, H. (2025). Integrating social data and engineering solutions for developing resilient water infrastructure against coastal climate change. *Scientific Reports*, *15*(1), 7241. Advance online publication. DOI: 10.1038/s41598-025-91147-0 PMID: 40021917

Santos, R. M., & Bakhshoodeh, R. (2021). Climate change/global warming/climate emergency versus general climate research: Comparative bibliometric trends of publications. *Heliyon*, *7*(11), e08219. DOI: 10.1016/j.heliyon.2021.e08219

Sargani, G. R., Wang, B., Leghari, S. J., & Ruan, J. (2025). Is digital transformation the key to agricultural strength? A novel approach to productivity and supply chain resilience. *Smart Agricultural Technology*, *10*, 100838. DOI: 10.1016/j.atech.2025.100838

Sasani, M. S. (2024). The importance of biosecurity in emerging biotechnologies and synthetic biology. *Avicenna Journal of Medical Biotechnology*, *16*(4), 223.

Sattar, A., Brown, C., Rounsevell, M., & Alexander, P. (2024). Typology analysis of Egyptian agricultural households reveals increasing income diversification and abandonment of agricultural activities. *Agricultural Systems*, *218*, 104000. DOI: 10.1016/j.agsy.2024.104000

Schachter, O. (2017). Towards a theory of international obligation. In Evans, M. D. (Ed.), *Sources of international law*. Routledge. (Original work published 2000) DOI: 10.4324/9781315087795-1

Schilling, J., Freier, K. P., Hertig, E., & Scheffran, J. (2012). Climate change, vulnerability and adaptation in North Africa with focus on Morocco. *Agriculture, Ecosystems & Environment*, *156*, 12–26. DOI: 10.1016/j.agee.2012.04.021

Schlenker, W., & Lobell, D. B. (2010). Robust negative impacts of climate change on African agriculture. *Environmental Research Letters*, *5*(1), 014010. Advance online publication. DOI: 10.1088/1748-9326/5/1/014010

Schlenker, W., & Roberts, M. J. (2009). Nonlinear temperature effects indicate severe damages to U.S. crop yields under climate change. *Proceedings of the National Academy of Sciences of the United States of America*, *106*(37), 15594–15598. DOI: 10.1073/pnas.0906865106 PMID: 19717432

Schmidt, D. (1998). Les définitions du contrôle d'une sociétés, in, La prise de contrôle d'une société. *Revue de jurisprudence commerciale,* (11).

Schmidt, R. C. (1997). Managing Delphi surveys using nonparametric statistical techniques. *Decision Sciences*, *28*(3), 763–774. DOI: 10.1111/j.1540-5915.1997.tb01330.x

Schreinemachers, P., Simmons, E. B., & Wopereis, M. C. (2018). Tapping the economic and nutritional power of vegetables. *Global Food Security*, *16*, 36–45. DOI: 10.1016/j.gfs.2017.09.005

Schwantes, R., Morales, Y., Pomp, E., Singer, J., Chavan, K., & Saravia, F. (2025). Thermally driven ultrapure water production for water electrolysis – A techno-economic analysis of membrane distillation. *Desalination*, *608*, 118848. Advance online publication. DOI: 10.1016/j.desal.2025.118848

Scoones, I. (1998). Sustainable rural livelihoods: A framework for analysis. IDS Working Paper 72.

Sée, A. (2025). Les participations minoritaires des personnes publiques au capital des sociétés de production d'énergie renouvelable. *L' Actualité Juridique Droit Administratif (AJDA)*(14).

Seidl, C., Wheeler, S. A., & Zuo, A. (2020). High turbidity: Water valuation and accounting in the Murray-Darling Basin. *Agricultural Water Management, 230*, 105929. Advance online publication. DOI: 10.1016/j.agwat.2019.105929

Semchuk, I., Mishchenko, V., Petrukha, N., Strochenko, N., & Kapelista, I. (2025). Agricultural practices' impact on achieving the sustainable development goal: Eliminating world hunger. *Grassroots Journal of Natural Resources*, *8*(3), 737–758. DOI: 10.33002/nr2581.6853.080131

Seo, S. N., & Mendelsohn, R. (2008). Measuring impacts and adaptations to climate change: A structural Ricardian model of African livestock management. *Agricultural Economics, 38*(2), 151–165. DOI: 10.1111/j.1574-0862.2007.00289.x

Serdeczny, O., Adams, S., Baarsch, F., Coumou, D., Robinson, A., Hare, W., & Reinhardt, J. (2017). Climate change impacts in Sub-Saharan Africa: From physical changes to their social repercussions. *Regional Environmental Change*, *17*(6), 1585–1600. DOI: 10.1007/s10113-015-0910-2

Servat, D., Perrier, E., Treuil, J. P., & Drogoul, A. (1998). When agents emerge from agents: Introducing multi-scale viewpoints in multi-agent simulations. In *Multi-Agent Systems and Agent-Based Simulation: First International Workshop, MABS'98, Paris, France, July 4-6, 1998.* [Springer Berlin Heidelberg.]. *Proceedings*, *1*, 183–198.

Sghaier, M., Frija, A., Postigo, J., Speelman, S., Alary, V., & Sghaier, M. (2025). Assessing pastoral reforms through the performance of agro-pastoral community-based organizations in South Tunisia. *Rangeland Ecology and Management*, *98*, 49–62. DOI: 10.1016/j.rama.2024.07.008

Shaheen, H. M. (2025). The Role of the Paris Agreement in Addressing Gender-Based Violence through Nationally Determined Contributions (NDC s): A Comparative Study of Egypt and Tunisia. *Studi Magrebini*, *22*(2), 201–217. DOI: 10.1163/2590034x-20240110

Sharma, M. P., Grover, M., Chourasiya, D., Bharti, A., Agnihotri, R., Maheshwari, H. S., & Bagyaraj, D. J. (2020). Deciphering the role of trehalose in tripartite symbiosis among rhizobia, arbuscular mycorrhizal fungi, and legumes for enhancing abiotic stress tolerance in crop plants. *Frontiers in Microbiology*, *11*, 509919.

Shen, Z., Wang, S., Boussemart, J.-P., & Hao, Y. (2022). Digital transition and green growth in Chinese agriculture. *Technological Forecasting and Social Change*, *181*, 121742. DOI: 10.1016/j.techfore.2022.121742

Shiva, V. (1991). *The violence of the Green Revolution: Third World agriculture, ecology, and politics*. Zed Books.

Shiva, V. (2016). *Who Really Feeds the World? The Failures of Agribusiness and the Promise of Agroecology*. North Atlantic Books.

SHOBANKE, M., BHATT, M. & SHITTU, E. (2025). Advancements and future outlook of Artificial Intelligence in energy and climate change modeling. *Advances in Applied Energy*, •••, 100211.

Sibhatu, K. T., & Qaim, M. (2017). Rural food security, subsistence agriculture, and seasonality. *PLoS One*, *12*(10), e0186406. DOI: 10.1371/journal.pone.0186406

Silva, E. R. O., Silva, T. L., Wei, M. C. F., Souza, R. A., & Molin, J. P. (2025). Spatial and temporal variability management for all farmers: A cell-size approach to enhance coffee yields and optimize inputs. *Plants*, *14*(2), 169. DOI: 10.3390/plants14020169

Simola, A., Boysen, O., Ferrari, E., Nechifor, V., & Boulanger, P. (2022). Economic integration and food security – The case of the AfCFTA. *Global Food Security*, *35*, 100651. DOI: 10.1016/j.gfs.2022.100651

Singh, R. (2025). *Industry 5. 0: Sustainable Business Practices for a Bright Future* (1st ed.). Nova Science Publishers, Incorporated. DOI: 10.52305/UYKO4001

Sissoko, K., van Keulen, H., Verhagen, J., Tekken, V., & Battaglini, A. (2011). Agriculture, livelihoods and climate change in the West African Sahel. *Regional Environmental Change*, *11*(S1), 119–125. DOI: 10.1007/s10113-010-0164-y

Skogly, S. I. (2007). Right to adequate food: National implementation and extraterritorial obligations. *Max Planck Yearbook of United Nations Law*, *11*(1), 339–358. DOI: 10.1163/18757413-90000013

Skulmoski, G. J., Hartman, F. T., & Krahn, J. (2007). The Delphi method for graduate research. *Journal of Information Technology Education*, *6*(1), 1–21. DOI: 10.28945/199

Slimani, H., & Aidoud, A. (2004). Desertification in the Maghreb: A case study of an Algerian high-plain steppe. In: Environmental Challenges in the Mediterranean 2000–2050: Proceedings of the NATO Advanced Research Workshop on Environmental Challenges in the Mediterranean 2000–2050 Madrid, Spain 2–5 October 2002 (pp. 93-108). Dordrecht: Springer Netherlands. DOI: 10.1007/978-94-007-0973-7_6

Slingo, J. M., Challinor, A. J., Hoskins, B. J., & Wheeler, T. R. (2005). Introduction: Food crops in a changing climate. *Philosophical Transactions of the Royal Society of London. Series B, Biological Sciences*, *360*(1463), 1983–1989. DOI: 10.1098/rstb.2005.1755

Sofia, G., Zaccone, C., & Tarolli, P. (2024). Agricultural drought severity in NE Italy: Variability, bias, and future scenarios. *International Soil and Water Conservation Research*, *12*(2), 403–418. DOI: 10.1016/j.iswcr.2023.07.003

Sola, I., Santana-Anticoy, C., Silva-García, R., Pérez-Hernández, G., Pereira-Rojas, J., Blanco-Murillo, F., Díaz, M. J., Sáez, C. A., & Rodríguez-Rojas, F. (2025). Evaluating physico-chemical and biological impacts of brine discharges for a sustainable desalination development on South America's Pacific coast. *Journal of Hazardous Materials*, *489*, 137464. Advance online publication. DOI: 10.1016/j.jhazmat.2025.137464 PMID: 39955987

Solaiman, Z. M., Blackwell, P., Shafi, M. I., Salman, N. D., Storer, P., & Babur, E. (2025). Co-application of biochar and phosphorus increases soil microbial biomass, mycorrhizal colonization, growth, and nutrition of subterranean clover. *AIMS Microbiology*, *11*(1), 59–73.

Soliman, N. F., & Yacout, D. M. (2016). Aquaculture in Egypt: Status, constraints and potentials. *Aquaculture International*, *24*(5), 1201–1227. DOI: 10.1007/s10499-016-9989-9

Sortais, J.-P. (1993). Protection des minoritaires: droit des sociétés. In Dalloz (Ed.), *Répertoire des sociétés* (Vol. P). Dalloz: Paris.

Sowers, J., Vengosh, A., & Weinthal, E. (2011). Climate change, water resources, and the politics of adaptation in the Middle East and North Africa. *Climatic Change*, *104*(3–4), 599–627. DOI: 10.1007/s10584-010-9835-4

State of the Climate in Africa 2023. (2024, November 26). World Meteorological Organization. https://wmo.int/publication-series/state-of-climate-africa-2023

Stern, N. (2007). *The economics of climate change*. https://doi.org/DOI: 10.1017/CBO9780511817434

Stevanović, M., Popp, A., Lotze-Campen, H., Dietrich, J. P., Müller, C., Bonsch, M., Schmitz, C., Bodirsky, B. L., Humpenöder, F., & Weindl, I. (2016). The impact of high-end climate change on agricultural welfare. *Science Advances*, *2*(8), e1501452. Advance online publication. DOI: 10.1126/sciadv.1501452

Stone, G. D. (2011). Field versus farm in Warangal: Bt cotton, higher yields, and larger questions. *World Development*, *39*(3), 387–398. DOI: 10.1016/j.worlddev.2010.09.008

Sulaiman, N., Zalilah, M. S., & Asnarulkhadi, A. S. (2010). Food security: Concepts and definitions. *Journal of Community Health*, *16*(2), 2–9. http://journalarticle.ukm.my/651/1/Vol16%282%29-Norhasmah.pdf

Sumberg, J., Thompson, J., & Woodhouse, P. (2021). Science, technology and innovation for agricultural development: Reframing the discourse. *Outlook on Agriculture*, *50*(2), 91–97. DOI: 10.1177/0030727021995701

Sunderland, T. C. H.SUNDERLAND. (2011). Food security: Why is biodiversity important? *International Forestry Review*, *13*(3), 265–274. DOI: 10.1505/146554811798293908

Supharatid, S., & Nafung, J. (2021, November 1). Projected drought conditions by CMIP6 multimodel ensemble over Southeast Asia. *Journal of Water and Climate Change*, *12*(7), 3330–3354. DOI: 10.2166/wcc.2021.308

Surminski, S., Bouwer, L. M., & Linnerooth-Bayer, J. (2016). How insurance can support climate resilience. *Nature Climate Change*, *6*(4), 333–334. DOI: 10.1038/nclimate2979

Swinburn, B., Sacks, G., Vandevijvere, S., Kumanyika, S., Lobstein, T., Neal, B., Barquera, S., Friel, S., Hawkes, C., Kelly, B., L'Abbé, M., Lee, A., Ma, J., Macmullan, J., Mohan, S., Monteiro, C., Rayner, M., Sanders, D., Snowdon, W., & Walker, C. (2013). INFORMAS (International Network for Food and Obesity/non-communicable diseases Research, Monitoring and Action Support): Overview and key principles. *Obesity Reviews*, *14*(S1), 1–12. DOI: 10.1111/obr.12087

Taabni, M., & Jihad, M. E. (2012). Eau et changement climatique au Maghreb : Quelles stratégies d'adaptation? *Les Cahiers d'Outre-Mer*, *65*(260), 493–518. DOI: 10.4000/com.6718

Tasumi, M., Trezza, R., Allen, R. G., & Wright, J. L. (2005). Operational aspects of satellite-based energy balance models for irrigated crops in the semi-arid U.S. *Irrigation and Drainage Systems*, *19*(4), 355–376. DOI: 10.1007/s10795-005-8138-9

Terneyre, P. (2003). Les nouveaux contrats de partenariat et la typologie des instruments du partenariat public-privé. *Revue de droit Immobilier,* (520).

Testu, F.-X. (2019). Titre 0 - Introduction. In Dalloz (Ed.), *Dalloz référence Contrats d'affaires*. Paris.

Tewathia, N., Kamath, A., & Ilavarasan, P. V. (2020). Social inequalities, fundamental inequities, and recurring of the digital divide: Insights from India. *Technology in Society*, *61*, 101251. DOI: 10.1016/j.techsoc.2020.101251

The Convention on Biological Diversity. (2025). *The Convention on Biological Diversity.* https://www.cbd.int

The Economics of Climate Change: The Stern Review - Grantham Research Institute on climate change and the environment. (2024, August 19). Grantham Research Institute on Climate Change and the Environment. https://www.lse.ac.uk/granthaminstitute/publication/the-economics-of-climate-change-the-stern-review/

The European Commission. (2025). *The European Commission.* https://environment.ec.europa.eu/index_en

The World Bank. (2018). Beyond scarcity: water security in the Middle East and North Africa. (2018). In *Washington, DC: World Bank eBooks*. https://doi.org/DOI: 10.1596/978-1-4648-1144-9

Thi, N. Q., Govind, A., Le, M. H., Linh, N. T., Anh, T. T. M., Hai, N. K., & Ha, T. V. (2023). Spatiotemporal characterization of droughts and vegetation response in Northwest Africa from 1981 to 2020. *The Egyptian Journal of Remote Sensing and Space Sciences*, *26*(3), 393–401. DOI: 10.1016/j.ejrs.2023.05.006

Thomsen, J., Abdulrazzaq, N. M., AlRand, H., Consortium, U. A. S., Elhag Ahmed, A., Yousef, A. F., & Enshasy, D. A. M. (2023). Surveillance of antimicrobial resistance in the United Arab Emirates: The early implementation phase. *Frontiers in Public Health*, *11*, 1247627. DOI: 10.3389/fpubh.2023.1247627

Tilman, D., Cassman, K. G., Matson, P. A., Naylor, R., & Polasky, S. (2002). Agricultural sustainability and intensive production practices. *Nature*, *418*(6898), 671–677. DOI: 10.1038/nature01014 PMID: 12167873

Tohit, N. F. M., Ya, R. M., & Haque, M. (2025). Unveiling the impacts of food security on community health: A scoping review. *Bangladesh Journal of Medical Science*, *24*, 307–330.

Tompkins, E. L., & Eakin, H. (2012). Managing private and public adaptation to climate change. *Global Environmental Change, 22*(1), 3–11. https://doi.org/DOI: 10.1016/j.gloenvcha.2011.09.010

Tompkins, E. L., & Amundsen, H.TOMPKINS. (2008). Perceptions of the effectiveness of the United Nations Framework Convention on Climate Change in advancing national action on climate change. *Environmental Science & Policy, 11*(1), 1–13. DOI: 10.1016/j.envsci.2007.06.004

Trenberth, K. E. (2018). Climate change caused by human activities is happening and it already has major consequences. *Journal of Energy & Natural Resources Law, 36*(4), 463–481. DOI: 10.1080/02646811.2018.1450895

Tripathi, S., Srivastava, P., Devi, R. S., & Bhadouria, R. (2020). Influence of synthetic fertilizers and pesticides on soil health and soil microbiology. In *Agrochemicals Detection, Treatment and Remediation* (pp. 25–54). Elsevier. DOI: 10.1016/B978-0-08-103017-2.00002-7

Tsan, M., Totapally, D. S., Hailu, D. M., & K Addom, B. (2019). *The digitalization of African agriculture report* (978-92-9081-657–7). The Technical Centre for Agricultural and Rural Cooperation (CTA).

Tubiello, F. N., Karl, K., Flammini, A., Gütschow, J., Obli-Laryea, G., Conchedda, G., Pan, X., Qi, S. Y., Halldórudóttir Heiðarsdóttir, H., Wanner, N., Quadrelli, R., Rocha Souza, L., Benoit, P., Hayek, M., Sandalow, D., Mencos Contreras, E., Rosenzweig, C., Rosero Moncayo, J., Conforti, P., & Torero, M. (2022). Pre- and post-production processes increasingly dominate greenhouse gas emissions from agri-food systems. *Earth System Science Data, 14*(4), 1795–1809. DOI: 10.5194/essd-14-1795-2022

Udry, C. (1995). Risk and saving in Northern Nigeria. *The American Economic Review, 85*(5), 1287–1300.

Umar, O. B., Ranti, L. A., Abdulbaki, A. S., Bola, A. U. L., Abdulhamid, A. K., Biola, M. R., & Victor, K. O. (2021). Stresses in plants: Biotic and abiotic. In *Current trends in wheat research*. IntechOpen. Khan, A. A., Wang, Y. F., Akbar, R., & Alhoqail, W. A. (2025). Mechanistic insights and future perspectives of drought stress management in staple crops. *Frontiers in Plant Science, 16*, 1547452.

UNCCD (2023). Report from Algeria. Performance review and assessment of implementation system. Seventh reporting process.

UNCCD. (2020). Neutralité en matière de dégradation des terres: Mettre le concept en pratique. https://www.unccd.int/sites/default/files/ldn_targets/algeria-ldn-country-report-final-draft.pdf

UNDP. (2025). *New Homes for Vulnerable Households in Yen Bai Ahead of Storm Season, Thanks to Post-Typhoon Yagi Recovery Support.* https://www.undp.org/vietnam/press-releases/new-homes-vulnerable-households-yen-bai-ahead-storm-season-thanks-post-typhoon-yagi-recovery-support

UNEP finance. (2023). *Adapting to a New Climate in the MENA Region: An assessment of physical risk management and climate adaptation finance in the MENA region.* https://www.uncclearn.org/wp-content/uploads/library/Adapting-to-a-new-climate-MENA.pdf

United Nations Development Programme (UNDP). (2023). Annual Report 2022. New York: UNDP. https://annualreport.undp.org/2022/assets/Annual-Report-2022-FR.pdf

United Nations Economic and Social Commission for Western Asia (ESCWA). (2024). *Arab sustainable development report 2024* (E/ESCWA/CL5.SDGS/2024/1). https://www.unescwa.org/publications/arab-sustainable-development-report-2024

United Nations Economic and Social Commission for Western Asia. United Nations Population Fund Arab States Regional Office, & Save the Children. (2022, November). *Through their eyes: Climate change and economic inequality in the MENA region* (E/ESCWA/CL2.GPID/2022/BOOKLET.1). https://www.unescwa.org/publications/through-their-eyes-climate-change-and-economic-inequality-mena-region

United Nations. (2025a). *Food.* https://www.un.org/ar/global-issues/food

United Nations. (2025b). *Human Rights Treaty Body Documents: International Covenant on Economic, Social and Cultural Rights.* https://tbinternet.ohchr.org/_layouts/15/treatybodyexternal/TBSearch.aspx?Lang=en&TreatyID=9&DocTypeID=11

USDA (2022). Food and agricultural import regulations and standards country report: Algeria. FAIRS Annual Country Report. AG2022-0013. USDA.

Usigbe, M. J., Asem-Hiablie, S., Uyeh, D. D., Iyiola, O., Park, T., & Mallipeddi, R. (2023). Enhancing resilience in agricultural production systems with AI-based technologies. *Environment, Development and Sustainability*, *26*(9), 21955–21983. DOI: 10.1007/s10668-023-03588-0

Uzundumlu, A. S., & Ateş, T. (2025). Olive Production Forecasts in the Leader Countries for 2023–2027. *Applied Fruit Science*, *67*(2), 56. DOI: 10.1007/s10341-025-01268-z

Vandôme, P. (2023). Rendre l'innovation technologique accessible aux systèmes irrigués: Co-conception, évaluation et implications de la low-tech numérique pour accompagner la gestion de l'eau en agriculture. Sciences agricoles, Institut national d'enseignement supérieur pour l'agriculture, l'alimentation et l'environnement. https://theses.hal.science/tel-04747653v1/file/23-0025_Vandome.pdf

Vatankhah, T., Moosavi, S. N., & Tabatabaei, S. M. (2019). The economic impacts of climate change on agriculture in Iran: A CGE model analysis. *Energy Sources. Part A, Recovery, Utilization, and Environmental Effects*, *42*(16), 1935–1949. DOI: 10.1080/15567036.2019.1604903

Vennetier, M. (2020). Forêts et changement climatique. Le constat en région méditerranéenne. *Sciences Eaux & Territoires*, *33*(3), 18–25. DOI: 10.3917/set.033.0018

Verhoef, P. C., Broekhuizen, T., Bart, Y., Bhattacharya, A., Qi Dong, J., Fabian, N., & Haenlein, M. (2021). Digital transformation: A multidisciplinary reflection and research agenda. *Journal of Business Research*, *122*, 889–901. DOI: 10.1016/j.jbusres.2019.09.022

Vermeulen, S. J., Aggarwal, P. K., Ainslie, A., Angelone, C., Campbell, B. M., Challinor, A. J., Hansen, J. W., Ingram, J. S. I., Jarvis, A., Kristjanson, P., Lau, C., Nelson, G. C., Thornton, P. K., & Wollenberg, E. (2012). Options for support to agriculture and food security under climate change. *Environmental Science & Policy*, *15*(1), 136–144. DOI: 10.1016/j.envsci.2011.09.003

Verner, D. (2010). *Reducing poverty, protecting livelihoods, and building assets in a changing climate: Social implications of climate change in Latin America and the Caribbean*. World Bank., DOI: 10.1596/978-0-8213-8238-7

Verner, D. (Ed.). (2012). *Adaptation to a changing climate in the Arab countries: A case for adaptation governance and leadership in building climate resilience*. World Bank., DOI: 10.1596/978-0-8213-9459-5

Verner, D., Treguer, D., Redwood, J., Christensen, J., McDonnell, R., Elbert, C., & Konishi, Y. (2018). *Climate variability, drought, and drought management in Tunisia's agricultural sector. World Bank*. World Bank.

Verpoorten, M., Arora, A., Stoop, N., & Swinnen, J. (2013). Self-reported food insecurity in Africa during the food price crisis. *Food Policy*, *39*, 51–63. DOI: 10.1016/j.foodpol.2012.12.006

Vial, G. (2019). Understanding digital transformation: A review and a research agenda. *The Journal of Strategic Information Systems*, *28*(2), 118–144. DOI: 10.1016/j.jsis.2019.01.003

Volschenk, T. (2020). Water use and irrigation management of pomegranate trees - a review. *Agricultural Water Management, 241*, 106375. DOI: 10.1016/j.agwat.2020.106375

Vorobey, N. A., Kukol, K. P., Kots, S. Y., Pukhtaievych, P. P., & Patyka, V. P. (2023). Viability of nodule bacteria Bradyrhizobium japonicum on soybean seeds by treatment with fungicides during extended storage.

Vurukonda, S. S. K. P., Vardharajula, S., Shrivastava, M., & SkZ, A. (2016). Enhancement of drought stress tolerance in crops by plant growth promoting rhizobacteria. *Microbiological Research, 184*, 13–24. DOI: 10.1016/j.micres.2015.12.003 PMID: 26856449

Wahab, A., Muhammad, M., Munir, A., Abdi, G., Zaman, W., Ayaz, A., & Reddy, S. P. P. (2023). Role of arbuscular mycorrhizal fungi in regulating growth, enhancing productivity, and potentially influencing ecosystems under abiotic and biotic stresses. *Plants, 12*(17), 3102.

Waha, K., Krummenauer, L., Adams, S., Aich, V., Baarsch, F., Coumou, D., Fader, M., Hoff, H., Jobbins, G., Marcus, R., Mengel, M., Otto, I. M., Perrette, M., Rocha, M., Robinson, A., & Schleussner, C.-F. (2017). Climate change impacts in the Middle East and Northern Africa (MENA) region and their implications for vulnerable population groups. *Regional Environmental Change, 17*(6), 1623–1638. DOI: 10.1007/s10113-017-1144-2

Wahid, N., Bouazzama, B., Yousfi, H. E., & Zascerinska, J. (2022). A HISTORICAL OVERVIEW OF EVOLUTION OF THE IRRIGATED AGRICULTURE SECTOR IN MOROCCO. *Journal of Regional Economic and Social Development, 14*, 109–118. DOI: 10.17770/jresd2022vol14.6896

Walz, A., Lardelli, C., Behrendt, H., Grêt-Regamey, A., Lundström, C., Kytzia, S., & Bebi, P. (2007). Participatory scenario analysis for integrated regional modelling. *Landscape and Urban Planning, 81*(1–2), 114–131. DOI: 10.1016/j.landurbplan.2006.11.001

Wang, H., Xue, L., Peng, J., Liu, G., & Chen, Z. (2022). Green Operations Management of Interbasin Water Transfer Project: An Extended Framework of Joint Pricing and Inventory Management. *Discrete Dynamics in Nature and Society, 2022*(1), 5927108. Advance online publication. DOI: 10.1155/2022/5927108

Wepner, B., Neuberger, S., Hörlesberger, M., Molin, E. M., Lampert, J., & Koch, H. (2025). How can digitalisation support transformation towards sustainable agrifood systems? Scenario development in Lower Austria. *Agricultural Systems, 224*, 104251. DOI: 10.1016/j.agsy.2024.104251

Westerman, G., Bonnet, D., & McAfee, A. (2014). *Leading digital: Turning technology into business transformation*. Harvard Business Review Press.

Wineman, A., Zanello, G., Degefaw, D. M., Samuel, F., Yates, J., Shankar, B., Adeyemi, O., Anderson, A. K., Awasthi, A., Bell, W., Chege, C. K., Choo, E., Degefaw, D. M., Deitchler, M., Freeman, M. C., Frongillo, E. A., Gelli, A., Havelaar, A. H., Guitian, J., & Zewdu, A.Nutritious Food Systems Consortium. (2024). A call to action for transformation towards nutritious food systems. *Nature Food*, *5*(1), 2–3. DOI: 10.1038/s43016-023-00896-6

WMO. (2023). *State of the Climate in Africa 2024. WMO-No. 1370*. World Meteorological Organization.

Wolfert, S., Ge, L., Verdouw, C., & Bogaardt, M. J. (2017). Big Data in Smart Farming – A review. *Agricultural Systems*, *153*, 69–80. DOI: 10.1016/j.agsy.2017.01.023

World Bank (2023a). Diagnostic sur la gestion des risques climatiques et de catastrophe en Algérie. Banque internationale pour la reconstruction et le développement, Groupe de la Banque mondiale.

World Bank (2023b). Note sur les forêts algériennes: Gestion durable des forêts pour lutter contre les feux de forêts. Banque internationale pour la reconstruction et le développement.

World Bank (2024a). Climatology: Libya. Climate Change Knowledge Portal (CCKP). The World Bank Group.

World Bank. (2018). *Climate change in MENA: Challenges and opportunities for the world's most water-stressed region*. World Bank Group. https://documents.worldbank.org/en/publication/documents-reports/documentdetail/894251519999525186/climate-change-in-mena-challenges-and-opportunities-for-the-world-s-most-water-stressed-region

World Bank. (2020). *Poverty and shared prosperity 2020: Reversals of fortune*. World Bank Group. https://www.worldbank.org/en/publication/poverty-and-shared-prosperity

World Bank. (2021). *Climate-smart agriculture in the Middle East and North Africa: Opportunities and constraints*. World Bank Group. https://openknowledge.worldbank.org/handle/10986/35296

World Bank. (2022) "Maroc: Rapport de suivi de la situation économique (Hiver 2022-23): Le Maroc face aux chocs d'offre" a. https://documents1.worldbank.org/curated/en/099332002132325417/pdf/IDU0e3742c55010a3044730a80a0ad6062acd9db.pdf (Accessed 25 October 2023)

World Bank. (2022). Beyond Scarcity: Water Security in the Middle East and North Africa.

World Bank. (2024). Rapport de suivi de la situation économique en Algérie: Investir dans les données pour une croissance diversifie. Banque Internationale pour la Reconstruction et le Développement. Banque Mondiale. Bekkis S. & Benmehaia M.A (2025). Céréales et lait en Algérie: Sujétions de service public et résultats économiques. *Maghreb, Machrek*, •••, 261.

World Bank. (2024b). *Trends and Significant Change against Natural Variability in Tunisia. Climate Change Knowledge Portal (CCKP).* The World Bank Group.

WorldClim. (2024). Records on the bioclimatic variables: Historical weather data. *WorldClim*, *2*, 1. worldclim.org/data/monthlywth.html

WTO. (1994). *Agreement on Trade-Related Aspects of Intellectual Property Rights.* Marrakesh Agreement Establishing the World Trade Organization.

Wu, G., Yang, W., Hou, X., & Tian, Y. (2023). Agri-food supply chain under live streaming and government subsidies: Strategy selection of subsidy recipients and sales agreements. *Computers & Industrial Engineering*, *185*, 109647. DOI: 10.1016/j.cie.2023.109647

Xie, H., Wen, Y., Choi, Y., & Zhang, X. (2021). Global Trends on Food Security Research: A Bibliometric Analysis. *Land (Basel)*, *10*(2), 119. DOI: 10.3390/land10020119

Xu, J., Wang, M., Xu, G., Sokolova, E., Vail, T., & Song, D. (2025). A critical review of ISO standards for seawater desalination: Classification, applications, and future development. *Separation and Purification Technology*, *367*, 132898. Advance online publication. DOI: 10.1016/j.seppur.2025.132898

Yang, D., & Choi, H. (2007). Are remittances insurance? Evidence from rainfall shocks in the Philippines. *The World Bank Economic Review*, *21*(2), 219–248. DOI: 10.1093/wber/lhm003

Yaqoob, N., Ali, S. A., Kannaiah, D., Khan, N., Shabbir, M. S., Bilal, K., & Tabash, M. I. (2023). The effects of Agriculture Productivity, Land Intensification, on Sustainable Economic Growth: A panel analysis from Bangladesh, India, and Pakistan Economies. *Environmental Science and Pollution Research International*, *30*(55), 116440–116448. DOI: 10.1007/s11356-021-18471-6

Yarullina, L. G., Kalatskaja, J. N., Cherepanova, E. A., Yalouskaya, N. A., Tsvetkov, V. O., Ovchinnikov, I. A., & Mardanshin, I. S. (2023). Approaches to improving the biological activity of agricultural formulations based on bacteria of the genus Bacillus and chitosan nanocomposites. *Applied Biochemistry and Microbiology, 59*(5), 549–560.

Zabihi, H., Vogeler, I., Mat Amin, Z., & Ramezani Gourabi, B. (2016). Mapping the sensitivity of citrus crops to freeze stress using a geographical information system in Ramsar, Iran. *Weather and Climate Extremes, 14*, 17–23. DOI: 10.1016/j.wace.2016.10.002

Zaied, Y. B., & Zouabi, O. (2016). Impacts of climate change on Tunisian olive oil output. *Climatic Change, 139*(3-4), 535–549. DOI: 10.1007/s10584-016-1801-3

Zaki, N. A., Kløve, B., & Haghighi, A. T. (2022). Expanding the irrigated areas in the MENA and Central Asia: Challenges or opportunities? *Water (Basel), 14*(16), 2560. DOI: 10.3390/w14162560

Zeng, H., Wu, B., Elnashar, A., & Fu, Z. (2024). Dryland dynamics in the Mediterranean region. In *Dryland Social-Ecological Systems in Changing Environments* (pp. 243–271). Springer Nature Singapore. DOI: 10.1007/978-981-99-9375-8_8

Zhang, H., Zhang, J., Lang, Z., Botella, J. R., & Zhu, J.-K. (2018). Genome editing—Principles and applications for functional genomics research and crop improvement. *Critical Reviews in Plant Sciences, 37*(4), 215–234.

Zhang, M., Riaz, M., Xia, H., Li, Y., Wang, X., & Jiang, C. (2022). Four-year biochar study: Positive response of acidic soil microenvironment and citrus growth to biochar under potassium deficiency conditions. *The Science of the Total Environment, 318*, 152515. DOI: 10.1016/j.scitotenv.2021.152515

Zhang, Y., Ahmad, M. I., Shen, Q., Boota, M. W., Zhao, C., & Ma, H. (2025). Unveiling the environmental impact of corn production in China: Evidence from panel ARDL approach. *Frontiers in Sustainable Food Systems, 8*, 1492262. Advance online publication. DOI: 10.3389/fsufs.2024.1492262

Zhong, J., Cheng, H., & Jia, F. (2024). Supply chain resilience capability factors in agri-food supply chains. *Operations Management Research.* DOI: 10.1007/s12063-024-00470-8

Zhou, H. C., Wu, L., He, B., & Shao, H. S. (2010). Study of agricultural water rights transfer price in Dalian. *Dalian Ligong Daxue Xuebao/Journal of Dalian University of Technology, 50*(2), 251-257.

Zhou, C. H., He, Y. L., Chen, Z., Fan, Y., Yimingniyazi, A., Ma, P. C., & Kadier, A. (2025). A novel and sustainable invasive plant wood-based solar interfacial evaporator for efficient brackish water desalination. *Desalination, 606*, 118760. Advance online publication. DOI: 10.1016/j.desal.2025.118760

Zhu, J. K. (2016). Abiotic stress signaling and responses in plants. *Cell, 167*(2), 313–324. DOI: 10.1016/j.cell.2016.08.029 PMID: 27716505

Zimmermann, A., Schmidt, R., Sandkuhl, K., El-Sheikh, E., Jugel, D., Schweda, C., Möhring, M., Wißotzki, M., & Lantow, B. (2016). Leveraging Analytics for Digital Transformation of Enterprise Services and Architectures. In El-Sheikh, E., Zimmermann, A., & Jain, L. C. (Eds.), *Emerging Trends in the Evolution of Service-Oriented and Enterprise Architectures* (Vol. 111, pp. 91–112). Springer International Publishing., DOI: 10.1007/978-3-319-40564-3_6

ZORBAS, C., RESNICK, D., JONES, E., SURI, S., IRUHIRIYE, E., HEADEY, D. D., MARTIN, W., VOS, R., ARNDT, C. & MENON, P. 2024. From promises to action: Analyzing global commitments on food security and diets since 2015, Intl Food Policy Res Inst.

Zuehlke, D., Geweniger, T., Heimann, U., & Villmann, T. (2009). Fuzzy Fleiss-kappa for comparison of fuzzy classifiers. In *Proceedings of the European Symposium on Artificial Neural Networks (ESANN)*.

Zurqani, H. A. (Ed.). (2021). *The soils of Libya*. Springer. DOI: 10.1007/978-3-030-66368-1

About the Contributors

Mohammed El Hadi El Maknouzi is an Associate Professor of Commercial Law and current Dean of the College of Law at the University of Al Dhaid, UAE, specialized in agriculture and sustainable development. He holds a Ph.D. in Commercial and Business Law and a Diplôme Advanced Studies Diploma from Mohammed V University in Rabat. Additionally, he earned an LL.M. in Enterprise Law from Toulouse University in France, jointly with ISCAE in Morocco. Dr. El Maknouzi has over 19 years of academic and professional experience. He has published more than 16 research papers in internationally recognized journals indexed in Scopus, including Q1 journals ranked in the top 1%, 2%, and 10% globally. He is also the author of six books and has participated in 11 international conferences, focusing on commercial law, financial markets, and emerging technologies' legal implications. Dr. El Maknouzi has supervised over 80 master's theses and 10 doctoral dissertations. He is a two-time recipient of the Abu Dhabi University Research Excellence Award and has contributed significantly to advancing academic research and legal education.

Hicham Sadok is a full professor of Economics and Finance at Mohammed V University of Rabat (UM5R-Rabat), Morocco. He is a visiting professor at three universities: Mons (Belgium), Tunis (Tunisia), and Toulouse (France). He is the director of the "Economic Analysis and Modeling Laboratory" at UM5- Rabat. Hicham Sadok holds a Master's degree in Finance and Capital Markets from the École Supérieure de Commerce de Paris (2001), a Postgraduate Diploma (DEA) in Management Sciences from Montesquieu-Bordeaux University (2003) in France, and a PhD in Financial Economics from Bordeaux IV University (2008). He is certified in the United Kingdom (OBPUK) as an Analyst in Geopolitics and International Finance (2022), in Financial Risk Management (2019), in Economic Analyst (2017), in Financial Investment Advisor (2013), and in Wealth Management (2019). He has been a financial analyst since 2010. Since 2009, he has taught Finance and economics at Mohammed V University in Rabat, Morocco. He has contributed to nine books, including three with CRC Press and two with IGI Global. He has also published over 63 research articles in internationally renowned peer-reviewed journals, including Q1 and Q2 journals, as well as conference proceedings from IEEE, Springer, and Elsevier. His research has been cited over 640 times, with an H-index of 14, an i-10 index of 20, and a cumulative JCR impact factor of over 100. His research focuses on financial economics, the digital economy, and economic development issues.

**

Imad Eddin Abdelhaye holds a distinguished position as a Professor in the field of Commercial Law. His research interests span various aspects of commercial law, including maritime law, aviation law, multimodal transport, commercial companies, and banking operations. Dr. Abdelhaye earned his Bachelor of Laws degree from Aleppo University in Syria in 1987. He pursued further studies and obtained a Master's degree in Maritime Law from the University of Nantes, France, in 1994. Later, he completed his Ph.D. in Commercial Maritime Law from the same university in 1997. Throughout his academic career, Dr. Abdelhaye has held teaching positions at Aleppo University in Syria, starting as a teaching assistant from 1988 to 1992. He then pursued his doctoral studies in France on a scholarship from the French government. Upon completion of his Ph.D., he continued his academic journey as an Assistant Professor at Aleppo University until 2006 when he was promoted to Associate Professor. Subsequently, he served as an Associate Professor at the College of Law, University of Ajman, UAE, from 2013 to 2015, and at the University of Sharjah, UAE, from 2015 onwards. Dr. Abdelhaye has received recognition for his scholarly contributions, including awards such as the University of Sharjah Encouragement Award for distinguished faculty members in scientific research and the College of Law Award for ranking first among faculty members in research productivity. Additionally, he has been honored for his academic excellence by the Aleppo University Teachers Syndicate.

Enas Mohammed Al-Qodsi is an Associate Professor of Private Law at the College of Law, United Arab Emirates University. She holds a Ph.D. in Private Law from Cairo University, awarded with high distinction and honors. With over 20 years of academic and professional experience, Dr. Al-Qodsi has made significant contributions to legal education and research in the region. Previously, she served as Faculty Coordinator at the College of Law, Abu Dhabi University, for nearly five years. She also worked as a Legal Advisor at the Supreme Court under the Supreme Judicial Council in Palestine, where she provided legal opinions and prepared legal memoranda on various judicial matters .Dr. Al-Qodsi has published 22 peer-reviewed research papers in high-ranking international journals indexed in Scopus, including journals ranked within the top 1%, 2%, 5%, and 9%. She holds an h-index of 8. Her research primarily focuses on civil law, tort and contract law, consumer protection, sustainability, environmental law, and artificial intelligence. Her outstanding contributions have earned her multiple awards and honors, including the Community Service Award (2019/2020) and the Teaching Excellence Award (2020/2021) from Abu Dhabi University. In 2025, she received the Research Excellence Award at the College of Law level from the United Arab Emirates University, in addition to several certificates of appreciation for research distinction from the university. Dr. Al-Qodsi has actively participated in numerous international conferences with peer-reviewed papers and has served on the scientific committees of various international conferences. She is a member of editorial boards of several Scopus-indexed journals and a certified peer reviewer for numerous reputable journals. She has reviewed a large number of research papers in her field. She has received substantial research funding from Abu Dhabi University, United Arab Emirates University, and external funding from the University of Sharjah. Her impactful research and dedication to legal scholarship continue to significantly contribute to the advancement of law both in the UAE and internationally.

Mariam Ahmed Al-Sandal is an Assistant Professor of Civil Procedure Law at Al Dhaid University, United Arab Emirates. Its an institution that places agriculture, veterinary medicine, and sustainable development at the core of its mission. Dr. Mariam has practical and professional experience in the academic field at the University of Sharjah prior to joining Al Dhaid University. She has published ten research papers and has had six other papers accepted in refereed scientific journals, all of which hold scientific value and discuss distinguished topics, with a focus on issues related to civil procedure law. I have also authored five scientific books on various subjects.

Saher Ibrahim Shukri Al-Waleed, a Palestinian national, graduated from the Faculty of Law at Al-Azhar University in Gaza, Palestine. Following his graduation, he was appointed as a teaching assistant at the same faculty in 1998. He then worked as a lecturer there after obtaining his master's degree in 2001 from the University of Jordan, and then as an assistant professor after obtaining his doctorate from Cairo University in 2008. Work Experience: 2024 - Present Assistant Professor of Criminal Law at Sultan Qaboos University, Sultanate of Oman. 2020-2024 Professor of Criminal Law at Al-Azhar University in Gaza, Palestine. 2014-2020 Associate Professor of Criminal Law at Al-Azhar University in Gaza, Palestine. 2008-2014 Assistant Professor of Criminal Law at Al-Azhar University in Gaza, Palestine.

2001-2008 Lecturer of Criminal Law at Al-Azhar University in Gaza, Palestine. 1998-2001 Teaching Assistant at the Faculty of Law, Al-Azhar University, Gaza, State of Palestine. Research Interests: Professor Saher Ibrahim Shukri Al-Walid's research interests focus on criminal law (penal law and criminal procedures law), with a particular focus on contemporary issues related to criminal justice, fair trial guarantees, and the rights of defendants during investigation and trial. His interests include analyzing comparative criminal legislation, developing penal policies that achieve a balance between deterrence and reform, and studying emerging crimes such as cybercrime. He is also interested in evaluating the performance of judicial bodies and enhancing the effectiveness of procedural systems in line with international human rights standards. He has published international research contributions that seek to develop criminal legal thought in both the Arab and international contexts.

Badraoui Ismail is an assistant professor of Operations Management at the College of Business at Zayed University (ZU). He received his Ph.D. in Operations Research and Logistics from Wageningen University and Research in the Netherlands, and his research focuses on inter-firm collaboration in agri-food supply chains. He specializes in empirical research that includes both quantitative and qualitative methods. Before joining ZU, Dr. Badraoui worked as an assistant professor of engineering management at Abu Dhabi University (ADU), and as an assistant professor of operations and supply chain management at the International University of Rabat. Before entering academia, Dr. Badraoui worked as a project manager at the Agricultural Development Agency in Morocco, where he managed international development projects funded by international institutions such as the World Bank and the African Development Bank.

Boubaker Dhehibi is an agricultural Economist in the Social, Economic and Policy Research Program (SEPRP) at the International Center for Agricultural Research in the Dry Areas (ICARDA). Boubaker has a PhD in quantitative economics from the University of Zaragoza (Spain) and an MSc in Marketing from the Mediterranean Agronomic Institute of Zaragoza (Spain) – CIHEAM. His research interests include production economics, climate change, the economics of water resources, the economics of natural resources management, applied micro-econometrics, food demand analysis, international trade, economic modeling, competitiveness and productivity analysis of the agriculture sector in the Middle East and North Africa (MENA) region, growth analysis, and the economics of development.

Syeda Maira Hamid is currently pursuing PhD degree at the College of Life Science and Technology, Tarim University, Alar, Xinjiang, China. She secured distinction in both her Master's degree in Industrial Biotechnology and a Bachelor's degree in Biotechnology. Her research interests focus on microbiology and nanotechnology with applications in sustainable agriculture, aquaculture, and environment. She is committed to addressing current scientific challenges by developing innovative and cost effective solutions to promote sustainability and health in medicine, agriculture and aquaculture management systems.

Abdallahi H'moide is currently a PhD candidate in Economics, specializing in climate change economics, at the Laboratory for Economic Analysis and Modeling (LEAM), Faculty of Legal, Economic, and Social Sciences, Mohammed V University of Rabat, Morocco. He holds a Master's degree in Economic Analysis and Modeling, as well as a Bachelor's degree in Public Economics. His research interests focus on the economics of climate change, food security, adaptation strategies, and the economic dynamics of the agricultural sector.

Iyad Jadalhaq holds a doctorate degree in Civil Law from University of Athens in Greece. His teaching experience in many universities is over 20 years. He have many publications in international journals indexed in SCOPUS, so he has been a recipient of many awards during his previous academic work, such as, the University Research Award, University Research Fellow Award, College of Law Teaching Award, and twice received the College of Law Research Award. In addition, he has 8 authorized books published in Palestine and UAE.

Houda Mahboub is an Associate Professor in SUPMTI in Rabat (Morocco) and has PhD in Economics and Business. She has served as a revewer in several indexed journals and international conferences. Her main research areas are digital transformation, organizational behavior, investment under uncertainty, digital economy and management, reflected in the publication of several indexed articles and more than 10 scientific conferences.

Achyuta Nanda Mishra is a Professor at School of Law, Christ University, Bangalore. He has 25 years of teaching experience. His research interests include Constitutional law, Legal theory, Democracy, Globalization and Global Justice. He has supervised a Postdoctoral research scholar and is currently guiding Ph.D. scholars under his supervision.

Amine Oulmane PhD in agroeconomics, researcher at CREAD, specializing in irrigation water governance and policies, agroecology, and value chain analysis.

Ananya Pandey is a Research Scholar in School of Law at Christ University, Bangalore. She has her research interests in the field of Constitutional Law,

Environment Law, Human Rights, Food Security, Intellectual Property Law and Jurisprudence.

Index

A

B

C

D

E

F

G

H

I

M

N

P

R

S

T

U

W

www.ingramcontent.com/pod-product-compliance
Lightning Source LLC
LaVergne TN
LVHW081318300926
849161LV00008B/1800